Chemistry: Principles, Techniques and Applications

Chemistry: Principles, Techniques and Applications

Edited by
Erica Young

WILLFORD PRESS

www.willfordpress.com

Published by Willford Press,
118-35 Queens Blvd., Suite 400,
Forest Hills, NY 11375, USA

ISBN: 978-1-68285-585-0

Cataloging-in-Publication Data

Chemistry : principles, techniques and applications / edited by Erica Young.
    p. cm.
Includes bibliographical references and index.
ISBN 978-1-68285-585-0
1. Chemistry. I. Young, Erica.
QD31.3 .C44 2019
540--dc21

For information on all Willford Press publications
visit our website at www.willfordpress.com

WILLFORD PRESS

Contents

Preface

This book aims to highlight the current researches and provides a platform to further the scope of innovations in this area. This book is a product of the combined efforts of many researchers and scientists, after going through thorough studies and analysis from different parts of the world. The objective of this book is to provide the readers with the latest information of the field.

Chemistry is the scientific study of the composition, structure, physical and chemical properties of compounds as well as their interactions with other compounds. Compounds are substances formed through the chemical bonding of atoms and molecules that share the same chemical properties. Chemistry studies in detail the chemical bonds between atoms and molecules to formulate new compounds. It branches out into multiple sub-fields like organic, inorganic, analytical, physical, nuclear chemistry among many others. This book traces the progress of this field and highlights some of its key concepts and applications. This book is a vital tool for all researching and studying the discipline of chemistry. Those who are interested in broadening the expanse of their knowledge will be immensely benefited by this book.

I would like to express my sincere thanks to the authors for their dedicated efforts in the completion of this book. I acknowledge the efforts of the publisher for providing constant support. Lastly, I would like to thank my family for their support in all academic endeavors.

Editor

Development and Utilization of Reversed Phase High Performance Liquid Chromatography Methods for a Series of Therapeutic Agents

Jason Olbrich* and Joel Corbett

Poly-Med, Inc., Anderson, SC, USA

Abstract

Capability to quantify drug release is vital in many areas of pharmaceutical research. Accurate, rapid and repeatable detection of pharmaceutical agents allows for comparisons between drug delivery constructs prior to *in vivo* studies as well as identification of therapeutic levels in samples both *in vitro* and *in vivo*. Perhaps the foremost scientific method for monitoring of therapeutics is High Performance Liquid Chromatography. This process allows for clear and simple quantification of therapeutic presence in a sample, with an inherent or induced chromophore, through the detection of bound specimen to the utilized column. Multiple drugs can be used through this process and exemplary methods are presented for the pharmaceutics; cefuroxime, clindamycin, dexamethasone, dicloxacillin, doxycycline, metronidazole, oxymetazoline, paclitaxel, tobramycin, and vancomycin. Each of these drugs is analyzed using Reversed-Phase High Performance Liquid Chromatography utilizing a hydrophobic column. Independent, repeatable methods are developed for each drug. Where necessary a pre-column derivatization is used to allow for visualization of otherwise undetectable tobramycin. In the case of each drug, a standard curve is presented to show the linearity of response for the proposed method within a range of absorbance. This work shows the ability to analyze multiple drugs with a single simple, quick, cost effective system and detection methods are evaluated for each drug with a R^2 value of greater than 0.99.

Keywords: Reversed-Phase High Performance Liquid Chromatography; Clindamycin; Metronidazole; Paclitaxel; Tobramycin; Gradient

Introduction

HPLC

High Performance Liquid Chromatography (HPLC) is a widely used method to quantify the presence of a pharmaceutical solute in samples. This analytical method involves the injection of a sample into the chromatographic column which then retains the desired sample from the mobile phase. The extracted and bound sample is increased and analyzed through a variety of methods, which involve the detection of a chromophore of the bound agent. HPLC analysis is superior to other forms of UV/Vis quantification due to the process ability to clearly quantitate material as well as remove interference problems inherent in other processes [1]. Despite this distinct advantage, HPLC is a very difficult process with variables such as temperature, column choice, mobile phase composition and mobile phase pH all being essential to consider [2].

As stated, column choice is a crucial variable to control the efficacy of an HPLC method. Columns are offered in a wide variety of types and sizes, but can generally be divided into normal phase and reverse phase columns. These column names relate to the manner in which the desired solute is removed from the mobile phase and drawn to the analytical column [3]. Reversed phase is by far the more prevalent technology at present and involves hydrophobic columns and aqueous based mobile phases. Commonly used columns in reversed-phase HPLC (RP-HPLC) include C8 and C18 columns. Column choice parameters have been covered in other areas of the literature [2,3].

Following the choice for column type, HPLC methods again fall into two groups based on analytical scheme. These two scheme types are referred to as isocratic and gradient methods. Isocratic methods refer to the use of a single mobile phase composition for the entirety of analysis, whereas gradient methods refer to methods which utilize a changing composition through the process. Isocratic analysis has been considered quicker until recently, however gradient analysis allows higher throughput [4], better resolution [4], and separation of multiple compounds in a single sample [5]. For these reasons gradient analysis is

often employed in analytical settings where applicable. When choosing a gradient method, the mobile phase is of great concern. Mobile phases in RP-HPLC usually involve water, acetonitrile, and/or methanol and sometimes a dopant agent such as triethanolamine or triflouroacetic acid is used [2]. Because of the large amounts of variables present in such systems, method development is quite difficult and results in many methods for analyzing the same drug [6]. For this reason, it is important to select and/or develop the best method for a given application's indication.

Discussed antibiotics and published HPLC methods

The following sections briefly outline the utilized drugs and their relation to HPLC analysis in the lab today. Each of these drugs was chosen for method development based on clinical need.

Cefuroxime: Cefuroxime is a member of the cephalosporin family which displays activity against both Gram-positive and Gram-negative bacteria [7]. It is a well characterized drug with many methods for quantification and has a prominent chromophore with peak UV absorbance in the 270-280 nm range, making it acceptable for HPLC analysis [8]. The drug is usually used in salt form, with cefuroxime sodium being prevalent. This form is very hydrophilic and for this reason RP-HPLC can be used to quantitate drug presence in solution samples.

Clindamycin: Clindamycin is a lincosamide that is finding a wider use in today's clinic due to the wide range of efficacy for the drug [9]. It can be used to treat infections from staphylococcal to anaerobic

*Corresponding author: Jason Olbrich, Poly-Med, Inc., Technology Center: 51 Technology Drive, Anderson, SC 29625, USA
E-mail: Jason.Olbrich@Poly-Med.com

to even some protozoal infections [10]. Clindamycin is supplied as a hydrophilic salt, traditionally as a hydrochloride. With this enhanced solubility, RP-HPLC is used with detection around 200 nm [9,10]. Clindamycin methods usually are performed isocratically [9,10], however, gradient methods are useful especially as clindamycin is often used in combination treatments [5].

Dexamethasone: Dexamethasone is a steroid of the glucocorticoid family [1]. These drugs have exhibited effectiveness with treatment of many ocular ailments [11]. Dexamethasone shows the most anti-inflammatory activity within this family and is therefore finding many applications within the ocular field [12]. However, excessive levels have been linked with undesirable side effects and so quantification is of the utmost importance [13]. Dexamethasone, like most steroids, is generally lipophilic but can be quantified through HPLC with reverse phase columns. Analysis is commonly performed on C18 columns [11,12,14] and detection is usually between 238 and 255 nm [1,11,12].

Dicloxacillin: Dicloxacillin is a modern member of the penicillin family. It is penicillinase resistant allowing for use in indications where other members of the group are not effective [15]. It is a highly polar compound which causes difficulty with extraction in many processes [16]. Previous work with this drug has involved reversed phase columns and utilized detection wavelengths between 220 and 240 nm [15-18].

Doxycycline: This common tetracycline drug is used widely due to its broad spectrum of activity [19]. It has been well defined pharmacokinetically with a reliable absorption and long half-life [20]. Recently, doxycycline has exhibited inhibition of enzymatic activity, such as that of the matrix metalloproteinase's (MMP's) [21]. For these reasons pared with wide patient tolerance, this drug has been used in many indications and HPLC methods have been developed for quantification [19-21]. These methods involve a wide range of detection wavelengths with 350 nm being the most common [19].

Metronidazole: This cytostatic drug has a wide range of uses within the clinic. Metronidazole is prescribed for conditions ranging from rosacea to bacterial vaginosis [22]. The drug has also been shown to sensitize tumors to radiotherapy, opening its use in oncology [23]. For these conditions, the drug is often carried within a gel [24], with release from the gel being an important experimental variable. For this reason, HPLC is often employed as the drug salt is water soluble, has a strong chromophore, and is used in the described wide variety of applications. Detection for this drug is performed over a wide range from 250 to 350 nm [22-24].

Oxymetazoline: Oxymetazoline is a widely used non-prescription drug which acts as a vasoconstrictor [25]. It is generally used in the nasal cavity as a decongestant [25,26] and can be used in combination therapies for a wide variety of ear, nose and throat disorders [25-27]. The drug has a relatively low molecular weight, much of which is contained within the chromophore structure itself. This drug is often used in ointments and sprays and therefore ability to quantify release through HPLC is of great value. Analysis is traditionally performed using a C18 column with detection at 230 nm [25-27].

Paclitaxel: Taxanes and paclitaxel in particular have taken a leading role in chemotherapy over the recent decade. This expansion of use is largely due to the efficacy of paclitaxel in treatment of a wide range of solid tumors [28]. This drug moves to block mitotic activity through interference with the formation of microtubules [29], thus preventing cancer cell division. However, due to the prevalence of significant side effects, paclitaxel is almost always used in a comparatively low dose to other bioactive agents [30]. This need for a precise window of treatment combined with the important nature of the drugs indications make paclitaxel an important drug for quantification. HPLC is uniquely suited for the low levels of drug used as well as the common use of paclitaxel with other drugs and excipients such as Chremophor EL® [28-31]. The hydrophobic nature of the drug does necessitate different mobile phase ratios than other drugs described, but hydrophobic columns can be used and detection is usually around 230 nm [28-31].

Vancomycin: Vancomycin is a glycopeptide drug which has come to be used for critical infections where drugs of lesser efficacy cannot be utilized or in instances of drug resistance [32]. Specifically, vancomycin is used clinically to treat methicillin resistant *Staphylococcus aureus* (MRSA) [33]. Vancomycin is a relatively large molecular weight drug with many chromophores and is also hydrophilic [34]. These characteristics lend themselves to RP-HPLC with many methods described in the literature with detection traditionally in the 220-240 nm range [32-34].

Tobramycin: Many serious infections within the body require specialized antibiotics. Tobramycin is often used to combat difficult infection such as osteomyelitis [35] and other serious gram negative bacillary infections [36]. This aminoglycoside chemically is similar to a sugar complex but presents challenges for HPLC analysis as it lacks a chromophore [35]. Because of this, pre-column derivatization techniques are often employed to make this drug visible to an associated UV detector [36,37]. The method utilized in this paper is similar to presented by Lai of Varian Chromatography Systems [37].

Materials and Methods

Materials

Pharmaceutical drugs and triflouroacetic acid (TFA) were acquired from Sigma-Aldrich (St. Louis, MO). Acetonitrile (ACN) was purchased from Fisher Chemical (Waltham MA). HPLC grade water was prepared in lab using a Barnstead E-pure system series 582.

HPLC system

A single HPLC instrument was utilized for the detection of all drugs, allowing for limited maintenance and ease of work flow interruption between projects. Column utilized was a Waters Symmetry® C18 5 um 4.6 × 150 mm column. Pump was a Waters 1525 binary HPLC pump. Pump and column were coupled with a Water 2996 Photodiode Array Detector (PDA). All mobile phases were first filtered through a Waters four channel in-line degasser prior to entrance into pump housing. Mobile Phases utilized were HPLC grade water with 0.1% TFA and ACN with 0.1% TFA. A Waters 717 plus autosampler was employed to allow multiple sample injection. Software for analysis was Waters Empower Pro. Flow rate was set at 1 ml/min. Methods herein are described by percentage of ACN mobile phase, omitting the complimentary percentage of HPLC composition in the mobile phase.

Method development: A general outline

Methods for each drug were generated specifically for the provided HPLC apparatus. A drug solution of greater than 0.1 mg/ml was prepared in mobile phase solution depending on the solubility of the given drug. Sample was then analyzed on HPLC apparatus through a gradient method of 5% ACN to 95% ACN over a period of 20 minutes. The 5% to 95% range was chosen so as to limit usage of the end ranges as this will enhance the lifetime of the column. This HPLC trace provides a general range for the determination of the second method attempt. Based on the elution time the original run the HPLC peak presents, a second run is prepared with a range of 15% change in ACN over a space of 10 minutes. Based on presentation of peak within this second method, mobile phase composition was adjusted until the major eluent peak presented more than 5 minutes after the solvent injection peak, to be assured of peak resolution from the solvent front. Analytical wavelength for each drug was determined as the wavelength which corresponded with the maximum absorbance response compared to baseline noise for the sample peak as provided by the attached PDA detector. Following development of method, process was analyzed for reproducibility through generation of a standard curve consisting of at least five varying injection quantities and the confirmation of consistency through a R^2 value of greater than 0.99 when comparing micrograms injected and absorbance reported.

Results and Discussion

Results are presented for each drug subsequently. For each drug, methodology, analytical wavelength and standard curve are provided.

Cefuroxime

Cefuroxime sodium salt was determined to be best analyzed through the gradient method of 20 to 35% ACN over a time of 10 minutes. With this method, a peak was found at approximately 7.25 minutes. This peak was also determined to have a best absorbance for analysis at 275 nm. A standard curve was then prepared of a sample solution injected at different injection volumes, resulting in an R^2 value of 0.9999. Prepared standard curve is seen in figure 1.

Clindamycin

Clindamycin hydrochloride analysis method was developed to be 20% to 35% ACN over a period of 10 minutes. With this method, a peak was found at approximately 8.15 minutes. This peak was also determined to have a best absorbance for analysis at 210 nm. A standard curve was then prepared of a sample solution injected at different injection volumes, resulting in an R^2 value of 0.9999. Prepared standard curve is seen in figure 2.

Dexamethasone

Dexamethasone sodium phosphate was determined to be best analyzed through the gradient method of 25 to 40% ACN over a time of 10 minutes. With this method, a peak was found at approximately 8.15 minutes. This peak was also determined to have a best absorbance for analysis at 240 nm. A standard curve was then prepared of a sample solution injected at different injection volumes, resulting in an R^2 value of 0.999. Prepared standard curve is seen in figure 3.

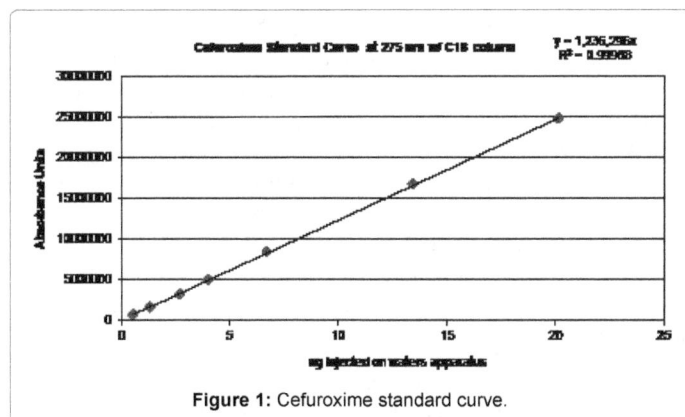

Figure 1: Cefuroxime standard curve.

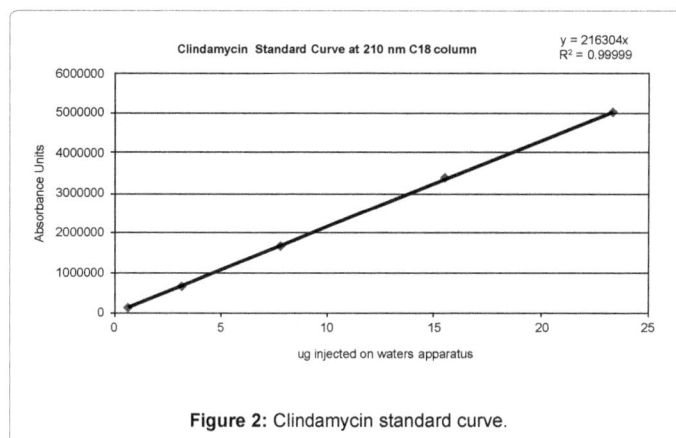

Figure 2: Clindamycin standard curve.

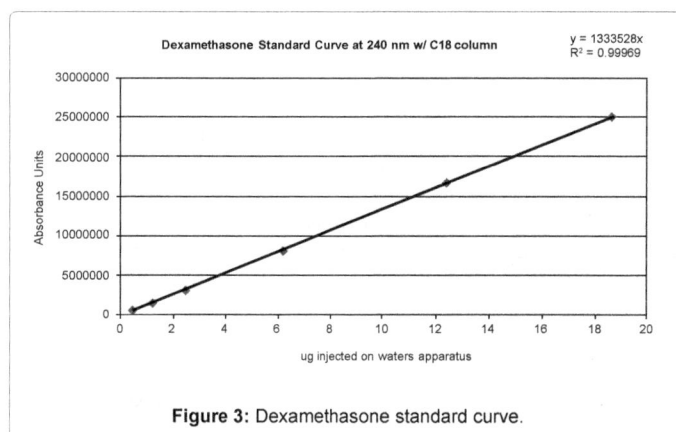

Figure 3: Dexamethasone standard curve.

Dicloxacillin

Dicloxacillin salt monohydrate was determined to be best analyzed through the gradient method of 45 to 60% ACN over a time of 10 minutes. With this method, a peak was found at approximately 7.33 minutes. This peak was also determined to have a best absorbance for analysis at 225 nm. A standard curve was then prepared of a sample solution injected at different injection volumes, resulting in an R^2 value of 0.999. Prepared standard curve is seen in figure 4.

Doxycycline

Doxycycline hyclate was determined to be best analyzed through the gradient method of 23 to 38% ACN over a time of 10 minutes. With this method, a peak was found at approximately 7.85 minutes. This peak was also determined to have a best absorbance for analysis at 350 nm. A standard curve was then prepared of a sample solution injected at different injection volumes, resulting in an R^2 value of 0.999. Prepared standard curve is seen in figure 5.

Metronidazole

Metronidazole free base was determined to be best analyzed through an isocratic method of 3% ACN over a time of 10 minutes. Gradient methods proved difficult due to the high water solubility of the drug, and a percentage outside of the 5 to 95% preferred range was required. This method exhibits the effectiveness of isocratic methods with the described HPLC apparatus. With this method, a peak was found at approximately 7.30 minutes. This peak was also determined to have a best absorbance for analysis at 275 nm. A standard curve was then prepared of a sample solution injected at different injection volumes, resulting in an R^2 value of 0.999. Prepared standard curve is seen in figure 6.

Oxymetazoline

Oxymetazoline hydrochloride was determined to be best analyzed through the gradient method of 25 to 40% ACN over a time of 10 minutes. With this method, a peak was found at approximately 8.10 minutes. This peak was also determined to have a best absorbance for analysis at 210 nm. A standard curve was then prepared of a sample solution injected at different injection volumes, resulting in an R^2 value of 0.999. Prepared standard curve is seen in figure 7.

Paclitaxel

Paclitaxel free base, free of excipients, was determined to be best analyzed through the gradient method of 50 to 65% ACN over a time of 10 minutes. With this method, a peak was found at approximately 6.55 minutes. This peak was also determined to have a best absorbance for analysis at 230 nm. A standard curve was then prepared of a sample solution injected at different injection volumes, resulting in an R^2 value of 0.999. Prepared standard curve is seen in figure 8.

Vancomycin

Vancomycin hydrochloride was determined to be best analyzed through the gradient method of 10 to 25% ACN over a time of 10 minutes. With this method, a peak was found at approximately 6.78 minutes. This peak was also determined to have a best absorbance for analysis at 210 nm. A standard curve was then prepared of a sample solution injected at different injection volumes, resulting in an R^2 value of 0.999. Prepared standard curve is seen in figure 9.

Tobramycin

Tobramycin sulfate salt was successfully prepared for HPLC

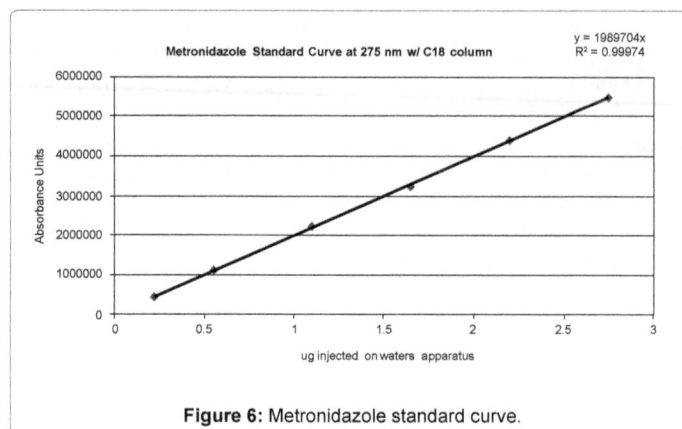

Figure 6: Metronidazole standard curve.

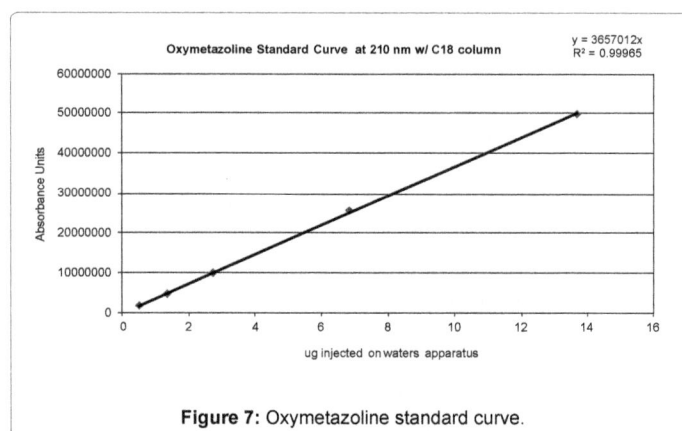

Figure 7: Oxymetazoline standard curve.

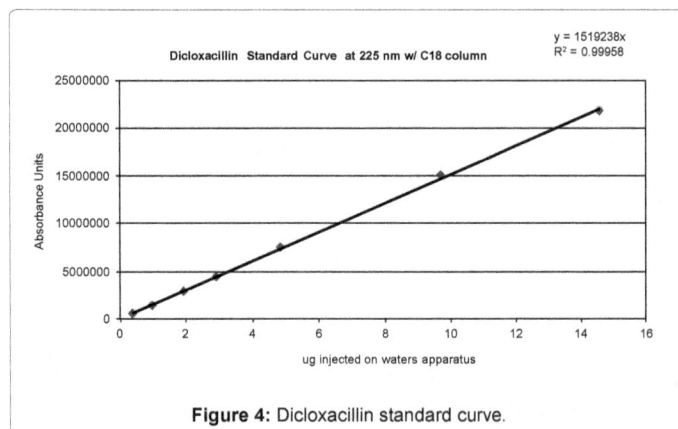

Figure 8: Paclitaxel standard curve.

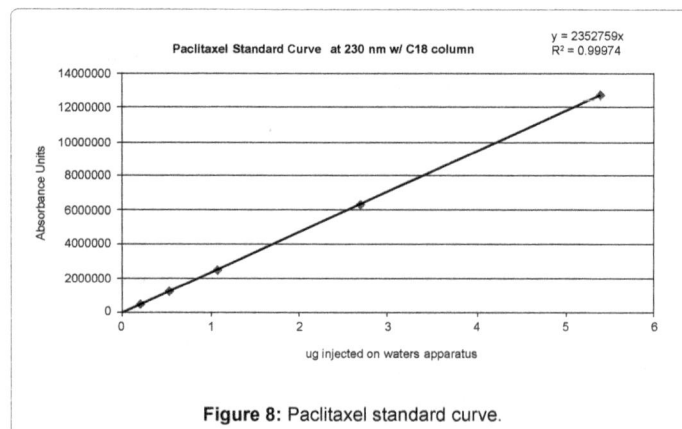

Figure 4: Dicloxacillin standard curve.

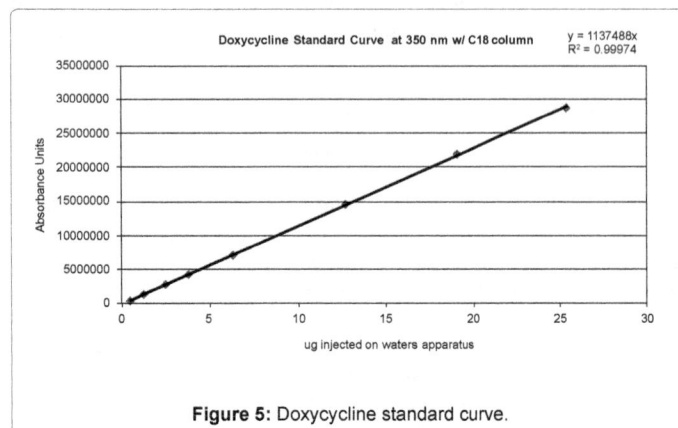

Figure 5: Doxycycline standard curve.

analysis through a pre-column derivatization as described [37]. A pre-column derivatization was required due to the lack of a chromophore on the tobramycin structure. Analysis of the tobramycin complex was determined to be best performed through the gradient method of 5 to 95% ACN over a time of 20 minutes. With this method, a peak was found at approximately 6.15 minutes. This peak was also determined to have a best absorbance for analysis at 254 nm. Other peaks present in chromatogram were omitted from standard curve analysis. A standard curve was prepared of a sample solution injected at different injection volumes, resulting in an R^2 value of 0.99. Prepared standard curve is seen in figure 10.

As the reaction utilized to prepare the HPLC visible complex is time sensitive, analysis of tobramycin visibility degradation was performed.

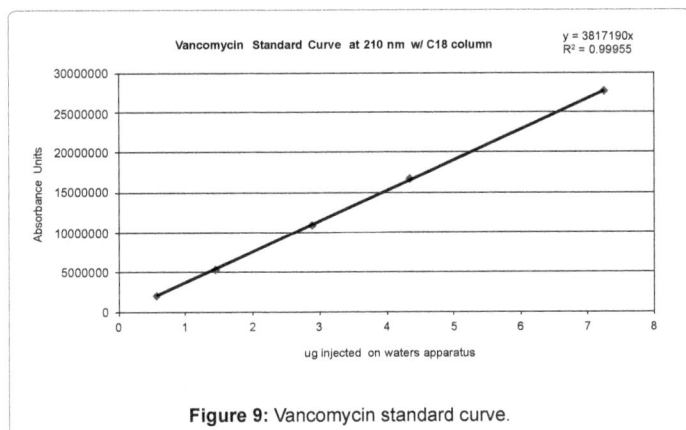

Figure 9: Vancomycin standard curve.

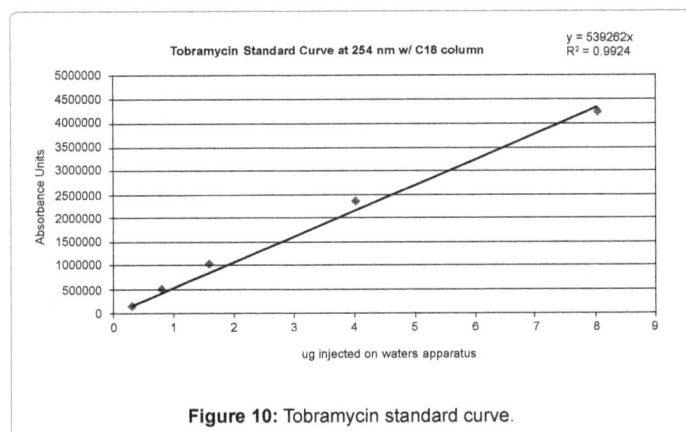

Figure 10: Tobramycin standard curve.

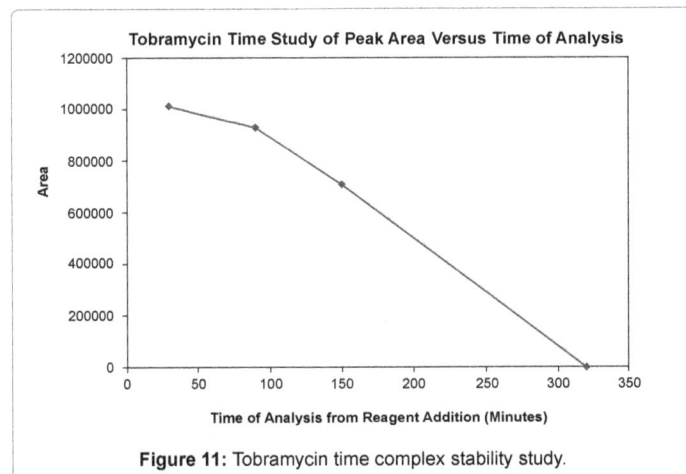

Figure 11: Tobramycin time complex stability study.

A sample was tested at 30, 90, 150 and 320 minutes for response. Results can be seen in figure 11. From this information it can be determined that analysis should take place in the first hour following the pre-column derivatization to achieve the best results and a repeatable method.

Conclusion

As can be seen from the provided data, multiple drugs can be analyzed from a single HPLC instrument not apparatus. Also, each drug has a well-defined and reproducible analytical method presented. These methods have been determined to be efficient for the drug at hand, and where applicable gradient methods allow for separation and detection of multiple compounds if called for by the experiment.

Drugs which are both hydrophobic and hydrophilic in nature have been quantified using a reverse-phase system. A method of detection for the aminoglycoside tobramycin is advanced and verified for the system. With a well-planned and executed methodology, HPLC can be a powerful tool for the detection of a wide range of antibiotics while requiring only a single apparatus.

References

1. Garcia CV, Breier AR, Steppe M, Schapoval EE, Oppe TP (2003) Determination of dexamethasone acetate in cream by HPLC. J Pharm Biomed Anal 31: 597-600.

2. Medenica M, Ivanovic D, Markovic S, Malenovic A, Misljenovic D (2003) Optimization of an RP-HPLC Method for Drug Control Analysis. Journal of Liquid Chromatography & Related Technologies 26: 3401-3412.

3. Ahuja S (2008) A Strategy for Developing HPLC Methods for Chiral Drugs. The Application Notebook. 70-79.

4. Joseph MJ (2003) Fast HPLC Analysis of Antibiotics Using Columns with Sub Two-Micron Particles. LC-GC North America, The Application Notebook 21: 41.

5. Thomas LA, Bizikova T, Minihan AC (2011) In vitro elution and antibacterial activity of clindamycin, amikacin and vancomycin from R-gel polymer. Vet Surg 40: 774-780.

6. Dejaegher B, Jimidar M, De Smet M, Cockaerts P, Smeyers-Verbeke J, et. al. (2006) Improving method capability of a drug substance HPLC assay. J Pharm Biomed Anal 42: 155-170.

7. Piva G, Farin D, Gozlan I, Kitzes-Cohen R (2000) HPLC method for determination of cefuroxime in plasma. Chromatographia 51: 154-156.

8. Vieira DC, Salgado HR (2011) Comparison of HPLC and UV spectrophotometric methods for the determination of cefuroxime sodium in pharmaceutical products. J Chromatogr Sci 49: 508-511.

9. Batzias GC, Delis GA, Athanasiou LV (2005) Clindamycin bioavailability and pharmacokinetics following oral administration of clindamycin hydrochloride capsules in dogs. Vet J 170: 339-345.

10. Batzias GC, Delis GA, Koutsoviti-Papadopoulou M (2004) A new HPLC/UV method for the determination of clindamycin in dog blood serum. J Pharm Biomed Anal 35: 545-554.

11. Gomez-Gaete C, Tsapis N, Besnard M, Bochot A, Fattal E (2006) Encapsulation of dexamethasone into biodegradable polymeric nanoparticles. Int J Pharm 331: 153-159.

12. Kumar V, Mostafa S, Kayo MW, Goldberg EP, Derendorf H (2006) HPLC Determination of dexamethasone in human plasma and its application to an in vitro release study from endovascular stents. Pharmazie 61: 908-911.

13. Song Y, Park J, Kim J, Kim C (2004) HPLC Determination of Dexamethasone in Human Plasma. Journal of Liquid Chromatography & Related technologies 27: 2293-2306.

14. Moya-Ortega M, Messner M, Jansook P, Nielsen T, Wintgens V, et al. (2011) Drug loading in cyclodextrin polymers: dexamethasone model drug. Journal of Inclusion Phenomena and Macrocyclic Chemistry 69: 377-382.

15. Dhoka MV, Sandage SJ, Dumbre SC (2010) Simultaneous determination of cefixime trihydrate and dicloxacillin sodium in pharmaceutical dosage form by reversed-phase high-performance liquid chromatography. J AOAC Int 93: 531-535.

16. Alderete O, Gonzalez-Esquivel DF, Del Rivero LM, Castro Torres N (2004) Liquid chromatographic assay for dicloxacillin in plasma. J Chromatogr B Analyt Biomed Life Sci 805: 353-356.

17. Samanidou VF, Nisyriou SA, Papadoyannis IN (2007) Development and validation of an HPLC method for the determination of penicillin antibiotics residues in bovine muscle according to the European Union decision 2002/657/EC. J Sep Sci 30: 3193-3201.

18. Samanidou VF, Giannakis DE, Papadaki A (2009) Development and validation of an HPLC method for the determination of seven penicillin antibiotics in veterinary drugs and bovine blood plasma. J Sep Sci 32: 1302-1311.

19. Kogawa AC, Salgado HR (2012) Quantification of Doxycycline Hyclate in Tablets by HPLC-UV Method. J Chromatog Sci.

20. Mitic S, Miletic G, Kostic D, Dokic D, Arisc B et, al. (2008) A rapid and

reliable determination of doxycycline hyclate by HPLC with UV detection in pharmaceutical samples. Journal of the Serbian Chemical Society 73: 665-671.

21. Skulason S, Ingolfsson E, Kristmundsdottir T (2003) Development of a simple HPLC method for separation of doxycycline and its degradation products. J Pharm Biomed Anal 33: 667-672.

22. Tashtoush BM, Jacobson EL, Jacobson MK (2008) Validation of a simple and rapid HPLC method for determination of metronidazole in dermatological formulations. Drug Dev Ind Pharm 34: 840-844.

23. Akay C, Ozkan SA, Senturk Z, Cevheroglu S (2002) Simultaneous determination of metronidazole and miconazole in pharmaceutical dosage forms by RP-HPLC. Farmaco 57: 953-957.

24. Miani PK, do Nascimento C, Sato S, Filho AV, da Fonseca MJ, et al. (2012) In vivo evaluation of a metronidazole-containing gel for the adjuvant treatment of chronic periodontitis: preliminary results. Eur J Clin Microbiol Infect Dis 31: 1611-1618.

25. Mahajan MK, Uttamsingh V, Gan LS, Leduc B, Williams DA (2011) Identification and Characterization of oxymetazoline glucuonidation in human liver microsomes: evidence of the involvement of UGT1A9. J Pharm Sci 100: 784-793.

26. Sudsakorn S, Kaplan L, Williams DA (2006) Simultaneous determination of triamcinolone acetonide and oxymetazoline hydrochloride in nasal spray formulations by HPLC. J Pharm Biomed Anal 40: 1273-1280.

27. Golubitskii GB, Basova EM, Ivanov VM (2008) Application of gradient high-performance liquid chromatography to the analysis of some multicomponent pharmaceutical preparations. J Anal Chem 63: 875-880.

28. Badea I, Ciutaru D, Lazar L, Nicolescu D, Tudose A (2004) Rapid HPLC Method for the determination of paclitaxel in pharmaceutical forms without separation. 34: 501-507.

29. Kim SC, Yu J, Lee JW, Park ES, Chi SC (2005) Sensitive HPLC Method for quantitation of paclitaxel (Genexol in biological samples with application to preclinical pharmacokinetics and biodistribution. J Pharm Biomed Anal 39: 170-176.

30. Yonemoto H, Ogino S, Nakashima MN, Wada M, Nakashima K (2007) Determination of paclitaxel in human and rat blood samples after administration of low dose paclitaxel by HPLC-UV detection. Biomed Chromatogr 21: 310-317.

31. Torne SJ, Ansari KA, Vavia PR, Trotta F, Cavalli R (2010) Enhanced oral paclitaxel bioavailability after administration of paclitaxel-loaded nanosponges. Drug Deliv 17: 419-425.

32. Jesus Valle MJ, Lopez FG, Navarro AS (2008) Development and validation of an HPLC method for vancomycin and its application to a pharmacokinetic study. J Pharm Biomed Anal 48: 835-839.

33. Liu M, Hu C (2006) Simultaneous Determination of the Purity and Potency of Vancomycin and Norvancomycin by HPLC. Chromatographia 65: 203-207.

34. Frick P, Nagy Z, Fekete J, Kettrup A, Gebefugi I (2001) Reverse Phase HPLC Method for Determination of Vancomycin in Influenza Vaccine. J Liq Chromatogr Relat Technol 24: 497-507.

35. Valentini F, Buldini PL, Landi E, Tampieri A, Tonelli D (2008) HPLC determination of tobramycin in a simulated body fluid. Microchemical Journal 90: 113-117.

36. He S, Chen Q, Sun Y, Zhu Y, Luo L, et al. (2011) Determination of tobramycin in soil by HPLC with ultrasonic-assisted extraction and solid phase-extraction. J Chromatogr B Analyt Tecnol Biomed Life Sci 879: 901-907.

37. Lai F, Sheehan T (1992) Enhancement of detection sensitivity and cleanup selectivity for tobramycin through pre-column derivatization. J Chromatogr 609: 173-179.

2

The Effects of Blanching, Harvest Time and Location (with a Minor Look at Postharvest Blighting) on Oleoresin Yields, Percent Curcuminoids and Levels of Antioxidant Activity of Turmeric (*Curcuma longa*) Rhizomes Grown in Jamaica

Cheryl E Green* and Sylvia A Mitchell

The Biotechnology Centre, Faculty of Science and Technology, University of the West Indies, Mona Campus, Kingston 7, Jamaica, West Indies

Abstract

Turmeric (*Curcuma longa*) grown in Jamaica was studied for its naturally occurring linear diarylheptanoid compounds namely curcumin, Bis-Demethoxy Curcumin (BDMC), Demethoxy Curucmin (DMC) and for its antioxidant activity. Evaluations were conducted on the basis of whether or not there were any potential effects of blanching, harvest time and location of growth on the quantity and quality of turmeric oleoresins. The highest antioxidant activity of 92.86% was obtained from turmeric rhizomes grown in the parish of Hanover while the highest turmeric oleoresin yields of 14.87% were obtained from the 15 minute blanched-treated turmeric rhizomes. With a new analog-selective RP-HPLC method, the curcumin, DMC, and BDMC were qualified and quantified. It was found that the highest yield of curcumin content of 22.69% was obtained from the 15 minute 'blanched' samples grown in the parish of Hanover from the 1st harvest period of the study. An analytical method validation with linear equations and correlation of regressions of $R^2=0.9991$, $R^2=0.9993$, $R^2=0.9998$ and $R^2=0.9992$ for inter-day precision analyses were performed to validate the HPLC method.

Keywords: *Curcuma longa*; Turmeric grown in Jamaica; Antioxidant activity; Blanching; Post-harvest; Effect of harvest time; Curcumin; Curcuminoids; Bis-demethoxy curcumin (BDMC); Demethoxy curcumin (DMC); DPPH (2, 2-diphenyl-1-picrylhydrazyl); Turmeric oleoresin; Blighting; Geographic effects on plant compounds; Spices

Introduction

In the midst of several 'consumer trends', culinary spices find themselves at a juncture in the food industry as they transition from conventional spices to 'nouveau healthy condiments'. Along with the bold flavour, ethnic flair and functional attributes, spices are healthy and natural, giving consumers an 'all-in-one' package deal. Culinary herbs and spices are enjoying this renaissance for the medicinal properties that they provide, as modern science proves the health promoting benefits that come from eating them [1].

Much of the focus about herbs and spices is in relation to their antioxidant content and how that relates to cancer prevention and their anti-inflammatory properties [2,3]. In the human body, metabolic reactions produce free radicals with unpaired electrons which make them reactive so they can interact with important cellular components such as DNA and cell membranes leading to cell damage, disease and ageing. Antioxidants stabilize these unpaired electron species by oxidative or reductive reactions and thus deter the propagation of damaging chemical chain reactions to take place within the important cellular matrix [4]. According to recent studies, it was found that 'salad dressings' that contain herbs and spices increase the antioxidant capacity of the vegetable salad itself [1]. The American Institute for Cancer Research suggests that herbs and spices should be used for their health-protective phytochemicals to ward off cancer and other diseases [1].

The Jamaican spice industry is well-known and respected for its high quality [5]; it is also of paramount importance to the island's economy [6]. Turmeric has been identified as one of the top ten commonly produced spices in Jamaica [7]. However, growing and processing of the Jamaican turmeric is done on an 'ad hoc' basis by individuals in the principal turmeric-growing parishes throughout the island of Jamaica [8]. The Ministry of Agriculture and Fisheries (MOAF) a government ministry of Jamaica has recommended a system for the orderly conduct and development of the Jamaican spice industry which includes: production of planting material; quality control at the farm-gate; dissemination of information to farmers about obtaining quality turmeric tubers; and the marketing, export and product promotion of the spice [9]. However, research to feed this improved system in Jamaica has been non-existent. This study was conceptualized out of the recognition that local Jamaican spice houses and Jamaican turmeric farmers will benefit from findings that will guide them to the best time to harvest and the best post-harvesting procedures to implement in order to acquire 'premium-grade' turmeric.

One step in the post-harvest processing of turmeric involves boiling or blanching the turmeric rhizomes [10]. Blanching (boiling in water for a set period of time) is carried out to facilitate the drying process of the rhizomes, to allow the starch granules to gelatinize and to inactivate enzymes [11,12]. The aim of this research was to evaluate a number of parameters that may impact the quality of turmeric during this post-harvest step. The parameters evaluated included blanching time, the time of year of harvest and the location where the turmeric was grown. The results from this study will provide information on the optimum time to harvest turmeric and how to further improve the post-harvest processing of turmeric. This information is beneficial to buyers and sellers of turmeric as this spice is deemed as an economically important value added product with proven health and anti-aging benefits [13,14].

*Corresponding author: Cheryl E Green, The Biotechnology Centre, Faculty of Science and Technology, University of the West Indies, Mona Campus, Kingston 7, Jamaica, West Indies, E-mail: greence100@yahoo.com

Materials and Methods

Chemicals

Commercial curcumin standard was purchased from Sigma Aldrich (St. Louis, MO). Turmeric standards kit containing curcumin, demethoxy curcumin and bis-demethoxy curcumin was purchased from Chroma Dex, Irvine, CA. The 1,1-diphenyl-2-picrylhdrazl (DPPH) was procured from Sigma Aldrich, St. Louis, MO. The 95% ethanol was purchased from Pharmco-AAper (Shelbyville, KY). HPLC grade isopropanol, acetonitrile and methanol and Fisher brand 25 mm 0.45 μm nylon syringe filters were purchased from Fisher Scientific (Fair Lawn, NJ). Whatman No. 1 filter paper was purchased from Whatman International Ltd. Maidstone, England.

Plant material

Turmeric was collected from two locations; the first location was from the traditional turmeric farming area in the western end of the island in the parish of Hanover (HAN) and the second from the eastern end of the island in the parish of St. Andrew (SA). There were a total of four harvesting periods: 1) February from both locations; 2) March from both locations; 3) the last week of April in Hanover and the first week of May in St. Andrew; and 4) June from both locations. Turmeric is a hardy, prolific crop that grows mostly wild in Jamaica all year round [9]. However when turmeric is cultivated in Jamaica, it is planted in September/October and harvested May/June of the following year, at full maturation in the 9th month of growth (as communicated by spice house, Betapac Ltd., Jamaica). Turmeric is reaped when the aerial leaves become 'dried-up' indicating readiness of 'harvest' for this spice.

Blanching

The harvested turmeric rhizomes were washed to remove debris and dirt. They were divided into three portions whereby the first portion was 'not blanched' and was assigned as the 'control' group. The other two portions were blanched by placing the turmeric rhizomes in pots of water so that all the rhizomes were submerged ensuring that no parts of the rhizomes were above the water level. The second portion of turmeric was boiled for 15 minutes and the third portion was boiled for 30 minutes.

After blanching, the turmeric rhizomes were sliced, steam dried and milled in an industrial miller with a mesh size of $^{1}/_{8}^{th}$ of an inch. Due to insufficiency of turmeric rhizomes in the first harvesting period, there was no 30 minute-blanching batch from Hanover. Subsequent to harvesting in the 3rd harvesting period, certain batches became blighted as they appeared moistened with an off-odour during the post-harvest procedure. Due to the nature of the study as a 'time-related' study, it was not possible to remedy the unforeseen blighting once the study

was under way. The blighted samples were kept, analysed and the data derived from them were used as a 'case study' so as to address the effects of 'deleterious issues' such as post-harvest blighting of turmeric rhizomes. Discussion and data for the blighted batches are assigned with an asterisk therein.

Classification of the samples

The turmeric samples were labelled according to location, harvesting period and blanch-treatment administered to the rhizomes. Hence the following codes: HAN1-0 represents samples from Hanover from the first harvest period with no blanching, HAN1-15 for those blanched for 15 minutes and HAN1-30 for 30 minutes. Similar codes were given for the samples from St. Andrew (SA1-0, SA1-15 and SA1-30). Then those for the second harvest are HAN2-0, HAN2-15 and so on.

Extraction of turmeric oleoresins

Each harvested batch of turmeric that was dried and milled was subjected to the following extraction procedure. A 50 gram portion of the dried milled turmeric was weighed and 700 mL of 95% ethanol was added to it. The mixture was allowed to macerate for 5 hours. After extraction, the remaining solution was filtered by gravity under suction. The residue was re-extracted with ethanol for 2 hours. The remaining solution was filtered by gravity under suction. Both filtrates were pooled and concentrated *in vacuo* using a rotary evaporator with a water-bath set at 50°C. The amount of oleoresin obtained, was weighed and the yield was calculated as percentage weight by weight (% w/w).

HPLC analysis of turmeric oleoresin

0.5 mg/mL turmeric oleoresin solution was prepared in HPLC grade methanol. A concentration of 0.5 mg/mL of commercial curcumin powder was prepared in HPLC-grade methanol. A stock solution of Bis-Demethoxy Curcumin (BDMC) and Demethoxy Curcumin (DMC) was prepared in methanol to yield 0.5 mg/mL solution. All prepared solutions were filtered through a 0.45 μm nylon syringe filter.

The HPLC settings

A relatively medium-polar quaternary mobile phase was prepared with isopropanol, acetonitrile, distilled water and acetic acid in a 3.0:1.5:5.0:0.5 v/v ratio. There was an injection volume of 20 μL, at a flow rate of 0.5 mL/minute, a run-time of 20 minutes, a wavelength setting at 420 nm, using an Agilent LiChrospher reversed phase RP C18 250 ×4 mm id column and an isocratic elution mode. Each sample preparation was injected in duplicates, separately onto the stationary phase column.

Preparation of a 4-point standard curve

A stock solution of 0.25 mg/mL curcumin standard was prepared

Hanover	Harvest 1	Harvest 2	Harvest 3	Harvest 4
HAN control	9.02 ± 0.10^{b}	12.00 ± 0.00^{b}	$^{*}11.07 \pm 0.07^{b}$	14.21 ± 0.46^{b}
HAN-15	7.22 ± 1.01^{b}	9.06^{a}	$^{*}8.83^{a}$	14.87^{a}
HAN-30	N D	10.58 ± 0.12^{c}	9.03 ± 0.95^{c}	12.22 ± 0.29^{b}
St. Andrew	**Harvest 1**	**Harvest 2**	**Harvest 3**	**Harvest 4**
SA-control	6.82^{a}	9.62 ± 0.07^{c}	8.20 ± 0.48^{c}	14.59 ± 0.44^{b}
SA-15	4.65^{a}	8.26^{a}	7.43 ± 0.07^{b}	14.34 ± 0.27^{b}
SA-30	6.78^{a}	10.29 ± 0.12^{b}	$^{*}5.89 \pm 0.02^{b}$	12.96 ± 0.60^{b}

[a]Mean values represent a single sample
[b]Mean values represent duplicates
[c]Mean values represent three replicates
[*] denotes the 'blighted' samples

Table 1: Mean percentage (%) turmeric oleoresin with standard error (±).

in methanol. From this stock solution, serial dilutions were prepared to yield 0.125, 0.0625, 0.0312 and 0.0156 mg/mL concentrations. A 4-point standard curve was generated by injecting each of the prepared serial dilutions of commercial curcumin standard onto the column. The linearity of the standard curve was assessed as correlation coefficient squared (R^2) where $R^2>0.995$ is considered to be linear, shown in Table 1.

An *in vitro* DPPH free radical-scavenging assay

The bioassay method described by Williams et al. was used as the antioxidant assay in this study [15]. A 50 µg/mL turmeric oleoresin solution was prepared in methanol. A blank was prepared by measuring 2.2 mL of methanol. A 0.02% DPPH stock solution was prepared in methanol. Accurately 400 µL of the 0.02 % (w/v) 1, 1-diphenyl-1-picrylhdrazl (DPPH) was removed and added to the 50 µg/mL oleoresin solution. This mixture was allowed to incubate for 20 minutes prior to determining the absorbance at 517 nm using a UV-vis spectrophotometer. From the 0.02% DPPH stock solution, 400 µL was removed and added to the methanol blank and was also allowed to be incubated along with the samples (above-mentioned) for 20 minutes prior to antioxidant activity measurements. Determination was conducted in duplicates.

DPPH is a stable free radical with a signature purple colour [14]. The turmeric oleoresins containing phenolic curcuminoids with radical scavenging activity reduced the 1, 1-diphenyl-2-picrylhdrazl (DPPH in the radical form) to the 2, 2-diphenyl-1-picrylhydrazine (the reduced DPPH form) which is colourless. The percentage antioxidant activity (% age AA) was determined by the following equation [15].

% Antioxidant activity calculation:

$$\%AA = \frac{Abs_{control} - Abs_{sample}}{Abs_{control}} \times 100$$

Statistical analysis

There were three independent variables in the studies; 1) the harvest season, an attribute variable; 2) the location, an attribute variable and 3) the blanching treatments, which is an 'active variable'. Both 'attribute' variables and the 'active' variable were evaluated on the basis of any potential effect they may have on the curcuminoid content, the oleoresin yields and the levels of antioxidant activities from the turmeric oleoresin.

There were three dependent variables in the study: namely the 1) curcuminoid content, 2) oleoresin yields and 3) antioxidant activity. These three dependent variables were assessed for whether or not they were affected by the independent variables: the time of harvest, location and blanching treatments.

Analysis of variance (ANOVA) was used to compare the variables with a significant difference at a *p-value*<0.05. The ANOVA was confined to data with corresponding values. With the application of a scatter plot for data points, the HPLC calibration curves with linear trend-lines, R-squared (R^2) values and line equations were generated.

Microsoft (MS) excel was used to generate all of the statistical analyses for this study. Standard error was calculated for oleoresin yields, % antioxidant activity and for the % curcuminoids and the results were expressed as the 'mean' % ± S.E. (standard error).

Results

Due to the relationship between curcuminoids and antioxidant activity and in order to accomplish the aim of the study, which is to assess the effects of blanching, harvesting and location on the quality and quantity of turmeric oleoresins, it was imperative to do the following tests: a gravimetric analysis to determine percent oleoresin yields; a HPLC analysis to determine curcuminoid content; and a bioassay to determine percent antioxidant activity of the turmeric oleoresins. The results from these analyses are discussed therein.

Oleoresin yields

In the parish of Hanover, the non-blanched rhizomes produced an increasing oleoresin yield over the period of study from 9.02% in the first harvesting period, 12.0% in the second and the highest yield of 14.21% in the fourth harvesting period (Table 1). From the parish of St. Andrew, the oleoresin yields also increased from 4.65% to 14.59% from the first to fourth harvest, respectively. Of the four harvesting periods, the highest oleoresin yields were obtained from the 4th harvesting period (highest was 14.87% from the parish of Hanover and turmeric rhizomes blanched for 15 minutes) (Table 1). *It must be noted that the blighted turmeric samples had unexpectedly the same levels of oleoresin yields as all the corresponding samples.

Curcuminoid content

Figure 1 represents the HPLC chromatogram for the turmeric oleoresin. Curcuminoid content was expressed as 'percent purity curcuminoid'. The order of retention times corresponding to each of the three curcuminoid derivatives was in accordance with the retention of the same derivatives observed from another HPLC method in the literature [16]. Curcumin was the main active of all three curcuminoids and its peak was the most abundant of the three derivatives, a finding that was in keeping with that reported in the literature [16]. It is notable that from all three curcuminoid derivatives, the highest quantity of curcumionoids occurred in the 1st harvest period with a noticeable decline in the other 3 ensuing harvest times (Table 2a, 2b and 2c).

The largest percent curcumin content of 22.96 ± 0.04% was obtained from HAN1-15 (the first harvesting period from Hanover blanched for 15 minutes). The demethoxy curcumin (DMC) content reflected the same trend with the highest DMC content (17.63 ± 0.14%) obtained from the first harvest, grown in Hanover (Table 2b), however unlike the curcumin derivative, the control group (HAN1-0) had the highest DMC content. For the third derivative, bis-demethoxy curcumin (BDMC), there was also a similar trend with the highest levels (13.09 ± 0.04%) obtained from HAN1-15 (Table 2c). *It must be noted that

% CURCUMIN

HANOVER	HARVEST 1	HARVEST 2	HARVEST 3	HARVEST 4
HAN CONTROL	20.47 ± 0.08	12.30 ± 0.00	*2.68 ± 0.01	11.32 ± 0.00
HAN-15	22.96 ± 0.04	11.36 ± 0.11	*7.30 ± 0.09	13.30 ± 0.02
HAN-30	N.D.	15.81 ± 0.16	12.14 ± 0.06	10.62 ± 0.00
ST. ANDREW	HARVEST 1	HARVEST 2	HARVEST 3	HARVEST 4
SA-CONTROL	13.19± 0.00	12.86 ± 0.02	9.44 ± 0.12	11.35 ± 0.08
SA-15	17.41 ± 0.15	16.56 ± 0.01	11.96 ± 0.12	9.47 ± 0.08
SA-30	13.74 ± 0.05	15.65 ± 0.03	*5.09 ± 0.01	9.92 ± 0.10

*Denotes the 'mean' values representing the 'blighted' samples from the 3rd harvesting period.

Table 2a: Mean percentage (%) curcumin content for four different harvested *Curcuma longa* rhizomes.

% DMC

HANOVER	HARVEST 1	HARVEST 2	HARVEST 3	HARVEST 4
HAN-CONTROL	17.63 ± 0.14	5.12 ± 0.01	*4.39 ± 0.01	5.06 ± 0.01
HAN-15	17.52 ± 0.07	4.89 ± 0.01	*4.78 ± 0.04	6.53 ± 0.01
HAN-30	N.D.	5.82 ± 0.01	7.02 ± 0.05	5.31 ± 0.01
ST. ANDREW	HARVEST 1	HARVEST 2	HARVEST 3	HARVEST 4
SA-CONTROL	11.60± 0.02	4.63 ± 0.02	6.25 ± 0.08	5.05 ± 0.01
SA-15	15.82 ± 0.07	5.27 ± 0.00	6.15 ± 0.02	4.61 ± 0.06
SA-30	11.75 ± 0.04	5.33 ± 0.01	*3.25 ±0.01	4.56 ± 0.00

Table 2b: Mean percentage (%) demethoxy curcumin (DMC) content for four different harvested *Curcuma longa* rhizomes.

HANOVER	HARVEST1	HARVEST 2	HARVEST3	HARVEST4
HAN-CONTROL	12.41 ± 0.10	2.55 ± 0.00	*1.71 ± 0.01	3.40 ± 0.06
HAN-15	13.09 ± 0.04	2.45 ± 0.01	*2.99 ± 0.03	4.26 ± 0.03
HAN-30	N.D.	3.05 ± 0.00	4.88 ± 0.02	3.20 ± 0.08
ST. ANDREW	HARVEST1	HARVEST 2	HARVEST3	HARVEST4
SA-CONTROL	9.71 ± 0.01	2.40 ± 0.00	4.60 ± 0.06	3.20 ± 0.12
SA-15	11.15 ± 0.04	2.98 ± 0.00	5.88 ± 0.01	3.27 ± 0.07
SA-30	8.86 ± 0.04	3.05 ± 0.01	*2.66 ± 0.01	3.12 ± 0.03

Table 2c: Mean percentage (%) Bis-demethoxy curcumin (BDMC) content for four different harvested *Curcuma longa* rhizomes.

Hanover	Harvest 1	Harvest 2	Harvest 3	Harvest 4
HAN control	92.86 ± 1.79	90.53 ± 0.42	'76.91 ± 0.36	90.60 ± 0.17
HAN-15	90.50 ± 0.43	90.27 ± 0.91	'79.47 ± 4.96	92.58 ± 0.04
HAN-30	N D	92.55 ±0.34	90.72 ± 0.95^c	91.31 ± 0.34
St. Andrew	**Harvest 1**	**Harvest 2**	**Harvest 3**	**Harvest 4**
SA-control	88.69 ± 2.20	92.11 ± 0.13	88.16 ± 0.47	91.78 ± 0.13
SA-15	92.77 ± 0.16	92.80 ± 0.16	89.38 ± 1.10	90.32 ± 0.30
SA-30	90.34 ± 1.79	92.11 ± 1.32	'81.78 ± 1.19	91.68 ± 0.27

'Denotes the 'mean' values representing the 'blighted' samples from the 3rd harvest period.

Table 3: Mean percentage (%) antioxidant activity with standard error (±).

the blighted samples had very low curcumin, BDMC and DMC content with levels as low as 2.68 ± 0.01% for curcumin, suggesting the adverse effects of blighting on the 'quality' of the turmeric oleoresin (Table 2a).

Antioxidant activity

The amount of antioxidant activity was expressed as percent antioxidant activity per gram extract. This bioassay uses the DPPH as the 'model free radical' which is scavenged by compounds such as the curcuminoids inherent in turmeric (*Curcuma longa*) rhizomes. As a result of the scavenging of electrons, there was a decrease in the absorbance of the DPPH solution which was quantified at an absorbance reading of 517 nm. The decrease in absorbance was inversely proportional to the antioxidant activity of the turmeric oleoresin (Table 3). The highest antioxidant activity of 92.86% came from Hanover, in the first harvesting period from the control samples, HAN1-0 (Table 3). All the samples tested were considered to be of 'high' antioxidant activity as compared to other plant materials in the literature [15]. *It must be noted that the blighted samples also had relatively 'high' antioxidant activity, for example there was 79.47% from the 15 minute blanched sample; however the blighted samples were consistently less than that of the untainted samples (Table 3). The turmeric oleoresins from this research were able to scavenge the free radicals (DPPH solution) at a concentration of 50 µg/mL.

An overall look at the effects between the 'observed' and 'manipulated' variables via ANOVA

A One-Way Analysis of Variance (ANOVA) can test three or more

"means" at a time by using variances [17]. ANOVA allows for the analysis of the interactive effects between variables and is ideal to test complex hypotheses [17]. The ANOVA was employed to give an 'overall effect' of the variables as it simultaneously compares the "means" in several groups as well as variances across groups [18](Table 4).

In this study the one-way analysis of variance (ANOVA) was used to evaluate the independent variables and their effects on the dependent, observed variables. The 'harvest period' had 4 levels (harvest 1, 2, 3, and 4); while 'location' had 2 levels (HAN [Hanover] and SA [St Andrew]) while the 'blanching treatments' had 3 levels (0, 15 and 30 minutes treatment). ANOVA was confined to data with corresponding values.

Overall effects of 'blanching' on turmeric oleoresin yields, curcuminoid content and antioxidant activity via ANOVA

There were 9 sets of ANOVA analyses conducted in respect to the 'effects of blanching' on oleoresin yields, curcuminoid content and antioxidant activity of turmeric extract (Table 4). The first set of ANOVA involved the comparison between the control and the 15 minute blanching treatments for the harvest periods of 1, 2 & 4 in Hanover (HAN). No statistical significance was found in terms of oleoresin yields, curcuminoid content and antioxidant activity. When the control and 30 minute-blanching treatment was compared, there was a significant difference for oleoresin yield and antioxidant activity. For the comparison between the 15 and 30 minute blanching treatments, there was a significance in the percent curcumin content, with the higher levels of curcumin obtained from the 30 minute

Variables compared	Factors incorporated	% Oleoresins	%Antioxidant	% Curcumin	% BDMC	% DMC
BLANCHING						
Control vs. 15 min	Hanover, Harvest 1, 2 & 4	NS	NS	NS	NS	NS
Control vs. 15 min	St Andrew, Harvest 1, 2 & 4	NS	NS	NS	NS	NS
Control vs. 15 min	HAN & SA, Harvest 1, 2 & 4	NS	NS	NS	NS	NS
Control vs. 30 min	Hanover, Harvest 2 & 4	Control >30 min	30 min >control	NS	NS	NS
Control vs. 30 min	St Andrew, Harvest 2 & 4	NS	NS	NS	NS	NS
Control vs. 30 min	HAN & SA, Harvest 2 & 4	NS	NS	NS	NS	NS
15 min vs. 30 min	Hanover, Harvest 2 & 4	NS	NS	30 min >15 min	NS	NS
15 min vs. 30 min	St Andrew, Harvest 2 & 4	NS	NS	NS	NS	NS
15 min vs. 30 min	HAN & SA, Harvest 2 & 4	NS	NS	NS	NS	NS
HARVESTING						
2nd vs. 4th harvest	Hanover - 0, 15 & 30	4th>2nd	NS	NS	4th>2nd	NS
2nd vs. 4th harvest	St Andrew - 0, 15 & 30	4th>2nd	NS	2nd>4th	4th>2nd	NS
2nd vs. 4th harvest	HAN & SA - 0, 15 & 30	4th>2nd	NS	2nd>4th	4th>2nd	NS
1st vs. 2nd harvest	HAN & SA - 0 & 15 min	2nd>1st	NS	1st>2nd	1st>2nd	1st>2nd
1st vs. 4th harvest	HAN & SA - 0 & 15 min	4th>1st	NS	1st>4th	1st>4th	1st>4th
2nd vs. 4th harvest	HAN & SA - 0 & 15 min	4th>2nd	NS	NS	4th>2nd	NS
1st vs. 2nd harvest	HAN - 0 & 15 min	NS	NS	1st>2nd	1st>2nd	1st>2nd
1st vs. 4th harvest	HAN - 0 & 15 min	4th>1st	NS	1st>4th	1st>4th	1st>4th
2nd vs. 4th harvest	HAN - 0 & 15 min	4th>2nd	NS	NS	4th>2nd	NS
1st vs. 2nd harvest	SA - 0 & 15 min	2nd>1st	NS	NS	1st>2nd	1st>2nd
1st vs. 4th harvest	SA - 0 & 15 min	4th>1st	NS	1st>4th	1st>4th	1st>4th
2nd vs. 4th harvest	SA - 0 & 15 min	4th>2nd	2nd>4th	2nd>4th	4th>2nd	NS
LOCATION						
HAN vs. SA	1st, 2nd & 4th harvest - 0 & 15 min	NS	NS	NS	NS	NS
HAN vs. SA	2nd & 4th harvest -0, 15 & 30 min	NS	NS	NS	NS	HAN>SA
HAN vs. SA	2nd harvest -0, 15 & 30 min	HAN>SA	NS	NS	NS	NS
HAN vs. SA	4th harvest -0, 15 & 30 min	NS	NS	HAN>SA	NS	HAN>SA
HAN vs. SA	Control - 1st , 2nd, & 4th harvest	NS	NS	NS	NS	NS
HAN vs. SA	15 min - 1st , 2nd, & 4th harvest	NS	NS	NS	NS	NS
HAN vs. SA	30 min - 1st , 2nd, & 4th harvest	NS	NS	NS	NS	NS

Table 4: ANOVA with *p-value*<0.05.

blanching treatment (Table 4).

Overall effect of 'harvesting time' on turmeric oleoresin yields, curcuminoid contents and antioxidant activities via ANOVA

Twelve sets of statistical ANOVA were conducted to assess the effects of harvesting periods on % oleoresin yields, % antioxidant activity levels, % curcumin, % BDMC and % DMC content (Table 4). The ANOVA analysis between the 2nd versus 4th harvest periods (incorporating all three blanching treatments) had a significant difference in 'oleoresin yields' and 'BDMC content' with the larger values occurring in the 4th harvesting period. In the case of the 'curcumin content' the 2nd harvest gave significantly higher yields. There was a significant difference between the harvesting periods except for antioxidant activity which only had one significant case of more antioxidant activity from the 2nd harvest (Table 4).

Overall effect of 'location' on turmeric oleoresin yield, curcuminoid content and antioxidant activity via ANOVA

There were four cases of significant differences as it relates to effect of location on oleoresin, curcuminoid and antioxidant activity. This confirmed that there was a significantly higher oleoresin yields, % curcumin and %DMC content from the parish of Hanover than from the parish of St Andrew (Table 4).

Discussion

Findings for the effects of blanching, harvesting and location on the % oleoresin yields

The maximum oleoresin yield for the turmeric grown in Jamaica was 14.87%; this is comparably high as it relates to the yields of "6 to 10%" from a similar solvent extraction procedure of turmeric powder recorded in the literature [19]. Another literature review records oleoresin yields varying from as low as 4% to as high as 24.3% depending on the cultivar of turmeric tested [20].

To account for the effects of 'blanching length of time' on oleoresin yields, a similar study which evaluated the effects of blanching on the yields of turmeric oleoresins was identified for comparison purposes. The results from their studies showed that at 15, 20, 25 and 30 minutes there was 4.99, 4.98, 4.86 and 4.81% oleoresin yields respectively. Their yields were low and there was no apparent difference between them [21]. It is notable that the present study on turmeric grown in Jamaica had variations in oleoresin yields between blanching treatments (Table 1) and some of these differences were significant (Table 4).

There was progressive increase in the oleoresin yields with respect to the 'time' of harvest. This was evident in both locations, and this effect was significant (Table 4). An example of the increase was noticed in St Andrew where oleoresin yields of 4.65% from the 1st harvest climbed to 14.34% in the 4th harvest period, of the same 15 minute blanched treatment (Table 2).

Turmeric oleoresins contain endogenous pigments and essential oils [22]. According to Shiyou et al. [23], the essential oils of turmeric reach their maximum quantity at about 7.5 and 8 months of plant maturity and this could account for the increase in the oleoresin yields (over time) as turmeric oleoresins are comprised of essential oils [23]. As such, it is postulated that the increase in the turmeric oleoresin yields is in relationship to the maturity of the turmeric plant.

The effect of location takes into account microclimatic differences in rainfall and temperature from one location to the other. Climatic changes have an effect on the production of plant chemicals even of the same plant variety and there can be varying amounts of phytochemicals when the plant is grown in different soils under different weather conditions [24]. The result from this research showed that there were significantly higher oleoresin yields obtained from the parish of Hanover for the first three harvesting periods of the study but the percent yields were similar in the 4th harvest period for both locations (Tables 1 and 4).

Findings for the effects of blanching, harvesting and location on the curcuminoids content

The 1st harvest period for this research took place in the month of February considered the 5th month of maturity for cultivated turmeric in Jamaica. There was a decrease of curcuminoid content between the 5th and 9th month of the growing season for turmeric in Hanover and it took place in the 6th to 9th month in St. Andrew, (Tables 2a-2c).This phenomenon of decrease of curcuminoid content with respect to plant maturity is in keeping with another study in literature by Shiyou et al.[23] whereby there was an increase in curcumin quantity with the highest yields occurring at 5 ½ month of maturity after which there was a noticeable decline of the pigments towards full maturity at the 9th month, mimicking the very same trend found in this study with the turmeric grown in Jamaica [23].

It has been suggested that curcuminoids reach their peak during a certain stage of plant maturation as a consequence of dynamic changes such as plant growth, rate of biosynthesis and the probable decrease in gene expression for these pigments due to loss of plant vigor over time. As such, production of these metabolites decline and they remain in situ. Knowledge of the precursors and biosynthetic pathways responsible for the formation of these secondary metabolites in the growth cycle of turmeric is not fully understood [25].

In the literature with respect to the effects of blanching on curcumin content, a study was conducted to evaluate the length of boiling time on the curcumin content from turmeric. The results from their studies showed that boiling turmeric at 15, 20, 25 and 30 minutes yielded curcumin contents of 4.23, 4.21, 3.91 and 2.29% respectively [21]. It must also be noted that there was a 'difference' between the 15 and 30 minutes of boiling with almost 54% more curcumin content obtained from the 30 minutes of boiling in comparison to the 15 minute boiling treatment [21]. In the case of the turmeric grown in Jamaica, there was overall insignificant difference observed between the blanched samples (Table 4). However, there was one noticeable difference with the yields of curcuminoids from the parish of St. Andrew in the 1st harvest period between the 15 minute and the 30 minute blanched treated samples

whereby the 15 minute blanched treatment gave yields of 17.41% while the 30 minute blanched treatment gave yields of 13.74%, (Table 2a); the same scenario was observed for the BDMC and the DMC (Table 2b and 2c). Although small, this variation is consistent with the literature [21] and deserves attention and further study.

With respect to effect of location on curcuminoid content, the highest curcumin content from Hanover was 22.96% while the highest curcumin content from St Andrew was 17.41% (Table 2a). The highest curcumin, the highest BDMC and the highest DMC contents all came from Hanover suggesting that this location was more favourable for yielding higher quantities of these metabolites from the turmeric rhizome (Tables 2a-2c).

Findings for the effects of blanching, harvesting and location on the antioxidant activity of turmeric oleoresins

The highest antioxidant activity of 92.8% came from the parish of St. Andrew, which was not significantly higher than that obtained from Hanover (Table 3, 4). It was however considerably higher than a comparable study in literature, conducted using a similar DPPH free radical methodology to determine the antioxidant activity of cultivated turmeric with results of 78.8% for organically grown turmeric and 67.5% for conventionally grown turmeric respectively [26]. These levels of antioxidant activity, though considerably high, were lower than those obtained from this research on turmeric grown in Jamaica, *even the blighted ones (Table 3).

In the literature a similar blanching study was conducted on white saffron (Curcuma mangga) and the antioxidant activity obtained from the non-blanched white saffron in comparison to the 5 minutes blanched treated in 0.05% citric acid solution were different with a higher antioxidant activity obtained from the blanched treated samples [27]. The studies conducted on turmeric grown in Jamaica, showed minor differences; for example in the 1st harvest period in St. Andrew the 15 minute blanched turmeric gave yields of 92.77% antioxidant activity while the 30 minute blanched treated turmeric gave yields of 90.34%, but there was no significant differences (Table 4).

It is said that the levels of antioxidant activity of medicinal plants is dependent on harvesting or rather on the seasonal variations and the time of the year that the plant is reaped [28]. However, this was not in accordance with the results on turmeric grown in Jamaica whereby there was no apparent effect of harvesting on the level of antioxidant activity (Table 4).

With respect to the effects of location on the antioxidant activity of medicinal plants, from the literature it is said that geographical regions of growth have some impact on the antioxidant activity of medicinal plants [28]. However, in the case of the research done on the turmeric grown in Jamaica there was no apparent effect of location on the level of antioxidant activity from the turmeric oleoresins (Table 4).

Inter-day Precision Analysis for HPLC validation

A linear curve analysis was generated to confirm the accuracy of the HPLC method. This improved HPLC method achieves excellent

Test day	Linear equation	Correlation of regression (R^2)
Day 1	y=702.03x – 44.52	R^2= 0.9991
Day 2	y=76617x – 691.23	R^2= 0.9993
Day 3	y=3076.2x + 827.45	R^2=0.9998
Day 4	y=20391x + 57.36	R^2=0.9992

Table 5: Calibration curves for the validation of the RP HPLC chromatograph method.

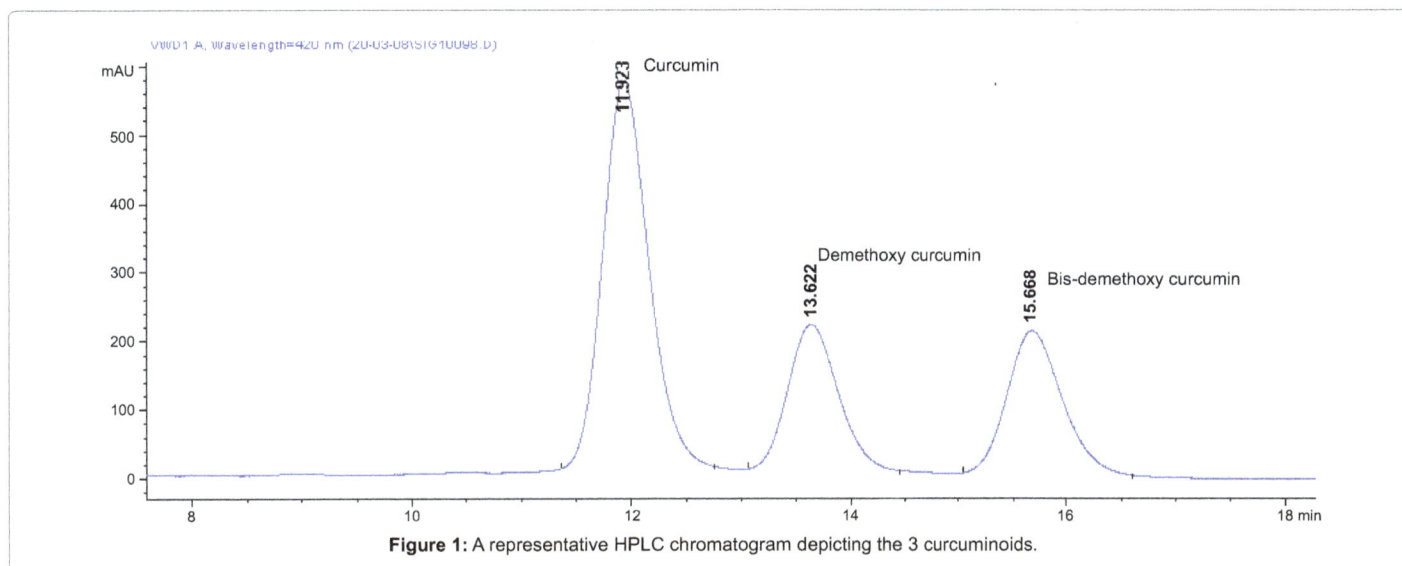

Figure 1: A representative HPLC chromatogram depicting the 3 curcuminoids.

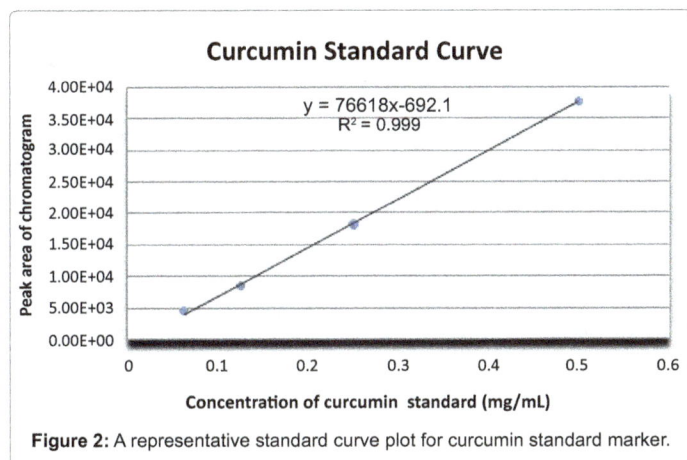

Figure 2: A representative standard curve plot for curcumin standard marker.

separation with a steady, noise-free baseline and with minimum shouldering between the chromatographic peaks. The method is considerably 'sensitive' as it requires relatively low microgram quantities of the turmeric oleoresin and is detected at a relatively short run-time with all 3 compounds fully eluted at about 17 minutes with all three curcuminoid derivatives completely eluted (Figure 1). In comparison to another HPLC method in the literature, the sensitivity and run-time parameters are markedly improved with this method [16]. With the use of the newly developed reversed phase HPLC method, the curcumin, demethoxy-curcumin and the bis-demethoxy curcumin were qualitatively and quantitatively determined. The analytical method was validated with the application of inter-day precision with correlations of $R^2=0.9991$, $R^2=0.9993$, $R^2=0.9998$ and $R^2= 0.9992$ for days 1, 2, 3, and 4 respectively. Along with the R^2 values, there was recovery data. For day one, there was an R^2 value of 0.9991 with a corresponding recovery data of $y=702.03x – 44.52$, see table 5 for the other consecutive days. All the correlation of regressions had high R^2 values indicating statistical soundness of the HPLC method, (Table 5). A representative standard curve for curcumin is shown in Figure 2. The inter-day precision method confirms the accuracy of the new optimized HPLC method as an ideal tool in the time analysis of endogenous pigments from turmeric oleoresins.

The matter of postharvest blighting

*There was an outstanding decrease in the curcuminoid content observed in the blighted turmeric samples in the 3rd harvest period (Tables 2a, 2b and 2c) and this occurrence is supported by research findings which have found that plant diseases such as 'root rot' suppress plant growth and plant quality [35]. The antioxidant activity of the turmeric oleoresin was also affected, with lower yields observed in the blighted samples (Table 3). Blighting is a cause of concern for tuberous spice rhizomes such as turmeric and ginger (both are from the Zingiberaceae plant family). A study was conducted to identify fungi and bacteria associated with the "postharvest rot" of ginger rhizomes (Zingiberofficinale Roscoe) in the Serrana region of Espírito Santo, Brazil. It was difficult to establish appropriate guidelines for the postharvest management of these ginger rhizomes in Brazil, due to the lack of information about the etiological agents associated with the rot [35]. The lack of information pertaining to the post-harvest pathology for turmeric requires further studies to curtail this problem.

Summary

Spices are often eaten daily and are one of the most promising and readily available sources of antioxidants. The findings in this research show that specific harvesting periods during the year produced the highest levels of curcumin, a compound recognized for its applications in the prevention of many degenerative diseases.

To fully appreciate the importance of the antioxidant activity of turmeric as evidenced in this study, leading to its application in the prevention of many degenerative diseases, the following information is relevant. There are important metabolic redox reactions that harness electrons and protons. This cellular activity is vital because proton movement is the driving force that provides the majority of ready energy (ATP) for the cell as nutrients are broken down. Reduction involves the addition of negatively charged electrons and it is an endergonic reaction as it takes the compound to a higher energy state. Redox reactions are important in metabolism as they relate to energy carrier compounds, nicotinamide adenine dinucleotide (NAD^+/NADH) and flavin adenine dinucleotide (FAD/$FADH_2$). These two molecules are specialized hydrogen carriers as they transport hydrogens and their associated energy to specialized metabolic pathways that generate energy (ATP) via the use of O_2 [40]. However during these metabolic reactions, reactive oxygen species or free radicals, such as superoxide anion are generated and these have detrimental effects on human health. Antioxidants can inhibit or retard oxidation either by scavenging the free radicals that initiate oxidation or by breaking the oxidative chain reactions [41]. Free radicals are implicated in cell aging, cell damage, damaged tissue, chronic diseases and overall poor health [15]. Research has indicated that there is a superfluous amount of active oxygen radicals produced in the human body causing stress at the cellular level and consequently inflammation, aging and diseases [29]. Biochemical processes such as reduction of oxygen (via enzyme oxidases), phagocytosis, and the formation of "advanced glyco-oxidation end products" (also known as AGEs) are constant sources that generate these free radicals. Environmental pollutants also add to the intracellular formation of Reactive Oxygen Species (otherwise known as ROS). Furthermore there are reactive nitrogen species such as nitric oxide (NO) and peroxynitrate ($\cdot$OONO$_2$) which when combined overwhelms the body's antioxidant defense system leading to a biochemical state called 'oxidative stress' [15]. There is growing evidence that acute overproduction of Reactive Oxygen Species (ROS) under pathophysiologic conditions is implicated in cardiovascular diseases [15]. A number of drugs termed 'antioxidants' such as β-blockers, inhibitors such as angiotensin-converting enzyme and calcium antagonists to name just a few have been administered to manage heart diseases [15]. There are another class of antioxidants namely the 'spice antioxidants' and these are important in the chemoprevention of lipid peroxidation, inflammation, cancer and the overall retardation of the aging process [29]. Curcumin, a spice antioxidant, has been confirmed as a complementary natural alternate medicine for cancer treatment [30].

Increase in consumer and food manufacturer interests in 'natural alternates' has led to an astronomical increase of 175% 'natural alternatives' between 1989 and 1990. During that time, the number of products that made claims to be 'without additives or preservatives' rose by 99% [31]. As a consequence, much emphasis has been given to the identification of natural antioxidants in food products. The research on natural antioxidants has developed enormously and there has been an enhanced public awareness of health issues too. Natural antioxidants have some advantages that their synthetic counterparts do not have. In the first place, natural antioxidants are more readily acceptable over the synthetic antioxidants; this is due to the fact that natural antioxidants are considered safe and many of them are identified on the Generally Recognized as Safe (GRAS) list. Turmeric is classified as a 'CPG Sec. 525.750 spice' on the GRAS list [32]. The rigorous criteria and procedures used to evaluate the "safety" of all substances on the GRAS list did not warrant genotoxicity analysis of the turmeric oleoresins in this study. This procedure includes many scientific studies on the genotoxicity of turmeric which have conclusively shown the safety of turmeric, curcumin and essential oil derivatives. Indeed, turmeric has been found to alleviate genotoxicity induced by other agents [42]. Also, a test for genotoxicity was not needed for this paper as it delves into the factors that may affect production of the curcumin in the plant and not on its effect in humans.

In the case of synthetic antioxidants there are concerns regarding the pathological effects of using them and this has prompted studies to find natural alternatives in the last few decades [33]. Butylated hydroxyanisole (BHA) and buytlated hydroxytoluene (BHT) are two well-known synthetic antioxidants used as food additives however they are restricted in some countries due to the possibility that they may have undesirable effects on the enzymes of human organs. Due to this controversy, there is growing interest in finding safe substitutes from natural resources such as edible plants, herbs and spices [33].

It is a 'paradox' that on the one hand 'oxygen' is needed to sustain life but on the other hand oxygen is the primary cause for oxidative stress in the human body and the degradation of foods. In order to address this issue there has been much focus on the involvement of free radicals responsible for aging and diseases [15]. Antioxidants are the defence system *in vivo* and the first line of defence to inhibit the formation of Reactive Oxygen Species (ROS) and free radicals which are implicated in many degenerative diseases such as cancer [34].

Conclusion

This study is the first analysis done on Jamaican turmeric and it has exposed their high quality. The maximum levels of curcuminoids occurred in the month of February and March, in Hanover and St. Andrew respectively. Maximum turmeric oleoresins occurred in the month of June in both locations. The overall anti-oxidant activity levels were high throughout the four harvesting periods, and in both locations, and are comparatively high in comparison to similar studies in literature [15]. These results warrant further more in-depth study with more samples and more locations to further analyze the factors that lead to high quality turmeric harvest. The high anti-oxidant activity for the turmeric extract should also lend leverage for the locally grown turmeric in Jamaica as an important spice with value-added benefits for its medicinal and health enhancing applications.

Recommendations

Despite the fact that there was minor impact of blanching on curcuminoid content, oleoresin yields and antioxidant activities (Table 4), information from the literature considers blanching a good post harvesting practice and an important preparative step hence it is recommended [36].

Curcumin is a 'quality indicator' as it is used to measure the colour value of turmeric as well as its antioxidant properties [37,38]. If the buyer of turmeric is interested in the 'quality' of turmeric oleoresin which is indicative of the abundance of curcumin content, then it is suggested that the turmeric rhizome be harvested about the 5th and 6th month of maturity when the curcuminoids were found to be at their maximum peak. However if the buyer of turmeric is more interested in 'quantity' of turmeric oleoresin, then the turmeric rhizome should be reaped at full maturity in accordance to the research results from this study. The opportunity to harness the highest levels of curcumin from turmeric is important as it is a highly therapeutic secondary metabolite [39].

Acknowledgment

This study was funded by an Environmental Foundation of Jamaica (EFJ) grant and a UWI research and publication grant. The utilization of certain laboratory instruments was made possible through the Scientific Research Council, Kingston Jamaica, and the Chemistry Department of the University of the West Indies, Mona campus, Kingston Jamaica. We are grateful and thankful to everyone who enabled the successful completion of this study. The research study is in the partial fulfilment for the PhD thesis for Cheryl E Green

References

1. Palmer S (2008) Food Product Design-Herbs and Spices: Hot and Healthy. Food Product Design.

2. Milner JA, Romagnolo DF (2010) Bioactive Compounds and Cancer. Humana Press, Springer, New York, NY, USA, pp 671.

3. Ninfali P, Mea G, Giorgini S, Rocchi M, Bacchiocca M (2005) Antioxidant capacity of vegetables, spices and dressings relevant to nutrition. Br J Nutr 93: 257-266.

4. Whitney E, Rady R (2012) Understanding Nutrition.Cengage Learning, Belmont, CA, USA, pp 361.

5. Jamaica Promotions Corporation (2012) Flavours of Jamaica, JAMPRO

publication, Kingston, Jamaica, p. 44.

6. Richardson MA (1993) A paper on spices and aromatic plants produced in Jamaica relative to roundtable on the development of spices and aromatics in the CARICOM countries 12th -15th October, 1993, Jamaica p. 1.

7. Export Centre and Business Information Point (2008) Market Brief-Sauces and Spices Bulletin. Jamaica Exporters Association, p. 5.

8. (1991) Report of the third meeting of the International Spice Group: Kingston, Jamaica, 18-23 November.

9. Collinder A (2012) 510 acres to be planted with ginger, turmeric. Business section, the Jamaica Gleaner.

10. Parthasarathy VA (2008) Organic Spices. New India Publishing Agency, New Delhi, India, pp 420.

11. Plotto A (2004) Turmeric: Post-Production Management, Food and Agriculture Organization of the United Nations (FAO), p 10.

12. Panda H (2010) Handbook on Spices and Condiments (Cultivation, Processing and Extraction). Asia Pacific Business Press Inc., New Delhi, India, pp 22

13. Comstock F (2010) Antiaging 101: Course Manual: A Proactive Preventive Health Care Program. AuthorHouse, Bloomington, USA, pp 89.

14. Jain PK, Agrawal RK (2008) Antioxidant and Free Radical Scavenging Properties of Developed Mono- and Polyherbal Formulations, Asian J Exp Sci 22: 213-220.

15. Williams LAD, Hibbert SL, Porter RBR,Bailey-Shaw YA, Green CE (2006) Jamaican plant with in vitro anti-oxidant activity. In Biologically active natural products for the 21st Century. Research Signpost, Kerala, India, Chapter 2. pp. 2, 3, 8, 11.

16. Green CE, Hibbert SL, Bailey-Shaw YA, Williams LA, Mitchell S, et al. (2008) Extraction, processing, and storage effects on curcuminoids and oleoresin yields from Curcuma longa L. grown in Jamaica. J Agric Food Chem 56: 3664-3670.

17. (2014) Introduction to ANOVA / MANOVA.

18. (2014) Statistical Advisor, How to Compare Means / Variances in Multiple Groups.

19. Weise EA (2002) Spice Crops, CABI Publishing, New York, USA, pp 350.

20. Peter KV (2007) Underutilized and Underexploited Horticultural Crops.Volume 4, New India Publishing, Delhi, India, pp 339.

21. Shinde GU, Kamble KJ, Harkari MG, More GR (2011) Process Optimization in Turmeric Heat Treatment by Design and Fabrication of Blancher.Proceedings of International Conference on Environmental and Agriculture Engineering, IACSIT Press, Singapore 15: 40.

22. Pruthi JS (1998) Quality Assurance in Spices and Spice Products, Modern Methods of Analysis.Alliend Publishers Limited, Mumbai, India, pp 505.

23. Li S, Yuan W, Deng G, Wang P, Yang P, et al. (2011) Chemical Composition and Product Quality Control of Turmeric (Curcuma longa L.). Pharmaceutical Crops 2: 28-54.

24. Bonakdar RA (2012) The H.E.R.B.A.L. Guide: Dietary Supplement Resources for the Clinician. Lippincott Williams & Wilkins, Philadelphia, PA, USA, pp 169.

25. Fraga CG (2009) Plant Phenolics and Human Health: Biochemistry, Nutrition and Pharmacology. Volume 1 of The Wiley-IUBMB Series on Biochemistry and Molecular Biology, John Wiley & Sons, Hoboken, NJ, USA, pp 31.

26. Roghelia V, Patel VH (2013) Antioxidant profile of organically and conventionally grown fresh turmeric (Curcuma longa L): A comparative study. J Cell and Tissue Research 13: 3749-3751.

27. Pujimulyani D, Raharjo S, Marsono Y, Santoso U (2013) The Phenolic Substances and Antioxidant Activity of White Saffron (Curcuma mangga Val.) as Affected by Blanching Methods.Intl J of BioVetAgric and Food Engineering 7: 606-609.

28. Škrovánková S, Mišurcová L, Machů L (2012) Antioxidant activity and protecting health effects of common medicinal plants. Adv Food Nutr Res 67: 75-139.

29. Shahidi F (1997) Natural Antioxidants: Chemistry, Health Effects, and Applications. AOCS Press, Champaign, IL, USA, pp 73.

30. Alaoui-Jamali M (2010) Alternative and Complementary Therapies for Cancer: Integrative Approaches and Discovery of Conventional Drugs. Springer Science & Business Media, New York, NY, USA, pp 63.

31. Madhavi DL, Deshpande SS, Salunkhe DK (1995) Food Antioxidants: Technological: Toxicological and Health Perspectives. CRC Press, Boca Raton, FL, USA, pp 73.

32. US FDA (2014) Inspections, Compliance, Enforcement, and Criminal Investigations. U.S. Food and Drug Administration

33. Charles DJ (2012) Antioxidant Properties of Spices, Herbs and Other Sources. Springer Science & Business Media, New York, USA, pages 55, 570.

34. Lobo V, Patil A, Phatak A, Chandra N (2010) Free radicals, antioxidants and functional foods: Impact on human health. Pharmacogn Rev 4: 118-126.

35. Moreira SI, Dutra DC, Rodrigues AC, de Oliveira JR, Dhingra OD, et al. (2013) Fungi and bacteria associated with post-harvest rot of ginger rhizomes in Espirito Santo, Brazil. Trop plant pathol 38: 218.

36. Elliott SM (2003) Rhizome rot disease of ginger. Crop and Plant Protection Unit, Ministry of Agriculture, Bodles Research Station, Jamaica.

37. Sforza S (2013) Food Authentication using Bioorganic Molecules. DEStech Publications, Inc. Lancaster, PA, USA, pp 299.

38. Dasgupta A, Hammett-Stabler C (2011) Herbal Supplements: Efficacy, Toxicity, Interactions with Western Drugs, and Effects on Clinical Laboratory Tests. John Wiley & Sons, Hoboken, NJ, USA, pp 198.

39. Hickey S, Saul AW (2010) The Vitamin Cure for Migraines. Basic Health Publications Inc., Laguna Beach, CA, USA, pp 147.

40. Brown SP, Miller WC, Eason JM (2006) Exercise Physiology: Basis of Human Movement in Health and Disease. Lippincott Williams & Wilkins Publishers, Baltimore, MD, USA, pp 55.

41. Venugopal V (2011) Marine Polysaccharides: Food Applications. CRC Press, Boca Raton, FL, USA, pp 167.

42. Nirmala K, Panpatil VV, Raja Kumar AK, Balakrishna N, Balansky R (2011) Turmeric alleviates benzo(a)pyrene induced genotoxicity in rats: Micronuclei formation in bone marrow cells and DNA damage in tissues. International Journal of Cancer Research 7: 114-124.

3

Chemical Sensor for Determination of Mercury in Contaminated Water

Ahmed Khudhair Hassan*

Environment and Water Research and Technology Directorate, Ministry of Science and Technology, Baghdad, Iraq

Abstract

In this research, we constructed chemical sensor for determining mercury in contaminated water because we needed fast, simple, low-cost, and accurate determination of mercury in different environmental systems. The constructed membrane composed of (Poly Vinyl Chloride) PVC as a matrix material, 1,5-diphenylthiocarbazone (dithizone) as electro active compound, and di-n-butyl phthalate (DBPH) as a plasticizer. The optimum membrane composition 30% PVC, 65% DBPH, 5% dithizone exhibited the better Nernstian response. The results showed that probe is high stability along the pH range from (3.5 to 8). The electrode displays a linear log [Hg^{2+}] versus Electromotive Force (EMF) response over a wide concentration range of (5×10^{-6} to 1×10^{-2}M) with Nernstian slope of 29.7 ± 0.5 mV decade^{-1} and limit of detection 3×10^{-6} M. The proposed sensor shows relatively high selectivity for mercury ion in different matrix solution, other ions had negligible interference effect on the reading.

Keywords: Ion selective electrodes; Polymeric membranes; Determination of mercury

Introduction

Mercury is generally found at very low concentration in the environment. Mercuric ion can be absorbed readily by humans and other organisms. It may cause kidney toxicity, neurological damage, paralysis, chromosome breakage, and birth defects [1,2]. Due to its serious hazardous effect on human health and toxicity in the environment, it is important to control its levels in natural and potable water.

Thus, it is very necessary to monitor the mercury levels in our environment. The common methods for the purpose that are being adopted are complexometry and spectrophotometry [3], Cold Vapor Atomic Absorption Spectrometry (CV-AAS) [4] inductively coupled plasma Atomic emission spectrophotometry (ICP-AES) [5] and Inductively Coupled Plasma Mass Spectrometry (ICP-MS) [6,7], X-Ray fluorescence [8] and Voltammetry [9]. But the potentiometric technique has advantages such as high selectivity, sensitivity, good precision, simplicity and low cost. There is a considerable attention has been given for drug analysis using Ion-Selective Electrodes (ISEs) [10,11].

A number of ISEs based on conventional polymeric membrane, and coated wire electrodes utilizing various neutral ionophores were made for determination of mercury ion [12-15].

The aim of this paper was therefore to evaluate a simple method for inorganic mercury determination in aqueous solution by PVC-membrane electrode based on 1,5-diphenylthiocarbazone (dithizone) (Figure 1).

Experimental

Reagents and materials

All the reagents were of analytical grade and were used as received. Solvent mediator (plasticizer) di-n-butyl phthalate (DBPH) was obtained from Merck (Germany). High molecular weight (PolyVinyl Chloride) (PVC) was obtained from Sigma-Aldrich (USA). AR grade tetrahydrofuran (THF), 1,5-diphenylthiocarbazone (dithizone), hydrochloric acid and sodium hydroxide were obtained from Fluka (Germany). Stock solutions (0.1 M) of Na$^+$, K$^+$, Mg^{2+}, Ca^{2+}, Hg^{2+}, Pb^{2+}, Zn^{2+},Cu^{2+}, Fe^{3+}, Cr^{3+} were prepared by direct dissolution of proper amounts of metal salts in doubly distilled water.

Apparatus

Electrochemical measurements were made pH/ION/ Cond 750 Ion analyzer WTW at 25°C in conjunction with a ceramic junction calomel electrode. The electrochemical cell assembly used for this study was as follows: Ag/AgCl | Internal solution (0.1 M) of Hg^{2+} | PVC membrane | Sample solution | Hg/Hg$_2$Cl$_2$, KCl (saturated). A pH meter WTW, inoLab' pH 720/7200 was used to check the pH of the solutions.

Construction of mercury (ll) membrane electrode

Membranes were prepared by the Moody-Thomas method [16]. Hg^{2+}-selective membrane was prepared by dissolving 25 mg of dithizone as ionophore, 150mg PVC and 325 mg of DBPH in 5 mL THF. The solution was poured into glass Petri dishes (5 cm diameter), and was allowed to evaporate overnight at room temperature. The thickness of the obtained membrane was about 0.3 mm. Membranes (10 mm diameter) were cut out and glued to the polished end of PVC tubes by means of a PVC-THF solution. The electrode bodies consisted of a glass tube, to which the PVC tube was attached at one end and filled with an internal solution (0.1 M of Hg^{2+}). The membrane was conditioned by immersing in a 0.1 M Hg^{2+} solution for 3 hours before measurements.

The electrochemical cell assembly used for this study was as follows: Ag/AgCl | internal solution (0.1 M) of Hg^{2+}| PVC membrane | sample solution | Hg/Hg$_2$Cl$_2$, KCl (saturated). A brief schematic diagram of the measuring cell is shown in Figure 2.

Direct potentiometric determination of mercury (II)

The electrode was calibrated by transferring 20 mL aliquots of $1\times10^-$

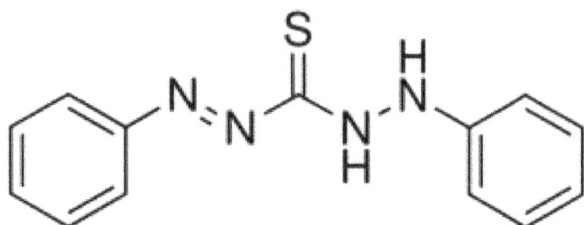

Figure 1: Chemical structure of dithizone.

***Corresponding author:** Ahmed Khudhair Hassan, Environment and Water Research and Technology Directorate, Ministry of Science and Technology, Baghdad, Iraq, E-mail: ahmedkhh71@yahoo.com

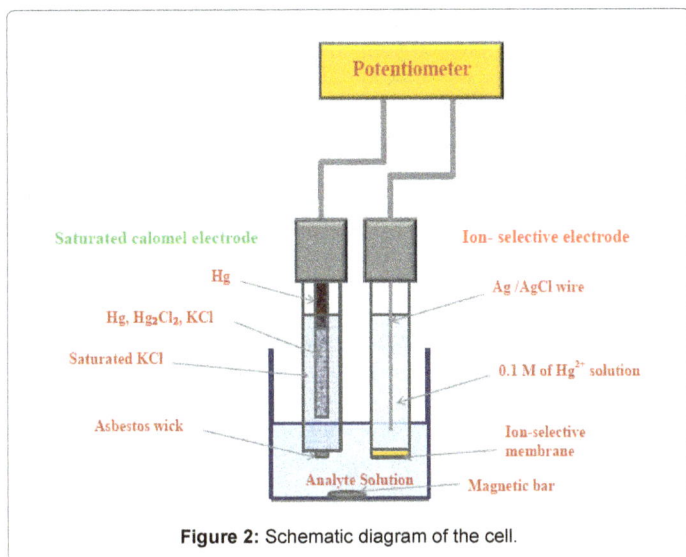

Figure 2: Schematic diagram of the cell.

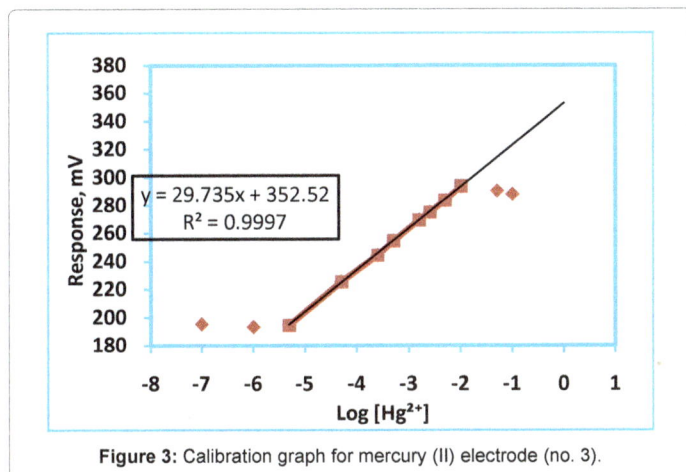

$y = 29.735x + 352.52$
$R^2 = 0.9997$

Figure 3: Calibration graph for mercury (II) electrode (no. 3).

^{7}-1×10^{-1}M aqueous solutions of mercury (II) to 50 mL beakers, followed by immersing the Ion-selective membrane electrode, together with a calomel reference electrode in the solution. The potential readings were recorded after stabilization to ± 0.5 mV, and the EMF was plotted as a function of the logarithm of the mercury (II) concentrations. The calibration graph was used for subsequent determinations of unknown mercury (II) concentrations. A typical calibration plot for electrodes No. 3 is shown in Figure 3.

Selectivity of the electrode

The selectivity coefficients over interfering species were evaluated by the separate solution method [17-20] at 1×10^{-3} M concentration of mercury (II) solution and interfering.

Results and Discussion

Optimization of the electrodes

Composition of the membranes: Five membranes of the different compositions were investigated with the DBPH as plastizer given in (Table 1). For each composition, the amount of polymer (Poly Vinyl Chloride) (PVC) was kept constant (30% w/w) and varying the percentage (w/w) of dithizone as ionophore and plasticizer [21]. The results reveal that the composition having the 5% dithizone leads to exhibit a better slope (29.7 ± 0.5 mV decade^{-1}); correlation coefficient

(0.9997) and wide concentration range (5×10^{-6} to 1×10^{-2} M). In all subsequent studies electrodes made of the membrane composition No.3 (PVC, 30%: plasticizer, 65%: ionophore, 5%) were used (Table 1).

The effect of pH

The influence of pH on the potential of the electrodes was investigated by measuring the Electro Motive Force (EMF) of the cell at 5×10^{-4} and 5×10^{-3} M of mercury (II) solutions. The pH values of the cell were adjusted by the addition of very small volumes of (0.01-0.1 M) Hydrochloric acid or sodium hydroxide. The results are shown in Figure 4 it is evident that the electrode does not respond to pH changes in the range (3.5-8.0). Under more acidic conditions, the ligand may be protonated thereby losing its capacity to complex with the metal ions. The drift in potential at pH 8.5 is attributed to formation of mercury (II) hydroxide [22] (Figure 4).

Construction of the calibration graphs

The electrodes were calibrated by transferring 20 mL aliquots of (1×10^{-7} to 1×10^{-1} M) aqueous solutions of mercury (II) to 50 mL beakers, followed by immersing the mercury-selective electrode in conjunction with a calomel reference electrode in the solution. The potential readings were recorded after stabilization to ± 0.5 mV and the EMF was plotted as a function of the logarithm of the mercury (II) concentration. The calibration graph was used for subsequent determination of unknown mercury (II) concentrations. A typical calibration plot for electrodes No.3 is shown in Figure 3.

Response time and life span

Long-term stability of electrode potential is an important parameter in practical applications of ion-selective electrodes. Large potential drift is a major drawback. A purpose of this work was to shade some light on stability of prepared electrode. Potential stability of mercury ISE

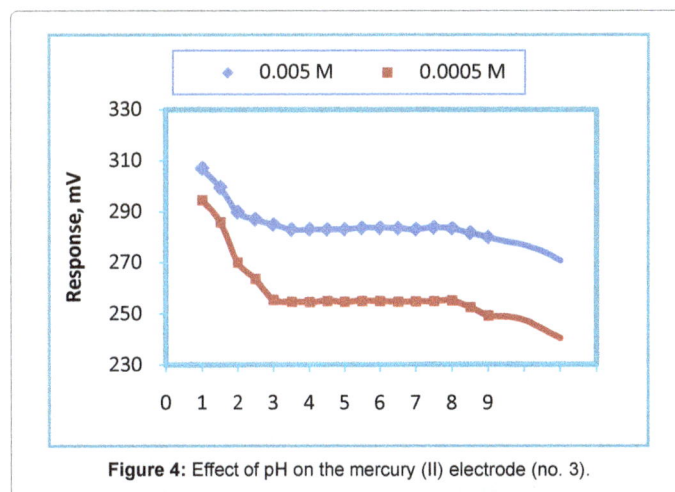

Figure 4: Effect of pH on the mercury (II) electrode (no. 3).

| No. | Composition % (w/w) | | | Slope (mV/ decade) | Correlation coefficient |
	PVC, %	DBPH, %	Dithizone, %		
1	30	69	1	27.3	0.986
2	30	67	3	28.6	0.989
3*	30	65	5	29.7	0.999
4	30	63	7	27.2	0.991
5	30	60	10	23.8	0.980

*Optimum composition

Table 1: Composition of different membranes and slopes of the corresponding calibration graphs.

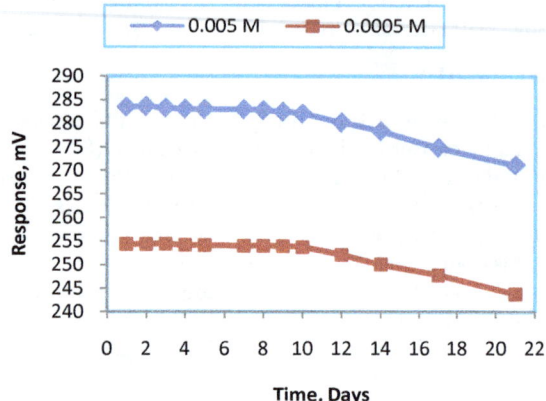

Figure 5: Stability of the mercury (II) electrode (no. 3) as a function of time.

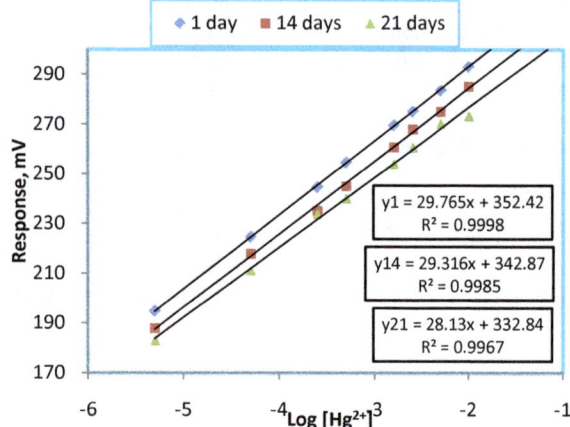

$$y1 = 29.765x + 352.42$$
$$R^2 = 0.9998$$

$$y14 = 29.316x + 342.87$$
$$R^2 = 0.9985$$

$$y21 = 28.13x + 332.84$$
$$R^2 = 0.9967$$

Figure 6: Calibration of the mercury (II) electrode (no. 3) as a function of time.

Parameter	
Slope (mV decade^{-1})	29.7 ± 0.5
Linear concentration range (M)	$5 \times 10^{-6} - 1 \times 10^{-2}$
Intercept (mV)	352.5
Correlation coefficient, r	0.9997
Lower detection limit (M)	3.0×10^{-6}
Response time for 1×10^{-3} (M) solution (sec)	20 ± 1
Working pH range	3.5 - 8.0
Life time (day)	14

Table 2: Summary of the response characteristic of the mercury sensor

electrodes was monitored over 21 days by measuring their potentials in 5×10^{-3} and 5×10^{-4} M standard mercury (II) solutions each day as show in Figure 5. The slope of each electrode in mV per decade was calculated and compared with the original slope obtained when this electrode was calibrated at the first time in the 1×10^{-6} to 1×10^{-2} M mercury (II) solutions. Electrode was considered no longer suitable for measurements when the slope differences exceeded 1.0mV per decade Figure 5.

In Figure 5, the potentials of electrode were shown to be significantly stable up to 10 days. A drift of <5mV per decade was observed after 14 days. It indicate that, electrode remains fully operational and the Ag/AgCl internal reference remains free from water transport for at least 14 days. It was observed that the investigated electrode (no. 3) exhibited

good stability in terms of slope in the linear domain of concentration and the electrode can be used continuously for about 14 days without considerable decrease in its slope value as show in Figure 6. A decrease in electrode stability after 14 days might be attributed to leaching the ionophore (dithizone) from the membrane. A 15-30 s response times were recorded for prepared electrode in the 1×10^{-2} to 1×10^{-5} M mercury (II) solutions. After 14 days, a decrease in electrodes stability is associated with increase in response times up to 1-2 min. (Table 2) showed summary of the response characteristic of the mercury sensor (Table 2).

Selectivity of the electrode

The selectivity coefficients are the most important characteristics of the membrane sensor, informing about the ability of the electrode in discriminating the primary ion against other ions of the same charge signs. Selectivity coefficients were determined by the separate solution method [17-20] in which the following equation was applied:

$$\log K^{pot}_{Hg^{2+},j^{z+}} = (E_2 - E_1)/S + \log\left[Hg^{2+}\right] - \log\left[j^{z+}\right]^{1/z} \quad (1)$$

E_1 is the electrode potential in a 1.0×10^{-3} M Hg^{2+} solution.

E_2 is the potential of the electrode in a 1.0×10^{-3} M solution of the interferent ion (j^{z+}).

S is the slope of the calibration plot.

The influence of some inorganic cations on the electrode response was investigated. This method is considered to be the simplest way to evaluate the degree of interference that might be taking place and is used to perform measurements in aqueous samples. The selectivity coefficients obtained by this method. Table 3 showed that the proposed electrode was highly selective toward Hg^{2+} ion. The inorganic cations did not interfere due to the differences in their mobility's and permeability's as compared with Hg^{2+} ion. As can be seen from Table 3, most ions have negligible interference; the ions Pb^{2+} show intermediate effect, which is common interfering ion on Hg (II) ion-selective electrode as they have comparable size and characteristics to those of mercury ions [22] (Table 3).

Analytical applications

The proposed sensor was found to work well under laboratory conditions. It is clear that the amount of Hg^{2+} ions can be accurately determined using the proposed sensor. To assess the applicability of the proposed sensor to real samples, Hg^{2+} was measured in treated tap water. Each sample was analyzed in triplicate, using this sensor by the direct method. The results in (Table 4) show an average recovery of 97% with Relative Standard Deviation (RSD) of 2% and indicate the utility of the proposed electrode (Table 4).

Foreign ion	$K^{pot}_{Hg2+,jz+}$
Na^+	1.2×10^{-3}
K^+	1.9×10^{-3}
Mg^{2+}	4.0×10^{-3}
Ca^{2+}	5.9×10^{-3}
Zn^{2+}	5.1×10^{-3}
Cu^{2+}	4.9×10^{-3}
Cr^{3+}	2.7×10^{-3}
Fe^{3+}	4.9×10^{-3}
Pb^{2+}	6.5×10^{-2}
Hg^{2+}	3.1×10^{-2}

Table 3: Selectivity coefficient values for mercury (II) electrode (no. 3).

Hg2+ added, (M)	Hg2+ Found*, (M)	RE, %	Rec., %	RSD, %
5.0×10^{-5}	4.83×10^{-5}	-3.4	96.6	1.8
1.0×10^{-4}	0.97×10^{-4}	-3.0	97.0	2.3
5.0×10^{-3}	4.88×10^{-3}	-2.4	97.6	1.9

*Average of three determinations

Table 4: Recovery of mercury ions from tap water

Conclusion

The proposed Hg^{2+} selective electrode based on 1,5-diphenylthiocarbazone (dithizone) as the electro active compound might be a useful analytical tool for the determinations of Hg(II) ions in the range from 5×10^{-6} to 1×10^{-2} M, and therefore an alternative to spectrophotometric methods. The proposed electrode was applied as indicator electrode and successfully used to determine mercury (II) in tap water samples with satisfactory results.

References

1. Wang J, Feng X, Anderson CW, Xing Y, Shang L (2012) Remediation of mercury contaminated sites-A review. J Hazard Mater 221-222: 1-18.

2. http://www.atsdr.cdc.gov/spl/

3. Hamza A, Bashammakh AS, Al-Sibaai AA, Al-Saidi HM, El-Shahawi MS (2010) Part 1. Spectrophotometric determination of trace mercury (II) in dental-unit wastewater and fertilizer samples using the novel reagent 6-hydroxy-3-(2-oxoindolin-3-ylideneamino)-2-thioxo-2H-1,3-thiazin-4(3H)-one and the dual-wavelength beta-correction spectrophotometry. J Hazard Mater 178: 287-292.

4. Shah AQ, Kazi TG, Baig JA, Afridi HI, Kandhro GA, et al. (2010) Determination of total mercury in chicken feed, its translocation to different tissues of chicken and their manure using cold vapour atomic absorption spectrometer. Food Chem Toxicol 48: 1550-1554.

5. Zhao L, Zhong S, Fang K, Qian Z, Chen J (2012) Determination of cadmium(II), cobalt(II), nickel(II), lead(II), zinc(II), and copper(II) in water samples using dual-cloud point extraction and inductively coupled plasma emission spectrometry. J Hazard Mater 239-240: 206-212.

6. Kenduzler E, Ates M, Arslan Z, McHenry M, Tchounwou PB (2012) Determination of mercury in fish otoliths by cold vapor generation inductively coupled plasma mass spectrometry (CVG-ICP-MS). Talanta 93: 404-410.

7. Rodrigues JL, de Souza SS, de Oliveira Souza VC, Barbosa F Jr (2010) Methylmercury and inorganic mercury determination in blood by using liquid chromatography with inductively coupled plasma mass spectrometry and a fast sample preparation procedure. Talanta 80: 1158-1163.

8. Alcalde-Molina M, Ruiz-Jimenez J, Luque de Castro MD (2009) Automated determination of mercury and arsenic in extracts from ancient papers by integration of solid-phase extraction and energy dispersive X-ray fluorescence detection using a lab-on-valve system. Anal Chim Acta 652: 148-153.

9. Giacomino A, Abollino O, Malandrino M, Mentasti E (2008) Parameters affecting the determination of mercury by anodic stripping voltammetry using a gold electrode. Talanta 75: 266-273.

10. Hassan AK, Ameen ST, Saad B (2013) Tetracaine-selective electrodes with polymer membranes and their application in pharmaceutical formulation control. Arabian Journal of Chemistry.

11. Hassan AK, Saad B, Ghani SA, Adnan R, Rahim AA (2011) Ionophore-based potentiometric sensors for the flow-injection determination of promethazine hydrochloride in pharmaceutical formulations and human urine. Sensors (Basel) 11: 1028-1042.

12. Abu-Shawish HM (2009) A mercury(II) selective sensor based on N,N'-bis(salicylaldehyde)-phenylenediamine as neutral carrier for potentiometric analysis in water samples. J. Hazard Mater 167: 602-608.

13. Ye G, Chai Y, Yuan R, Dai J (2006) A mercury(II) ion-selective electrode based on N,N-dimethylformamide-salicylacylhydrazone as a neutral carrier. Anal Sci 22: 579-582.

14. Hassan SS, Saleh MB, Abdel Gaber AA, Mekheimer RA, Abdel Kream NA (2000) Novel mercury (II) ion-selective polymeric membrane sensor based on ethyl-2-benzoyl-2-phenylcarbamoyl acetate. Talanta 53: 285-293.

15. AfkhamiA, Madrakian T, Sabounchei SJ, Rezaei M, Samiee S, Pourshahbaz M (2012) Construction of a modified carbon paste electrode for the highly selective simultaneous electrochemical determination of trace amounts of mercury(II) and cadmium(II). Sensors and Actuators B: Chemical 161: 542-548.

16. Craggs A, Moody GJ, Thomas JDR (1974) PVC matrix membrane ion-selective electrodes. Construction and laboratory experiments. J Chem Educ 51: 541-544.

17. Buck RP, Lindner E (1994) Recommendations for nomenclature of ion-selective electrodes. Pure Appl Chem 66: 2527-2536.

18. Umezawa Y, Umezawa K, Sato H (1995) Selectivity coefficients for ion-selective electrodes: Recommended methods for reporting values. Pure Appl Chem 67: 507-518.

19. Umezawa Y, Buhlmann P, Umezawa K, Tohda K, Amemiya S (2000) Potentiometric selectivity coefficients of ion-selective electrodes, Part I. Inorganic cations. Pure Appl Chem 72: 1851-2082.

20. Lindner E, Umezawa Y (2008) Performance evaluation criteria for preparation and measurement of macro and micro fabricated ion-selective electrodes. Pure Appl Chem 80: 85-104.

21. Ensafi AA, Meghdadi S, Allafchian AR (2008) Highly selective potentiometric membrane sensor for Hg(II) based on bis(benzoyl acetone) diethylene triamine. Sensors Journal, IEEE 8: 248-254.

22. Mahajan RK, Sood P, Pal Mahajan M, Marwaha A (2007) Mercury(II) ion-selective electrodes based on heterocyclic systems. Ann Chim 97: 959-971.

4

Synthesis and Characterization of 3s,5s,7s-Adamantan-1-Amine Complexes with Metals of Biological Interest

Najma Sultana[1], Saeed Arayne M[2]*, Amir Haider[2] and Hina Shahnaz[3]

[1]Research Institute of Pharmaceutical Sciences, Faculty of Pharmacy, University of Karachi, Karachi, Pakistan
[2]Department of Chemistry, University of Karachi, Karachi, Pakistan
[3]Department of Environmental Science, Sind Madressatul Islam University, Karachi, Pakistan

Abstract

3s,5s,7s-adamantan-1-amine, tricyclo[3.3.1.13,7]decan-1-amine, 1-adamantanamine, 1-aminoadamantane, 1-aminotricyclo[3.3.1.13,7]decane, 1-adamantylamine or amantadine with a tricyclic amine with cage like structure is an antiviral and antiparkinsonian compound. It also is used to prevent and treat respiratory infections caused by influenza A virus. Eleven metal complexes of amantadine with metals of biological interest as Mg^{II}, Ca^{I}, Cr^{II}, Mn^{II}, Fe^{II}, Fe^{III}, Co^{I}, Ni^{II}, Cu^{II}, Zn^{II} and Cd^{I} have been synthesized and characterized by spectroscopic techniques IR, 1H NMR, elemental analysis and atomic absorption spectroscopy. Prior to synthesis condutometric titrations were carried out to determine the molo ratios of drug metal interactions. In all complexes, amantadine acted as a monodentate ligand, two molecules of which are bound to the metal through the amino nitrogen showing a square planar geometry.

Keywords: Amantadine; Metal complexes; Conductance; FT-IR; ^{1}H-NMR

Introduction

Tricyclic amines have a great potential in the treatment and prevention of influenza A of which the most significant is amantadine, a synthetic alicyclic antiviral agent with an unusual cage like structure (Figure 1) [1]. Amantadine is an orally active antiparkinsonian and antiviral agent [2,3] discovered by workers at DuPont via an empiric screening program [1].

Amantadine hydrochloride possesses a unique, rigid, relatively unstrained ring system that is composed of three fused cyclohexane rings in the chair conformation [4]. Amantadine is considered to be the smallest repeating unit of the diamond lattice [5]. The symmetrical cage structure causes the infrared, nuclear magnetic resonance and mass spectra to be comparatively simple, as will be illustrated later.

Several metal complexes of amantadine are reported with iron [6], platinum [7], while other polyoxometalates containing Ce, W, Pr, Ni, V and Mn were reported by Liu and others [8]. Compounds of molybdenum and amantadine with the formulae, $(C_{10}H_{18}N)_5PMo_{12}O_{40}Cl_2.5H_2O$, $(C_{10}H_{18}N)_6As_2Mo_{18}O_{62}.6CH_3CN.6H_2O$ [9], $(C_{10}H_{18}N)_4Mo_8O_{26.6}$ $(CH_3)_2SO$ [10] and trans- $(AdNH)_2Mo(OSiMe_3)_4$ [11] have also been reported. At physiological pH amatadine forms complex with sodium molybdate, as the amino group of the drug is free for its function as antiviral, this study

suggests that the co-administration of amantadine with molybdenum supplements should be avoided [8].

Physical properties as solubility in water, size and ionic nature which in turn are dependent on the pH of the medium [12-14] effect the absorption of a drug through gastrointestinal tract. As only the free and unchanged drug can function at the active site in the body, if molybdate reacts with amantadine or when the two are administered together, the solubility and absorption can be affected. Owing to the fact that molybdnium is neither present in the body nor is administered as drug. Alternatively, essential and trace elements can be studied for complexation with amantadine, as being the metals of biological interest. In this paper we report synthesis and characterization of eleven metal complexes of amantadine with Mg^{II}, Ca^{II}, Cr^{II}, Mn^{II}, Fe^{II}, Fe^{III}, Co^{II}, Ni^{II}, Cu^{II}, Zn^{II} and Cd^{II}, which were characterized by FT-IR, ^{1}H NMR, atomic absorption spectroscopy and elemental analysis.

Materials and Methods

Chemicals

Amantadine was purchased from Sigma-Aldrich USA. The essential and trace elements used were in the form of their hydrated chloride salts of magnesium, calcium, chromium, manganese, ferric, cobalt, nickel, copper, zinc and cadmium, all of analytical grade. Methanol (TEDIA', USA), hydrochloric acid, sodium hydroxide from Merck, Darmstadt, Germany. Deionized water was freshly prepared in the laboratory and all glasswares were washed with chromic acid and then thoroughly rinsed with deionized water.

Instruments

Electrical balance [Mettler Toledo AB54], pH meter [Mettler Toledo

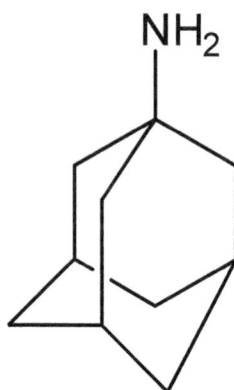

Figure 1: Amantadine.

*Corresponding author: Saeed Arayne M, Department of Chemistry, University of Karachi, Karachi-75270, Pakistan, E-mail: msarayne@gmail.com

MP220], UV-Visible double beam spectrophotometer (Shimadzu 1601), 1 cm rectangular quartz cells, ground glass distillation assembly, (Quickfit), de-ionizer (Stedec CSW -300), water distillation unit (GFL type 2001/2, No. 10793600G), melting point apparatus (Gallenkamp), FT-IR spectrophotometer (Nicolet Avatar 330) and proton NMR (Bruker); CHN elemental analysis were carried on Carlo Erba 1106, HSF-254. The reactions were monitored on TLC plates coated with Silica gel HF-254 and compounds were observed under UV lamp (254 nm). Perkin-Elmer AAnalyst 700 atomic absorption spectrometer used for atomic absorption studies.

Conductometric titration of amantadine metal complexes

Prior to synthesis conductometric titrations were performed to study the stoichiometric ratio of amantadine metal interactions in aqueous medium using conductivity/TDS meter. In individual experiments metal solutions of 1 mM were prepared and titrated with 1 mM ligand solution at 25°C. Conductivity of reacting mixtures were recorded after each addition of metal solution aliquots.

Synthesis of amantadine metal complexes

Amantadine (2 mM) was dissolved in 0.1 N HCl and 1 mM of each of these element salts were individually dissolved in 10 mL of methanol. Both of these solutions were mixed together and refluxed for three hours; the solution was concentrated and filtered while hot and then kept undisturbed for crystal growth at room temperature. The growth of crystals had a different time of crystallization. Crystals of magnesium, calcium, and chromium and manganese complexes with amantadine appeared in 15 days while the iron complex took one month for crystallization. On the other hand, cobalt, nickel, copper, zinc and cadmium complexes took 25~30 days for their growth. These

Metal	Concentration of standard (ppm)	Absorbance of standard	Absorbance of sample	Concentration of sample/50 mL	%	S.D
Mg	0.3	0.198	0.37	23800	23.8	0.000
Ca	4	0.191	5.3	30430	30.43	0.021
Cr	4	0.192	0.78	38780	38.78	0.012
Mn	2.5	0.19	1.86	39890	39.89	0.013
Fe	5	0.198	1.54	40100	40.10	0.025
Co	7	0.199	0.34	42340	42.34	0.020
Ni	7	0.198	5.35	42220	42.22	0.020
Cu	4	0.197	0.23	46120	46.12	0.012
Zn	1	0.204	1.81	44790	44.79	0.005
Cd	1.5	0.196	1.22	58230	58.23	0.004

Table 1: Atomic absorption analysis of amantadine metal complexes.

Sample	Melting point °C	Mole ratio	Solubility $H_2O:HCl$	Color
Amantadine	180		1:1	White
Amn+Mg	210	1:2	1:2	White
Amn+Ca	192	1:2	1:2	White
Amn+Cr	200	1:2	1:2	Green
Amn+Mn	228	1:2	1:2	White
Amn+FeCl$_3$	112	1:2	1:2	Yellow
Amn+FeSO$_4$	210	1:2	1:2	Light yellow
Amn+Co	218	1:2	1:2	Blue
Amn+Ni	212	1:2	1:2	Light green
Amn+Cu	210	1:1	1:2	Blue
Amn+Zn	202	1:1	1:2	Off white
Amn+Cd	232	1:1	1:2	White

Table 2: Physicochemical parameters of amantadine and its metal complexes.

metal complexes were recrystallized in absolute methanol, filtered, dried and physical characteristics were recorded.

Results and Discussion

Synthesis of complexes

A venture has been made to synthesize metal complexes of amantadine with various essential and trace elements in equimolar ratio in a mixture of hydrochloric acid and methanol. These complexes were than studied for their physicochemical parameters and characterized using techniques as IR, NMR and elemental analysis. Metals in all amantadine complexes were determined by using Pye-Unicam atomic absorption spectrometer (Table 1). Melting points were recorded on Gallenkamp melting point apparatus, while solubilities of all the complexes were checked and are present in Table 2.

IR Studies

All the synthesized complexes were studied spectroscopically in the IR region 4000 to 400 wavenumbers (cm^{-1}). The infrared spectrum of amantadine and their metal complexes were recorded as a potassium bromide disc method on a Nicolet Avatar 330 IR spectrophotometer. The main peak assignments are given in Table 3.

N–H stretch

The N-H stretching band of the amino group expected for amantadine is in the region 2961- 3087 [15,16] was observed for pure amantadine at 3038 cm^{-1} which shifts to 2900-3600 cm^{-1}.

For magnesium complex, N-H stretch band appeared at 3400 cm^{-1} as small band where as for calcium complex, a broad band was observed in the range of 3600-3100 cm^{-1} due to N-H stretching. In chromium and cadmium complexes, N-H stretch showed a short band at 3400 cm^{-1}, similarly in case of manganese, ferric chloride, cobalt, nickel, copper, zinc complexes a medium band was observed in the range of 3600-3100 cm^{-1} due to N-H stretching. While in ferrous ammonium citrate complex, N-H stretching showed a weak band in the region of 3400-3100 cm^{-1}.

The second incredibly significant peaks were asymmetric and symmetric stretching and their scissoring wagging and rocking which is the evidence of the attachment of metal with nitrogen of amine group. The CH$_2$ symmetric stretching was recorded at 2860 cm^{-1}, scissoring at 1449 cm^{-1}, rocking at 1263 cm^{-1} and wagging was at 1143 cm^{-1}. The comparison of the IR spectra of amantadine with metal complexes divulged that in case of all complexes the N-H stretching peak was shifted to 2914-3395 cm^{-1} in doublet and triplet weak bands.

CH$_2$ stretching (antisymmetric and symmetric)

Amantadine showed a very sharp band at 2900 cm^{-1} due to CH$_2$ antisymmetric and 2850 cm^{-1} due to CH$_2$ symmetric stretching whereas the reported peaks for antisymmetric is 2923 cm^{-1} and symmetric stretching at 2900 cm^{-1} [5]. For magnesium and calcium complex antisymmetric peaks appeared at 2900 cm^{-1}, 2950 cm^{-1} and symmetric very short band and sharp bands were observed at 2850 cm^{-1} for magnesium and calcium complexes. This band disappeared in chromium complex, whereas manganese, ferric chloride and ferric ammonium citrate nickel complexes showed suppressed bands of CH$_2$ stretching (antisymmetric and symmetric) at 2900 cm^{-1}. In cobalt complex very sharp symmetric band of CH$_2$ stretching was observed at 2850 cm^{-1} and in zinc and cadmium complexes very sharp asymmetric bands were observed at 2900 cm^{-1} and incredibly very short symmetric bands appeared between 2850-2860 cm^{-1}.

Compounds	υ(N–H) stretching	υ(N–H) bending	υ(N–H) oop bending	υ(N–H) stretching	$υ_{asym}(CH_2)$ stretch	$υ_{sym}(CH_2)$ stretch	$υ(CH_2)$ scissoring	$υ(CH_2)$ rocking	$υ(CH_2)$ wagging
Amantadine	2894	1608	811	1364	2860	-	1449	1263	1143
Amn+Mg	2920	1515	914	1080	2854	2711	1496	1313	1200
Amn+Ca	3300, 2914	1517	898	1090	2857	2708	1499	1366	1203
Amn+Cr	3123, 2923	1585	815	1090	2844	2658	1487	1329	1197
Amn+Mn	3212, 3073, 3046	1575	951	1092	2837	-	1479	1363	1203
Amn+FeCl₃	3173, 3116, 3050	1595	961	1080	2850	-	1472	1366	1193
Amn+FeSO₄	3176, 3119, 2954	1578	957	1077	2854	-	1476	1366	1197
Amn+CO	3365, 2906	1570	912	1094	2867	-	1480	1324	1206
Amn+Ni	3179, 2920-2904	1589	963	1080	2854	-	1472	1378	1203
Amn+Cu	3212, 3010-2910	1569	918	1090	2904	-	1476	1373	1207
Amn+Zn	3084, 2904	1574	964	1088	2850	-	1476	1311	1200
Amn+Cd	2914	1588	974	1097	2860	2728	1456	1366	1200

Table 3: Infrared assignments of amantadine metal complexes.

Drug & metal complex	$(CH_2)2$	J Values (Hz)	$(CH_2)_3$	J Values (Hz)	$(CH_2)_{10}$	J Values (Hz)	NH
Amantadine	1.71, 1.42	2.13, 11.48	1.18, 1.56	2.79, 4.21	1.18, 1.78	7.85, 6.78	1.45, 1.48
Mg complex	1.59, 1.49	2.52, 11.57	1.23, 1.61	2.90, 4.35	1.23, 1.67	7.90, 6.30	-
Ca complex	1.68, 1.38	2.65, 11.64	1.20, 1.67	2.89, 4.41	1.45, 1.61	7.71, 6.27	-
Cr complex	1.65, 1.32	2.58, 11.76	1.21, 1.60	2.91, 4.56	1.43, 1.65	7.69, 6.22	-
Mn complex	1.64, 1.45	2.65, 11.63	1.01, 1.63	2.93, 4.77	1.32, 1.90	7.67, 6.31	-
Fe⁺³complex	1.68, 1.49	2.76, 11.60	1.29, 1.61	2.67, 4.29	1.29, 1.96	7.75, 6.36	-
Fe⁺²complex	1.68, 1.37	2.31, 11.59	1.27, 1.70	2.7, 4.34	1.31, 1.82	7.93, 6.40	-
Co complex	1.62, 1.35	2.71, 11.62	1.21, 1.61	2.65, 4.43	1.38, 1.94	7.69, 6.29	-
Ni complex	1.60, 1.38	2.77, 11.48	1.23, 1.63	2.64, 4.67	1.23, 1.98	7.90, 6.30	-
Cu complex	1.78, 1.49	2.22, 11.68	1.09, 1.65	261, 4.33	1.29, 1.90	7.91, 6.22	-
Zn complex	1.62, 1.47	2.38, 11.71	1.09, 1.66	2.87, 4.38	1.44, 1.83	7.67, 6.24	-
Cd complex	1.55, 1.44	2.21, 11.56	1.22, 1.61	2.81, 4.41	1.27, 1.89	7.70, 6.29	-

Table 4: Chemical shifts (ppm) in 1H NMR spectra of amantadine and its metal complexes.

N–H overtones, N–H deformation and CH_2 deformation

Amantadine showed N–H overtones at 2000 cm⁻¹, N–H deformation at 1608 cm⁻¹, 1535 cm⁻¹, 1364 cm⁻¹ and CH_2 deformation at 1449 cm⁻¹ while reported for N–H overtones at 2000 cm⁻¹, N–H deformation at 1608 cm⁻¹, 1364 cm⁻¹ and CH_2 deformation at 1452 cm⁻¹ [5] are observed. For magnesium complex, N–H deformation was observed at 1515 cm⁻¹, 1510 cm⁻¹, 1365 cm⁻¹ as suppressed bands and CH_2 deformation suppressed band at 1496 cm⁻¹. N–H overtones observed at 2000 cm⁻¹ for all amantadine complexes. For calcium complex, N–H deformation shifted to 1517-20 cm⁻¹, 1499 cm⁻¹ and 1366 cm⁻¹, N–H overtones at 2000 cm⁻¹ and CH_2 deformation suppressed at 1452 cm⁻¹ have been recorded. In all metal-amantadine complexes the N–H bending and C-N stretching were at a lower wavelength at 1515-1589 cm⁻¹ and 1097-1077 cm⁻¹, respectively.

CH_2 wag and fingerprint region

Amantadine showed CH_2 wag sharp band at 1320 cm⁻¹ and fingerprint region at 1300 cm⁻¹ and below while reported CH_2 wag at 1307 cm⁻¹ [5]. For magnesium complex, CH_2 wag at 1320 cm⁻¹ (suppressed and doublet band), for calcium and manganese, ferrous ammonium citrate, ferric chloride, nickel, copper, zinc and cadmium complexes complexes suppressed bands were observed 1307-1310 cm⁻¹.

Due to the coordination of the metal ion with amantadine the N–H, C-N bands shifted to higher frequencies and overlapped [15]. It may be inferred that metal ions were strongly coordinated with amantadine through direct association with primary amine group [16,17].

¹NMR studies

By comparing main peaks of amantadine with its complexes (Table 4), it is observed that amine resonance was absent in the spectra of all metal complexes. The reported NH_2 (for C–1), β–CH_2 (for C–2, C–9), δ–CH_2 (for C–4, C–6, C–10), δ–CH (C–3, C–5, C–7) groups of actual drug, showed their presence at δ1.35, δ1.55, δ1.62, δ2.05 [5]. While the ¹NMR of the reference drug in D_2O at 600 MHz showed signal at 1.4 (d,2H, J=12.27) assigned for C–1 inferred the presence of NH_2 group. A doublet at δ1.5 having coupling constant J=11.48 confirmed the presence of β–CH_2 groups for C–2, C–8, C–9 positions. A doublet appeared at δ1.78 assigned for δ–CH_2 groups at C–4, C–6, C–10 positions (J=7.85) and the presence of CH groups at C–3, C–5, C–7 showed singlet at s, H δ1.56.

The magnesium complex, in $CD_3OD + CDCl_3$ at 600 MHz, showed a doublet at δ1.49 having coupling constant J=2.52 confirmed the presence of β–CH_2 groups for C–2, C–8, C–9 positions. A doublet appeared at δ1.67 (J=7.9) assigned for δ–CH_2 groups at C–4, C–6, C–10 positions and the presence of γ– CH groups at C–3, C–5, C–7 showed singlet at δ1.61. Calcium complex confirmed the complex formation, and a doublet at δ1.68 having coupling constant J=11.64 confirmed the presence of β–CH_2 groups for C–2, C–8, C–9 positions. While γ–CH_2 groups at C–4, C–6, C–10 positions (J=2.65) showed doublet at δ1.61 and the presence of γ–CH groups at C–3, C–5, C–7 showed singlet at δ1.2. Chromium complex showed and confirmed the complex formation signal at δ1.65 confirmed the presence of β–CH_2 groups for C–2, C–8, C–9 positions. The γ–CH_2 groups at C–4, C–6, C–10 positions and γ–CH groups at C–3, C–5, C–7 showed singlet at δ1.6 and δ1.43. Manganese complex showed signals at δ1.65 having coupling constant J=2.65 confirmed the presence of β–CH_2 groups for C–2, C–8, C–9 positions. γ–CH_2 groups at C–4, C–6, C–10 positions and γ–CH groups at C–3, C–5, C–7 showed singlet at δ1.9 and 1.32.

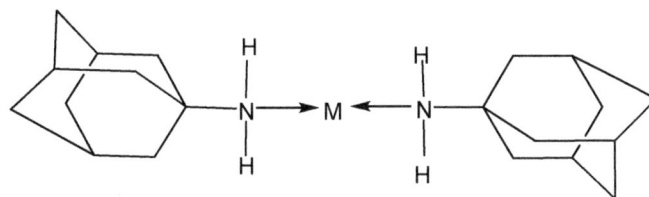

Figure 2: Proposed structure of amantadine metal complexes.

S.No	Compound	Chemical formula		C	H	N	Cl/S	Metal
1	Amantadine	$C_{10}H_{17}N$	Found	79.41	11.33	9.26		
			Calculated	79.75	11.9	8.79		
2	Amn + Mg	$C_{10}H_{17}Cl_2Mg_2NO$	Found	41.88	5.98	4.88		16.50
			Calculated	40.91	4.78	5.46	24.73	16.02
3	Amn + Ca	$C_{10}H_{15}Cl_2Ca_2N$	Found	40.0	5.03	4.66		26.69
			Calculated	40.45	5.1	3.79	23.61	26.59
4	Amn + Cr	$C_{20}H_{32}Cl_2CrN_2$	Found	56.47	8.06	6.59	16.67	12.22
			Calculated	56.42	8.10	6.53	16.63	12.20
5	Amn + Mn	$C_{20}H_{32}Cl_2MnN_2$	Found	56.08	8.00	6.54	16.55	12.83
			Calculated	56.05	8.01	6.58	16.50	12.88
6	Amn + Fechloride	$C_{20}H_{32}ClFeN_2$	Found	55.96	7.98	6.53	16.52	13.01
			Calculated	55.91	7.93	6.58	16.57	13.09
7	Amn + Fe sulfate	$C_{20}H_{32}FeN_2O_4S$	Found	53.10	7.13	6.19		12.34
			Calculated	52.51	7.55	5.54	7.09	11.93
8	Amn + Co	$C_{20}H_{32}ClCoN_2$	Found	55.56	7.93	6.48	16.40	13.63
			Calculated	55.59	7.99	6.42	16.47	13.68
9	Amn + Ni	$C_{20}H_{32}ClN_2Ni$	Found	55.59	7.93	6.48	16.41	13.58
			Calculated	55.61	7.99	6.50	16.47	13.60
10	Amn + Zn	$C_{20}H_{32}ClN_2Zn$	Found	54.75	7.81	6.38	16.16	14.90
			Calculated	54.70	7.88	6.42	16.20	14.96
11	Amn + Cd	$C_{20}H_{32}CdClN_2$	Found	49.45	7.05	5.77	14.60	23.14
			Calculated	49.50	7.10	5.81	14.66	23.17

Table 5: CHN microanalysis of amantadine and its metal complexes.

Ferrous sulfate complex showed a doublet at δ1.68 having coupling constant J=11.59 confirmed the presence of β–CH₂ groups for C–2, C–8, C–9 positions. δ–CH₂ groups at C–4, C–6, C–10 positions and γ–CH groups at C–3, C–5, C–7 showed singlet at δ1.82 and δ2.7. Similarly in ferric chloride complex doublet was observed at δ1.68 (J=11.6) and at δ1.49 indicated the NH₂ group for C–1, confirmed the complex formation, and β–CH₂ groups for C–2, C–8, C–9 positions.

The γ–CH₂ groups (C–4, C–6, C–10) and γ– CH groups (C–3, C–5, C–7) appeared at δ1.96 (singlet) and at δ1.29 (singlet). In the cobalt complex, the ¹NMR in CD₃OD at 600 MHz, confirmed the complex formation by showing doublet at δ1.35 having coupling constant J=2.71 confirmed the presence of β–CH₂ groups for C–2, C–8, C–9 positions. At δ1.8 (d, 6H, J=1.84, H–4, 6, 10) and singlet at δ1.61 (C–3, C–5, C–7) showed the presence of δ–CH₂ groups and γ– CH groups. For nickel complex, a doublet at δ1.6 having coupling constant J=2.7 confirmed the presence of β–CH₂ groups for C–2, C–8, C–9 positions. A singlet appeared at δ1.98 assigned for δ–CH₂ groups (C–4, C–6, and C–10) and γ–CH groups (C–3, C–5, and C–7) showed singlet at δ1.63 confirmed by signals δ1.78 (d, 2H, J=12.5) for β–CH₂ groups at C–2, C–8, C–9 positions. While δ–CH₂ groups at C–4, C–6, C–10 positions and CH groups at C–3, C–5, C–7 showed singlet at δ1.65 and at δ1.9. In zinc complex, a doublet at δ1.69 having coupling constant J=11.71 confirmed the presence of β–CH₂ groups for C–2, C–8, C–9 positions. A doublet appeared at δ1.83 (J=2.1) for δ–CH₂ groups at C–4, C–6, C–10 positions and singlet for CH groups (C–3, C–5, C–7) at δ1.66. Similarly in cadmium complex, the complex confirmed by the signals

of β–CH₂ groups (C–2, C–8, C–9), a doublet appeared at δ1.89 (J=7.7) for δ–CH₂ groups for C–4, C–6, C–10 positions and for γ–CH groups (C–3, C–5, C–7) singlet at δ1.27.

In case of all complexes the shifting of protons was observed at C2, C3 and C10 (Table 4). All the complexes showed resonance of methylene protons and other spectroscopic studies also account that there is an attachment of metal with the nitrogen of amine in amantadine molecule.

Structure of amantadine metal complexes

On the basis of above studies the amantadine nitrogen binds with metals and their proposed structures are shown as Figure 2. From the results obtained, it is proposed that amantadine forms complexes in the ratio of 2: 1 (drug: metal) [18]. The crystals of the complexes were very thin and we did not cope to obtain their X-ray crystallographs. The proposed formulae is established on the basis of spectroscopic and elemental analysis (Table 5) [6].

Conclusion

Complexes of metals of biological interest were synthesized with amantadine. The results from the elemental analysis, conductometric titration, AA spectroscopy, proton nuclear magnetic resonance and infrared studies reveals that in all complexes, amantadine acted as a monodentate ligand, two molecules of which were bound to the metal through the amino nitrogen showing a square planar geometry.

References

1. Davies W, Grunert RR, Haff RF, Mcgahen JW, Neumayer EM, et al. (1964) Antiviral Activity Of 1-Adamantanamine (Amantadine). Science 144: 862-863.

2. Whitley RJ, Alford CA (1978) Developmental aspects of selected antiviral chemotherapeutic agents. Annu Rev Microbiol 32: 285-300.

3. Tilley JW, Kramer MJ (1981) Aminoadamantane derivatives. Prog Med Chem 18: 1-44.

4. Fort RC (1976) Adamantane: The Chemistry of Diamond Molecules. Studies in Organic Chemistry, Volume 5, Marcel Dekker, New York, USA.

5. Fort RC, Schleyer PVR (1964) Adamantane: Consequences of the Diamondoid Structure. Chem Rev 64: 277-300.

6. Mustafa AA, Abdel-Fattah SA, Toubar SS, Sultan MA (2004) Spectrophotometric determination of acyclovir and amantadine Hydrochloride through metals complexation. Journal of Analytical Chemistry 59: 33-38.

7. Zák F, Turánek J, Kroutil A, Sova P, Mistr A, et al. (2004) Platinum(IV) complex with adamantylamine as nonleaving amine group: synthesis, characterization, and in vitro antitumor activity against a panel of cisplatin-resistant cancer cell lines. J Med Chem 47:761-763.

8. Liu SX, Wang CL, MiaoYU, Li YX, Wang EB (2005) Synthesis and Anti-influenza Virus Activity of Polyoxometalates Containing Amantadine. Acta Chim Sinica 63: 1069-1074.

9. Liu SX, Wang CM, Zhai HJ, Li DH, Wang EB (2004) Synthesis and crystal structures of novel Keggin and Dawson polyoxometalates containing amantadine. Chinese Chemical Letters 15: 216-219.

10. Li J, Qi Y, Wang E, Li J, Wang H, et al. (2004) Synthesis, structural characterization and biological activity of polyoxometallate-containing protonated amantadine as a cation. Journal of Coordination Chemistry 57: 715-721.

11. Sridevi N, Yusuff, Mohammed KK (2008) Synthesis, characterization and kinetic studies on complex formed between amantadine hhdrochloride and sodium molybdate at physiological pH. Indian Journal of Chemistry 47A: 836-842.

12. Porter RS (2011) The Merck Manual of Diagnosis and Therapy. 19th Edition, Section 22, Clinical Pharmacology, Merck & Co Inc, New Jersey, USA.

13. Leon S (1990) Pharm Review. John Wiley & Sons, New York, USA.

14. Gringauz A (1997) Introduction to Medicinal Chemistry: How Drugs Act and Why. Wiley-VCH, New York, USA.

15. Nakamoto K (1963) Infrared spectra of inorganic and coordination compounds. J. Wiley and Sons, USA.

16. Shishkin OV, Atroschenko Yu M, Gitis SS, Alifanova EN, Shakhkeldyan IV (1998) 3- Methyl-1,5-dinitro-3-azabicyclo[3.3.1]non-7-ene Acta Cryst 54: 271-273.

17. Bettahara N, Salmana SR (1996) X-ray powder diffraction data for 1-adamantanol. Powder Diffraction 11: 24-25.

18. Balode D, Valters R (1980) PSR Zinat Latv Vestis Ahad Kim Ser. 227; CA. 93: 168175e.

5

Non-Edible *Vernonia galamensis* Oil and Mixed Bacterial Cultures for the Production of Polyhydroxyalkanoates

Adrian D Allen[1,4]*, Folahan O Ayorinde[2] and Broderick E Eribo[3]

[1]*Department of Comprehensive Sciences, Howard University, Washington, DC, USA*
[2]*Department of Chemistry, Howard University, Washington, DC, USA*
[3]*Department of Biology, Howard University, Washington, DC, USA*
[4]*NOAA Center for Atmospheric Sciences, Howard University, Washington, DC, USA*

Abstract

Since the oil crisis of the 1970s much attempt has been made, albeit with varying degrees of success, to source the ideal substrate and bacteria for the production of PHA. The non-edible, naturally epoxidized seed oil from *Vernonia galamensis* and mixed cultures consisting of *Alcaligenes latus* (ATCC 29712), *Cupriavidus necator* (ATCC 17699), *Escherichia coli* (DH5α) and *Pseudomonas oleovorans* (ATCC 29347), were evaluated for PHA production under batch and fed-batch fermentations. PHA production, optimized by the mixed culture of *E. coli* and *C. necator*, was 0.4-19% (%wt/wt, cdw) for batch and fed-batch fermentations. Analyses of PHA by Matrix Assisted Laser Desorption Ionization-Time of Flight Mass Spectrometry (MALDI-TOF MS) and Gas Chromatography Mass Spectrometry (GC/MS) identified the 3-hydroxybutyrate (3HB) monomeric unit. The PHA ester bond stretching vibration (C=O), was confirmed at absorption 1740.66 cm^{-1}, using Fourier Transform Infrared Spectroscopy (FTIR). Gel Permeation Chromatography (GPC) indicated peak molecular weights between 3.8×10^3-1.12×10^6 Da with melting points (T_m), 60-90°C. The data further illustrates that inedible oils could be the ideal carbon source for the production of PHA.

Keywords: Polyhydroxyalkanoate; Mixed cultures; *Vernonia galamensis*; Batch/fed-batch fermentation; Mass spectrometry

Introduction

The exponential growth of the human population compounded with increased consumption of non-renewable resources and pollution have reinvigorated the interest in polyhydroxyalkanoates (PHA), a class of bacterial biodegradable and biocompatible elastomers. PHA accumulates intracellularly as inclusion bodies with a diameter of 0.2 to 0.5 µm, and may be utilized by bacteria from 75 genera as energy/carbon storage materials [1-4]. These materials may contain about 120 different (R)-3-hydroxy acid (3HA) monomeric units with the most common being poly(3-hydroxybutyrate), P(3HB). Blends of P(3HB) and other hydroxyacids are classified as copolymers, which may be short chain length (scl, 3 to 5 carbon atoms) or medium chain length (mcl, 6 to 16 carbon atoms). Scl-PHAs are often stiff and brittle, whereas, mcl-PHAs are elastomeric with lower crystallization. PHAs may have molecular weights ranging from 50 to 1000 kDa with a polydispersity (PD) of ≈ 2, and melting points (T_m) of 60 to 180°C [5-7].

The current production of PHAs is primarily done with conventional substrates such as glucose and organic acids in conjunction with pure cultures. Although these substrates have generated PHA contents of 48 to 77% of cell dry weight (cdw), they are costly, edible and are therefore not competitive for industrial-scale production of PHA [8-11]. With high gross operating expenses and the resulting cost of PHA at ca. US\$ 16/kg versus polyethylene or polypropylene (US\$ 1/kg), there is an urgent need to develop and optimize strategies to economically produce the material [12].

The utilization of substrates such as plant oils may provide the solution since they contain more carbon atoms (compared to glucose), generate PHA with higher molecular masses and may produce a theoretical yield coefficient of 0.65 to 0.98 g PHA/g plant oil, versus 0.32-0.48 g PHA/g glucose [13-17]. Similarly, a high percentage of PHA has been generated using edible oils; therefore, non-edibles such as Vernonia galamensis oil may be a competitive alternative [14,18-20]. *V. galamensis* is a tropical hardy plant which requires low rainfall, marginal conditions, is widely available and can be cheaply produced [21]. The unique property of *V. galamensis* lies in the seed oil (35-42% of seed), which contain 72-80% low viscosity, vernolic acid (cis-12, 13

epoxy-cis-9-octadecenoic or 18:1 epoxy); and fatty acids: linoleic (12-14%), oleic (4 to 6%), steric (2 to 3%), palmitic (2 to 3%), and trace amount of arachidic acid [21]. Previously, 42.8 wt% P(3HB) and P(3HB-co-3HV) were produced from saponified *V. galamensis* under batch conditions using *Cupriavidus metallidurans* (formerly *Alcaligenes eutrophus*) [1].

The utilization of pure cultures for PHA production has been associated with several advantages and disadvantages. For example, *Cupriavidus necator* can utilize organic acids such as lactate, acetate and butyrate, but not glucose, fructose or xylose [22]. The bacterium produces PHA when grown on emulsified plant oil medium with gum arabic [23], and can generate a PHA concentration and content of 8.37 g/L and 39.52% respectively when grown on condensed corn substrates [24]. The isolate will accumulate PHB with mass range 6×10^5 to 1×10^6 Da [4]. Nevertheless, only a few isolates can utilize the unmodified triacylgycerides (TAG), and oils must therefore be saponified [25,26]. Unlike *C. necator*, *A. latus* is a growth associated producer of PHA [27], utilize sucrose and can produces a P(3HB) content and productivity of 98.7 g/L (83%) and 4.94 g P(3HB)/L/h respectively [28]. However, the performance of *A. latus* may be susceptible to extreme temperature, pH, carbon to nitrogen ratio in the feed, concentration of substrates, trace elements, ionic strength, agitation intensity and dissolved oxygen level [29,30]. Likewise, wild type *Pseudomonas oleovorans*, is the best characterized mcl-PHA producer; however, it does not utilize substrates such as fructose or glucose, but grows on substrates such as fatty acids

*Corresponding author: Adrian Douglas Allen, Department of Comprehensive Sciences and NOAA Center for Atmospheric Sciences, Howard University, 415 College St. NW, Washington, DC, 20059, USA
E-mail: adriandallen@yahoo.com

or carbon sources which can undergo fatty acid *de novo* synthesis (n-alkanoic acids, n-alkanals and n-alkanes) [14,31-34]. *Pseudomonas* spp. could produce PHA with molecular weights in range $5×10^4$ to $6×10^4$ Da [35]. PHA can only be produced by recombinant *Escherichia coli* (transformed with the PHA operon). The isolate affords a rapid turnover and the production of 90% (wt/wt, cdw) [36,37].

To further maximize substrate usage and optimize PHA production, mixed cultures could be an ideal alternative. Mixed culture fermentations, involving two bacteria, are characterized by the conversion of a given substrate by one bacterium to an intermediate metabolite which is subsequently used by the other to generate PHA. For example, *Lactobacillus delbruckii* and *Cupriavidus necator* can produce PHB concentrations of 12 g/L by converting glucose to lactate, and lactate to P(3HB) [38-42]. Mixed cultures may confer minimal costs, simpler facility construction, and easy recovery of material from wastes, higher growth rate of cells, decrease culture contamination and PHA accumulation ≥62% cdw [43-45].

In this study, we investigate the suitability of saponified *V. galamensis* oil and mixed microbial cultures under batch and fed-batch fermentations, for PHA production.

Materials and Methods

Microbial culture and media

Cultures of *A. latus* (ATCC 29712), *Cupriavidus necator* (formerly *R. eutropha*, ATCC 17699), and *P. oleovorans* (ATCC 29347) were obtained from the American Type Culture Collection (ATCC, Manassas, VA, USA), and *E. coli* DH5α (recombinant strain with PHA operon, from Genetic Stock Center (New Heaven, Ct, USA)). These bacteria were sub-cultured in Trypticase Soy Broth (TSB) and stock cultures of each maintained on trypticase soy agar (TSA) at 4°C. Media were prepared as outlined by the manufacturer.

The Mineral Salt Medium (MSM) used for PHA production was prepared as described elsewhere [38]. Briefly, the medium contained the following: 1.1 g $(NH_4)_2HPO_4$, 5.8 g K_2HPO_4, 3.7 g KH_2PO_4, 10 mL (100 mM) $MgSO_4$ solution, 1 mL microelement solution, and distilled water to make a final volume of one liter. The following were the constituents of the micronutrient solution (g/L): 2.78 g $FeSO_4.7H_2O$, 1.98 g $MnCl_2.4H_2O$, 2.81 g $CoSO_4.7H_2O$, 1.67 g $CaCl_2.2H_2O$, 0.17 g $CuCl_2.2H_2O$ and 0.29 g $ZnSO_4.7H_2O$. The medium was sterilized for 15 min at 121°C/15 psi, and stored at 4°C until required. Saponified *V. galamensis* oil (0.98 g/L) was utilized as the sole carbon source. Saponification of *V. galamensis* oil was done as mentioned elsewhere with modifications [1]. About 100 mL methanol and 4.95 g (0.124 mol) sodium hydroxide were added to a 500 mL round bottom flask equipped with a magnetic stirring bar and a water-jacketed condenser. The mixture was refluxed until the sodium hydroxide dissolved. A mass of 10 g of *V. galamensis* oil (0.011 moles, based on molecular weight of vernolic triacylglycerol) was added to the alkaline solution and refluxed with continuous stirring for 30 min. The resulting hot mixture was then slowly added to 50 g ice:50 g water and the resulting off-white semi-solid crushed, filtered and air dried.

Batch/fed-batch fermentation, extraction and purification of PHA

The isolates *A. latus*, *C. necator*, *P. oleovorans* and *E. coli* DH5α (recombinant strain with PHA operon) were utilized under batch and fed-batch fermentation. Under batch fermentation, aliquots of 100 mL Mineral Salt Medium (MSM) were inoculated with 5 mL of $2×10^8$ CFU/mL of each bacterium and fermentation conducted at 25°C, 120 rpm

for 96 h on a Lab-line Environ* incubator (USA) or a gyrotory shaker (New Brunswick, N.J., USA). Cells were harvested via centrifugation, lyophilized, weighed, and the polymer extracted and analyzed. Fed-batch fermentation was carried out as previously outlined for 24 h with subsequent additions of 20 mL MSM every 12 h for 96 h and, the resulting fermentation broths treated as before. All batch and fed batch fermentations were repeated five times. Furthermore, the effect of pH (4.5 and 8.5) and degree of agitation (90 and 240 rpm) on productivity was determined. Fermentations were carried out as mentioned previously under batch and fed-batch conditions and the resulting fermentation broths treated accordingly.

Throughout fermentations, 500 μl broth was evaluated for PHA accumulation using Nile blue A and the hypochlorite assay [46-48]. Smears were stained with Nile blue A (1%) at 55°C for 10 min and slides washed with water and 8% acetic acid for 1 min then evaluated. Similarly, for the hypochlorite assay, a 200 μl broth was mixed with 4.8 mL of 5.25% sodium hypochlorite (Clorox) then incubated at 38°C for 40 min. Lipids in solution were determined from the optical density at 436 nm.

Fermentation broths were centrifuged (Sorvall RC-5, Dupont Instruments, Newtown, CT) for 15 min at 15, 000 rpm, 4°C and pellets re-suspended by washing once with 20 ml Tris-HCl buffer (pH 7.2). Pellets were lyophilized (-50°C and 25 mmHg) (LABCONCO Freeze Dryer 5, Kansas City, Missouri), and 100 mg refluxed with 100 mL chloroform for 3 h, then filtered (Whatman cellulose extraction thimble, 43×123 mm single thickness, Aldrich, Milwaukee, WI, USA). The resulting oily film was purified using methanol and chloroform and the precipitate dried and stored (- 4°C) for further analysis.

The purity of the polymer was determined as mentioned previously [46]. Briefly, 0.1 mg polymer was mixed with 10 mL concentrated sulphuric acid (98%) and the mixture incubated at 97°C for 10 min. Optical density was measured between 220 to 250 nm, and melting point determined by using a Fisher Johns melting point apparatus (Fisher Scientific Company, USA).

PHA analyses

The purified material was analyzed using Matrix Assisted Laser Desorption-Time of Flight Mass Spectrometry (MALDI-TOF MS), Gas Chromatography/Mass Spectrometry (GC/MS), Fourier Transform Infrared Spectroscopy (FTIR) and Gel Permeation Chromatography (GPC).

MALDI-TOF MS

All samples (mixed and commercial) were base transesterified by reacting 1 mg sample with 100 μl dichlormethane, vortexing for 1 min, and 4 μl of 25% sodium methoxide in methanol and 5 μl glacial acetic acid added. The reaction was allowed to proceed for 30 min. 2,5-dihydrobenzoic acid (2,5 DHB, 99%), the matrix, was prepared by dissolving 162 mg 2,5 DHB in 1 mL tetrahydrofuran (THF). Cationized oligomers were analyzed by mixing 20 μl of the mixture with 50 μl matrix, and applying 1 μl to MALDI target (Bruker Reflex III MALDI-TOF mass spectrophotometer). Runs were performed with the following parameters: accelerating voltage, 20 kV; grid voltage, 56 to 99%; guide wire voltage 0 to 0.3%; positive high resolution reflective mode with summation of 150 transients. Three runs were performed for each sample.

GC/MS

GC/MS analysis of samples was done by base transesterification to their monomers as outlined before [1]. An Agilent Technologies 6890N Network GC System (CA, USA) interfaced directly to an Agilent Technologies 5973 Inert Mass Selective Detector was used to generate data. Electronic pressure through a capillary column was used to maintain a constant helium flow of 35 cm/s. The following parameters were included in the GC program: temperature was held at 50°C for 2 min then increased to 300°C at a rate of 20°C/min, then held at 300°C for a total of 14 min. The auxiliary temperature was set at 250°C and the column flow rate at 1.2. A solvent delay time of 4 min was applied to each sample. About 1 µl of the transesterified sample was injected into the GC/MS instrument and a Hewlett-Packard PC integration program used to calculate the peak areas and percentages of the monomeric components of the polymer. Data analysis was performed using the NIST02 database.

Fourier Transform Infrared Spectroscopy (FT-IR)

FTIR (Perkin Elmer Spectrum 100, Version 6.3.2) was used for analysis of PHA. Samples for analysis were purified as mentioned before, dissolved in 100 µl chloroform and 20-30 µl added to sample crystal window. The following parameters were included in the instrument program: resolution, 4.0 cm^{-1}, scan speed 0.2 cm/s, range 4000-650 cm^{-1} and 4 scans for each run. Runs were done in duplicate.

Gel Permeation Chromatography (GPC)

Gel permeation chromatography was used to ascertain the apparent number-average (M_n), weight-average (M_w), molecular masses and polydispersity (PD, M_w/M_n) of the PHA. Samples (1 mg) were dissolved in 50:50 tetrahydrofuran (THF, HPLC grade) chloroform (HPLC grade), and filtered (0.2 µm). A Waters 2690 GPC separation module equipped with a Waters 2410 refractive detector and Polymer laboratories C-Linear mixed-bed size exclusion columns (2×300 mm/7.5 mm) was used. The degasser was set at continuous with a pressure of 0.5 psi, and the lines at 100°C. Polystyrene standards A with peak molecular weight range of 5460 to 96,000 and B, 2930 to 50,400 (Polyscience Corp. Warrington, PA, USA) with low polydispersity were used to generate a chromatogram and calibration curve. THF was used as the eluant at a flow rate of 0.7 mL/min at 35°C.

Results

Fermentation, extraction and purification of PHA

The resulting dry cell weight and PHA yield under standard conditions, variations in pH and dissolved oxygen, and the resulting melting points and peak molecular weights are shown in Table 1. Higher yields were observed under batch compared to fed-batch fermentations.

The Nile blue A and hypochlorite assays indicated the accumulation of PHA during fermentations. For the former assay, the characteristic

orange fluorescence was visible at a wavelength of 460 nm (data not shown). Furthermore, a maximum absorbance at 235 nm (data not shown), was shown for the crotonic acid assay of the purified PHA.

PHA Analyses

MALDI-TOF MS: MALDI-TOF mass spectrometry was used to identify the oligomers resulting from base trans-esterification of mixed culture polymers (Figure 1). Generally, mass spectra showed mostly sodium [M+Na]$^+$ and potassium [M+K]$^+$ adducts attached to the oligomeric chains of the PHA. The mass difference between adjacent peaks was an average of 86 (Da) and indicated the presence of the PHA repeat unit [-OCH(CH$_3$)CH$_2$CO-]. These are clusters of isotopically resolved peaks of the same oligomer, but consisting of different end groups. The ion at m/z 1156.58 is identified as a sodiated dodecamer with an olefinic end group i.e. [CH$_3$CH=CHC(O)OCH$_3$(CH$_3$) CH$_2$C(O)$_{12}$OCH$_3$-Na]$^+$. The potassium adduct of this oligomer is shown at m/z 1174.1.

Gas chromatography/ mass spectrometry (GC/MS): GC/MS was used for the identification of the monomeric/oligomeric composition of the isolated polymers. Overall, the gas chromatograms illustrate the various signature peaks for the methyl-β-hydroxybutyrate produced from base hydrolysis of the polymer (Figure 2). The electron impact mass spectrum shows ions at m/z 85, 100 and 117 (+1). These were identified as {CH$_3$CHCH$_2$COO-}, {CH$_3$CHCH$_2$COOCH$_3$} and {CH$_3$CH{OH}CH$_2$COOCH$_3$} respectively. These signatures were identical to those produced from the base hydrolyzed commercial P(3HB) with average elution time of 6.6 min. Identity was confirmed by the NIST02 database.

Fourier Transform Infrared Spectroscopy (FTIR): FTIR was utilized for the determination of functional and/or other groups found within PHA (Figure 3). The observed infrared absorption for the mixed cultures of *A. latus* and *Cupriavidus necator* were 3019.98 (=C-H and =CH$_2$), 2917.41, 2849.44 (CH$_3$, CH$_2$ and CH), 1740.66 (C=O); *P. oleovorans* and *Cupriavidus necator*, 2849.32 and 2917.26 (CH$_3$, CH$_2$, CH), 1261.41 (O-C); and pure cultures of *P. oleovorans*, 3019.83 (=C-H and =CH$_2$), 2921.43, 2850.85 (CH3, CH$_2$ and CH), 1728.85 (C=O), 1459.24 (CH$_2$, CH$_3$); *Cupriavidus necator*, 2849.89, 2918.07 (CH$_3$, CH$_2$ and CH), 1740.66 (C=O), 1261.14 (O-C) and 1017.89 (-OH). The FTIR spectrum for the commercial P(3HB), (-COCH$_2$CHCH$_3$)$_n$, identified infrared absorption at cm^{-1} 3020.12 (straight, =CH), 1724.07 (saturated aldehyde, C=O), 1281.25 (acids straight, O-C), 1133.04, 1057.16, 980.40 (OH).

Gel-Permeation Chromatography (GPC): GPC was used to determine the weight number average mass (M_n), peak molecular weight (MP), molecular weight (M_w) and polydispersity (M_w/M_n) of the polymer produced from saponified *V. galamensis* oil (Table 1).

	A. latus + C. necator cdw(mg/L) PHA(%wt/wt, cdw)		P. oleovorans + C. necator cdw(mg/L) PHA(%wt/wt, cdw)		E. coli + C. necator cdw(mg/L) PHA(%wt/wt, cdw)	
pH 7	920/820	11/12	1160/860	10.3/11.6	1320/620	19/16
pH4	382/520	7.8/3	264/460	2.3/2.2	286/560	0.3/2.7
pH8	496/500	2/2	536/560	2.6/2.3	444/540	2.7/18
DO 90	240/240	2/3	280/320	3.2/2.2	180/320	3.3/3
DO 240	302/220	0.5/0.4	18/10	6/10	304/280	1.6/1.7
^{d}PM (Da)	1.2 × 10^5		3.7 × 10^5		3.9 × 10^4	
$^e T_m$ (°C)	80-90		61-65		61-70	

^{a}Culture, *A. latus* (ATCC 29712), *C. necator* (ATCC 17699), *E. coli* (DH5α) and *P. oleovorans* (ATCC 29347); ^{b}batch fermentation, cell dry weight, cdw; ^{c}Fedbatch fermentation; ^{d}PM(Da), Peak molecular weight in Daltons; $^e T_m$ (°C), melting temperature of the polymer in Celsius. Polystyrene standards A, with molecular weight 1300-377,400 Da, and standard B, with molecular weight 500-210,500 Da, were used to derive standard curve

Table 1: Yields and properties for PHA produced by mixed bacterial cultures grown on saponified *V. galamensis* oil under batch and fed-batch fermentation.

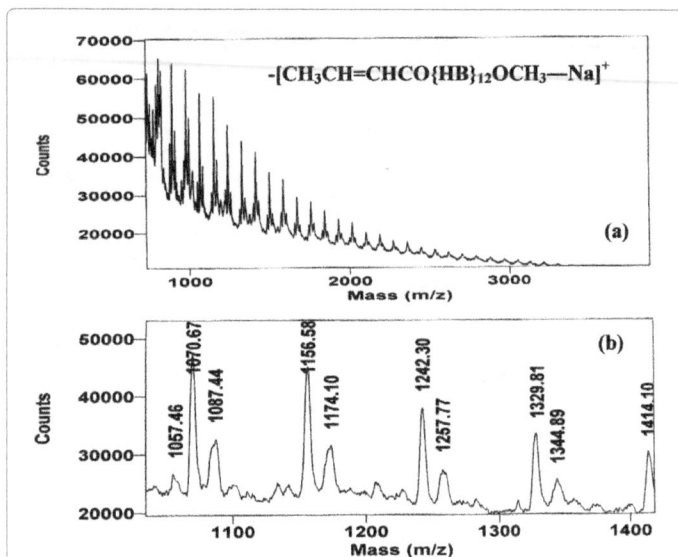

Figure 1: (a) Representative positive ion MALDI-TOF mass spectra of partially transesterified saponified *V. galamensis* oil PHA produced by mixed culture of *Escherichia coli* and *Ralstonia eutropha* (b) Expanded view of 'a' between *m/z* 1057 and 1400.   Peak at *m/z* 1156.58 indicate the sodiated 3-HB oligomer *m/z* 1156.58 [CH$_3$CH=CHCO{HB}$_{12}$OCH$_3$—Na]$^+$ and the potassium adduct of this oligomer is shown at 1174.10 [CH$_3$CH=CHCO{HB}$_{12}$OCH$_3$—K]$^+$. Runs were done in reflection mode with 2,5-dihydrobenzoic acid (2,5 DHB, 99%) and 120 transients. Analysis was accomplished using a Bruker Reflex III MALDI-TOF mass spectrophotometer with accelerating voltage, 20 kV, positive high resolution reflective mode with 150 transients.

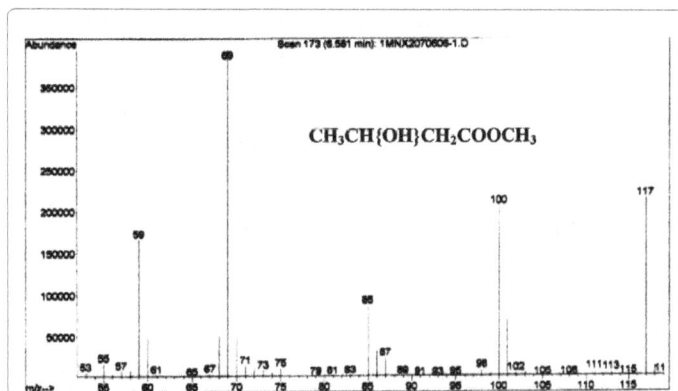

Figure 2: Electron-impact mass spectrum of trans esterified polymer produced by mixed culture cultivated on saponified *Vernonia galamensis* oil. The methylester of hydroxybutyric acid is indicated by the ion at *m/z* 117{CH$_3$CH{OH}CH$_2$COOCH$_3$}. The ions at *m/z* 100 {CH$_3$CHCH$_2$COOCH$_3$} and 86 {CH$_3$CHCH$_2$COO-} are fragment from the ion at *m/z* 117. Data was ascertained using an Agilent Technologies 6890N Network GC System (CA, USA) interfaced directly to an Agilent Techonologies 5973 Inert Mass Selective Detector with program: temperature was held at 50°C for 2 min then increased to 300°C at a rate of 20°C/min, then held at 300°C for a total of 14 min. The auxiliary temperature was set at 250°C and the column flow rate at 1.2.

The peak molecular weights for polystyrene standard A ranged from 1300 to 377 400 Da while that of standard B ranged from 580 to 210 500 Da were automatically used to generate a standard curve (data not shown). Analysis of all chromatogram illustrates peak molecular weight between 3.8×10^3-1.12×10^6 Da.

Discussion

In the current study, saponified *V. galamensis* oil was utilized by mixed bacterial cultures under batch and fed-batch fermentations to produce 19% (wt/wt, cdw) PHA with peak molecular weight in range 10^3 to 10^6 Da. This yield shows some improvement over previous studies wherein between 5.7 to 34.4% PHA was produced by pure culture of *Pseudomonas spp.* cultivated on rice, canola, sunflower, soybean, corn and hydrolyzed linseed oils [6,18]. Polymer accumulation, indicated by Nile blue A (Nile blue sulphate, basic blue) is consistent with previous reports and was a good indicator since stained PHA granules are easily identifiable by their characteristic orange fluorescence [16,47]. Neutral lipids do not affect identification of PHA since these lipids are liquids at the staining temperature (55°C) and have no affinity for Nile red (the oxidized form of Nile blue); therefore, cell membranes and other lipid containing cell components do not absorb enough dye to yield a detectable fluorescence at a wavelength of 460 nm [46]. Although this data was supported by the hypochlorite assay, the assay has been associated with several disadvantages which include severe degradation of polymer during digestion; only polymer in the native form of the lipid can be measured; the temperature and incubation time of the cell/hypochlorite mix must be precisely controlled and the suspension must be homogeneous [49-52].

Nevertheless, the methods were good predictors for the optimal points for PHA accumulation and cell dry weight (cdw). The resultant cdw was inversely proportional to PHA yield under both batch and fed-batch fermentation for all mixed cultures, and was dependent on *Cupriavidus necator* which is known to accumulate PHA when nutrients such as nitrogen and phosphorus are completely depleted from the environs [44]. The resulting lower PHA contents under batch and fed-batch modes indicates that bacterial growth rate under alkaline conditions is characterized by a logarithmic increase in viable cells [50]; therefore, an initial pH of 6.0 to 7.5 is ideal for microbial growth [52]. Similarly, variations in agitation had a significant effect on the growth of all mixed cultures and subsequently the PHA yields under batch and fed-batch fermentations. This further confirms the observation that high bacteria growth was inversely proportional to PHA yield. The consumption of oxygen is related to CO_2 production and ATP generation; therefore, manipulating the availability of oxygen

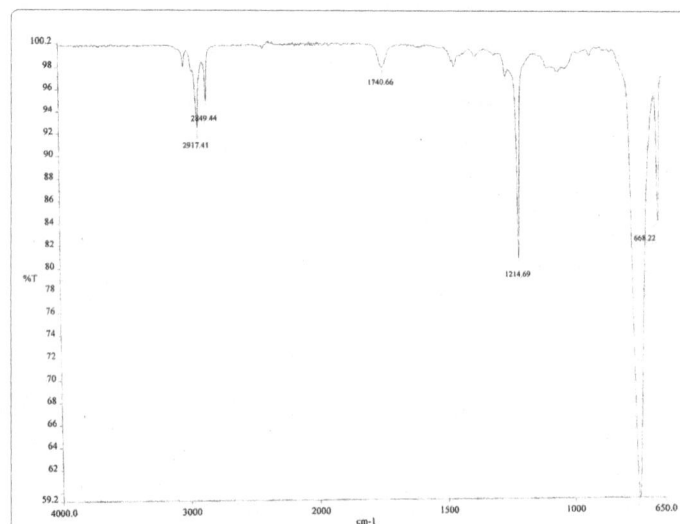

Figure 3: Representative Fourier Transform Infrared Radiation (FTIR) spectrum of PHA, produced by mixed culture of *A. latus* (ATCC 29712) + *R. eutropha* (ATCC 17699), cultured with saponified *Vernonia galamensis* oil as sole carbon source. The following parameters were included in the instrument program: resolution, 4.0 cm^{-1}, scan speed 0.2 cm/s, range 4000-650 cm^{-1} and 4 scans for each run. Runs were done in duplicate at RTP.

will contribute to overall fermentation yields i.e. PHB concentration, content and productivity, but not cell concentration [53,54]. Oxygen limitation (which can be generated by increasing the rpm) was previously shown to increase the PHB contents in fed-batch cultures containing recombinant *E. coli* with a PHB content of 80% [55]. Likewise, a cdw of 54 g/L (46% PHB) was produced by *Azotobacter chroococcum* under oxygen limitation, and 71 g/L dcw (20% PHB) under non-limiting oxygen conditions [56]. Deficiencies in oxygen, i.e., low oxidative capacity, could increase the rate of P(3HB) production in *Cupriavidus necator* and other microbes [53]. Assimilative activities such as protein and glycogen synthesis, and other cellular components are enhanced when oxygen supply is sufficient; however, if oxygen supply is curtailed, such activities will be minimized, thereby, contributing to increase production of PHA [5]. Suppression of assimilative activities contributes to the accumulation of NADH which inhibits the enzymes citrate synthase and isocitrate dehydrogenase in the tricarboxylic acid cycle (TCA) [35]. As a result, acetyl-CoA no longer enters the TCA cycle at the same rate and is instead converted to acetoacetyl-CoA by 3-ketothiolase. Therefore, increasing agitation speeds will limit the dissolved oxygen in the fermentor and contributes to increased PHB concentration, content and productivity, but not cell concentration [55].

The identity and purity of the extracted polymeric material was initially confirmed by spectrometric measurement (235 nm) of the α, β-unsaturated bond of crotonic acid (2-butenoic acid), produced by the reaction of PHA with sulphuric acid [46]. Further analyses indicated T_m between 61 to 90°C for all mixed culture PHA, compared to that derived from commercial material, 135 to 150°C. It has been suggested that the usual T_m for a PHA consisting primarily of the homopolymer, P(3HB), occurs between 174 to 179°C; however, this decreases dramatically if the homopolymer is coupled with varying percentages of other hydroxyacids such as hydroxyvalerate (HV), hydroxyhexanoate (HHx), among others [4,35]. For example, a copolymer containing 30 mol% 3HV had a T_m of 143°C [21,39]; and T_m range of 163 to 174°C (maximum at 172°C), for the copolymer, P(3HB-co-1mol% HV), produced using *C. necator* on saponified *V. galamensis* oil [1].

Further analysis of mixed culture PHA using MALDI-TOF MS identified only the HB oligomeric unit at *m/z* 1156.40 [CH$_3$CH=CHCO{HB}$_{12}$OCH$_3$-Na]$^+$ with isotopic fragments containing sodium and potassium adducts at 1174.1 [CH$_3$CH=CHCO{HB}$_{12}$OCH$_3$-K]$^+$ respectively. This observation was further confirmed by GC/MS i.e., only the 3-HB oliogomer was identified in the polymer. This finding is unusual, since *P. oleovorans* is a known producer of medium chain length PHA. Thus, copolymers, if produced, may have been utilized by the isolate for growth and/or concentrations were beyond the threshold of the instrument. FTIR identified the PHA marker absorption band at 1740.66 cm^{-1} i.e., the carbonyl (C=O) ester bond stretching vibration [57]. However, absorption at 1740.66 cm^{-1} and 1738.28 were previously shown to be representative of medium chain length hydoxyacids (mclHA), while the absorption at 1728.85 and 1724.07 cm^{-1}, observed for the commercial homopolymer (HB) in this study, for the HB unit and was similar to the previous study [58]. The evaluation of polymers by GPC indicated a peak molecular weight range of 1.2×10^5 to 1.12×10^6 Da. It was not possible to determine the uniformity of the polymers, since their polydispersities were not generated by the instrument. Other fragments eluted above 20 min had peak molecular weights in range 19 to 1000 Da and polydispersities between 1 and 1.4. Although immediate evidence is lacking, these fragments may be impurities, the lower molecular weight polymer called complexed PHA or other fragments from polymer

deterioration. The environmental conditions, method of isolation and the microbe used for fermentation, can adversely affect the molecular weight of polymers [35]. For example, *Cupriavidus necator* has been shown to accumulate PHB with peak molecular weight in the range 6×10^5 to 1×10^6 Da, while *Pseudomonas* spp. 5×10^4 to 6×10^4 Da [5,36]. The biosynthesis of mcl-PHA or other PHA copolymers is problematic for *E. coli*, due to the difficulty of drawing PHA precursors from the fatty acid metabolic pathways. Nevertheless, the current data suggests that the molecular weight of resultant polymers was dependent on bacteria used and culture conditions.

The current data illustrate that non-edible oils such as *V. galamensis* and mixed bacterial cultures could be optimized to generate PHAs with ideal melting points and molecular weights. Further studies are underway to evaluate other non-edible oils with these mixed cultures.

Acknowledgement

We would like to express our sincere gratitude to Dr. Emmanuel O. Akala and Dr. Simeon Adesina, Howard University School of Pharmacy and Ms. Nikki Handy, Howard University, Department of Chemistry for their help with data acquisition using Gel Permeation Chromatography (GPC) and Fourier Transform Infrared Spectroscopy (FTIR) respectively.

References

1. Ayorinde FO, Saeed KA, Price E, Morrow A, Collins WE, et al. (1998) Production of poly-(β-hydroxybutyrate) from saponified *Vernonia galamensis* oil by *Alcaligenes eutrophus*. Journal of Industrial Microbiology and Biotechnology 21: 46-50.

2. Alias Z, Tan IK (2005) Isolation of palm oil-utilising, polyhydroxyalkanoate (PHA)-producing bacteria by an enrichment technique. Bioresour Technol 96: 1229-1234.

3. Luengo JM, García B, Sandoval A, Naharro G, Olivera ER (2003) Bioplastics from microorganisms. Curr Opin Microbiol 6: 251-260.

4. Madison LL, Huisman GW (1999) Metabolic engineering of poly(3-hydroxyalkanoates): from DNA to plastic. Microbiol Mol Biol Rev 63: 21-53.

5. Salehizadeh H, Van Loosdrecht MC (2004) Production of polyhydroxyalkanoates by mixed culture: recent trends and biotechnological importance. Biotechnol Adv 22: 261-279.

6. Silva-Queiroz SR, Silva LF, Pradella JG, Pereira EM, Gomez JG (2009) PHA(MCL) biosynthesis systems in *Pseudomonas aeruginosa* and *Pseudomonas putida* strains show differences on monomer specificities. J Biotechnol 143: 111-118.

7. Sudesh K, Abe H, Doi Y (2000) Synthesis, structure and properties of polyhydroxyalkanoates: biological polyesters. Progress in Polymer Science 25: 1503-1555.

8. Nath A, Bhat S, Devle J, Desai AJ (2005) Enhanced production of 3-hydroxybutyric acid (3-HB) by *invivo* depolymerization of polyhydroxybutyric acid in 3-HB dehydrogenase mutants of *Methylobacterium sp.* ZP24. Ann Microbiol 55: 107-111.

9. Ren Q, Grubelnik A, Hoerler M, Ruth K, Hartmann R, et al. (2005) Bacterial poly(hydroxyalkanoates) as a source of chiral hydroxyalkanoic acids. Biomacromolecules 6: 2290-2298.

10. Shahid S, Mosrati R, Ledauphin J, Amiel C, Fontaine P, et al. (2013) Impact of carbon source and variable nitrogen conditions on bacterial biosynthesis of polyhydroxyalkanoates: evidence of an atypical metabolism in *Bacillus megaterium* DSM 509. J Biosci Bioeng 116: 302-308.

11. Sun Z, Ramsay J, Guay M, Ramsay B (2009) Enhanced yield of medium-chain-length polyhydroxyalkanoates from nonanoic acid by co-feeding glucose in carbon-limited, fed-batch culture. J Biotechnol 143: 262-267.

12. Scheller J, Conrad U (2005) Plant-based material, protein and biodegradable plastic. Curr Opin Plant Biol 8: 188-196.

13. Chen GQ, Zhang G, Park SJ, Lee SY (2001) Industrial scale production of poly(3-hydroxybutyrate-co-3-hydroxyhexanoate). Appl Microbiol Biotechnol 57: 50-55.

14. He W, Tian W, Zhang G, Chen GQ, Zhang Z (1998) Production of novel polyhydroxyalkanoates by *Pseudomonas stutzeri* 1317 from glucose and soybean oil. FEMS Microbiology Letters 169: 45-49.

15. Kahar P, Tsuge T, Taguchi K, Doi Y (2004) High yield production of polyhydroxyalkanoates from soybean oil by *Ralstonia eutropha* and its recombinant strain. Polymer Degradation and Stability 83: 79-86.

16. Tan IKP, Kumar KS, Theanmalar M, Gan SN, Gordon III B (1997) Saponified palm kernel oil and its major free fatty acids as carbon substrates for the production of polyhydroxyalkanoates *Pseudomonas putida* PGA1. Applied Microbiology and Biotechnology 47: 207-211.

17. Tsuge T (2002) Metabolic improvements and use of inexpensive carbon sources in microbial production of polyhydroxyalkanoates. J Biosci Bioeng 94: 579-584.

18. Casini E, de Rijk TC, de Waard P, Eggink G (1997) Synthesis of poly(hydroxyalkanoate) from hydrolyzed linseed oil. Journal of environmental polymer degradation 5: 153-158.

19. Saeed KA, Eribo BE, Ayorinde FO, Collier L (2002) Characterization of copolymer hydroxybutyrate/hydroxyvalerate from saponified vernonia, soybean, and "spent" frying oils. J AOAC Int 85: 917-924.

20. Thakor N, Trivedi U, Patel KC (2005) Biosynthesis of medium chain length poly(3-hydroxyalkanoates) (mcl-PHAs) by *Comamonas testosteroni* during cultivation on vegetable oils. Bioresour Technol 96: 1843-1850.

21. Baye T, Becker HC (2005) Genetic variability and interrelationship of traits in the industrial oil crop *Vernonia galamensis*. Euphytica 142: 119-129.

22. Shi H, Shiraishi M, Shimizu K (1997) Metabolic flux analysis for biosynthesis of poly(β-hydroxybutyric acid) in *Alcaligens eutrophus* from various carbon sources. Journal of Fermentation and Bioengineering 84: 579-587.

23. Budde CF, Riedel SL, Hübner F, Risch S, Popović MK, et al. (2011) Growth and polyhydroxybutyrate production by *Ralstonia eutropha* in emulsified plant oil medium. Appl Microbiol Biotechnol 89: 1611-1619.

24. Chakraborty P, Muthukumarappan K, Gibbons WR (2012) PHA productivity and yield of *Ralstonia eutropha* when intermittently or continuously fed a mixture of short chain fatty acids. J Biomed Biotechnol 2012: 506153.

25. Cromwick AM, Foglia T, Lenz RW (1996) The production of polyhydroxyalkanoates from tallow. Applied Microbiology and Biotechnology 46: 464-469.

26. Solaiman DK, Ashby RD, Foglia TA (2001) Production of polyhydroxyalkanoates from intact triacylglycerols by genetically engineered *Pseudomonas*. Appl Microbiol Biotechnol 56: 664-669.

27. Hrabak O (1992) Industrial production of poly-β-hydroxybutyrate. FEMS Microbiology Letters 103: 251-255.

28. Lee SY, Choi J (1999) Production and degradation of polyhydroxyalkanoates in waste management. Waste Management 19: 133-139.

29. Wang F, Lee SY (1997) Poly(3-Hydroxybutyrate) Production with High Productivity and High Polymer Content by a Fed-Batch Culture of *Alcaligenes latus* under Nitrogen Limitation. Appl Environ Microbiol 63: 3703-3706.

30. Yamane T, Chen X, Ueda S (1996) Growth-Associated Production of Poly(3-Hydroxyvalerate) from n-Pentanol by a Methylotrophic Bacterium, *Paracoccus denitrificans*. Appl Environ Microbiol 62: 380-384.

31. Fritzsche K, Lenz RW, Fuller RC (1990) Production of unsaturated polyesters by *Pseudomonas oleovorans*. Int J Biol Macromol 12: 85-91.

32. Preusting H, Nijenhuis A, Witholt B (1990) Physical characteristics of poly(3-hydroxyalkanoates) and poly(3-hydroxyalkenoates) produced by *Pseudomonas oleovorans* grown on aliphatic hydrocarbons. Macromolecules 23: 4220-4224.

33. Sandoval A, Arias-Barrau E, Bermejo F, Cañedo L, Naharro G, et al. (2005) Production of 3-hydroxy-n-phenylalkanoic acids by a genetically engineered strain of *Pseudomonas putida*. Appl Microbiol Biotechnol 67: 97-105.

34. Timm A, Steinbüchel A (1990) Formation of polyesters consisting of medium-chain-length 3-hydroxyalkanoic acids from gluconate by *Pseudomonas aeruginosa* and other fluorescent pseudomonads. Appl Environ Microbiol 56: 3360-3367.

35. Anderson AJ, Dawes EA (1990) Occurrence, metabolism, metabolic role, and industrial uses of bacterial polyhydroxyalkanoates. Microbiol Rev 54: 450-472.

36. Lee SY, Chang HN (1995) Production of poly(3-hydroxybutyric acid) by recombinant *Escherichia coli* strains: genetic and fermentation studies. Can J Microbiol 41 Suppl 1: 207-215.

37. Li R, Zhang H, Qi Q (2007) The production of polyhydroxyalkanoates in recombinant *Escherichia coli*. Bioresour Technol 98: 2313-2320.

38. Ganduri VSRK, Ghosh S, Patnaik PR (2005) Mixing control as a device to increase PHB production in batch fermentations with co-cultures of *Lactobacillus delbrueckii* and *Ralstonia eutropha*. Process Biochemistry 40: 257-264.

39. Katoh T, Yuguchi D, Yoshii H, Shi H, Shimizu K (1999) Dynamics and modeling on fermentative production of poly (beta-hydroxybutyric acid) from sugars via lactate by a mixed culture of *Lactobacillus delbrueckii* and *Alcaligenes eutrophus*. J Biotechnol 67: 113-134.

40. Tohyama M, Takagi S, Shimizu K (2000) Effects of controlling lactate concentration and periodic change in DO concentration on fermentation characteristics of a mixed culture of *Lactobacillus delbrueckii* and *Ralstonia eutropha* for PHB production. Journal of Bioscience and Bioengineering 89: 323-328.

41. Delgenes JP, Escare MC, Laplace JM, Moletta R, Navarro JM (1998) Biological production of industrial chemicals, i.e. xylitol and ethanol, from lignocelluloses by controlled mixed culture systems. Industrial Crops and Products 7: 101-111.

42. Fang HHP, Liu Y (2000) Intracellular polymers in aerobic sludge of sequencing batch reactors. Journal of Environmental Engineering 126: 732-738.

43. Iwamoto Y, Satoh H, Mino T, Matsuo T (1994) Production of biodegradable plastic PHA by excess sludge from anaerobic–aerobic activated sludge process. Proc Environ Eng Res 31: 305-314.

44. Lemos PC, Viana C, Salgueiro EN, Ramos AM, Crespo JPSG, et al. (1998) Effect of carbon source on the formation of polyhydroxyalkanoates (PHA) by a phosphate-accumulating mixed culture. Enzyme and Microbial Technology 22: 662-671.

45. Saito Y, Soejima T, Tomozawa T, Doi Y, Kiya F (1995) Production of biodegradable plastics from volatile acids using activated sludge. J Environ Syst Eng 521: 145-154.

46. Ostle AG, Holt JG (1982) Nile blue A as a fluorescent stain for poly-beta-hydroxybutyrate. Appl Environ Microbiol 44: 238-241.

47. Wendlandt KD, Geyer W, Mirschel G, Al-Haj Hemidi F (2005) Possibilities for controlling a PHB accumulation process using various analytical methods. J Biotechnol 117: 119-129.

48. Law JH, Slepecky RA (1961) Assay of poly-beta-hydroxybutyric acid. J Bacteriol 82: 33-36.

49. Serafim LS, Lemos PC, Levantesi C, Tandoi V, Santos H, et al. (2000) Methods for detection and visualization of intracellular polymers stored by polyphosphate-accumulating microorganisms. J Microbiol Methods 51: 1-18.

50. Padan E, Bibi E, Ito M, Krulwich TA (2005) Alkaline pH homeostasis in bacteria: new insights. Biochim Biophys Acta 1717: 67-88.

51. Braunegg G, Lefebvre G, Genser KF (1998) Polyhydroxyalkanoates, biopolyesters from renewable resources: physiological and engineering aspects. J Biotechnol 65: 127-161.

52. Kim HM, Ryu KE, Bae K, Rhee YH (2000) Purification and characterization of extracellular medium-chain-length polyhydroxyalkanoate depolymerase from *Pseudomonas sp.* RY-1. J Biosci Bioeng 89: 196-198.

53. Kim BS (2000) Production of poly(3-hydroxybutyrate) from inexpensive substrates. Enzyme Microb Technol 27: 774-777.

54. Kim BS, Chang HN (1998) Production of poly (3-hydroxybutyrate) from starch by *Azotobacter chroococcum*. Biotechnology letters 20: 109-112.

55. Randriamahefa S, Renard E, Guérin P, Langlois V (2003) Fourier transform infrared spectroscopy for screening and quantifying production of PHAs by *Pseudomonas* grown on sodium octanoate. Biomacromolecules 4: 1092-1097.

56. Hong K, Sun S, Tian W, Chen GQ, Huang W (1999) A rapid method for detecting bacterial polyhydroxyalkanoates in intact cells by Fourier transform infrared spectroscopy. Applied Microbiology and Biotechnology 51: 523-526.

57. Khanna S, Srivastava AK (2005) Recent advances in microbial polyhydroxyalkanoates. Process Biochemistry 40: 607-619.

58. Grothe E, Moo-Young M, Chisti Y (1999) Fermentation optimization for the production of poly(β-hydroxybutyric acid) microbial thermoplastic. Enzyme and Microbial Technology 25: 132-141.

6

On the Primary Reaction Pathways in the Photochemistry of Nitro-Polycyclic Aromatic Hydrocarbons

Carlos E Crespo-Hernández*, Aaron Vogt R and Briana Sealey

Department of Chemistry and Center for Chemical Dynamics, Case Western Reserve University, 10900 Euclid Avenue, Cleveland, Ohio 44106, USA

Abstract

The primary reaction pathways in the photochemistry of nitro-polycyclic aromatic hydrocarbons under specific laboratory conditions are briefly summarized. In addition, photochemical data is presented for 2-nitronaphthalene, 1-nitronaphthalene, and 2-methyl-1-nitronaphthalene in cyclohexane and acetonitrile solutions under aerobic and anaerobic conditions. It is shown that molecular oxygen significantly reduces the photodegradation quantum yield of 1-nitronaphthalene and 2-methyl-1-nitronaphthalene by 63% and 81%, respectively, whereas 2-nitronaphthalene is photoinert in both solvents under aerobic and anaerobic conditions. In addition, it is proposed that recombination of the aryl and nitrogen (IV) dioxide geminate radical pair within the solvent cage or internal conversion of an initially formed intramolecular charge transfer state may play an important role in the fraction of excited molecules that returns to the ground state in 1-nitronaphthalene and 2-methyl-1-nitronaphthalene. Scavenging of radical species by molecular oxygen and the generation of singlet oxygen in high yield are proposed to contribute to the photochemistry of these nitro-naphthalene derivatives in solution.

Keywords: Photochemistry of nitro-polycyclic aromatic compounds; Radical pair recombination; Excited-state dynamics; Environmental photochemistry; Ultrafast spectroscopy; Computational chemistry

Introduction

Nitro-polycyclic aromatic hydrocarbons (NPAHs) constitute one of the most troubling classes of environmental pollutants. Concern about these compounds arises partly from their ubiquity: NPAHs are released to the environment directly from a variety of incomplete combustion processes [1] and are also formed in situ by atmospheric reactions of polycyclic aromatic hydrocarbons (PAHs) [2-5]. They have been identified as components of grilled food as well as in diesel, gasoline, and wood smoke emissions, and are also commonly found in aquatic systems and in sediments [5-9].

Despite the potential negative impact of NPAHs to human health, the emission of NPAHs to the environment continues, primarily from municipal incinerators [10], motor vehicles (particularly from diesel exhausts) [11,12], and industrial power plants [13] among other sources. These emissions are likely to rise with the increasing consumption of petroleum and coal particularly in developing countries. It is estimated that NPAH in ambient air, either as vapor or particle accounts for ~50% of the total direct mutagenicity [14]. Epidemiological studies show that exposure to diesel exhaust and urban air pollution is associated with an increased risk of lung cancer [9]. The increase in lung cancer risk that results from human exposure to particulate air pollution and the detection of NPAH compounds in lungs of non-smokers with lung cancer have led to continued interest in assessing their potential risk to humans and also in monitoring their concentrations and fate in the environment [15].

Photochemical transformation is thought to be one of the main routes of natural removal of NPAHs in the environment [8,16-19]. Importantly, NPAHs are frequently more toxic than their parent PAHs [8,20], and photochemical degradation of a number of NPAHs leads, in turn, to photoproducts, some of which are more toxic than their parent compounds [8,20,21]. However, the light-induced transformation mechanisms of NPAHs are still under debate and knowledge of their fates in the environment is of current interest [22-39]. Fundamental gaps remain in our knowledge of the elementary steps that lead to product formation with regard to the competition among available reaction pathways, the intermediate species involved in the photochemical transformations, and the degradation rates as a function of compound structure, added co-solutes, and micro-environment [26,34,35,40-43]. Consequently, development of laboratory models enabling accurate and quantitative measurements of these photochemical properties is essential for modeling the concentration and persistence of NPAHs in the environment and to understand their overall contribution to air quality [44-46]. Moreover, laboratory studies can facilitate the development of quantitative structure-activity relationships (QSARs), which connects photophysical and photochemical data to the physicochemical and structural properties of NPAHs [20,22,23,30,32,38,39,47]. In recent years, QSARs have been successfully developed [19,48,49], relating the toxicity, carcinogenicity, and mutagenicity of NPAHs with their physicochemical and structural properties.

The present work summarizes the primary reaction pathways that have been identified in the photochemistry of NPAHs under specific laboratory conditions. In addition, photochemical results of 2-nitronaphthalene, 1-nitronaphthalene, and 2-methyl-1-nitronaphthalene in cyclohexane and acetonitrile solutions are presented that provide new insights about the photochemical fate of these nitro-aromatic compounds in the environment.

Materials and Methods

Chemicals

Cyclohexane (99.9%) was obtained from Fisher Scientific. Acetonitrile (99.6%) was obtained from Acros. Both solvents were used as received. 2-Nitronaphthalene, 1-nitronaphthalene and 2-methyl-1-nitronaphthalene were obtained from Sigma-Aldrich (99.7%, 99% and 99%, respectively). 2-nitronaphthalene, 1-nitronaphthalene and 2-methyl-1-nitronaphthalene are moderately toxic compounds.

---

*Corresponding author: Carlos E Crespo-Hernandez, Frank Hovorka Assistant Professor, Department of Chemistry, Case Western Reserve University, Cleveland, Ohio, USA, E-mail: carlos.crespo@case.edu

Proper safety precautions were taken at all times to limit health risks. Due to the presence of a minor impurity in the commercial sample of 2-nitronaphthalene [39], recrystallized samples of 2-nitronaphthalene from methanol were used in the experiments reported in this work.

Steady-state measurements

Steady-state absorbance measurements were performed using a Cary 100 UV/Vis spectrometer (Varian, Inc.). Photodegradation experiments were performed using a 150 W Xe lamp (Newport-Oriel, Apex Source Arc, source model 66453, lamp model 6255), as reported recently [39]. The wavelength range of 275-375 nm was selected for irradiation by using a FGUVS11S colored glass filter (Schott), except for the determination of photodegradation rates, where a Pyrex filter was also used to restrict the irradiation wavelengths to the 300 to 375 nm range. The polychromatic light was focused through a 450 mm lens placed at 5 cm from the front of the lamp source. The sample was placed at 41 cm from the front of the lens. The beam width at the sample was 0.95 cm. Solutions were contained in a 1 cm quartz cuvette (Starna, Inc.) and stirred continuously with a magnetic stir bar (Starna, Inc.) to assure a homogenous irradiation of the solutions at all times.

For the determination of the polychromatic photodegradation quantum yields, the change in concentration of the nitronaphthalene derivatives as a function of irradiation time was obtained using a HPLC (Shimadzu LC-20AD) with an amide column (Ascentis RP-Amide, 5 μm, 25 cm×4.6 mm). The HPLC was used in order to separate the parent compounds from the photoproducts. Calibration curves were obtained for each nitronaphthalene derivative. At least 5 data points were used for each of the calibration curves, giving area under the chromatographic fraction versus concentration, as described recently [39]. An isocratic elution was used with a solvent composition of 80% acetonitrile and 20% water. A photodiode array detector (Shimadzu SPD-M20A) was used to measure the absorbance of the eluting compounds. Potassium ferrioxalate was used as an actinometer to measure the lamp intensity in photons/s [50]. Photodegradation yields were measured using the method recently developed by Dodson et al. [51] under O_2- and air-saturated conditions.

Results and Discussion

Brief summary of the primary photochemical pathways in NPAHs

In 1966, Chapman et al. correlated the probability of photoreactivity of NPAHs with the out-of-plane arrangement of the nitro group relative to the plane of the aromatic moiety [47]. NPAH compounds with parallel nitro substituents were proposed to be more stable to light degradation than those with perpendicular nitro substituents. This correlation is known as the Chapman's orientation-photoreactivity relationship. In general, nitro groups peri to two hydrogen atoms are expected to be in a perpendicular orientation (Figure 1a), while nitro groups peri only to one hydrogen atom are expected to stay in a parallel orientation relative to the aromatic rings (Figure 1b) [48,52-54]. The preferred perpendicular orientation of the nitro group peri to two hydrogen atoms is due to steric forces acting between the hydrogen atoms on the aromatic rings and the oxygen atoms on the nitro group, which restrict the torsion angle of the nitro group to an out-of-plane conformation.

Chapman proposed that the first step after excitation of compounds with nitro groups that are perpendicular to the aromatic rings is rearrangement of the nitro group to a nitrite intermediate (Figure 2 (2)). It was further proposed that perpendicular orientations increased

probability of overlap between the half-vacant, non-bonding orbital of the nitro group and the adjacent orbital of the aromatic ring in the excited state [47]. This orbital overlap leads to the formation of an oxaziridine-type transition state (Figure 2 (1)) that can collapse to the nitrite intermediate (Figure 2 (2)). A dissociation reaction then produces nitrogen (II) oxide (NO•) and aryloxy (ArO•) radicals (Figure 2 (3)). Oxidation products such as quinones and hydroxyl compounds are expected to be formed subsequently from secondary reactions of the aryloxy radical intermediate. In parallel orientations, orbital overlap between the π-orbitals of the aromatic ring and the half-vacant, non-bonding orbital of the nitro group is unfavorable in the excited state, precluding the formation of the oxaziridine-type transition state (Figure 2).

However, it is still unclear which excited state is responsible for the formation of the nitrite intermediate and of the NO and ArO radicals. Figure 3 summarizes the primary electronic-energy relaxation pathways that have been proposed in the photochemistry of NPAH compounds. A generic NPAH compound (NO_2-Ar) is used as example. A direct dissociation-recombination mechanism from the excited singlet state was first proposed by Brown [55]. In this mechanism, the C-N bond is weakened in the excited singlet state leading to the formation of the aryl (Ar•) and nitrogen (IV) dioxide geminate radical pair. This radical pair was suggested to rearrange to a nitrite intermediate, which then forms the aryloxy radical (pathway 2b in Figure 3). Subsequently, Chapman proposed that an upper nπ* triplet state (pathway 2a in Figure 3) is another plausible candidate for the formation of nitrogen (II) oxide and the aryloxy radicals but did not provide spectroscopic evidence of its participation [47].

Spectroscopic studies, mainly by Hamanoue [56-60], and Testa and their co-workers [61,62] were subsequently performed in an effort to detect the proposed excited states and reactive intermediate species participating in the photochemistry of NPAH compounds. Studies by Hamanoue et al. using 9-nitroanthracene and several meso-substituted nitroanthracene derivatives suggested that NPAH compounds decay through two simultaneous pathways after light absorption [29,56-60]. Singlet-to-triplet intersystem crossing was proposed as one of the relaxation pathways (path 1 in Figure 3). A second pathway involves the formation of nitrogen (II) oxide and the aryloxy radicals through a photoinduced dissociation reaction of the NPAH compound from an intermediate triplet state (path 2a in Figure 3). Electron spin resonance and transient absorption studies have been used to detect the radical

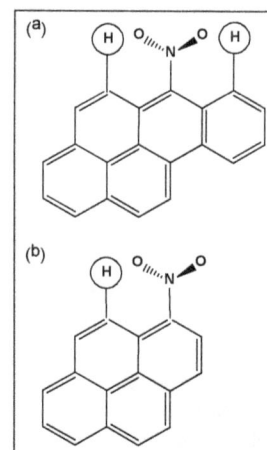

Figure 1: The nitro-group in (a) 6-nitrobenzo[a]pyrene is *peri to* two hydrogen atoms while in (b) 1-nitropyrene is *peri* to one hydrogen atom.

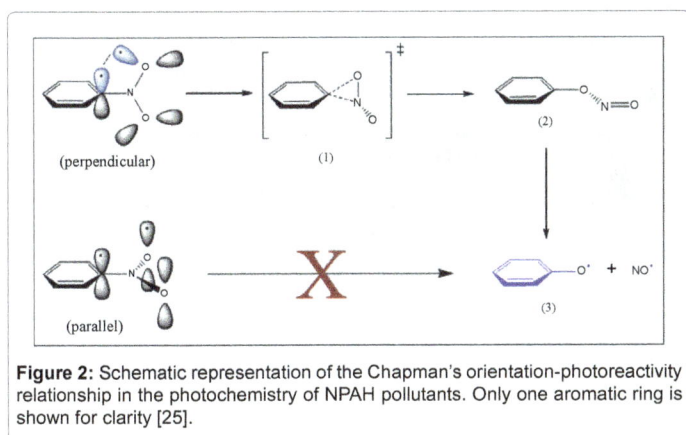

Figure 2: Schematic representation of the Chapman's orientation-photoreactivity relationship in the photochemistry of NPAH pollutants. Only one aromatic ring is shown for clarity [25].

intermediates formed from the dissociation reaction [57,59,63]. The lowest-energy excited triplet state has been detected using transient absorption, Raman spectroscopy, and steady state phosphorescence at low-temperature [26,34,56,59,64,65]. However, direct evidence of the population of the intermediate excited triplet state and of the excited state responsible for the formation of the nitrogen (II) oxide and the aryloxy radicals were not obtained because of the limited time-resolution and detection sensitivity used in those works. It has also been suggested that the lowest-energy excited triplet state in NPAH compounds may undergo intermolecular electron transfer or hydrogen abstraction depending on the solvent used (Figure 3) [34,35,61,65,66], but direct spectroscopic evidence of these reactive intermediates is lacking (Figure 3).

Several groups have recently challenged Chapman's orientation-reactivity relationship [23,25,26,29,30]. Warner et al. studied the photodegradation of several NPAH pollutants in solution and solid particles [23]. The authors grouped the NPAHs in three different categories based on their photoreactivity in solution. While several NPAH compounds followed Chapman's relationship others do not. It was concluded that the nitro group orientation is not the only factor controlling the photoreactivity of the NPAH pollutants [23], and recent works support this idea [25,29,32,34,38,39]. For instance, femtosecond broadband transient absorption spectroscopy in combination with computational methods has been recently used to study the excited state dynamics in 1-nitropyrene [25]. Ground state optimization of 1-nitropyrene results in a nitro group torsion angle of 32.8° relative to the aromatic rings. Optimization of the lowest-energy excited singlet state resulted in a nitro group torsion angle of 0.07°. The transient absorption spectra revealed that the rotation of the nitro group in the excited singlet state occurs in ~100 femtoseconds [25]. Based on Chapman's relationship (Figure 2), the nitro to nitrite rearrangement should not be favored in the photochemistry of 1-nitropyrene. However, 1-hydroxypyrene is the major product in the photochemistry of 1-nitropyrene in nonpolar solvents and in the absence of oxygen [8,26,34,35]. The formation of 1-hydroxypyrene has been proposed to originate from the formation of the pyrenyloxyl radical, followed by hydrogen atom abstraction from the solvent [26]. This mechanism is inconsistent with Chapman's hypothesis if the nitro group of 1-nitropyrene relaxes to a planar conformation in the excited state. A dissociation-recombination mechanism was thus proposed as a potential precursor of 1-hydroxypyrene (pathway 2b in Figure 3) [25]. However, direct evidence of the radical pair intermediate was not presented for 1-nitropyrene [25], but has been recently documented in 9-nitroanthracene [31]. Furthermore, experimental and computational data is now amassing in support of a direct

photodissociation mechanism from an intramolecular charge transfer state as a competitive degradation channel in the photochemistry of NPAHs under anaerobic conditions [25,29,31,35,38,39].

We have recently focused part of our effort toward investigating the electronic structure, steady-state UVA (320 to 400 nm) photochemistry, and excited-state dynamics of 1-nitronaphtalene, 2-nitronaphthalene, and 2-methyl-1-nitronaphthalene in solution [29,38,39]. Quantum-chemical calculations predict that the torsion angle in these nitro-PAHs is modulated by electrostatic forces acting between the lone pairs of the oxygen atoms on the nitro group and the neighboring atoms on the aromatic moiety. The calculations also predict that there is a distribution of torsion angles in the ground state at room temperature for each NPAH. Upon excitation, the distribution of torsion angles accesses different regions of configurational space in the excited singlet state potential energy surface in each molecule, which is thought to modulate the time-resolved and steady-state photochemistry of these nitronaphthalene derivatives [29,38,39]. Furthermore, the calculations predict a branching of the Franck-Condon excited singlet state to populate two main relaxation pathways in these nitronaphthalene derivatives. The first, main decay channel connects the excited singlet state with a receiver, high-energy triplet state (T_n) that has significant $n\pi^*$ character. The second channel involves conformational relaxation, primarily of the nitro group, to populate an intramolecular charge transfer state in 1-nitronaphtalene and in 2-methyl-1-nitronaphthalene, whereas a ca. 5 kcal/mol energy barrier prohibits the population of the intramolecular change transfer state in 2-nitronaphthalene [38]. This energy barrier explains the lack of photoreactivity in 2-nitronaphthalene in the solvent used [39]. The dissociative channel is proposed to be responsible for the dissociation-recombination mechanism proposed originally by Brown [55], leading to the formation of the aryl radical (Ar•) and nitrogen (IV) dioxide geminate radical pair (Figure 3). This radical pair rearranges to a nitrite intermediate (2 in Figure 2), which then form the nitrogen (II) oxide (NO) and aryloxy (ArO) radicals (3 in Figure 2 and pathway 2b in Figure 3). The formation of the NO and ArO radicals has been proposed to be the primary photochemical pathway leading to the observed photochemistry in 1-nitronaphthalene and 2-methyl-1-nitronaphthalene under anaerobic conditions [39].

Role of aryl and nitrogen (IV) dioxide geminate radical pair recombination in the photochemical fate of nitronaphthalene derivatives

An important aspect in the development of the kinetic model discussed above for the nitronaphthalene derivatives [29,38,39] is the fact that internal conversion from the lowest-energy excited singlet state to the ground state is negligible, as time-resolved anisotropy measurements for 1-nitronaphthalene indicated [24]. This is further supported by recent CASPT2//CASSCF(12/12) calculations in which conical intersections between the S_0 and excited singlet states were not found [37]. These nitronaphthalene derivatives also exhibit negligible fluorescence quantum yields of ca. 10^{-4} or less [38,39]. However, the

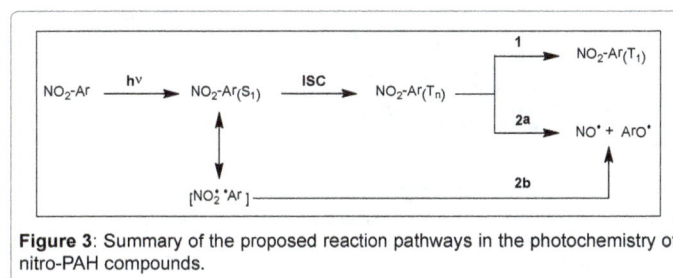

Figure 3: Summary of the proposed reaction pathways in the photochemistry of nitro-PAH compounds.

reported photodegradation yield of 1-nitronaphthalene is ca. 10^{-3} and its triplet yield is only 64% under anaerobic conditions [39], leaving a large fraction of the initial excited singlet state population unaccounted for. This result seems counterintuitive because it implies that close to 36% of the excited singlet state population in 1-nitronaphthalene should return back to the ground state. Similarly, the photodegradation yield of 2-nitronaphthalene is practically zero and the reported triplet yield is only 93% [39]. This implies that 7% of the excited singlet state population should decay back to the ground state. In the case of 2-methyl-1-nitronaphthalene, the triplet yield is 33% and the photodegradation yield is 12% [39]. This leaves 55% of the excited-state population returning to the ground state unaccounted. A potential explanation is that the triplet quantum yields reported for these molecules might be underestimated by as much as 10-20%, due to the energy transfer method used for their determination [39]. However, even under such an assumption, approximately 16% and 29% of the initial excited singlet state population returning to the ground state remain unaccounted for in 1-nitronaphthalene and 2-methyl-1-nitronaphthalene, respectively (Scheme 1).

To reconcile the fact that the time-resolved and the photochemical results available to date cannot account for the decay of all of the excited singlet state population, we propose that a large fraction of the aryl and nitrogen (IV) dioxide geminate radical pair decays back to the ground state by radical-pair recombination within the solvent cage, while only a small fraction dissociates to form the ArO and NO radicals (Scheme 1). An alternative explanation is that a fraction of the initially populated intramolecular charge transfer state internally converts back to the ground state before the geminate radical pair is formed. Either process can provide a reasonable explanation for the missing fraction of the excited singlet state population that returns back to the ground state. As explained above, a sizable energy barrier in the pathway leading to the intramolecular charge transfer state explains the triplet yield of close to unity in 2-nitronaphthalene [39], as well as its lack of photoreactivity of under anaerobic conditions [39]. The question arises as to whether this kinetic model can also explain the photoreactivity of the nitronaphthalene derivatives under aerobic conditions or if additional photochemical pathways are involved. For instance, the recently reported [39] high triplet yield in these nitronaphthalene derivatives may result in the efficient photosensitization of singlet oxygen, which can in turns lead to the formation of new oxidation products.

Photodegradation rates and quantum yields in nitronaphthalene derivatives under anaerobic and aerobic conditions

Figure 4 shows the changes in the absorption spectra of 1-nitronaphthalene as a function of irradiation time in the spectral

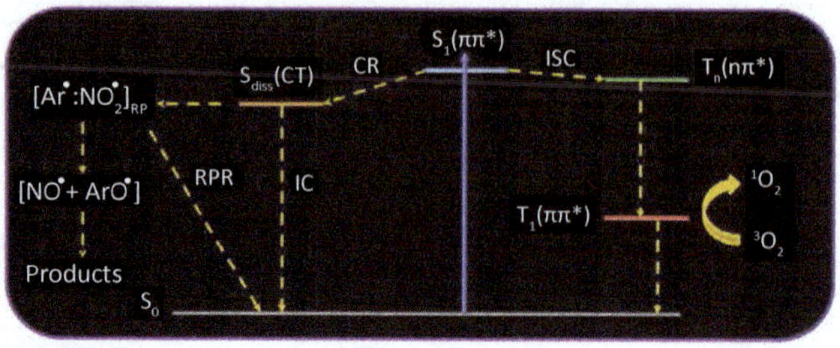

Scheme 1: Proposed reaction pathways in the photochemistry of 1-nitronaphthalene and 2-methyl-1-nitronaphthalene in cyclohexane and acetonitrile solutions. UVA excitation leads to an initial branching of the S_1 state into two relaxation pathways: (1) ultrafast intersystem crossing (ISC) to the triplet manifold and (2) conformational relaxation (CR) of the nitro group to an intramolecular charge transfer state with dissociative character, S_{diss}(CT). The S_{diss}(CT) state population can dissociate to form the aryl (Ar) and nitrogen (IV) dioxide (NO$_2$·) geminate radical pair (RP) or decay non radiatively by internal conversion (IC) to the ground state. The geminate radical pair can either recombine within the solvent cage (RPR) or rearrange to form the aryloxy (ArO·) and nitrogen (II) oxide radical (NO·), resulting in product formation. It is clear that molecular oxygen reduces the photodegradation yields significantly, likely forming singlet oxygen, whereas it is unclear whether or not the triplet state participate in the photochemistry of these nitronaphthalene derivatives [38].

region from 275 to 375 nm using a Xe-lamp as described in the methods section. These experiments were performed in cyclohexane under air-saturated conditions. The absorption spectra of the major products are also shown in Figure 4 under air-saturated conditions, using a photodiode array detector after HPLC separation of the reaction mixture. In particular, we highlight herein the absorption spectrum of the photoproduct with elution time of 4.23 min. The absorption spectrum of this product shows a red-shifted absorption band with maximum at ~445 nm relative to 1-nitronaphthalene. This is indicative of the formation of an oxidation product. This oxidation product might result from the aryloxy radical after hydrogen abstraction from the solvent or from a reaction of singlet oxygen with 1-nitronaphthalene. However, more work is needed to characterize the photoproducts of 1-nitronaphthalene and those resulting from the photolysis of 2-methyl-1-nitronaphthalene in different solvents and in the presence of different additives. While this is out of the scope of the present work, below we explore what can be learned regarding the primary photochemical pathways in these nitronaphthalene compounds from the determination of photodegradation rates and quantum yields in the presence or absence of molecular oxygen (Figure 4).

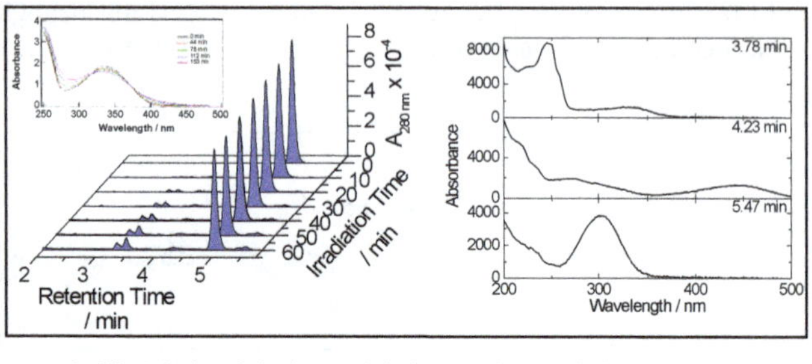

The photodegradation rates for 2-nitronaphthalene, 1-nitronaphthalene, and 2-methyl-1-nitronaphthalene were estimated from the changes in absorbance as a function of irradiation time for each nitronaphthalene derivative using absorption spectroscopy under air-saturated conditions. A Pyrex filter was used to block the radiation wavelengths below 300 nm (see the methods section for details). Precautions were taken to ensure that the integrated area of the initial absorption band above 300 nm (i.e., before the irradiation experiments began) were similar for the three nitronaphthalene derivatives. The changes in absorbance were converted to changes in concentration using the extinction coefficients reported in the literature for each compound [39]. A kinetic, zero-order rate equation was used to determine the photodegradation rates because the percent of degradation of each nitronaphthalene derivative was kept at 15% or less. In the case of 2-nitronaphthalene, no detectable changes in absorption spectrum were observed even after more than 6 hours of continuous irradiation in cyclohexane or acetonitrile solutions. Hence, it is assumed to be photoinert under the air- and O$_2$-saturated conditions used in this

Figure 4: Photolysis of 1-nitronaphthalene using a Xe-lamp in cyclohexane solution and identification of the absorption spectra of the major products under air-saturated conditions.

work. Similar results were obtained recently for the photochemistry of 2-nitronaphthalene in acetonitrile under anaerobic conditions [39] (Figure 5).

Figure 5 shows the estimated photodegradation rates as a function of the calculate ground-state torsion angle for each compound. The torsion angles were obtained from density functional calculations in vacuum at the B3LYP/6-31++G(d,p) level of theory, as reported previously [29,39]. The photodegradation rates increase with an increase in the nitro-group torsion angle of the nitronaphthalene derivatives independent of the solvent used. This is in general agreement with Chapman's orientation-reactivity relationship (Figure 2) [47] and in good agreement with the photodegradation rates previously measured in the gas phase [22]. While the photodegradation rate of 1-nitronaphthalene does not vary significantly in cyclohexane versus acetonitrile, the photodegradation rate of 2-methyl-1-nitronaphthalene is twofold higher in cyclohexane than in acetonitrile. The latter observation correlates with the smaller rate of intersystem crossing reported for this compound in cyclohexane (2.7×10^{12} s^{-1}) relative to acetonitrile (4.8×10^{12} s^{-1}) [38]. In other words, assuming that the solvent-dependent intersystem crossing rates equate to a decrease in the triplet yield of 2-methyl-1-nitronaphthalene in cyclohexane versus acetonitrile, an increase in the yield of the dissociative pathway would be expected (Scheme 1), which can lead to the observed increase in the photodegradation rate in cyclohexane. Another possibility is that there is a higher formation of singlet oxygen in cyclohexane than in acetonitrile for 2-methyl-1-nitronaphthalene [67], which leads to a significant increase in the formation of oxidative products. However, the participation of singlet oxygen in the formation of oxidation products seems to be insignificant. A more quantitative analysis of the potential participation of singlet oxygen in the photochemistry of 1-nitronaphthalene and 2-methyl-1-nitronaphthalene requires the determination of photodegradation yields in the presence and absence of molecular oxygen, which are reported next (Table 1).

Table 1 collects the photodegradation yields of 2-nitronaphthalene, 1-nitronaphthalene, and 2-methyl-1-nitronaphthalene in acetonitrile under N$_2$-, O$_2$-, and air-saturated conditions. The photodegradation quantum yields were obtained using a HPLC to separate the photoproducts from the parent compound, as explained in the method section. Analogous measurements in cyclohexane were impractical because the higher volatility of this solvent. As in the case of the photodegradation rate reported above, no photodegradation was observed in the irradiation experiments for 2-nitronaphthalene in the presence or absence of molecular oxygen and thus its

Molecule	Nitrogen [39]	Air	Oxygen
2-nitronaphthalene	~0	~0	n.d.[c]
1-nitronaphthalene	0.0035 ± 0.0004[a]	0.0015 ± 0.0004	0.0013 ± 0.0002
2-methyl-1-nitronaphthalene	0.123[b] ± 0.008	0.043 ± 0.002	0.023 ± 0.002

[a]errors are equal to two standard deviations from three independent measurements; [b]value did not change significantly with freshly distilled acetonitrile (0.126); [c]n.d. = not determined.

Table 1: Photodegradation yields of the nitronaphthalene derivatives in acetonitrile (initial solute concentration of ca. 150 µM).

photodegradation yield is reported as zero in Table 1. On the other hand, the photodegradation quantum yield of 1-nitronaphthalene and 2-methyl-1-nitronaphthalene decreases with an increase of molecular oxygen concentration. This might be indicative of quenching of the triplet state in the presence of molecular oxygen, leading to large yields of singlet oxygen generation in 1-nitronaphthalene [67] and 2-methyl-1-nitronaphthalene solutions. However, this will also suggest that singlet oxygen is unable to oxidize 1-nitronaphthalene or 2-methyl-1-nitronaphthalene in acetonitrile, as has been observed in nitropyreno derivatives [68].

The suggestion above that singlet oxygen cannot oxidize the nitronaphthalene derivatives would also imply at first glance that the triplet state should play a large role in the photochemistry of 1-nitronaphthalene and 2-methyl-1-nitronaphthalene in acetonitrile, contrary to what has been recently proposed [39]. However, the reduction in the photodegradation yield under aerobic conditions can also be explained by assuming that molecular oxygen can scavenge one or more of the radicals formed in the intramolecular charge transfer pathway (Scheme 1). In fact, Garcia-Berrios and Arce have recently proposed that molecular oxygen can scavenge the pyrenyloxyl radical of 1-nitropyrene efficiently [35], thus reducing by 67% the formation yield of 1-hydroxypyrene, one of the major photoproducts [26]. If molecular oxygen can also scavenge one or more of the radicals formed in 1-nitronaphthalene and 2-methyl-1-nitronaphthalene, then this would explain part or all of the reduction in the photodegradation yield reported in Table 1 under aerobic conditions. For instance, recent experiments in aqueous solutions have shown that the radical anion of 1-nitronaphthalene can be scavenged by molecular oxygen [69]. Clearly, however, the experiments presented herein cannot prove or rule out the participation of the triplet state in the photoreactivity of these nitronaphthalene derivatives or the direct scavenging by molecular oxygen of one or more of the radical species proposed in Scheme 1. Altogether, it is clear that the photochemistry of these nitronaphthalene derivatives is more complex than previously thought and further steadstate and time-resolved photochemical investigations in the presence of triplet quenchers and radical/electron scavengers are warranted.

Experiments aimed at measuring and quantifying the generation of singlet oxygen in the photochemistry of 2-nitronaphthalene and 2-methyl-1-nitronaphthalene are essential to evaluate their potential phototoxicity, as has been recently shown for other NPAHs [20]. A high yield of singlet oxygen has been reported for 1-nitronaphthalene in both acetonitrile and cyclohexane solutions [67]. In particular, 2-nitronaphthalene might prove to be an excellent singlet oxygen generator upon UVA excitation, which can explain its high toxicity [19]. If so, 2-nitronaphthalene has the potential to be used in photosensitization and therapeutic applications, where the use of a photochemically-robust, photoinert sensitizer is required.

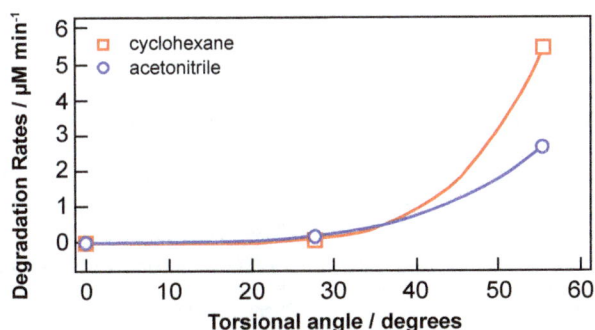

Figure 5: Correlation between the nitro-group orientation of the nitronaphthalene derivatives in the ground state with the photodegradation rates of 2-nitronaphthalene (0.0°), 1-nitronaphthalene (27.7°), and 2-methyl-1-nitronaphthale (55.4°). Symbols are connected by a solid line to guide the eye.

Conclusions

The major reaction pathways in the photochemistry of NPAHs under specific laboratory conditions have been presented. Readers interested in a broader discussion about the photochemistry of nitro-aromatic compounds are referred to previous monographs [4,8,19,70,71]. Despite recent progress at understanding the photodegradation mechanism of NPAHs, fundamental gaps remain in our knowledge of the elementary steps that lead to product formation. This is particularly evident regarding the competition among available reaction pathways, the intermediate species involved in the photochemical transformations, and the degradation rates as a function of compound structure, added co-solutes, and micro-environment.

New insights regarding the primary reaction pathways participating in the photochemistry of 2-nitronaphthalene, 1-nitronaphthalene, and 2-methyl-1-nitronaphthalene in cyclohexane and acetonitrile solutions under aerobic and anaerobic conditions were presented. It was shown that molecular oxygen significantly reduces the photodegradation quantum yield of 1-nitronaphthalene and 2-methyl-1-nitronaphthalene, whereas 2-nitronaphthalene is essentially photoinert. Recombination of the aryl and nitrogen (IV) dioxide geminate radical pair within the solvent cage was suggested to play an important role in the fraction of excited singlet state molecules that returns to the ground state in 1-nitronaphthalene and 2-methyl-1-nitronaphthalene. An alternative ground-state recovery pathway was proposed, where a fraction of the intramolecular charge transfer state internally converts to the ground state before the dissociation-recombination process occurs. Scavenging of radical species by molecular oxygen and the generation of singlet oxygen in high yield were proposed to contribute to the photochemistry of these nitro-naphthalene derivatives in solution. Singlet oxygen did not seem to oxidize the nitronaphthalene derivatives under the experimental conditions used in this work. The triplet state was suggested to play a minor role in the photochemistry of 1-nitronaphthalene and 2-methyl-1-nitronaphthalene. Further studies are needed to corroborate or refute this latter hypothesis using triplet quenchers other than molecular oxygen.

Acknowledgments

This research was partially supported by the donors of the American Chemical Society Petroleum Research Fund. The authors also acknowledge the support to B.S. from the American Chemical Society and the Project SEED endowment.

References

1. World Health Organization (1989) Diesel and gasoline engine exhausts and some nitroarenes. International Agency for Research on Cancer 46: 41.

2. Atkinson R, Arey J (1994) Atmosphere chemistry of gas-phase polycyclic aromatic hydrocarbons: formation of atmospheric mutagens. Environ Health Perspect 102: 117-126.

3. Arey J, Zielinska B, Harger WP, Atkinson R, Winer AM (1988) The contribution of nitrofluoranthenes and nitropyrenes to the mutagenic activity of ambient particulate organic matter collected in southern California. Mutat Res 207: 45-51.

4. Arey J, Atkinson R (2003) Photochemical reactions of PAHs in the atmosphere. PAHs: An ecotoxicological perspective. John Wiley & Sons, Limited, UK.

5. Tokiwa H, Ohnishi Y (1986) Mutagenicity and carcinogenicity of nitroarenes and their sources in the environment. Crit Rev Toxicol 17: 23-60.

6. Möller L (1994) In vivo metabolism and genotoxic effects of nitrated polycyclic aromatic hydrocarbons. Environ Health Perspect 102: 139-146.

7. Purohit V, Basu AK (2000) Mutagenicity of nitroaromatic compounds. Chem Res Toxicol 13: 673-692.

8. Yu H (2002) Environmental carcinogenic polycyclic aromatic hydrocarbons: photochemistry and phototoxicity. J Environ Sci Health C Environ Carcinog Ecotoxicol Rev 20: 149-183.

9. Arlt VM (2005) 3-nitrobenzanthrone, a potential human cancer hazard in diesel exhaust and urban air pollution: a review of the evidence. Mutagenesis 20: 399-410.

10. Kamiya A, Ose Y (1988) Isolation of dinitropyrene in emission gas from a municipal incinerator and its formation by a photochemical reaction. Sci Total Environ 72: 1-9.

11. Henderson TR, Sun JD, Royer RE, Harvey TM, Hunt DH, et al. (1983) Triple-quadrupole mass spectrometry studies of nitroaromatic emissions from different diesel engines. Environ Sci Technol 17: 443-449.

12. Handa T, Yamauchi T, Sawai K, Yamamura T, Koseki Y, et al. (1984) In situ emission levels of carcinogenic and mutagenic compounds from diesel and gasoline engine vehicles on an expressway. Environ Sci Technol 18: 895-902.

13. Hönor A, Arnold M, Hüsers N, Kleiböhmer W (1995) Monitoring polycyclic aromatic hydrocarbons in waste gases. Journal of Chromatography A 710: 129-137.

14. Schauer C, Niessner R, Pöschl U (2004) Analysis of nitrated polycyclic aromatic hydrocarbons by liquid chromatography with fluorescence and mass spectrometry detection: air particulate matter, soot, and reaction product studies. Anal Bioanal Chem 378: 725-736.

15. Tokiwa H, Sera N, Horikawa K, Nakanishi Y, Shigematu N (1993) The presence of mutagens/carcinogens in the excised lung and analysis of lung cancer induction. Carcinogenesis 14: 1933-1938.

16. Kamens RM, Zhi-Hua F, Yao Y, Chen D, Chen S, et al. (1994) A methodology for modeling the formation and decay of nitro-PAH in the atmosphere. Chemosphere 28: 1623-1632.

17. Fan Z, Chen D, Birla P, Kamens RM (1995) Modeling of nitro-polycyclic aromatic hydrocarbon formation and decay in the atmosphere. Atmospheric Environment 29: 1171-1181.

18. Fan Z, Kamens RM, Hu J, Zhang J, McDow S (1996) Photostability of nitro-polycyclic aromatic hydrocarbons on combustion soot particles in sunlight. Environ Sci Technol 30: 1358-1364.

19. Janet K, Ulrich W, Inge M (2003) Selected nitro and nitro-oxy-polycyclic aromatic hydrocarbons. Environmental health criteria. World Health Organization, Geneva.

20. Xia Q, Yin JJ, Zhao Y, Wu YS, Wang YQ, et al. (2013) UVA photoirradiation of nitro-polycyclic aromatic hydrocarbons-induction of reactive oxygen species and formation of lipid peroxides. Int J Environ Res Public Health 10: 1062-1084.

21. Victorin K (1994) Review of the genotoxicity of nitrogen oxides. Mutat Res 317: 43-55.s

22. Phousongphouang PT, Arey J (2003) Rate constants for the photolysis of the nitronaphthalenes and methylnitronaphthalenes. Journal of Photochemistry and Photobiology A: Chemistry 157: 301-309.

23. Warner SD, Farant JP, Butler IS (2004) Photochemical degradation of selected nitropolycyclic aromatic hydrocarbons in solution and adsorbed to solid particles. Chemosphere 54: 1207-1215.

24. Morales-Cueto R, Esquivelzeta-Rabell M, Saucedo-Zugazagoitia J, Peon J (2007) Singlet excited-state dynamics of nitropolycyclic aromatic hydrocarbons: direct measurements by femtosecond fluorescence up-conversion. J Phys Chem A 111: 552-557.

25. Crespo-Hernandez CE, Burdzinski G, Arce R (2008) Environmental photochemistry of nitro-PAHs: direct observation of ultrafast intersystem crossing in 1-nitropyrene. J Phys Chem A 112: 6313-6319.

26. Arce R, Pino EF, Valle C, Agreda J (2008) Photophysics and photochemistry of 1-nitropyrene. J Phys Chem A 121: 10294-10304.

27. Mohammed OF, Vauthey E (2008) Excited-state dynamics of nitroperylene in solution: solvent and excitation wavelength dependence. J Phys Chem A 112: 3823-3830.

28. Zugazagoitia JS, Collado-Fregoso E, Plaza-Medina EF, Peon J (2009) Relaxation in the triplet manifold of 1-nitronaphthalene observed by transient absorption spectroscopy. J Phys Chem A 113: 805-810.

29. Reichardt C, Vogt RA, Crespo-Hernández CE (2009) On the origin of ultrafast nonradiative transitions in nitro-polycyclic aromatic hydrocarbons: Excited-state dynamics in 1-nitronaphthalene. J Chem Phys 131: 224518.

30. Vyas S, Onchoke KK, Rajesh CS, Hadad CM, Dutta PK (2009) Optical spectroscopic studies of mononitrated benzo[a]pyrenes. J Phys Chem A 113: 12558-12565.

31. Plaza-Medina EF, Rodríguez-Córdoba W, Morales-Cueto R, Peon J (2011)

Primary photochemistry of nitrated aromatic compounds: excited-state dynamics and NO· dissociation from 9-nitroanthracene. J Phys Chem A 115: 577-585.

32. Plaza-Medina EF, Rodríguez-Córdoba W, Peon J (2011) Role of upper triplet states on the photophysics of nitrated polyaromatic compounds: S(1) lifetimes of singly nitrated pyrenes. J Phys Chem A 115: 9782-9789.

33. Murudkar S, Mora AK, Singh PK, Nath S (2011) Origin of ultrafast excited state dynamics of 1-nitropyrene. J Phys Chem A 115: 10762-10766.

34. Arce R, Pino EF, Valle C, Negrón-Encarnación I, Morel M (2011) A comparative photophysical and photochemical study of nitropyrene isomers occurring in the environment. J Phys Chem A 115: 152-160.

35. García-Berríos ZI, Arce R (2012) Photodegradation mechanisms of 1-nitropyrene, an environmental pollutant: the effect of organic solvents, water, oxygen, phenols, and polycyclic aromatics on the destruction and product yields. J Phys Chem A 116: 3652-3664.

36. Healy RM, Chen Y, Kourtchev I, Kalberer M, O'Shea D, et al. (2012) Rapid formation of secondary organic aerosol from the photolysis of 1-nitronaphthalene: role of naphthoxy radical self-reaction. Environ Sci Technol 46: 11813-11820.

37. Orozco-Gonzalez Y, Coutinho K, Peon J, Canuto S (2012) Theoretical study of the absorption and nonradiative deactivation of 1-nitronaphthalene in the low-lying singlet and triplet excited states including methanol and ethanol solvent effects. J Chem Phys 137: 054307.

38. Vogt RA, Reichardt C, Crespo-Hernández CE (2013) Excited-state dynamics in nitro-naphthalene derivatives: Intersystem crossing to the triplet manifold in hundreds of femtoseconds. J Phys Chem A.

39. Vogt RA, Crespo-Hernández CE (2013) Conformational control in the population of the triplet state and photoreactivity of nitro-naphthalene derivatives.

40. Feilberg A, Kamens RM, Strommen MR, Nielsen T (1999) Modeling the formation, decay, and partitioning of semivolatile nitro-polycyclic aromatic hydrocarbons (nitronaphthalenes) in the atmosphere. Atmospheric Environment 33: 1231-1243.

41. Feilberg A, Kamens RM, Strommen MR, Nielsen T (1999) Photochemistry and partitioning of semivolatile nitro-PAH in the atmosphere. Polycyclic Aromatic Compounds 14: 151-160.

42. Feilberg A, Nielsen T (2000) Effect of aerosol chemical composition on the photodegradation of nitro-polycyclic aromatic hydrocarbons. Environ Sci Technol 34: 789-797.

43. Feilberg A, Nielsen T (2001) Photodegradation of nitro-PAHs in viscous organic media used as models of organic aerosols. Environ Sci Technol 35: 108-113.

44. Vione D, Maurino V, Minero C, Pelizzetti E, Harrison MA, et al. (2006) Photochemical reactions in the tropospheric aqueous phase and on particulate matter. Chem Soc Rev 35: 441-453.

45. Vione D, Minero C, Hamraoui A, Privat M (2007) Modelling photochemical reactions in atmospheric water droplets: an assessment of the importance of surface processes. Atmospheric Environment 41: 3303-3314.

46. Miet K, Le Menach K, Flaud PM, Budzinski H, Villenave E (2009) Heterogeneous reactions of ozone with pyrene, 1-hydroxypyrene, and 1-nitropyrene adsorbed on particles. Atmospheric Environment 43: 3699-3707.

47. Chapman OL, Heckert DC, Reasoner JW, Thackaberry SP (1966) Photochemical studies on 9-nitroanthracene. J Am Chem Soc 88: 5550-5554.

48. Fu PP, Herrero-Saenz D (1999) Nitro-polycyclic aromatic hydrocarbons: A class of genotoxic environmental pollutants. Journal of Environmental Science and Health, Part C: Environmental Carcinogenesis and Ecotoxicology Reviews 17: 1-43.

49. Vogt RA, Rahman S, Crespo-Hernandez CE (2010) Structure-activity relationships in nitro-aromatic compounds. Practical Aspects of Computational Chemistry 217-240.

50. Calvert JG, Pitts JN (1966) Photochemistry. Wiley: New York, USA.

51. Dodson LG, Vogt RA, Marks J, Reichardt C, Crespo-Hernández CE (2011) Photophysical and photochemical properties of the pharmaceutical compound salbutamol in aqueous solutions. Chemosphere 83: 1513-1523.

52. Fu PP, Chou MW, Beland FA (1988) Effects of nitro substitution on the in vitro metabolic activation of polycyclic aromatic hydrocarbons. In: Yang SK, Silverman BD (editions) polycyclic aromatic hydrocarbons carcinogenesis: structure-activity relationships. CRC Press, Boca Raton, USA.

53. Fu PP (1990) Metabolism of nitro-polycyclic aromatic hydrocarbons. Drug Metab Rev 22: 209-268.

54. Fu PP, Qui FY, Jung H, Von Tungeln LS, Zhan DJ, et al. (1997) Metabolism of isomeric nitrobenzo[a]pyrenes leading to DNA adducts and mutagenesis. Mutat Res 376: 43-51.

55. Brown HW, Pimentel GC (1958) Photolysis of nitromethane and methyl nitrite in an argon matrix; infrared detection of nitroxyl, HNO. J Chem Phys 29: 883-888.

56. Hamanoue K, Hirayama S, Nakayama T, Teranishi H (1980) Nonradiative relaxation process of the higher excited states of meso-substituted anthracenes. J Phys Chem 84: 2074-2078.

57. Hamanoue K, Amano M, Kimoto M, Kajiwara Y, Nakayama T, et al. (1984) Photochemical reactions of nitroanthracene derivatives in fluid solutions. J Am Chem Soc 106: 5993-5997.

58. Hamanoue K, Nakayama T, Ushida K, Kajiwara K, Yamanaka S (1991) Photophysics and photochemistry of nitroanthracenes. Part 1-Primary processes in the photochemical reactions of 9-benzoyl-10-nitroanthracene and 9-cyano-10-nitroanthracene studied by steady-state photolyis and nanosecond laser photolysis. J Chem Soc Faraday Trans 87: 3365-3371.

59. Hamanoue K, Nakayama T, Kajiwara K, Yamanaka S (1992) Photophysics and photochemistry of nitroanthracenes. Part 2-Primary process in the photochemical reaction of 9-nitroanthracene studied by steady-state photolysis and laser photolysis. J Chem Soc Faraday Trans 88: 3145-3151.

60. Hamanoue K, Nakayama T, Amijima Y, Ibuki K (1997) A rapid decay channel of the lowest excited singlet state of 9-benzoyl-10-nitroanthracene generating 9-benzoyl-10-anthryloxy radical and nitrogen (II) oxide. Chemical Physics Letters 267: 165-170.

61. Hurley R, Testa AC (1968) Triplet-state yield of aromatic nitro compounds. J Am Chem Soc 90: 1949-1952.

62. Testa AC, Wild UP (1986) Holographic photochemical study of 9-nitroanthracene. J Phys Chem 90: 4302-4305.

63. Fukuhara K, Kurihara M, Miyata N (2001) Photochemical generation of nitric oxide from 6-nitrobenzo[a]pyrene. J Am Chem Soc 123: 8662-8666.

64. Anderson Jr RW, Hochstrasser RM, Lutz H, Scott GW (1974) Measurements of intersystem crossing kinetics using 3545 Å picosecond pulses: nitronaphthalenes and benzophenone. Chemical Physics Letters 28: 153-157.

65. Fournier T, Tavender SM, Parker AW, Scholes GD, Phillips D (1997) Competitive energy and electron-transfer reactions of the triplet state of 1-nitronaphthalene: a laser flash photolysis and time-resolved Raman study. J Phys Chem A 101: 5320-5326.

66. Görner H (2002) Photoreduction of nitronaphthalenes in benzene by N,N-dialkylanilines and triethylamine: a time-resolved UV-vis spectroscopic study. J Chem Soc Perkin Trans 2: 1778-1783.

67. Wilkinson F, McGarvey DJ, Olea AF (1994) Excited Triplet state interactions with molecular oxygen: Influence of charge transfer on the bimolecular quenching rate constants and the yields of singlet oxygen [O·2(1.delta.g)] for substituted naphthalenes in various solvents. J Phys Chem 98: 3762-3769.

68. Arce R personal communication.

69. Brigante M, Charbouillot T, Vione D, Mailhot G (2010) Photochemistry of 1-nitronaphthalene: a potential source of singlet oxygen and radical species in atmospheric waters. J Phys Chem A 114: 2830-2836.

70. Malkin J (1992) Photochemistry of Nitro Compounds. In: Photophysical and Photochemical Properties of Aromatic Compounds. CRC Press Inc., 426 pages.

71. Dopp D (1995) Photochemical Reactivity of the Nitro Group. In: CRC Handbook of Organic Photochemistry and Photobiology. CRC Press, Inc., Boca Raton, Florida, USA.

7

QSAR and Synthesis of a Novel Biphenyl Carboxamide Analogue for Analgesic Activity

Prasanna Datar*

Sinhgad Institute of Pharmacy, Narhe, Pune, Maharashtra, India

Abstract

A compound 2-(anilino)-1-(4-phenylphenyl)ethanone was synthesized by reaction between biphenyl acid chloride and aniline. The synthesized compound was screened for analgesic activity. The selection of the compound was on the basis of QSAR study performed on the biphenyl analogs having anti-inflammatory activity. 2D QSAR study using topological descriptors revealed the important features required for the design of new potent anti-inflammatory agent. QSAR equation found in present study had r^2 of 0.842 and internal predictivity of 0.69. Using QSAR results a molecule was designed to have analgesic activity.

Keywords: MLR; Electrotopological indices; Biphenyl; Carboxamide; Analgesic; Anti-inflammatory

Introduction

An analgesic is used to relieve pain either by acting at the peripheral or central nervous systems; they include paracetamol (acetaminophen) and Non-Steroidal Anti-Inflammatory Drugs (NSAIDs) such as the salicylates, aryl acetic acids, anthranilic acid derivatives and propionic acid derivatives. In choosing analgesics, the severity of the pain and the type of pain such as neuropathic pain are taken into consideration. The WHO pain ladder, originally developed in cancer-related pain, is widely applied to find suitable analgesic drugs in a stepwise manner [1].

The exact mechanism of action of paracetamol/acetaminophen is uncertain, but it appears to be acting centrally rather than peripherally in the brain rather than in nerve endings. Aspirin and other Non-Steroidal Anti-Inflammatory Drugs (NSAIDs) inhibit cyclooxygenases, leading to a decrease in prostaglandin production. This reduces pain and also inflammation. Paracetamol and aspirin have one aromatic center and other carboxyl group center as pharmacophoric feature [2-5].

Paracetamol has few side effects and is regarded as safe, although intake above the recommended dose can lead to liver damage, which can be severe and life-threatening, and occasionally kidney damage [6-8]. NSAIDs predispose to peptic ulcers, renal failure, allergic reactions, and occasionally hearing loss, and they can increase the risk of hemorrhage by affecting platelet function. The ulceration was thought to be due to presence of carboxyl group function [9,10].

There are various biphenyl analgesics present in the market such as diflunisal and flurbiprofen. Diflunisal acts by inhibiting the production of prostaglandins [11], while flurbiprofen reduces the hormone that causes inflammation and pain in the body. Biphenyl-4-carboxylic acid has been reported for anti-inflammatory activity [12], antimicrobial activity [13,14], insecticidal [15], antidiabetic [16], cytotoxicity, leishmanicidal, trypanocidal and antimycobacterial activities [17]. 3D QSAR [9,10], docking and pharmacophore modeling studies have been performed earlier on biphenyl compounds. The common scaffold in most of the NSAID's had two aromatic ring centers and one acidic center as pharmacophoric element. The presence of carboxyl group or any other acidic center was also found to be the cause of ulcerogenic property of NSAID's. Recently nonacidic molecules have been found to be effective NSAID that lack acidic function such as carboxyl group e.g. nabumetone.

Introduction of third aromatic center has lead to the development of COX-2 inhibitors for almost two decades. Recently in 2004, rofecoxib and valdecoxib were withdrawn from the market due to their report of cardiovascular side effects. Thus introducing third aromatic center has made the molecule to interact with other biological receptors leading to severe side effects.

In order to stand with the known importance of two aromatic ring centers, a series of compounds containing biphenyl nucleus, substituted with carboxamide linkage at position-2 was designed using QSAR study and synthesized to evaluate analgesic activity.

Materials and Methods

2D QSAR studies

Training set was selected that spans a large chemical domain. The domain can be limited or diverse with respect to chemical to develop and evaluate methodology for increasing the reliability of QSAR predictions. Regardless of the diversity of the training data used it is important to realize that empirical QSAR models are only valid in the domain space in which they were trained and validated. Extrapolation of empirical models is dangerous and can lead to grossly erroneous model predictions [18]. Twenty five substituted direct analogues of flurbiprofen [4'-methylbiphenyl-2-(substituted phenyl) carboxamide derivatives] reported by Shah et al. [19] for anti-inflammatory activity, as shown in Table 1. Training set was selected by sphere exclusion method and manual selection using chemical diversity. Data set was distributed manually to find mean and was further divided into training set and test set such that test set distribution lies within the limits of training set. Training set consists of a total of seventeen molecules such as, 1, 5, 6, 7, 8, 10, 11, 12,15, 16, 17, 18, 20, 21, 22, 23, 24. While test set consists of eight molecules such as 2, 3, 4, 9, 13, 14, 19, 25.

Biological data

Anti-inflammatory activity of these molecules was found reported [19] as percent inhibition required inhibiting rat paw edema induced by carrageenan. The activity was converted to logarithmic value (Log

***Corresponding author:** Prasanna Datar, Sinhgad Institute of Pharmacy, Narhe, Pune, 411041, Maharashtra, India, E-mail: d_pras_anna@rediffmail.com

BA) as shown in Table 1.

Experimental

Computational details

All molecular modeling studies were carried out using VLife MDS 3.5 [20]. Structures were constructed and partial charges were assigned using the MMFF. The molecules were subsequently minimized using MMFF force field until a root mean square deviation of 0.001 kcal/mol Å was achieved.

Calculation of descriptors

Various molecular descriptors were calculated such as molecular weight, dipole moment, partition coefficient (ClogP), surface area descriptors, H-bond donor count, H-bond acceptor count, ionization potential, electron affinity, electrotopological indices and topological distance indices.

The topological distance indices captures the bond distances between any two atoms in a molecule. This bond distances (T) are useful to find distance between heteroatoms (H) and any other (O) atoms in number of bond path distances (D) indicated as T_H_O_D. One can find such distances commonly occurred in active molecules. A comparison can be made of such records with inactive molecules. Electrotopological indices gives the sum of E-state of atom type occurred in a molecule [21]. Electrotopological indices can be calculated easily in less time and can be used for virtual screening of databases to find useful hits that comply with the requirement of query substructure. The query substructure can be represented as sequence of atom types and path distances. This approach for finding novel entities is much simpler than earlier 3D QSAR studies reported [19]. The reproducibility of 3D QSAR study requires alignment of molecules and occupying grid space. In comparison our study using electrotopological indices generates search key which would be applicable without limitations of alignment and 3D space.

Multiple Linear Regression (MLR) models were generated to find the relationship between response variable and various physicochemical, structural descriptors (as independent variables). Significant descriptors were chosen on the basis of statistical data of analysis. Inter-correlation between these descriptors was not beyond 0.5 to confirm the independence of the variables.

The predictive power of equations was validated by Leave-One-Out (LOO) cross-validation method. Standard deviation based on predicted residual sum of squares (SPRESS) and Standard Deviation of Error of Prediction (SDEP) were used to guide selection of variables.

The statistical quality of the developed equations was judged by the parameters like coefficient of determination (r^2), standard error of estimate (s), cross-validation r^2 (q^2), SPRESS and SDEP.

The QSAR model generated by any method should be predictive as well as statistically significant and robust. Therefore randomization test was performed for 100 runs by shuffling the activity column in each run. The r^2 generated from randomization was compared with the r^2 of equation 1. The validity of the best model can be further checked by carrying out the effects of outlier detection by prediction diagnostics. The outlier can be identified by Z score value. Z score value is calculated by the following formula.

Z score = (x - x_mean)/s

Where, initially the mean is subtracted from every value, then the mean-shifted values are divided by the standard deviation, s. Z score is a value that estimates in terms of the number of standard deviations the value is above or below the mean of a data set.

Chemistry

Melting points were determined on a VEEGO Melting Point digital (VMPD) apparatus and are uncorrected. The FT-Infrared spectra were recorded in KBr on Jasco 6100 FT-IR spectrometer. The TLC was performed for the intermediate compounds using the solvent system (Benzene (4 ml) : n-Hexane (1ml)). Eddy's hot plate (Ugo Basile 7250) was used for analgesic activity. Acetic acid, CMC, syringe, stop watch, weighing pan, Eddy's hot plate method device.

Step 1: Synthesis of bromoacetyl chloride

Acetic acid 6 ml (0.1 mol) and thionyl chloride 12 ml (0.4 mol) are placed in 250 ml flask equipped with magnetic stirrer and condenser with drying tube. The reaction mixture was stirred and heated at 70°C using heating mental for half an hour [22]. The flask was then cooled to room temperature and to this reaction mixture were added finely powdered N-bromo succinamide 13 g (0.2 mol) along with thionyl chloride 12 ml (0.1 mol) and 4 drops of concentrated hydrochloric acid. The flask was again heated to 85°C for 1.25 hours and then the solvent was removed under reduced pressure at 61.2°C (uncorrected).

Step 2: Preparation of 1-biphenyl-4-yl-2-bromoethanone

In a 250 ml three necked flask provided with a dropping funnel, a mechanical stirrer and a reflux condenser, 1.54 g (0.01mol) of biphenyl, 1.33 g (0.01mol) of finely powdered anhydrous aluminum chloride and 35 ml of anhydrous carbon disulphide was placed. The dropping funnel was charged with 0.8 ml (0.01mol) of pure bromoacetyl chloride and closed with a calcium chloride guard tube. The mixture was heated on a water bath until gentle reflux commenced and bromoacetyl chloride was added drop wise, the addition product made its appearance as a curdy mass when about three quarters of the bromoacetyl chloride was added. The reaction mixture was refluxed gently for an hour. The reaction mixture was then cooled and poured slowly with stirring on to crushed ice to which hydrochloric acid was added and stirred. The product was filtered and washed with water to remove traces of hydrochloric acid and dried. It was recrystallized from methanol.

Step 3: Preparation of 2-(anilino)-1-(4-phenylphenyl) ethanone

Aniline 5 ml (0.02) and 10% sodium hydroxide 5 ml solution was taken in volumetric iodine flask and biphenyl acyl bromide 2 ml (0.5 ml at a time) was added with constant shaking and cooling in water. Then warm the reaction mixture and shake for 20 min till odour of acid disappears. pH was checked after every 5 min. Reaction mixture was made alkaline using alcoholic 10% sodium hydroxide (pH 7-14) to get the solid product. The product was filtered and then washed with water. The product was recrystallized from aqueous ethanol (1 ml water + 2 ml ethanol).

Biological Methods

Animals

Swiss albino mice of either sex weighing 20-25 g and wistar rats of 175-200 g were obtained from National Toxicological Centre, Pune, India which is approved breeder of laboratory animals. They were housed under standard environmental conditions of temperature (24 ± 1°C) and relative humidity of 30-70%. A 12:12 h light dark cycle was followed. All animals had free access to water and standard pelleted laboratory animal diet. All the experimental procedures and protocols

S. No	R	%Reduction in paw volume[a] (100 mg /kg)[b]	Log BA[c]
1		16.32	1.212
2		31.63	1.500
3'		19.59	1.292
4		35.50	1.550
5		36.88	1.566
6		15.8	1.198
7		44.11	1.644
8		41.17	1.614
9		39.70	1.598
10		35.29	1.547
11		17.64	1.246
12		22.05	1.343
13		29.42	1.468
14		24.90	1.396
15		22.05	1.343
16		30.88	1.489
17		40.48	1.607

No.	Structure		
18	HO— (phenyl)—N H	23.80	1.376
19	O— (phenyl)—N H	19.52	1.290
20	H_2N—(phenyl)—N H	23.86	1.377
21	HOOC—(phenyl)—N H	41.42	1.617
22	Cl (phenyl)—N H	26.12	1.416
23	F (phenyl)—N H	39.55	1.597
24	F (phenyl)—N H	33.21	1.521
25	O—(phenyl) (phenyl)—N H	25.75	1.410

[a]Carrageenan-induced rat paw edema model, using six animal group of male wistar rats
[b]Data analyzed by ANOVA followed Dunnetts test. $P < 0.05$,
[c]Logarithm of percent reduction in paw volume

Table 1: *In vivo* anti-inflammatory activity of 4'-methylbiphenyl-2-carboxamide analogues

used in this study were reviewed and approved by the Institutional Animal Ethical Committee (IAEC) of College, Pune, constituted in accordance with the guidelines of the Committee for the Purpose of Control and Supervision of Experiment on Animals (CPCSEA), Government of India.

Analgesic activity by hot plate method

In hot plate method various responses (jumping and licking) were recorded. The control group of mice (n= 6) received vehicle (1% w/v CMC, 0.5 ml p.o.). The test group mice received synthesized compound at 10, 20 and 30 mg/kg p.o. and diclofenac (9 mg/kg i.p.), respectively [23,24]. One hour following the test compound or diclofenac administration, the mice were individually placed on Eddy's hot plate (Ugo Basile 7250) maintained at 50-55°C. The latency period of 20 sec was defined as complete analgesia. To avoid injury, the measurement was terminated if latency period exceeded 20 sec.

Writing method

The analgesic activity was determined by acetic acid induced writhing method using six albino mice (25-30 g) of either sex selected by random sampling technique [25]. Standard drug Aspirin (100 mg/kg) and synthesized compound, 2-(anilino)-1-(4-phenylphenyl) ethanone (20 mg/kg) were given intraperitoneally 30 min prior to the administration of the writhing agent (0.6% v/v aqueous acetic acid, 10 ml/kg). The number of writhing and stretching produced in the animal was observed for 30 min. The numbers of writhing records were compared with the control drug.

Results and Discussion

QSAR study

The stepwise method during MLR method yielded significant QSAR equation. The statistically significant equation is given in Table 2. The randomization r^2 is much lower than the coefficient of determination (r^2), which indicates that the resultant QSAR equation is not by chance. The internal predictivity of the model is significant. Figure 1 gives the fitness plot of the equation 1 for predicted activity vs. actual activity. Predictivity of QSAR equation was assessed by predicting the activities of the test set molecules and the difference between predicted and actual activity values of the test set is given in Table 3.

Design of molecule

Lipinski's rule of five suggests important features the molecule should have for its bioavailability. The series under consideration had CH_3 group common on one side of biphenyl part which imparts bulk to the molecule and increases its size. Hence newly designed molecule did not have CH_3 group at this end. The other side of the biphenyl has ortho substitution amide bond arm. This ortho substitution mimics the pharmacophore of COX-2 inhibitor. Therefore instead of ortho, substitution was made at para position. Since the amide bond is liable to break by amidases, it limits the duration of action of compound hence it was eliminated. Instead non-amide link was introduced to connect the third aromatic ring. MLR equation showed descriptor T_N_O_4, that gives the importance of distance between nitrogen and oxygen by 4 bond path length. This is true in reported compounds 4 and 18.

Equation	N	r^2	CVr^2 (q^2)	F	Std error (s)	Ran_r^2	Zscore r^2
Activity = [0.1022(±0.0089)]H-AcceptorCount + [0.2307(±0.0466)] SsClcount + [0.2260(±0.0546)] T_N_O_4 + [0.1479(±0.0733)] T_N_F_3 1.0665	17	0.842	0.696	16.09	0.07	0.217	4.123

Table 2: Stepwise regression result using MLR method

Compound	Actual activity	Predicted activity	Difference
02	1.500	1.270	0.229
03	1.292	1.496	-0.204
04	1.55	1.927	0.463
09	1.598	1.373	0.224
13	1.468	1.270	0.197
14	1.396	1.270	0.125
19	1.290	1.373	-0.083
25	1.410	1.373	0.036

Table 3: The prediction of test set

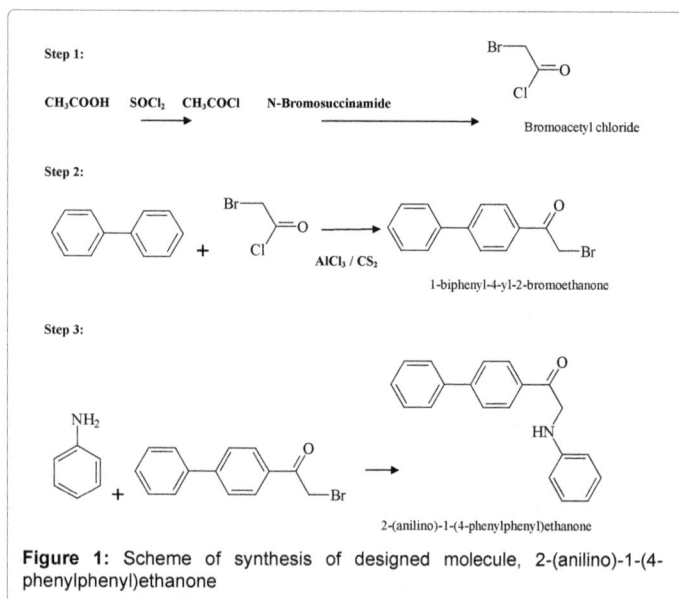

Figure 1: Scheme of synthesis of designed molecule, 2-(anilino)-1-(4-phenylphenyl)ethanone

MLR equation also showed the importance of substituted F and Cl on the aromatic nucleus and their distance from nitrogen in descriptors SsClcount and T_N_F_3. But chlorine atom has higher van der Waals radius than fluorine. In the newly designed molecule chlorine and fluorine, both were eliminated to study the effect of prominence of the parameter. Instead of chlorine or fluorine, electronically rich and bulky aniline functional group was introduced (Figure 3). The MLR equation gives the importance of carbonyl group as hydrogen bond acceptor count. Hence looking at these input features the molecule was designed with para substitution keeping the distance between carbonyl oxygen and nitrogen by four and predicted from MLR equation 1. The molecular modeling studies using systematic search method for identification of conformation showed that the third aromatic ring tilts away from axis that passes through biphenyl nucleus. The predicted value for the designed molecule was 1.5, which is higher than lower active molecules.

Characterization

The designed molecule was synthesized by using scheme given in Figure 2 with percentage yield 68%. The melting point of the compound (2-(anilino)-1-(4-phenylphenyl)ethanone) was 108°C (uncorrected). The FTIR gave the peak at 1680 for carbonyl group and NH peak at 3520. The proton NMR gave the δ ppm (m, 7.4-7.8) for aromatic ring and δ ppm (s, 2.4) for NH group.

Anti-inflammatory activity: *In vivo* activity

Hot plate method: The responses of drug at 55°C on mice: In Eddy's hot plate method, the response time was noted as the time at which animals reacted to the pain stimulus either by paw licking or jump response, whichever appeared first as shown in Table 4.

The synthesized compound closely matches structurally to the drugs such as flurbiprofen and diflunisal mentioned as NSAIDs, therefore the standard utilized for the study was diclofenac. The synthesized compound is made into 1 mg/ml concentration as a stock solution and 0.1, 0.2, 0.3 ml of this solution was injected peritonially to the mice. The licking and jumping responses showed that the drug with lowest concentration had maximum latency period than the dose of higher concentration. As seen in standard drug diclofenac administration, the response is of higher latency period.

Writhing Method

After administration of 0.6% v/v acetic acid the mice showed writhing action after 5 min in control group (Figure 3). It was found that the action of acetic acid is very irritating to mice. The records of observation of writhing are mentioned in the Table 5. After administration of Aspirin to mice as per dose 10 mg/kg of mice and then after 10 min, the count of writhing was observed following administration of 0.6% acetic acid. The writhing movements were less than that of control acetic acid as shown in the Table 5.

The second group of six mice of average weight 42 g were administered with synthetic compound in the dose of 20 mg/kg by oral route and the mice were kept for one hour. The responses after administration of 2-(anilino)-1-(4-phenylphenyl) ethanone (20 mg/kg) are also given in counts. The synthesized compound showed writhing response after 15 min.

Conclusion

QSAR study has revealed useful equation in terms of topology and E-indices. The novelty of the designed molecule lies in absence of acidic character which was meant to be essential pharmacophoric feature for analgesic and anti-inflammatory activity. The synthesized compound showed moderate effect on central nervous system while in case of writhing method it was found that the synthesized compound has retention of analgesic activity.

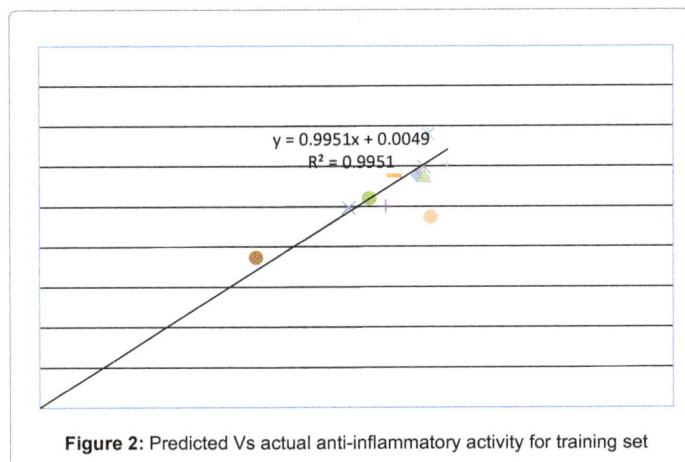

Figure 2: Predicted Vs actual anti-inflammatory activity for training set

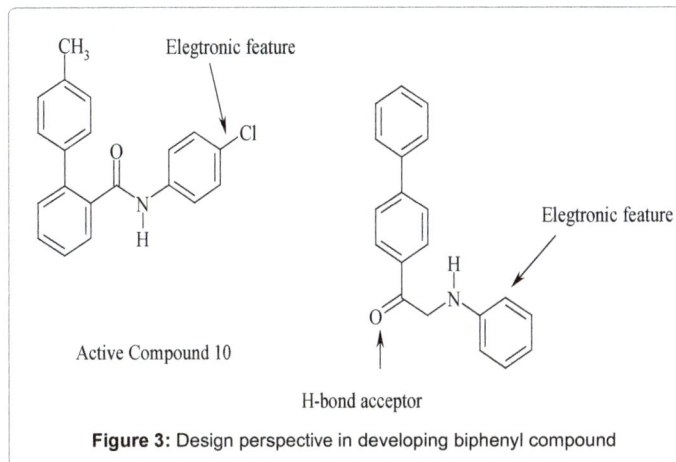

Figure 3: Design perspective in developing biphenyl compound

Animal number	Weight of animal	Response for control group mice (in second)	Response for compound injected mice (in second)	Response for Diclofenac injected mice (in second)
1	44 g	1 sec	3 sec	5 sec
2	52 g	2 sec	4 sec	4 sec
3	48 g	1 sec	3 sec	5 sec
4	41 g	1 sec	3 sec	4 sec
5	47 g	1 sec	3 sec	4 sec
6	44 g	2 sec	3 sec	5 sec

Table 4: The response of mice before and after administration of compound [2-(anilino)-1-(4-phenylphenyl)ethanone] (1 mg/ml), Diclofenac (9 mg/kg)

Animal number	Weight of animal	No. of writhes of mice for control group (Acetic acid)		No. of writhes of mice (Aspirin)		No. of writhes of mice (compound*)	
		0-10 min	10-20 min	0-10 min	10-20 min	0-10 min	10-20 min
1	44 g	1	10	0	3	0	5
2	48 g	0	12	0	2	0	7
3	52 g	1	13	0	3	0	6
4	41 g	2	18	0	3	0	5
5	44 g	2	15	0	3	0	5
6	45 g	2	17	0	2	0	6

Compound*: Responses after administration of 2-(anilino)-1-(4-phenylphenyl) ethanone at oral dose of 20 mg/kg

Table 5: Response of writhing response of mice to acetic acid, standard drug Aspirin and synthesized compound

Acknowledgement

Author thanks Mr. Anand Datar for providing facility and Sinhgad Institute of Pharmacy for organizing approval from animal ethical committee.

References

1. (1990) Cancer pain relief and palliative care. Report of a WHO Expert Committee. World Health Organ Tech Rep Ser 804: 1-75.

2. Dworkin RH, Backonja M, Rowbotham MC, Allen RR, Argoff CR, et al. (2003) Advances in neuropathic pain: diagnosis, mechanisms, and treatment recommendations. Arch Neurol 60: 1524-1534.

3. Futaki N, Yoshikawa K, Hamasaka Y, Arai I, Higuchi S, et al. (1993) NS-398, a novel non-steroidal anti-inflammatory drug with potent analgesic and antipyretic effects, which causes minimal stomach lesions. Gen Pharmacol 24: 105-110.

4. Gans KR, Galbraith W, Roman RJ, Haber SB, Kerr JS, et al. (1990) Anti-inflammatory and safety profile of DuP 697, a novel orally effective prostaglandin synthesis inhibitor. J Pharmacol Exp Ther 254: 180-187.

5. Penning TD, Talley JJ, Bertenshaw SR, Carter JS, Collins PW (1997) Synthesis and biological evaluation of the 1,5-diarylpyrazole class of cyclooxygenase-2 inhibitors: identification of 4-[5-(4-methylphenyl)-3-(trifluoromethyl)-1H-pyrazol-1-yl]benze nesulfonamide (SC-58635, celecoxib). J Med Chem 40: 1347-1365.

6. Habeeb AG, Praveen Rao PN, Knaus EE (2000) Design and syntheses of diarylisoxazoles: Novel inhibitors of cyclooxygenase-2 (COX-2) with analgesic-antiinflammatory activity. Drug Dev Res 51: 273-286.

7. Puig C, Crespo MI, Godessart N, Feixas J, Ibarzo J, et al. (2000) Synthesis and biological evaluation of 3,4-diaryloxazolones: A new class of orally active cyclooxygenase-2 inhibitors. J Med Chem 43: 214-223.

8. Kalgutkar AS, Crews BC, Rowlinson SW, Marnett AB, Kozak KR, et al. (2000) Biochemically based design of cyclooxygenase-2 (COX-2) inhibitors: facile conversion of nonsteroidal antiinflammatory drugs to potent and highly selective COX-2 inhibitors. Proc Natl Acad Sci U S A 97: 925-930.

9. Woods KW, McCroskey RW, Michaelides MR, Wada CK, Hulkower KI, et al. (2001) Thiazole analogues of the NSAID indomethacin as selective COX-2 inhibitors. Bioorg Med Chem Lett 11: 1325-1328.

10. Kalgutkar AS, Marnett AB, Crews BC, Remmel RP, Marnett LJ (2000) Ester and amide derivatives of the nonsteroidal antiinflammatory drug, indomethacin, as selective cyclooxygenase-2 inhibitors. J Med Chem 43: 2860-2870.

11. Wallace JL (2008) Prostaglandins, NSAIDs, and gastric mucosal protection: why doesn't the stomach digest itself? Physiol Rev 88: 1547-1565.

12. Deep A, Jain S, Sharma PC (2010) Synthesis and anti-inflammatory activity of some novel biphenyl-4-carboxylic acid 5-(arylidene)-2-(aryl)-4-oxothiazolidin-3-yl amides. Acta Pol Pharm 67: 63-67.

13. Deep A, Jain S, Sharma PC, Verma P, Kumar M, et al. (2010) Design and biological evaluation of biphenyl-4-carboxylic acid hydrazide-hydrazone for antimicrobial activity. Acta Pol Pharm 67: 255-259.

14. Kannappan MAN, Deep A, Kumar P, Kumar M, Verma P (2009) Synthesis and antimicrobial studies of biphenyl-4-carboxylic acid 2-(aryl)-4-oxo-thiazolidin-3-yl –amide. International Journal of Chem Tech Research 1: 1376-1380.

15. Plummer EL (1985) Insecticidal 2,2'-bridged(1,1'-biphenyl)-3-ylmethyl carboxamides. US patent 4493844.

16. Sachan N, Thareja S, Agarwal R, Kadam SS, Kulkarni (2009) Substituted biphenyl ethanones as antidiabetic agents: Synthesis and in-vivo screening. International Journal of PharmTech Research 1: 1625-1631.

17. de Souza AO, Hemerly FP, Busollo AC, Melo PS, Machado GM, et al. (2002) 3-[4'-bromo-(1,1'-biphenyl)-4-yl]-N, N-dimethyl-3-(2-thienyl)-2-propen-1-amine: synthesis, cytotoxicity, and leishmanicidal, trypanocidal and antimycobacterial activities. J Antimicrob Chemother 50: 629-637.

18. Eriksson L, Johansson E (1996) Multivariate design and modeling in QSAR. Chemom Intell Lab Syst 34: 1-19.

19. Shah UA, Wagh NK, Deokar HS, Kadam SS, Kulkarni VM (2010) 3D-QSAR of biphenyl analogues as anti-inflammatory agents by genetic function approximation (GFA) [part-II]. International Journal of Pharma and Bio Sciences 1: 512-522.

20. http://www.vlifesciences.com/products/VLifeMDS/Product_VLifeMDS.php

21. Furniss BS, Hannaford AJ, Smith PWG, Tatchell AR (1989) Vogel's Textbook of Practial Organic chemistry. Fifth edition, Longman Group, UK, 723 pages.

22. Kulkarni SK (1999) Handbook of Experimental Pharmacology. 3rd revised edition, Vallabh Prakashan, New Delhi, India, 123-125.

23. Eddy NB, Leimbach D (1953) Synthetic analgesics. II. Dithienylbutenyl- and dithienylbutylamines. J Pharmacol Exp Ther 107: 385-393.

24. Witkin LB, Heubner CF, Galdi F, O'Keefe E, Spitaletta P, et al. (1961) Pharmacology of 2-amino-indane hydrochloride (Su-8629): a potent non-narcotic analgesic. J Pharmacol Exp Ther 133: 400-408.

Synthesis and Characterization of α, α-Dimethyl-4-[1-Hydroxy-4-[4-(Hydroxyldiphenyl- Methyl)-1-Piperidinyl]Butyl] Benzeneacetic Acid Metal Complexes of Biological Interest

Saeed M Arayne[1]*, Najma Sultana[2], Hina Shehnaz[3] and Amir Haider[4]

[1]Department of Chemistry, University of Karachi, Karachi, Pakistan
[2]Research Institute of Pharmaceutical Sciences, Faculty of Pharmacy, University of Karachi, Karachi-75270, Pakistan
[3]Department of Environmental Science, Sind Madressatul Islam University, Karachi, Pakistan
[4]Arysta Life Science Pakistan, Horizon Vista - 3rd Floor, Commercial 10, Block 4, Clifton, Karachi-75600, Pakistan

Abstract

α,α-Dimethyl-4-[1-hydroxy-4-[4-(hydroxydiphenylmethyl)-1-piperidinyl]butyl]benzene acetic acid generically known as fexofenadine or carboxyterfenadine is a second generation H_1-receptor antagonist, much widely used due to its non-sedative effects. Metal complexes of fexofenadine with various essential and trace elements of biological interest, have been synthesized and characterized by FT-IR, ^{1}H-NMR, UV, CHN elemental analysis. Conductometric titration was carried out to determine the mole ratio of interaction of the drug and metal. Spectral data clearly showed the complexation of fexofenadine with nitrogen of the piperidine ring in case of magnesium, calcium, chromium and manganese complexes while oxygen of carboxylato group is involved in complexation with iron, cobalt nickel, copper, zinc and cadmium. Elemental analysis reveals monodentate character of these complexes.

Keywords: Fexofenadine; H_1-receptor antagonist; Essential and trace elements; FT-IR; ^{1}H-NMR

Introduction

Fexofenadine (carboxyterfenadine) (Figure 1), is α,α-Dimethyl-4-[1-hydroxy-4-[4-(hydroxydiphenylmethyl)-1-piperidinyl]butyl] benzeneacetic acid is a non-anticholinergic, non sedating-type histamine H -receptor antagonist [1, 2] and an active metabolite of terfenadine [3-5]. Bioavailability of fexofenadine decreases when the drug is co-administered with grapefruit juice, orange juice and apple juice which potentially reduced pharmacokinetic effects of fexofenadine [6] and also lowers the plasma drug concentrations by reducing drug absorption catalyzed by the Organic Anion Transporting Polypeptide (OATP) [7].

Literature survey reveals few studies on the interaction of metal ions with fexofenadine [8]. Metals are considered essential to a human body being an integral part of an organic structure in performing physiologically important and vital functions in the body [9]. It seems that the role of metal ions is imperative for the way of function of fexofenadine. Synthesis and characterization of new metal complexes with fexofenadine are of great importance for better understanding of drug-metal ion interactions [10]. A thorough survey of the literature has revealed that no work has been reported on the synthesis of fexofenadine metal complexes, though our research group has reported a number of metal complexes with other co-administered drugs as

levofloxacin [11], enoxacin [12,13], sparfloxacin [14], gatifloxacin [15], gliquidone [16] and atenolol [17].

In this paper, we present the synthesis of a series of fexofenadine–metal complexes with metals of biological interest in an attempt to find the mode of coordination as well as to study drug–metal interactions. More specifically, the complexes have been synthesized and characterized by elemental analysis and diverse spectroscopic techniques (IR, UV-Vis and NMR techniques).

Experimental

Materials

Fexofenadine was a kind gift from Getz Pharmaceuticals (Pvt.) Ltd. Karachi, Pakistan. The metal salts and other chemicals were of analytical grade (Merck Germany). All the glassware's were washed with chromic acid followed by a thorough washing with deionized water which was freshly prepared in the laboratory.

Instrumentation

Conductometric studies were carried on Vernier LabPro. Data acquisition and analysis were carried out by using the Logger pro 3.2 software. Thin Layer Chromatography (TLC) was performed on a HSF-254 TLC plate and the samples were visualized under a UV lamp, melting point of complexes was recorded on a Gallenkamp electrothermal melting point apparatus and is uncorrected. The characterization of fexofenadine metal complexes was carried out by FT-IR spectrophotometer (Shimadzu Prestige-21 200 VCE), coupled to a P IV-PC and loaded with IR resolution software. The disks were

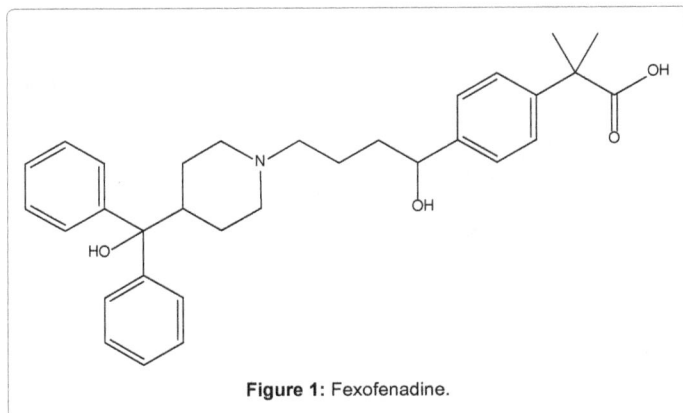

Figure 1: Fexofenadine.

*Corresponding author: Saeed M Arayne, Department of Chemistry, University of Karachi, Karachi-75270, Pakistan, E-mail: msarayne@gmail.com

placed in the holder directly in the IR beam. Spectra were recorded at a resolution of $2 cm^{-1}$, and 50 scans were accumulated. Stability of these complexes was studied 4.5 CF stability chamber from PARAMETER Generation and Control USA with constancy control $\pm 0.5\%$.

Proton NMR studies of metal complexes were carried out on a Brucker AMX 500MHz instrument in deuterated methanol (CD_3OD) using TMS as an internal standard. CHN elemental studies were carried out by standard micro methods using Carlo Erba 1106.

Conductometric titrations

Prior to the synthesis of the metal complexes, the stoichiometry of complexes was determined using conductometric titration technique [18]. 0.1 mM metal solution was titrated against the drug solution at 298 K with constant stirring and change in conductivity was measured until the addition resulted in no change of conductivity values of the resulting solution. Similar process was repeated for all metals.

Synthesis of metal complexes

Metal complexes of fexofenadine were synthesized with magnesium, calcium, chromium, manganese, ferric, ferrous, cobalt, nickel, copper, zinc and cadmium hydrated salts. The synthesis of magnesium, calcium, chromium and manganese complexes with fexofenadine were carried out in M:L ratio of 1:2 while for rest of the metals M:L ratio was 1:1. These ratios were determined earlier by conductometric measurements. Drug and metal salts were dissolved separately in hot methanol (20 mL) and deionized water (10 mL). Both solutions were mixed with constant stirring and then refluxed for about three hours. The reactions were monitored by thin layer chromatography. The volume of the reaction mixture was then reduced by evaporation, filtered while hot and then kept undisturbed for crystal growth at room temperature.

The growth of crystals depended upon many factors like type of crystal, environment, concentration of solution, temperature etc. Due to this all these metal complexes had variable time of crystallization. Crystals of magnesium, calcium, chromium and manganese complexes were obtained within 10 to 15 days, while the iron complex took one month for crystallization. Cobalt, nickel, copper, zinc and cadmium complexes were crystallized in 20~25 days. Crystals of different colors for different complexes were obtained. Finally, these were filtered, washed with water and methanol, and vacuum dried. Their melting points were recorded and were recrystallized from absolute methanol till constant melting point. Their physicochemical parameters were noted.

Physical data

Fexofenadine: Color white, m.p. 190°C, UV nm (ε): 1.775(218 nm), IR (KBr) vmax: 3400-3350 db, s, 3058.04 s, 1700 s, 1600 and 1475sm 1447.60 m, 1403.55 m, 1279 m, 1166.41 m,1099 s, 1068m, 983, 965 db, sm, 834 ms, 747, 702,638, 577, 525.53 dp, s. 1H NMR (MeOD, 400 MHz) : 7.19 (s, J=1.16 Hz), 7.11 (s, J=0.5 Hz) and 7.06 aromatic, 2.0 N-H, 1.52 CH_3, 2.29, 2.19 CH_2-N, 2.34 (t, J=12.2 Hz) -CH_2-CH_2-N, 2.36 (t, J=14.0 Hz), 1.39, 1.77 and 4.5 (s, J=1.0 Hz) -CH_2 open chain, 4.95 (s) alcoholic. Calculated for $C_{32}H_{39}NO_4$: C, 49.14; H, 5.96; N, 3.18. Found: C, 49.01; H, 5.74; N, 4.33.

Fexofenadine magnesium complex: Colorless, Yield 50% , m.p. 112°C, UV nm (ε): 0.4615(207 nm), IR (KBr) vmax: 3421 s, 2967 db, s, 1680 dp, s, 1446.45 sm, 1320 db,sm 0.4615(207 nm) 1300-1320 db, sm, 1100 sm, 1090 sm, 1410, 1637 sm, 1151 m, 750 sm, 707 sm, 1H NMR (MeOD, 400 MHz) δ: 7.10, 7.06 (s, J=0.69Hz) aromatic, 1.65 MgOOC-C-CH_2, 2.36, 1.19 (m, J=0.77Hz)CH_2-CH_2-N-Mg, 1.94 (t, J=0.99 Hz) CH_2-N-Mg, -$(CH_2)_3$-N-Mg, 1.79 1.69 (t, J=4.5Hz) Mg-N-CH_2, 2.42,

1.64, 1.44 -CH_2 open chain, 1.76 (s) alcoholic, 3.49, 1.58 (t) -CH_2-OH. Calculated for $C_{32}H_{39}ClMgNO_4$: C, 68.46; H, 7.00; N, 2.49. Found: C, 68.42; H, 7.07; N, 2.43.

Fexofenadine calcium complex: Colorless Yield 42% , m.p. 72°C, UV nm (ε): 0.5173(209nm), IR (KBr) vmax: 3650 sm, 3411 m, 2967 db, sm, 1716.58 db, m 1619.35 0.5173(209nm) sm, -$CH_2$1446.84 m, 1383.80 sm, 1100-1140 db, 1490 m, 750, sm, 706 sm, 668.14 sm, 602.08 sm. 1H NMR (MeOD, 400 MHz) δ: 7.34, 7.07, 7.23, 7.34 (s, J=0.63Hz) aromatic, 1.65 CaOOC-C-CH_2 2.92 (t,J=3.5 Hz) CH_2-NH, 1.85 (m, J=3.5 Hz) -CH_2-NH, 1.94 (t, J=14.0Hz), 1.64, 1.44 -$(CH_2)_3$-NH open chain, 4.28 (J=9.0Hz) (piperidine)(Ph)-CH-OH, 1.90 (J=3.5Hz) NH-$(CH_2)_2$-CH, 1.43 (s) alcoholic. Calculated for $C_{32}H_{39}ClCaNO_4$: C, 66.59; H, 6.81; N, 2.43. Found: C, 66.63; H, 6.85; N, 2.40.

Fexofenadine chromium complex: Color light green Yield 39%, m.p. 72°C, UV nm (ε): 0.28(257nm), IR (KBr) vmax: 3650 sm, 2967 db, sm 1725 db, 1691 sm, 1490 m, 1447 m, 1310 db, sm, 1146 m, 1471.97 db, m, 750 sm, 706 sm. 1H NMR (MeOD, 400 MHz) δ: 7.13 (s, J=7.67Hz) 7.18 (s, J=2.12Hz) aromatic, 4.16 (J=0.31Hz) -CH_2-COOCr, 1.91, 1.88 (t, J=12.2 Hz) CH_2-N-Cr, 1.35 (m, J=2.0 Hz) (m, J=7.69 Hz), -CH_2-CH_2-N, 2.40 (t, J=14.0Hz), 1.64, 1.50 (t, J=0.5 Hz) $(CH_2)_3$-N-Cr, 3.32 (J=4.8Hz) (piperidine)(Ph)-CH-OH, 1.76 (s) alcoholic. Calculated for $C_{32}H_{39}ClCrNO_4$: C, 65.24; H, 6.67; N, 2.38. Found: C, 65.22; H, 6.69; N, 2.41.

Fexofenadine manganese complex: Colorless Yield 34%, m.p. 128°C, UV nm (ε): 0.7 (215 nm), IR (KBr) vmax: 3647 sm , 3420 sm, 2969.91 db, m, 1717 db, 1653 trp, 1490 sm, 1458 m, 1397 db, sm, 1365 db, 1146-1099.36 trp, 840 m, 750 sm, 706 sm. 1H NMR (MeOD, 400 MHz) : 7.13 (s, J=0.31Hz) aromatic, 7.18 (s, J=0.77Hz) 1.46, 3.12 -CH_2-COOMn, 2.27, 1.79, 1.69 CH_2-N-Mn, 1.69 (t, J=0.77 Hz) 1.46 (q, J=2.0 Hz) (t, J=12.2 Hz) -CH_2-CH_2-N, $(CH_2)_3$-N-Mn open chain at 1.39, 3.39, 1.44 (t,J=6.8 Hz) 4.5 (J=6.8Hz) -CH-O-Mn, 3.49 (J=4.8Hz) (piperidine) (Ph)-CH-OH, 2.0 (s) alcoholic. Calculated for $C_{32}H_{39}ClMnNO_4$: C, 64.92; H, 6.64; N, 2.37. Found: C, 64.97; H, 6.62; N, 2.31.

Fexofenadine ferric chloride complex: Color Yellow Yield 19%, m.p. 112°C, UV nm (ε): 0.7 (215 nm), IR (KBr) vmax: 3425 m, 3055 db, m, 2968 db, 1716 m, 1591 sm, 481-1415 trp, sm, 1373 db, 1151 s, 790 s, 750sm, 636sm. 1H NMR (MeOD, 400 MHz) δ: 7.34 (s, J=0.69Hz), 7.23 (s, J=0.63Hz), 7.38 (s, J=7.21Hz) 7.07 (s, J=7.21Hz) (s, J=1.11Hz) 7.18 (s,J=0.77Hz) aromatic, 1.64 -$(CH_3)_2$-C-COOFe, 2.93, 2.92 CH_2-N, 1.73 1.85 -CH_2-CH_2-N, 2.41 1.52 1.64 $(CH_2)_3$-N, open chain at 4.61 (J=4.8Hz) (piperidine)(Ph)-CH-OH, 1.36 (s) alcoholic. Calculated for $C_{36}H_{51}FeNO_4$: C, 70.01; H, 8.32; N, 2.27. Found: C, 70.06; H, 8.37; N, 2.31.

Fexofenadine ferric ammonium citrate complex: Color light yellow Yield 19%, m.p. 210°C, UV nm (ε): 0.348 (254 nm), IR (KBr) vmax: 3405 sm, 2980 db, m, 1717 db, sm, 1600 s, 1491 sm, 1447 m, 1689 db, m, 1254 sm, 1154 sm, db, 1066 db, sm, 845 sm, 750 sm, 704 sm, 637 sm. 1H NMR (MeOD, 400 MHz) δ: 7.34 (s, J=0.69Hz), 7.23 (s, J=0.63Hz), 7.38 (s, J=7.21Hz) 7.07 (s, J=7.21Hz) (s, J=1.11Hz) 7.18 (s,J=0.77Hz) aromatic, 1.64 -$(CH_3)_2$-C-COOFe, 2.93, 2.92 CH_2-N, 1.73 1.85 -CH_2-CH_2-N, 2.41 1.52 1.64 $(CH_2)_3$-N, open chain at 4.61 (J=4.8Hz) (piperidine)(Ph)-CH-OH, 1.36 (s) alcoholic. Calculated for $C_{36}H_{51}FeNO_4$: C, 70.01; H, 8.32; N, 2.27. Found: C, 70.06; H, 8.36; N, 2.22.

Fexofenadine ferrous sulfate complex: Color yellow Yield 15%, m.p. 137°C, UV nm (ε): 0.517(216nm), IR (KBr) vmax: 3450 sm, 3055 db, sm, 2969 db, sm, 1718 s db, 1448 m, 1264 sm, 1150-1098 trp, 840 sm, 749 sm, 708 sm, 618 sm. 1H NMR (MeOD, 400 MHz) δ: 7.11, 7.06 (s, J=7.21Hz), (s, J=0.63 Hz) aromatic, 1.46, 1.52 -CH_3, 11.0 COOH,

2.19, 2.92 CH_2-N, 1.34, 1.59 -CH_2-CH_2-N, $(CH_2)_3$-N open chain at 2.36, 1.39, 1.77, 4.50 (t, J=0.5 Hz) (J=4.8Hz) 3.49 (piperidine)-CH-OH, 2.0 (s) (J=11.9Hz) alcoholic. Calculated for $C_{36}H_{51}FeNO_4$: C, 70.01; H, 8.32; N, 2.27. Found: C, 70.06; H, 8.35; N, 2.26.

Fexofenadine cobalt complex: Color blue Yield 20%, m.p. 118°C, UV nm (ε): 0.6(202nm), IR (KBr) vmax: 3420 m, 2969 db, m, 1717 m db, 1447 m, 1380 db, sm, 1151 sm, 1180-1100 db, m, 790 sm, 750 sm, 706 m. 1H NMR (MeOD, 400 MHz) δ: 7.06 (s, J=2.31Hz) 7.28 (s, J=2.12Hz) aromatic, 1.46, 1.64 –$(CH_3)_2$-C-COOCo, 2.24 (t, J=7.5 Hz) CH_2-N-Co, 1.50 (t, J=7.5 Hz) -CH_2-CH_2-N, 1.39 (t, J=7.0 Hz), 3.39, 1.77 (t, J=1.0 Hz) protons, 4.5 (J=1.0Hz) $(CH_2)_3$-N-Co open chain, 2.0 (s) alcoholic. Calculated for $C_{36}H_{51}CoNO_4$: C, 69.66; H, 8.28; N, 2.26. Found: C, 69.61; H, 8.228; N, 2.30.

Fexofenadine nickel complex: Color light green Yield 33%, m.p. 212°C, UV nm (ε): 0.78(204nm), IR (KBr) vmax: 3390 m, 2969 db, m, 1718 s db, 1447 m ,1380 db, sm, 1300 sm, 1150-1098 db, 950-990 trp, sm, 850 sm, 750 sm, 707 m, 637-619 db, m. 1H NMR (MeOD, 400 MHz) δ: 7.19 (s, J=7.12Hz), 7.08 (s, J=0.5 Hz), 7.38, 7.06, 7.11 aromatic, 1.64 –$(CH_3)_2$-C-COONi, 2.29 2.19 (t, J=2.0 Hz), CH_2-N, 1.59, 1.34 -CH_2-CH_2-N, 3.2, 1.39, 1.77 (t, J=7.0 Hz) 4.5$(CH_2)_3$-N open chain, 4.49 (J=4.8Hz) (piperidine)(Ph)-CH-OH, 2.0 (s) alcoholic. Calculated for $C_{36}H_{51}NNiO_4$: C, 69.68; H, 8.28; N, 2.26. Found: C, 69.64; H, 8.23; N, 2.29.

Fexofenadine copper complex: Color light blue Yield 44%, m.p. 110°C, UV nm (ε): 1.16(244nm), IR (KBr) vmax: 3390sm, 2968db,m, 1717 sm db, 1447 m, 1388 db, sm, 1300 sm, 1150 sm, 1150-1100 db, m, 1000- 1050 db, sm, 850 sm, 790 m, 706 sm, 750 sm. 1H NMR (MeOD, 400 MHz) : 6.99, 7.07 aromatic, 4.45 –CH_2-COOCu, 2.29, 2.19 (t, J=3.5 Hz) CH_2-N, 1.59, 1.34 -CH_2-CH_2-N, 2.36 (t, J=14.0 Hz) 1.39 (t, J=7.0 Hz) 1.77 (t, J=6.3 Hz) 4.5 $(CH_2)_3$-N open chain, 3.49 (J=4.8Hz) Ph-CH_2-OH, 3.9 (s) alcoholic. Calculated for $C_{36}H_{51}CuNO_4$: C, 69.14; H, 8.22; N, 2.24. Found: C, 69.10; H, 8.29; N, 2.27.

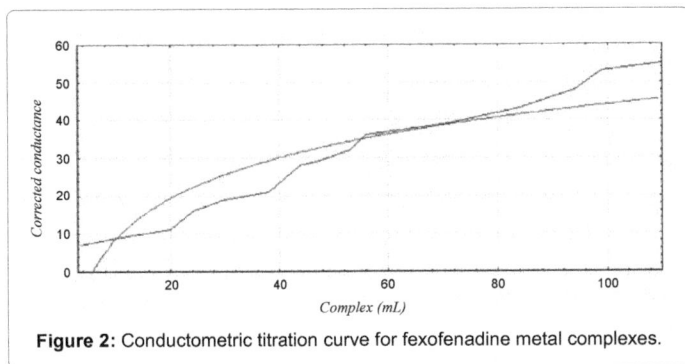

Figure 2: Conductometric titration curve for fexofenadine metal complexes.

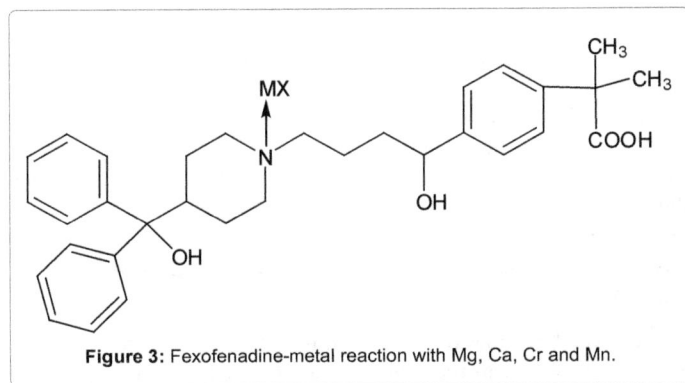

Figure 3: Fexofenadine-metal reaction with Mg, Ca, Cr and Mn.

Fexofenadine zinc complex: Color dirty white Yield 46.64%, m.p. 102°C, UV nm (ε): 0.86(217nm), IR (KBr) vmax: 3618 sm, 3376 db, 2924 m, db, 1717, 1697 db m, 1605 sm, 1596-1508 trp, m, 1480 s, 1310 db, sm, 1220 s, 1115 db, sm, 1033-1007 db, 753 m, 537 sm, 468 sm. 1H NMR (MeOD, 400 MHz) δ: 6.96, 7.08, 7.21 aromatic, 3.49 –CH_2-COOZn, 2.29, 2.19 (t, J=2.0 Hz), CH_2-N, 1.59, 1.34 (t, J=1.34 Hz), -CH_2-CH_2-N, 3.9 (t, J=14.0 Hz), 1.39 (t, J=8.0 Hz), 1.77 (t, J=7.0 Hz) 4.5 $(CH_2)_3$-N open chain, 2.51 (J=4.8Hz) (piperidine)(Ph)-CH-OH, 5.5 (s) alcoholic. Calculated for $C_{36}H_{51}NO_4Zn$: C, 68.94; H, 8.20; N, 2.23. Found: C, 68.91; H, 8.26; N, 2.28.

Fexofenadine cadmium complex: Colorless Yield 42.5%, m.p. 132°C, UV nm (ε): 1.95(221nm), IR (KBr) vmax: 3629 sm, 3450 s, 3070 db, m, 1710-1725 db, sm, 1491-1972, db, sm, 1447 m, 1410 db, sm, 1270 db sm, 1154 m, 1099-1017 db, m, 950 db, sm, 860 sm, 940 sm, 750 sm. 1H NMR (MeOD, 400 MHz) δ: 6.99, 7.30 aromatic, 3.6 –CH_2-COOCd, 2.29, 2.19 (t, J=3.75 Hz)(t, J=7.69 Hz) CH_2-N, 1.59, 1.34 (t, J=5.75 Hz), -CH_2-CH_2-N, 2.36 (t, J=14.0 Hz), 1.39 (t, J=8.0 Hz), 1.77 (t, J=7.0 Hz) 4.5 $(CH_2)_3$-N open chain, 3.49 (J=5.75Hz) piperidine-CH-OH, 4.4 (s) alcoholic. Calculated for $C_{36}H_{51}CdNO_4$: C, 64.13; H, 7.62; N, 2.08. Found: C, 64.17; H, 7.67; N, 2.11.

Results and Discussion

Fexofenadine is a zwitterion having good solubility in stomach and small intestinal fluids. Grapefruit juice decreases the dissolution of fexofenadine because of its interaction with excipients included in formulation to impair this process [19]. In the light of this study was initiated to explore fexofenadine interactions with metals of biological interest.

Transition metal complexes of fexofenadine were synthesized in mole ratios as described earlier in methanol. Prior to synthesis complexation studies between fexofenadine and metals were investigated by conductometric titrations. The effects on conductivity caused by the increased volume of reaction medium were compensated by correcting it for dilution by correction term [(v + V)/ V] and equivalence point were calculated. Molar conductivity values (Λ_m) were then calculated by correcting conductivity values (µS) using the formula Λ_m=K/C, where, K is the measured conductivity and C is the electrolyte concentration.

From the graph it is found that magnesium, calcium, chromium and manganese metal ions bind to ligand in the 2:1 ratio, whereas nickel, copper, zinc, cadmium and iron complexes are formed in equimolar ratio (Figure 2).

Metal complexes of fexofenadine were synthesized by refluxing fexofenadine with metal salt solutions in a mixture of methanol and water in equimolar ratios and then crystallizing them at room temperature. Solubility and melting points were noted; all these complexes are 50% soluble in methanol and 25% soluble in ethanol, ethyl acetate and acetone.

Stability of these complexes was studied by taking their melting points at an interval of 24 hours for seven days according to Q1A (R2) compliant [20]. Samples were stored at 25°C/60% RH, 30°C/60% RH, 30°C/65% RH, 40°C/75% RH, 30°C/75% RH and 25°C/40% RH. Melting points of the complexes were recorded after 24 hours for 7 days. No appreciable changes in the melting points were observed, and the estimated error was ± 1°C. It was concluded that the complexes were stable according to ICH guidelines [21]. Their structures were established from the elemental analyses, which agree well with their proposed formulae.

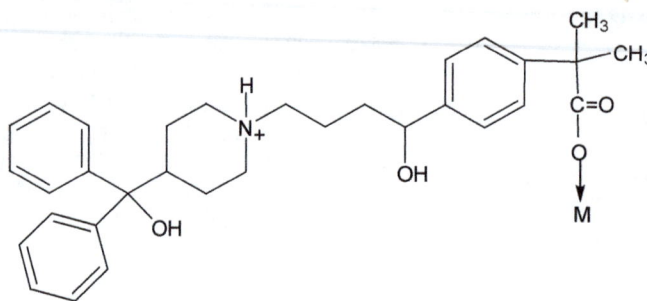

Figure 4: M=Fe, Co, Ni, Cu, Zn, Cd complexes of fexofenadine.

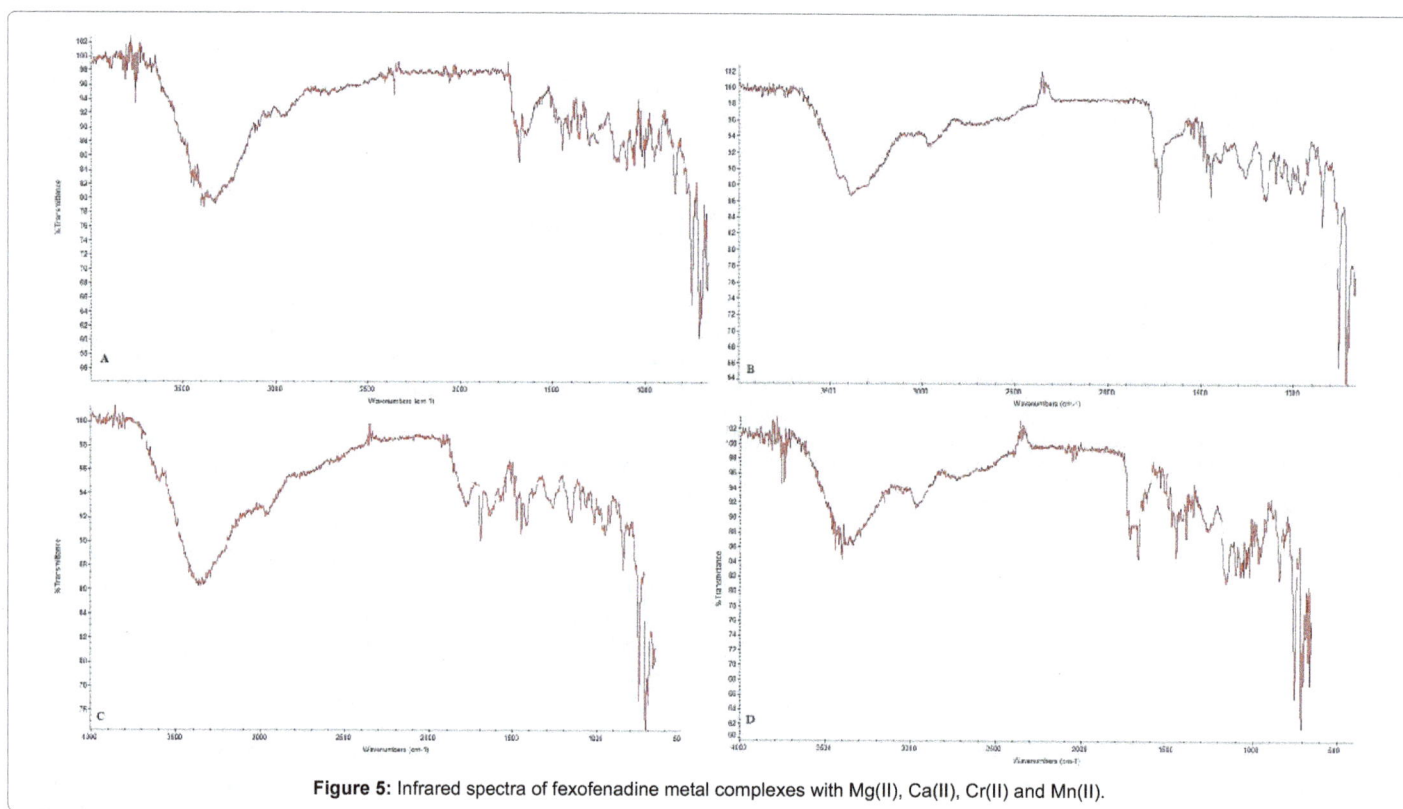

Figure 5: Infrared spectra of fexofenadine metal complexes with Mg(II), Ca(II), Cr(II) and Mn(II).

Infrared studies

Fexofenadine metal complexes were characterized by studying the most typical vibrations in the IR spectrum in the region 400-4000 cm^{-1}. In the IR spectra of metal complexes, some very prominent peak shifting has been observed along with a change in intensities of several important peaks indicating complexation of fexofenadine with metals (Figures 3 and 4).

The major infrared absorption bands of fexofenadine for -OH stretching occurred in the region 3300-3370 cm^{-1} and out of plane bending at 965 cm^{-1}, C-H stretching at 3058.04 cm^{-1} due to strained molecule because of the substitution of aromatic rings, a sharp band of C=O stretching at 1700 cm^{-1}, at 1600 and 1475 cm^{-1}, C=C aromatic medium weak bands, at 1447.60 cm^{-1} bending absorptions of –CH$_2$, C-O medium stretching vibration at 1279 cm^{-1}, at 1099 cm^{-1} C-N stretching region and at 1167 and 747 cm^{-1} four or more CH$_2$ groups.

On comparing the IR spectrum of fexofenadine with its magnesium,

manganese, iron and cobalt complexes it is found that the major doublet band of –OH group shifted to 3420 cm^{-1} as the singlet sharp band which was due to –O-metal and N-metal stretching [22,23]. In calcium complex two very weak bands appeared at 3650 cm^{-1} and another at 3411 cm^{-1} as sharp band, in chromium complex as a sharp band it was at 3400 cm^{-1}, in case of manganese complex a very weak band was at 3647 cm^{-1}. In case of iron ammonium citrate this band appeared as broadband in the region 3440-3380 cm^{-1} which was due to the N-H stretching of ammonium ion. In iron sulfate complex it was at 3400-3390 cm^{-1} as a medium band. In nickel and copper complex a sharp stretching band appeared at 3390 cm^{-1}. Fexofenadine zinc complex showed three consecutive small singlet at 3618 and sharp doublet at 3376. In cadmium complex a small band at 3629 cm^{-1} and 3450 cm^{-1} was recorded. Water bands are also appeared above 3000 cm^{-1} (Figures 5 and 6).

Fexofenadine showed C=O stretching at 1700 cm^{-1} due to carboxylic group which shifted to 1680 cm^{-1} as doublet strong band in magnesium

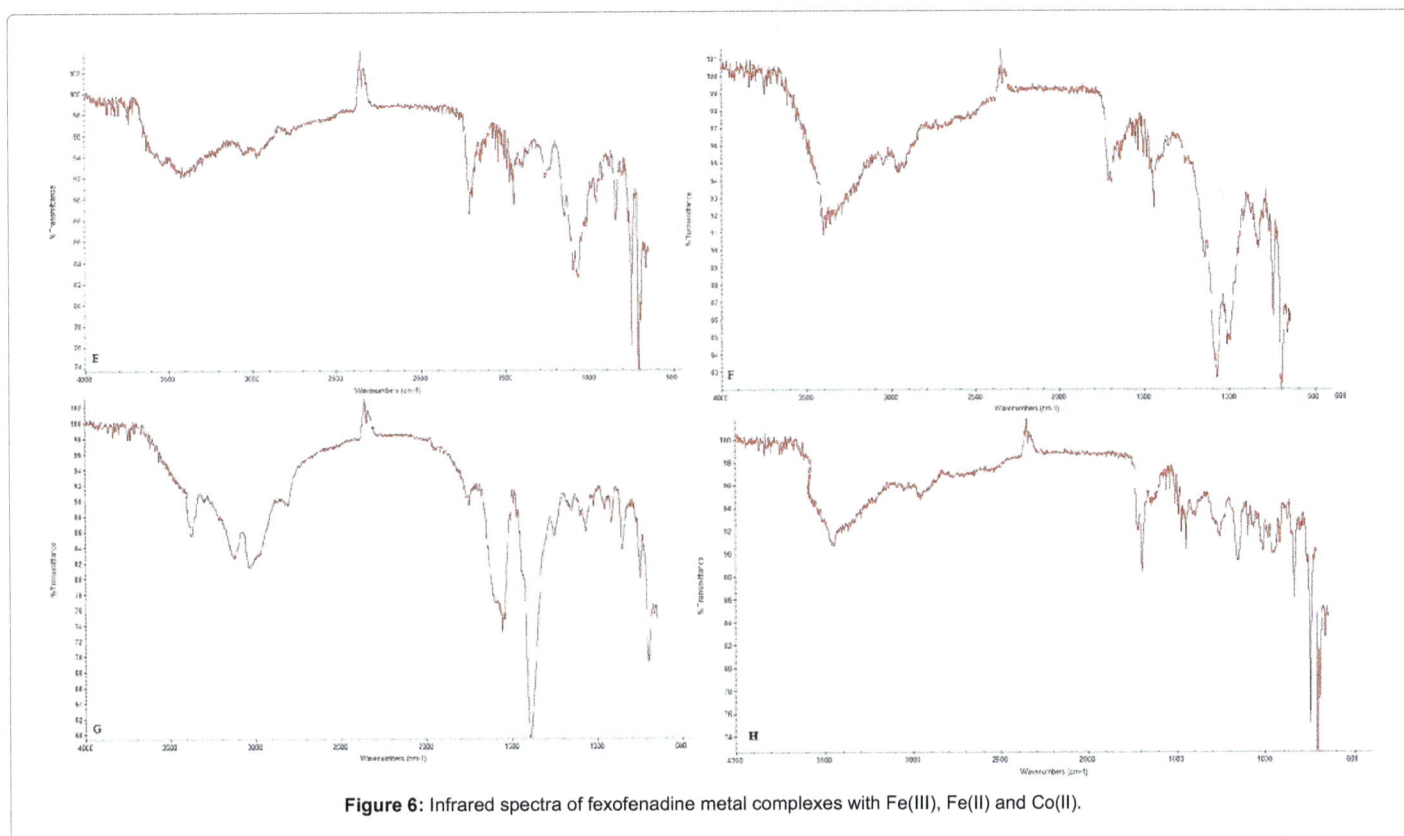

Figure 6: Infrared spectra of fexofenadine metal complexes with Fe(III), Fe(II) and Co(II).

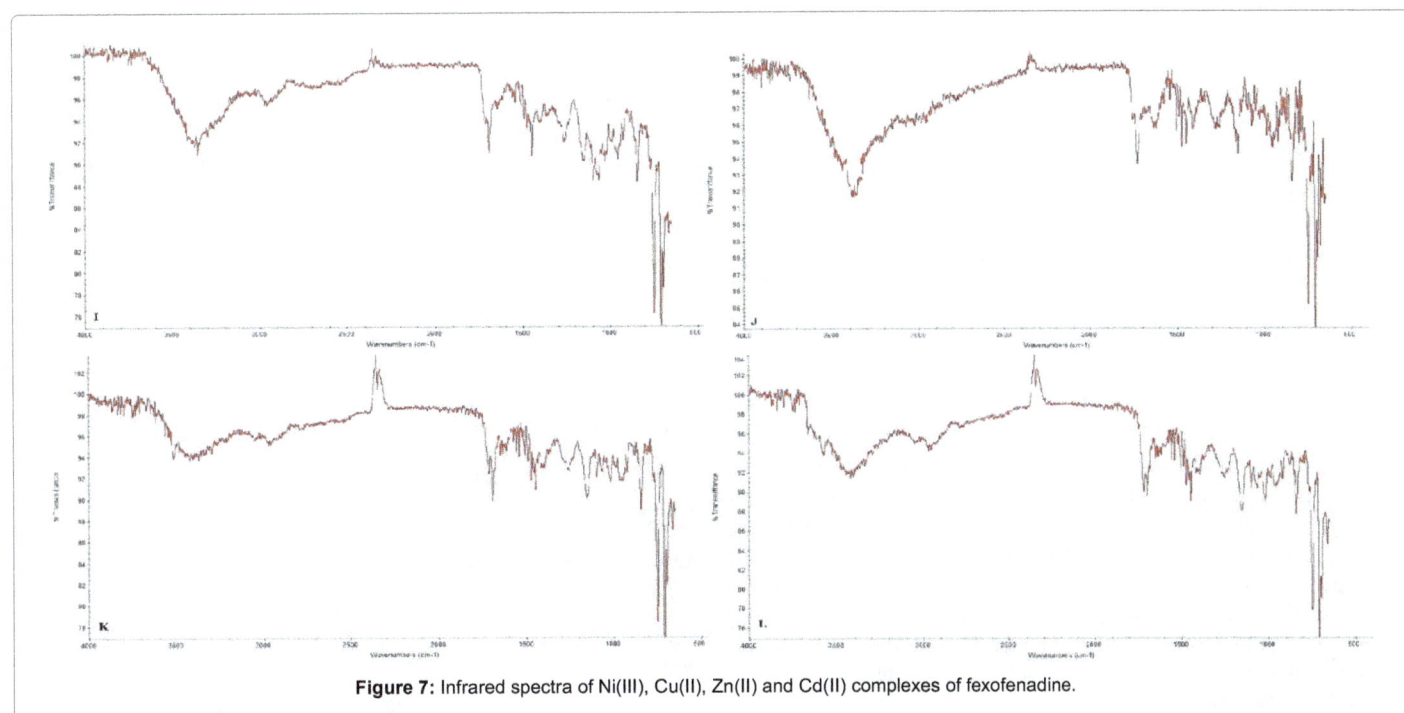

Figure 7: Infrared spectra of Ni(III), Cu(II), Zn(II) and Cd(II) complexes of fexofenadine.

complex whereas in cadmium, calcium, chromium, manganese, iron chloride, iron sulfate, iron ammonium citrate, cobalt, nickel, copper and zinc complexes this band appeared in the region 1715-1720 cm^{-1} [24,25] complexes indicating the coordination of this moiety to the metal ions [26] (Figure 7).

Fexofenadine C-O stretching appeared at 1279 cm^{-1} as single band of medium stretching, while in magnesium, calcium and cobalt complex this band appeared at 1300 cm^{-1}, in chromium it was in the form of broadband, whereas in manganese and iron chloride it was in the region of 1300-1280 cm^{-1}, iron sulfate it was at 1264 cm^{-1} as weak band, in iron ammonium citrate it appeared at 1253 cm^{-1} as a medium intensity band. In nickel complex at 1300 cm^{-1} as medium band whereas in copper it was a doublet of weak intensity, in zinc complex a weak doublet appeared at 1248 cm^{-1} and in cadmium complex at 1260 cm^{-1}. The change in the C-O stretching showed that the carboxylate ion also coordinated to a metal [27].

The C-N stretching in the reference compound was recorded at 1167 cm^{-1} as medium stretching but in magnesium, chromium, manganese, ferric chloride, cobalt, copper, zinc and cadmium complex this appeared in the region 1149-1151 cm^{-1}, a sharp doublet appeared at 1100-1140 cm^{-1} in calcium, whereas in nickel complex it was at 1120 cm^{-1} as a medium band. In iron sulfate a doublet band was found at 1114 cm^{-1} because of metal to nitrogen bonding and was absent in iron ammonium citrate [28-30].

The medium intensity bands of metal nitrogen absorptions were recorded in the region 1370-1395 cm^{-1} as medium bands whereas in the range 670-580 cm^{-1} as low intensity bands. From the above stretching modes of metal nitrogen stretching in fexofenadine metal complexes, there is no coordination of metal with nitrogen as there are no strong bands in these regions in the spectrum of fexofenadine [26,31].

Metal oxygen stretching was recorded at 1510-1490, weak bands of metal oxygen appeared in low intensity region in between 500-600 cm^{-1} as sharp and weak singlets and medium doublets [24,29]. When atoms of higher atomic masses were attached to the oxygen of carboxylate ion the stretching frequency lowered and another reason is that if metal oxygen bond was weaker the stretching frequency will also lower. That's why weaker bonds are recorded at lower intensity.

The metal chloride stretching frequencies were recorded in the region 690-640 cm^{-1} as weaker bands in magnesium, chromium and manganese but prominent bands in the rest of the complexes [22].

^{1}H NMR studies

The NMR spectra of fexofenadine and its metal complexes were recorded in deuterated methanol on 400 MHz Avance 500 spectrophotometer.

On comparing main peaks of fexofenadine with its complexes, it is observed that all the signals of the free ligand are present in the 1HNMR spectra of the complexes. Fexofenadine showed the aromatic proton resonating at δ7.19 (s, J=1.16 Hz), 7.11 (s, J=0.5 Hz) and 7.06. The proton resonating at δ2.0 is assigned to N-H proton, at δ1.52 are assigned for two methyl on tertiary carbon. Protons of piperidine ring CH$_2$-N and -CH$_2$-CH$_2$-N resonate at δ 2.29, 2.19 and 2.34 (t, J=12.2 Hz), -CH$_2$ open chain at 2.36 (t, J=14.0 Hz), 1.39, 1.77 and 4.5 (s, J=1.0 Hz) respectively, at δ4.95 (s) are alcoholic protons.

Two tertiary methyl groups have their chemical shift at 1.52 ppm in reference compound which shifted to 1.64 ppm in iron sulfate, cobalt and nickel complexes. The protons of methyl substituent with carboxylic group were found at 1.61 ppm in chromium complex, 1.46 ppm and 3.12 ppm in manganese complex, 1.60 ppm in copper complex, 1.63 ppm in zinc complex and 1.62 ppm in cadmium complex respectively. The protons of adjacent carbon of carboxylic group in magnesium complex shifted to 1.65 ppm, 1.64 ppm in calcium complex and in iron ammonium citrate this shift appeared at 2.8 and 2.55 ppm.

In iron sulfate, nickel, copper, zinc and cadmium complexes -CH$_2$ adjacent to piperidine nitrogen have peaks at 2.29 ppm and 2.19 ppm (t, J=2.0 Hz), (t, J=3.5 Hz) (t, J=3.75 Hz) (t, J=7.69 Hz). The methylene proton peak in magnesium complex was found at 1.94 ppm (t, J=0.99 Hz), in calcium complex at 2.92 ppm (t, J=3.5 Hz), in chromium complex at 1.91 ppm 1.88 ppm (t, J=12.2 Hz), in manganese complex at 2.27 ppm, 1.79 ppm and 1.69 ppm, in iron chloride at 2.93 ppm,

2.92 ppm, in iron sulfate 2.19 ppm, 2.92 ppm and in cobalt complex the peak was observed at 2.24 ppm (t, J=7.5 Hz). In iron ammonium citrate the methylene protons were at 4.79, 2.8, 2.55, 3.49 ppm and aromatic protons were at 6.99 and 7.07 ppm.

The set of signals of most of the groups were almost similar except for piperidine nitrogen, its adjoining protons and a carboxylic group because both these functional groups were taking part in the coordination.

The methylene protons adjacent to carboxylic group of magnesium, calcium, chromium and manganese complex the methyl protons of tertiary carbon shifted in the region 1.46-3.12 ppm. The methylene protons of piperidine ring shifted in all these complexes which confirmed the coordination of metal with nitrogen of piperidine ring and carboxylic group as evidenced from ^{1}H NMR data.

At 1.42, 1.52 and 1.64 ppm shifting of methyl protons of tertiary carbon adjacent to carboxylic group were observed in fexofenadine iron chloride and iron sulfate complex, whereas there was a shift to 2.92 ppm for iron chloride and 2.19 for iron sulfate complex which confirms that the coordination of metal with nitrogen of piperidine ring and carboxylic group was evidenced from ^{1}H NMR data. The shifting of methylene protons adjacent to carboxylic group to 2.8 and 2.55 ppm confirmed that the coordination of metal with carboxylic group was evidenced from ^{1}H NMR data.

The methyl protons of tertiary carbon adjacent to carboxylic group shifted in the region 1.46-1.64 ppm and methylene protons of piperidine ring shifted to 2.24 ppm confirmed that the coordination of metal with nitrogen of piperidine ring and carboxylic group was evidences from ^{1}H NMR data of cobalt complex.

Spectroscopic studies as UV, IR, ^{1}H NMR, and CHN analysis fully supports the formation of fexofenadine metal complexes in which fexofenadine is coordinated to the metal through piperidine nitrogen, oxygen of carboxylic group and oxygen of alcoholic group (Figures 2 and 3).

Conclusion

Fexofenadine metal complexes with magnesium, calcium, chromium, manganese, iron (ferric and ferrous); cobalt nickel, copper, zinc and cadmium were synthesized and characterized by FT-IR, ^{1}H-NMR and elemental analysis. Fexofenadine coordinate to metal ions to form stable monodentate complexes. The coordination being through piperidine nitrogen in case of magnesium, calcium, chromium and manganese complexes and with oxygen of carboxylato group in complexation with iron, cobalt nickel, copper, zinc and cadmium.

Disclosure Statement

There are no actual or potential conflicts of interest in this paper.

References

1. O'Neil, Patricia E Heckelman, Cherie B Koch, Kristin J. Roman (2006) The Merck Index. An encyclopedia of Chemicals Drugs, and Biologicals, Fourteenth, Merck & Co., Inc., NJ, USA.

2. Caballero E, Ocaña I, Azanza JR, Sádaba B (1999) [Fexofenadine: a antihistaminic. Review of its practical characteristics]. Rev Med Univ Navarra 43: 93-97.

3. Amon U, Amon S, Gibbs BF (2000) In vitro studies with fexofenadine, a new nonsedating histamine H1 receptor antagonist, on isolated human basophils. Inflamm Res 49 Suppl 1: S13-14.

4. Stone BM, Turner C, Mills SL, Nicholson AN (1999) Studies into the possible central effects of the H-1 receptor antagonist, fexofenadine. Int Arch Allergy Immunol 118: 338.

5. Turncliff RZ, Hoffmaster KA, Kalvass JC, Pollack GM, Brouwer KL (2006) Hepatobiliary disposition of a drug/metabolite pair: Comprehensive pharmacokinetic modeling in sandwich-cultured rat hepatocytes. J Pharmacol Exp Ther 318: 881-889.

6. Greenblatt DJ (2009) Analysis of drug interactions involving fruit beverages and organic anion-transporting polypeptides. J Clin Pharmacol 49: 1403-1407.

7. Shimizu M, Fuse K, Okudaira K, Nishigaki R, Maeda K, et al. (2005) Contribution of OATP (organic anion-transporting polypeptide) family transporters to the hepatic uptake of fexofenadine in humans. Drug Metab Dispos 33: 1477-1481.

8. Triggiani M, Gentile M, Secondo A, Granata F, Oriente A, et al. (2001) Histamine induces exocytosis and IL-6 production from human lung macrophages through interaction with H1 receptors. J Immunol 166: 4083-4091.

9. Castillo O, Luque A, Román P, Lloret F, Julve M (2001) Syntheses, crystal structures, and magnetic properties of one-dimensional oxalato-bridged Co(II), Ni(II), and Cu(II) complexes with n-aminopyridine (n = 2-4) as terminal ligand. Inorg Chem 40: 5526-5535.

10. Cuevas A, Kremer C, Suescun L, Russi S, Mombrú AW, et al. (2007) Synthesis, crystal structure and magnetic properties of novel heterobimetallic malonate-bridged MIIReIV complexes (M = Mn, Fe, Co and Ni). Dalton Trans 7: 5305-5315.

11. Sultana N, Arayne MS, Rizvi SBS, Haroon U, Mesaik MA (2012) Synthesis, spectroscopic and biological evaluation of some levofloxacin metal complexes. Medicinal Chemistry Research 22: 1371-1377.

12. Sultana N, Humza E, Arayne MS, Haroon U (2011) Effect of metal ions on the in vitro availability of enoxacin, it's in vivo implications, kinetic and antibacterial studies. Quim Nova 34: 186-189.

13. Arayne S, Sultana N, Haroon U, Mesaik MA (2009) Synthesis, characterization, antibacterial and anti-inflammatory activities of enoxacin metal complexes. Bioinorg Chem Appl.

14. Sultana N, Arayne MS, Gul S, Shamim S (2010) Sparfloxacin–metal complexes as antifungal agents – their synthesis, characterization and antimicrobial activities Journal of Molecular Structure 975: 285-291.

15. Sultana N, Arayne MS, Naz A, Mesaik MA (2010) Synthesis, characterization, antibacterial, antifungal and immunomodulating activities of gatifloxacin-metal complexes. Journal of Molecular Structure 969: 17-24.

16. Arayne MS, Sultana N, Mirza AZ (2009) Preparation and spectroscopic characterization of metal complexes of gliquidone. Journal of Molecular Structure 927: 54-59.

17. Sultana N, Arayne MS, Iftikhar B, Nawaz M (2008) A new RP-HPLC method for monitoring of atenolol: Application to atenolol metal interaction studies. J Chem Soc Pak 30: 113-118.

18. Lingane JJ (1958) Electroanalytical Chemistry Second edition, Interscience Publishers, New York, USA.

19. Bailey DG (2010) Fruit juice inhibition of uptake transport: a new type of food-drug interaction. Br J Clin Pharmacol 70: 645-655.

20. Ford R, Schwartz L, Dancey J, Dodd LE, Eisenhauer EA, et al. (2009) Lessons learned from independent central review. Eur J Cancer 45: 268-274.

21. (2003) Stability testing of new drug substances and products. International Conference on Harmonisation of Technical Requirements for Registration of Pharmaceuticals for Human Use.

22. Iwamoto R, Matsuda T (2004) Characterization of infrared and near-infrared absorptions of free alcoholic OH groups in hydrocarbon. Appl Spectrosc 58: 1001-1009.

23. Kasumov VT, Medjidov AA, Yayli N, Zeren Y (2004) Spectroscopic and electrochemical characterization of di-tert-butylated sterically hindered Schiff bases and their phenoxyl radicals. Spectrochim Acta A Mol Biomol Spectrosc 60: 3037-3047.

24. Korshin EE, Leitus G, Shimon LJ, Konstantinovski L, Milstein D (2008) Silanol-based pincer Pt(II) complexes: synthesis, structure, and unusual reactivity. Inorg Chem 47: 7177-7189.

25. Wille U, Tan JC, Mucke EK (2008) A computational study of multicomponent orbital interactions during the cyclization of silyl, germyl, and stannyl radicals onto C-N and C-O multiple bonds. J Org Chem 73: 5821-5830.

26. García-Terán JP, Castillo O, Luque A, García-Couceiro U, Román P, et al. (2004) One-dimensional oxalato-bridged Cu(II), Co(II), and Zn(II) complexes with purine and adenine as terminal ligands. Inorg Chem 43: 5761-5770.

27. Garrone E, Bulánek R, Frolich K, Otero Aréan C, Rodríguez Delgado M, et al. (2006) Single and dual cation sites in zeolites: theoretical calculations and FTIR spectroscopic studies on CO adsorption on K-FER. J Phys Chem B 110: 22542-22550.

28. Cossío FP, Arrieta A, Sierra MA (2008) The mechanism of the ketene-imine (staudinger) reaction in its centennial: still an unsolved problem? Acc Chem Res 41: 925-936.

29. Martin R, Buchwald SL (2008) Palladium-catalyzed Suzuki-Miyaura cross-coupling reactions employing dialkylbiaryl phosphine ligands. Acc Chem Res 41: 1461-1473.

30. Tye JW, Weng Z, Johns AM, Incarvito CD, Hartwig JF (2008) Copper complexes of anionic nitrogen ligands in the amidation and imidation of aryl halides. J Am Chem Soc 130: 9971-9983.

31. Martínez-Lillo J, Delgado FS, Ruiz-Pérez C, Lloret F, Julve M, et al. (2007) Heterotrimetallic oxalato-bridged ReIV2MII complexes (M=Mn, Co, Ni, Cu): synthesis, crystal structure, and magnetic properties. Inorg Chem 46: 3523-3530.

9

Effect of Environmental Factors on Phenolic Compounds in Leaves of *Syzygium jambos* (L.) Alston (Myrtaceae)

Wilma P Rezende[1], Leonardo L Borges [1*], Danillo L Santos[1], Nilda M Alves[2] and José R Paula[1]

[1]*Natural Products Research Laboratory, Federal University of Goiás, Brazil*
[2]*School of Health Sciences, Department of Pharmacy and Biochemistry, University of Rio Verde, Brazil*

Abstract

Background: *Syzygium jambos* (L.) Alston, Myrtaceae, is a plant widely used for the treatment of toothache, mouth scores, cough, wound dressing and infectious diseases. Among the metabolite groups identified in leaves of *S. jambos* are the polyphenols, highlighting tannins and flavonoids, related to the pharmacological properties of this plant. Studies on the influence of environmental factors over production of secondary metabolites in *S. jambos* are important because they contribute with knowledge for its cultivation and harvest, besides establish quantitative parameters of secondary metabolites in the plant drug. The aim of this paper was to evaluate the effects of environmental factors on levels of phenolic compounds in *S. jambos* leaves.

Materials and Methods: Total phenols, tannins, flavonoids and mineral nutrients were quantified in leaves, while soil fertility was also analyzed in two different sites and in two months (January and July) from ten specimens (five from each locality).

Results: The data were statistically analyzed and the results have shown that the levels of phenolic compounds in *S. jambos* leaves were influenced by environmental factors, particularly some foliar nutrients (P_l, K_l Ca_l, Na_l, Fe_l, Co_l and Mo_l), soil nutrients (Al_s, K_s, S_s, Na_s and Mn_s) and climatic factors (temperature and rainfall).

Conclusion: The results obtained in this work will be useful for knowledge of the best conditions for leaves collection from *S. jambos*, besides the data analyzed suggested that environmental factors can affect the levels of tannins in this species.

Keywords: Myrtaceae; Secondary metabolism; Environmental factors; Seasonality

Introduction

Syzygium jambos (L.) Alston, also known as *Eugenia jambos* or *Jambosa jambos*, is a medicinal plant traditionally used for the treatment of toothache, mouth scores, cough, wound dressing and infectious diseases [1]. Anti-inflammatory activity was also reported for *S. jambos* leaves extract and its isolated flavonoid glycosides and these properties is closely related to analgesic activity [2-4].

Environmental factors, such as soil composition, rainfall, temperature and humidity, can influence the levels of phenolic compounds in medicinal plants [5-7]. The development of the plant can affect the tannin amounts due a response to the environmental changes [8-10]. Environmental influence over secondary metabolites and studies about chemical variability were investigated in some species from Myrtaceae family [11-15]. In *S. jambos* species, the chemical variability and environmental factors which can influence essential oils were also investigated [16].

Regarding the lack of data on the influence of environmental factors on the production of phenolic metabolites in leaves of *S. jambos*, this study was carried out in order to obtain new data that would inform the appropriate cultivation and sampling of this plant.

Materials and Methods

Plant material

The plant material was collected from ten wild specimens of the plant located in two different municipalities in Goiás state, Brazil: Rio Verde (17°48′33.9″ S; 50°56′ 39,1″ W; 710 m), (17°46′33.6″ S; 50°54′13.2″ W; 688 m), (17°46′27.1″ S; 50°54′52.2″ W; 750 m), (17°46′9.6″ S; 50°54′52.6″ W; 781 m), (17°46′41.2″ S; 50°56′43.1″ W; 758 m) and Nova América (15°01′12″ S; 49°52′33.4″ W; 782 m), (15°01′48.7″ S; 49°51′27.9″ W; 657 m), (15°01′48.6″ S; 49°51′29.9″

W; 652 m), (15°02′58.5″ S; 49°51′53.4″ W; 614 m). The samples were collected in January and July 2011 and received botanic identification by Prof. José Realino de Paula. A voucher specimen has been deposited at the Herbarium of Federal University of Goiás under code number 47579. The samples were air-dried in a chamber at 40°C and ground into a powder.

Colorimetric assays

Total phenolics assay (TP): Ferric chloride was added to each extract under alkaline conditions to result a colored complex with phenols, which was read at 510 nm, following the Hagerman and Butler [17], adapted by Waterman and Mole [18]. All solutions were analyzed in triplicate. The standard curve was constructed with tannic acid at the following dilutions: 0.10, 0.15, 0.20, 0.25 and 0.30 mg/mL.

Protein Precipitation assay (PP): Hagerman and Butler [17] adapted by Waterman and Mole [19] uses Bovine Serum Albumine solution (BSA, 1.0 mg/mL) in 0.2M acetate buffer (pH 4.9). The extract solutions were precipitated with BSA and after centrifugation, the precipitate was dissolved in sodium dodecyl sulfate/triethanolamine solution and the tannins were complexed with ferric chloride and the colored complex was read at 510 nm. All solutions were analyzed in triplicate. The standard curve was

*Corresponding author: Leonardo Luiz Borges, Natural Products Research Laboratory, Faculty of Pharmacy, Federal University of Goiás, Brazil
E-mail: leonardoquimica@gmail.com

constructed with tannic acid at the following dilutions: 0.10, 0.20, 0.30, 0.40 and 0.50 mg/mL.

Total flavonoids assay (Fv): The methanolic extract was directly read at 361 nm [20]. All solutions were prepared in triplicate. The standards curves were prepared with rutin at the dilutions: 0.010, 0.015, 0.020, 0.025, 0.030 mg/mL.

Climatic data

Average temperature and average daily precipitation for the periods were collected from the official site of the National Institute for Space Research (Instituto Nacional de Pesquisas Espaciais) [21].

Chemical analysis of leaves and soil

Chemical analysis of soil (500 g) and leaf samples (15 g) was performed at the Solocria Agricultural Laboratory, following standard procedures [22]. The nitrogen (N) was extracted by digestion with H_2SO_4 and catalysts. The minerals Phosphorus (P), Potassium (K), Calcium (Ca), Magnesium (Mg), Sulfur (S), Copper (Cu), Iron (Fe), Manganese (Mn) and Zinc (Zn) were extracted by digestion with $HClO_4$ and HNO_3.

The soil samples were collected at a depth of 0-20 cm in four locations around each specimen of *S. jambos*, subsequently homogenized and then air dried. The pH was determined in a volume of water-soil at 1:1. Ca, Mg and Al were extracted with KCl 1M, and P, K, Zn, Cu, Fe and Mn were extracted with Mehlich's solution. Organic matter (OM), Cation Exchange Capacity (CEC), potential acidity (H+Al), base saturation (V) and aluminum saturation (m) were determined by standard methods [22].

The quantitative determination of minerals in leaves and soil was performed according to the methodology described by Silva [22]. Nitrogen was determined by distillation (semi-micro Kjeldahl method), phosphorus by colorimetry, potassium by flame photometry and sulfur by turbidimetry. Calcium, magnesium, copper, iron, manganese and zinc were determined by atomic absorption.

Statistical analyses

The relationship between phenolic compounds found in leaves of *S. jambos* and environmental variables were analyzed by stepwise Multiple Regression and Pearson's Correlation Analysis implemented using SAS GLM (General Linear Models) and SAS CORR (Correlation test) procedure, respectively [23]. Cluster Analysis was also applied to study the similarity of samples on the basis of constituent distribution. The hierarchical clustering was performed according to the Ward's variance minimization method [24]. For these analyses were employed the software's SAS (Statistical Analysis System) and Statistica 7.

Results

Tables 1-4 present the environmental data. The phenolic compounds of leaves are shown in Table 5.

The following equations were obtained by stepwise Multiple Regression with significant variables (*p*-values less than 0.10 to entry and stay variables in model):

$$TP(\%)=15.03+0.0045\ Fe_s+0.0444\ Na_l\ (R^2=0.4540;\ R=0.6738)\ (1)$$

$$PP(\%)=4.209+0.2817\ Cu_s+0.0022\ Fe_s+0.3349\ N_l\ (R^2=0.5555;\ R=0.7453)\ (2)$$

$$Fv(\%)=-0.3819+0.0503\ Cu_s+0.1324\ Ca_l+0.2889\ Mg_l+0.0989\ Cu_l+0.018\ Zn_l+3.52\ Co_l\ (R^2=0.8736;\ R=0.9347)\ (3)$$

Discussion

The Table 6 showed that potassium levels were negatively correlated with PP and Fv (weak correlation; $R<0.5$), this result can be attributed to fact that the levels of this macronutrient possess consistent positive capacity to reducing the incidence of diseases, suggesting a mechanism of compensation for the lack of K, capable to increasing resistance to pathogens by synthesis of phenolic compounds [25,26]. This correlation was also observed to *Myrcia tomentosa* species, which belongs to the same family of *S. jambos* [12].

Multiple coefficient of determination (R^2) means the proportion of the total variation that is explained by the regression model, so when R^2 is higher, the model fits better to data [27]. The value of R^2 is 0.8736 for Equation 3, showing that there are 87.36% changes in response variables (total phenols). By comparing models, the Equation 3 is better model fits to the data than the others models.

Multiple correlation coefficient (R) is employed to verify how far the relationship between one dependent variable and a set of independent variables [28]. The value of R is 0.9347 for Equation 1 and shows the multiple correlation strength between flavonoids and Cu_s, Ca_l, Mg_l, Cu_l, Zn_l and Co_l. The foliar nutrients were the principal set of environmental variables that can influence the levels of flavonoids in leaves of *S. jambos*. The total phenols and tannins by protein precipitation presented strong multiple correlation ($R>0.7$) with its sets of independent variables.

The tannins presented negative correlation with temperature (Table 6), with $R=-0.41778$. The increased levels of phenolic compounds in the leaves may be related to increased activity of phenylalanine ammonialyase (PAL) at lower temperatures, given the fact that PAL is an important enzyme in the biogenesis of various phenolic compounds, including tannins, which could explain the negative correlation found [29,30]. The same behaviour was observed in bananas that were stored in different temperatures,

Sample	Precipitation (mm)	Temperature (°C)
NA01/Jan/2011	11.93	24.48
NA02/Jan/2011	11.93	24.48
NA03/Jan/2011	11.93	24.48
NA04/Jan/2011	11.93	24.48
NA05/Jan/2011	11.93	24.48
RV01/Jan/2011	8.29	23.9
RV02/Jan/2011	8.29	23.9
RV03/Jan/2011	8.29	23.9
RV04/Jan/2011	8.29	23.9
RV05/Jan/2011	8.29	23.9
NA01/Jul/2011	-	21.62
NA02/Jul/2011	-	21.62
NA03/Jul/2011	-	21.62
NA04/Jul/2011	-	21.62
NA05/Jul/2011	-	21.62
RV01/Jul/2011	-	21.5
RV02/Jul/2011	-	21.5
RV03/Jul/2011	-	21.5
RV04/Jul/2011	-	21.5
RV05/Jul/2011	-	21.5

NA: Nova América; RV: Rio Verde

Table 1: Climate data for the collection sites in the period of January 2011 and July 2011. Mean precipitation (mm) and mean temperature (°C).

Sample	Cu mg/dm³	Fe mg/dm³	Mn mg/dm³	Zn mg/dm³	P mg/dm³	K mg/dm³	Ca mg/dm³	Mg mg/dm³
NA01/Jan/2011	6.00	310.00	34.00	20.00	1.00	10.00	5.00	1.70
NA02/Jan/2011	5.00	286.00	25.00	23.00	0.90	8.00	5.20	1.60
NA03/Jan/2011	5.00	276.00	24.00	22.00	1.20	10.60	4.90	1.50
NA04/Jan/2011	4.00	240.00	20.00	21.00	1.30	9.80	5.40	1.70
NA05/Jan/2011	7.00	243.00	18.00	19.00	1.20	8.40	4.70	2.20
RV01/Jan/2011	8.00	370.00	25.00	20.00	1.00	8.00	5.20	1.60
RV02/Jan/2011	5.00	305.00	22.00	21.00	1.00	8.00	5.50	1.60
RV03/Jan/2011	6.00	228.00	47.00	16.00	1.10	9.20	6.00	1.40
RV04/Jan/2011	6.00	202.00	61.00	16.00	1.40	9.60	6.20	1.60
RV05/Jan/2011	5.00	294.00	29.00	18.00	1.00	8.80	5.80	1.80
NA01/Jul/2011	4.00	241.00	29.00	11.00	1.00	7.60	5.20	1.70
NA02/Jul/2011	2.00	154.00	20.00	10.00	1.10	6.80	4.20	1.80
NA03/Jul/2011	3.00	367.00	19.00	11.00	1.00	8.00	4.40	1.60
NA04/Jul/2011	3.00	141.00	22.00	14.00	1.10	8.80	6.20	1.50
NA05/Jul/2011	3.00	148.00	24.00	14.00	1.20	8.60	6.50	1.60
RV01/Jul/2011	4.00	270.00	28.00	13.00	1.00	4.80	5.00	2.50
RV02/Jul/2011	4.00	363.00	22.00	14.00	1.10	6.40	6.00	1.80
RV03/Jul/2011	4.00	367.00	45.00	11.00	1.20	8.60	5.20	1.70
RV04/Jul/2011	6.00	345.00	58.00	11.00	1.20	8.00	5.40	1.60
RV05/Jul/2011	2.00	311.00	53.00	13.00	1.40	8.60	5.10	1.80

NA: Nova América; RV: Rio Verde

Table 2: Levels of mineral nutrients and fertility parameters of soil from each sample collection site.

Sample	H+Al cmolc/dm³	Al cmolc/dm³	CEC cmolc/dm³	O.M. %	M %	V %	Ca/CEC %	Mg/CEC %	K/CEC %
NA01/Jan/2011	2.1	0.0	4.54	7.00	0.00	53.66	33.04	8.81	11.23
NA02/Jan/2011	2.3	0.0	6.70	14.00	0.00	65.63	43.28	16.42	5.67
NA03/Jan/2011	1.3	0.0	7.58	33.00	0.00	76.22	47.49	17.15	11.35
NA04/Jan/2011	1.8	0.0	11.07	67.00	0.00	88.27	73.17	9.03	5.87
NA05/Jan/2011	2.7	0.0	5.87	8.00	0.00	54.03	32.37	17.04	4.26
RV01/Jan/2011	4.0	0.1	6.82	8.00	3.44	41.39	26.39	13.20	1.61
RV02/Jan/2011	2.9	0.0	8.35	12.00	0.00	65.24	47.90	15.57	1.56
RV03/Jan/2011	1.9	0.0	8.26	11.00	0.00	76.95	61.74	10.90	4.00
RV04/Jan/2011	2.6	0.0	6.49	10.00	0.00	59.96	40.06	16.95	2.62
RV05/Jan/2011	2.8	0.0	9.30	14.00	0.00	69.85	50.54	13.98	5.05
NA01/Jul/2011	2.6	0.1	6.52	23.00	2.50	60.15	33.74	10.74	15.34
NA02/Jul/2011	2.5	0.4	4.46	14.00	17.17	43.86	29.15	6.73	7.40
NA03/Jul/2011	2.0	0.0	7.46	18.00	0.00	73.21	45.58	16.09	11.26
NA04/Jul/2011	1.7	0.0	9.62	25.00	0.00	82.30	49.90	19.75	12.47
NA05/Jul/2011	2.0	0.0	10.33	28.00	0.00	80.60	46.47	19.36	14.52
RV01/Jul/2011	2.7	0.0	6.34	12.00	0.00	57.39	42.59	12.62	2.05
RV02/Jul/2011	2.7	0.0	7.11	14.00	0.00	62.07	47.82	12.66	1.41
RV03/Jul/2011	1.7	0.0	7.72	12.00	0.00	78.00	60.88	10.36	6.48
RV04/Jul/2011	2.2	0.0	5.55	13.00	0.00	60.41	50.45	7.21	2.52
RV05/Jul/2011	2.2	0.0	5.45	18.00	0.00	59.61	45.87	7.34	6.24

NA: Nova América; RV: Rio Verde

Table 3: Levels of mineral nutrients and fertility parameters of soil from each collection site.

and PAL activity was increased in lower temperatures [31]. In other study with tomato plants (*Lycopersicon esculentum* L.), the thermal stress caused highest phenylalanine ammonia-lyase activity, and this results are in agreement with the trend found in our paper [32]. To other hand, this sensibility to climatic changes was also observed in phenolic compounds present in grape skins, in which warm temperatures presented positive correlation with the levels of phenolic compounds [33]. Also, related to PAL, the micronutrient Cu is linked to production of phenolic compounds in plants, because it is capable to activate the PAL pathway, and this may explain the positive correlation between Cu present in soil with tannins and flavonoids observed in Table 6. Another explanation suggested of the increased levels of phenolic compounds in these tissues, is that this is associated with a mechanism of tolerance to Cu, since Cu is a catalyst for redox reactions that can generate free radicals harmful to the plant; consequently, increased levels of phenolic compounds may have two goals: to decrease the concentration of free Cu in plant tissue by the reaction of this with phenols, and to minimize the deleterious effects of free radicals formed, through the antioxidant reactions of the phenolic compounds [34].

Sample	N	P	K	Ca	Mg	S	Cu	Fe	Mn	Zn
NA01/Jan/2011	12.00	1.00	10.00	5.00	1.70	1.00	6.00	310.00	34.00	20.00
NA02/Jan/2011	12.20	0.90	8.00	5.20	1.60	1.30	5.00	286.00	25.00	23.00
NA03/Jan/2011	12.00	1.20	10.60	4.90	1.50	1.00	5.00	276.00	24.00	22.00
NA04/Jan/2011	12.80	1.30	9.80	5.40	1.70	1.20	4.00	240.00	20.00	21.00
NA05/Jan/2011	12.00	1.20	8.40	4.70	2.20	1.10	7.00	243.00	18.00	19.00
RV01/Jan/2011	13.60	1.00	8.00	5.20	1.60	1.00	8.00	370.00	25.00	20.00
RV02/Jan/2011	12.50	1.00	8.00	5.50	1.60	1.10	5.00	305.00	22.00	21.00
RV03/Jan/2011	13.20	1.10	9.20	6.00	1.40	1.10	6.00	228.00	47.00	16.00
RV04/Jan/2011	14.00	1.40	9.60	6.20	1.60	1.20	6.00	202.00	61.00	16.00
RV05/Jan/2011	12.60	1.00	8.80	5.80	1.80	1.10	5.00	294.00	29.00	18.00
NA01/Jul/2011	14.00	1.00	7.60	5.20	1.70	1.20	4.00	241.00	29.00	11.00
NA02/Jul/2011	13.00	1.10	6.80	4.20	1.80	1.40	2.00	154.00	20.00	10.00
NA03/Jul/2011	12.60	1.00	8.00	4.40	1.60	1.60	3.00	367.00	19.00	11.00
NA04/Jul/2011	13.40	1.10	8.80	6.20	1.50	1.40	3.00	141.00	22.00	14.00
NA05/Jul/2011	12.80	1.20	8.60	6.50	1.60	1.50	3.00	148.00	24.00	14.00
RV01/Jul/2011	14.20	1.00	4.80	5.00	2.50	1.60	4.00	270.00	28.00	13.00
RV02/Jul/2011	14.40	1.10	6.40	6.00	1.80	1.80	4.00	363.00	22.00	14.00
RV03/Jul/2011	13.00	1.20	8.60	5.20	1.70	1.90	4.00	367.00	45.00	11.00
RV04/Jul/2011	13.20	1.20	8.00	5.40	1.60	1.60	6.00	345.00	58.00	11.00
RV05/Jul/2011	15.00	1.40	8.60	5.10	1.80	1.80	2.00	311.00	53.00	13.00

NA: Nova América; RV: Rio Verde

Table 4: Levels of macronutrients (N_l, P_l, K_l, Ca_l, Mg_l, S_l in g/kg) and micronutrients (Cu_l, Fe_l, Mn_l, Zn_l in mg/kg) in the leaves of *Syzygium jambos* from each collection site in January 2011 to April 2011.

Sample	TP	PP	Fv
NA01/Jan/2011	21.151 ± 0.003	9.985 ± 0.010	2.151 ± 0.012
NA02/Jan/2011	21.134 ± 0.002	10.133 ± 0.011	1.914 ± 0.006
NA03/Jan/2011	19.766 ± 0.004	9.068 ± 0.008	1.910 ± 0.029
NA04/Jan/2011	20.750 ± 0.001	8.421 ± 0.008	2.006 ± 0.013
NA05/Jan/2011	23.784 ± 0.003	9.675 ± 0.015	2.218 ± 0.018
RV01/Jan/2011	20.143 ± 0.003	9.897 ± 0.005	2.364 ± 0.014
RV02/Jan/2011	18.614 ± 0.002	9.818 ± 0.013	2.153 ± 0.003
RV03/Jan/2011	20.582 ± 0.002	9.524 ± 0.005	2.272 ± 0.005
RV04/Jan/2011	18.147 ± 0.002	9.675 ± 0.008	2.165 ± 0.008
RV05/Jan/2011	19.453 ± 0.005	9.259 ± 0.010	2.216 ± 0.004
NA01/Jul/2011	21.748 ± 0.003	11.647 ± 0.008	2.265 ± 0.043
NA02/Jul/2011	23.047 ± 0.005	9.535 ± 0.006	1.815 ± 0.005
NA03/Jul/2011	20.397 ± 0.003	10.269 ± 0.005	2.061 ± 0.007
NA04/Jul/2011	22.912 ± 0.003	10.147 ± 0.010	2.156 ± 0.008
NA05/Jul/2011	21.629 ± 0.003	11.111 ± 0.022	2.448 ± 0.006
RV01/Jul/2011	18.860 ± 0.002	10.239 ± 0.012	2.372 ± 0.726
RV02/Jul/2011	19.949 ± 0.003	10.721 ± 0.024	2.480 ± 0.005
RV03/Jul/2011	21.591 ± 0.004	10.109 ± 0.011	2.169 ± 0.003
RV04/Jul/2011	20.658 ± 0.002	8.341 ± 0.013	2.200 ± 0.040
RV05/Jul/2011	20.329 ± 0.005	9.780 ± 0.013	2.134 ± 0.012

NA: Nova América; RV: Rio Verde; TP: total phenols content; PP: tannins by protein precipitation assay; Fv: total flavonoids

Table 5: Amounts of phenolic compounds in g/100 g dry weight (± standard deviation) of leaves of *Syzygum jambos*.

The nutrient Fe_s showed positive correlation with TP and PP, and according Jin et al. [35] phenolic compounds may complex with Fe^{3+} and be transported to other tissues, facilitating its mobilization between different tissues, and also participating in reduction reactions of Fe^{3+} to Fe^{2+}, aiding reductase-type enzymes.

The Hierarchical Cluster Analysis employing Ward's variance minimizing method showed a highly variability within phenolic compounds of *S. jambos* leaves. From Figure 1 it can be seen the similarities of the samples on the basis of the distribution of the constituents and this may indicate that the main factor responsible in chemical variability is the collection time, due the similarities of

samples of a same site and the collection site would have less influence in phenols variability between the two different localities.

Conclusion

This paper suggests that there is a relevant influence of environmental factors over production of phenolic compounds in the leaves of *S. jambos*, with foliar nutrients (P_l, K_l Ca_l, Na_l, Fe_l, Co_l and Mo_l) soil nutrients (Al_s, K_s, S_s, Na_s and Mn_s) and climatic factors (temperature and rainfall), being the main variables that may change the levels of phenolic compounds in this plant tissue. From the data obtained, the best conditions of collection can be established for leaves of *S. jambos*.

	TP	PP	Fv
Cas	-0.10705	-0.27518	0.005473
Ms	-0.02181	0.23104	0.19327
Als	0.37711	0.036006	-0.3875**
HAl	-0.23139***	0.17572	0.40273**
Ks	0.40544**	0.37155***	-0.03278
Ss	-0.38337**	0.14683	0.37144***
Nas	0.33869***	0.042196	-0.21794
Zns	-0.09854	-0.26323	0.10575
Bs	0.020883	-0.25372	0.00098
Cus	-0.19116	0.49054*	0.36616***
Fes	0.58551*	0.42248**	-0.34742***
Mns	0.50044*	0.39947**	-0.17427
Nl	-0.27365	0.2824	0.43661**
Pl	-0.03023	-0.35143***	-0.06747
Kl	0.02338	-0.38467**	-0.34668***
Cal	-0.25612	0.15188	0.55497*
Mgl	0.036881	0.10184	0.23637
Sl	0.057513	0.2213	0.19748
Nal	0.17358	0.18636	0.40286**
Cul	-0.19142	-0.2768	0.23941
Fel	-0.38217**	-0.10146	0.16464
Mnl	-0.32657	-0.25279	0.099418
Znl	-0.22435	-0.30987	-0.1938
Col	0.2682	0.40508**	0.22787
Mol	0.24594	0.43497**	0.24723
Temperature	-0.17664	-0.41778**	-0.27652
Rainfall	-0.1406	-0.42803**	-0.30053

*Significant at: *1%, **5% and ***15%. TP: total phenols content; PP: tannins by protein precipitation assay; Fv: total flavonoids

Table 6: Values of Pearson's coefficient between environmental variables and phenolic compounds found in leaves of *Syzygum jambos*.

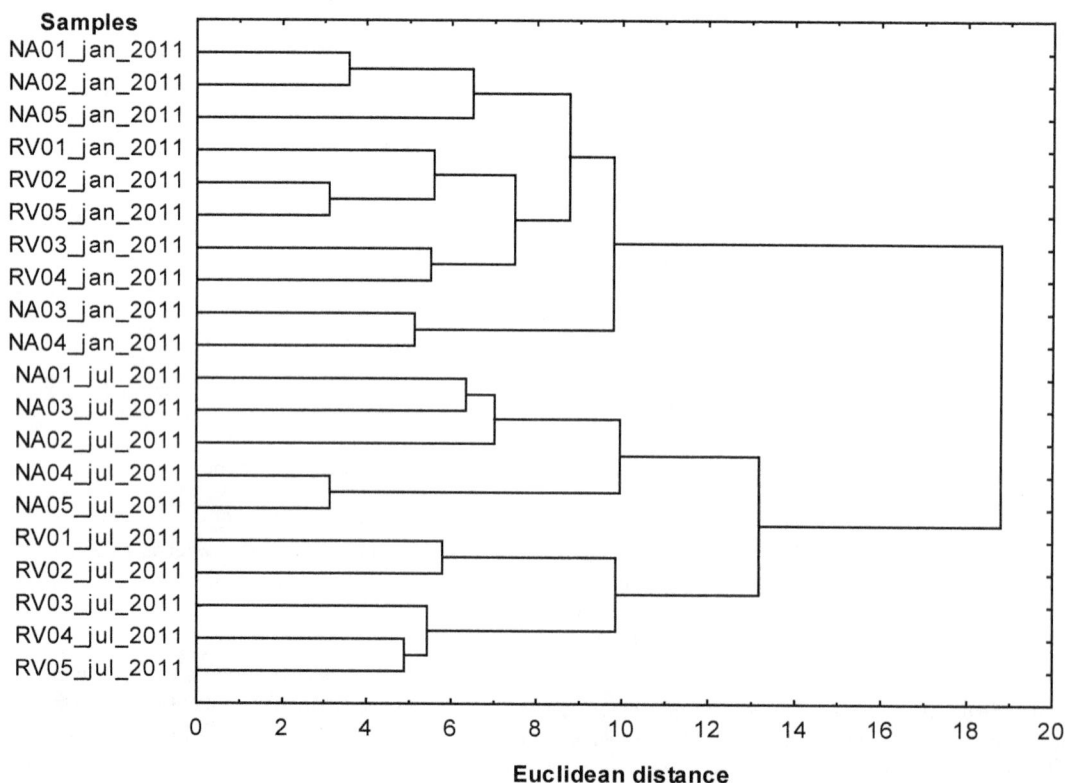

Figure 1: Dendrogram representing chemical composition similarity relationships among leaves of *S. jambos*, linking the climatic data, soil nutrients, foliar nutrients and phenolic compounds according to Ward's variance minimization method.

References

1. Iwu M (1993) Handbook of African Medicinal Plants. CRC Press, Florida, USA.

2. Slowing K, Söllhuber M, Carretero E, Villar A (1994) Flavonoid glycosides from *Eugenia jambos*. Phytochemistry 37: 255-258.

3. Slowing K, Carretero E, Villar A (1994) Anti-inflammatory activity of leaf extracts of *Eugenia jambos* in rats. J Ethnopharmacol 43: 9-11.

4. Slowing K, Carretero E, Villar A (1996) Anti-inflammatory compounds of *Eugenia jambos*. PTR 10: 8126-8127.

5. Kouki M, Manetas Y (2002) Resource availability affects differentially the levels of gallotannins and condensed tannins in *Ceratonia siliqua*. Biochem Syst Ecol 30: 631-639.

6. Monteiro JM, Albuquerque UP, Neto EMFL, Araújo EL, Albuquerque MM, et al. (2006) The effects of seasonal climate changes in the Caatinga on tannin level. Rev Bras Farmacogn 16: 338-344.

7. Avila-Peña D, Peña N, Quintero L, Suárez-Roca H (2007) Antinociceptive activity of *Syzygium jambos* leaves extract on rats. J Ethnopharmacol 112: 380-385.

8. Hatano T, Kira R, Yoshizaki M, Okuda T (1986) Seasonal changes in the tannins of *Liquidanbar formosana* reflecting their biogenesis. Phytochemistry 25: 2787-2789.

9. Salminen JP, Ossipov V, Haukioja E, Pihlaja K (2001) Seasonal variation in the content of hydrolysable tannins in leaves of *Betula pubescens*. Phytochemistry 57: 15-22.

10. Sampaio BL, Bara MT, Ferri PH, Santos SC, Paula JR (2011) Influence of environmental factors on the concentration of phenolic compounds in leaves of *Lafoensia pacari*. Rev Bras Farmacogn 21: 1127-1137.

11. Sá FAS, Borges LL, Paula JAM, Sampaio BL, Ferri PH, et al. (2012) Essential oils in aerial parts of *Myrcia tomentosa*: composition and variability. Rev Bras Farmacogn 22: 1233-1240.

12. Borges LL, Alves SF, Sampaio BL, Conceição EC, Bara MTF, et al. (2013) Environmental factors affecting the concentration of phenolic compounds in *Myrcia tomentosa* leaves. Rev Bras Farmacogn 23: 230-238.

13. Borges LL, Alves SF, Carneiro FM, Conceição EC, Bara MTF, et al. (2012) Influence of environmental factors on the concentration of phenolic compounds in barks of *Myrcia tomentosa* (Aubl.) DC. J Pharm Res 5: 1323-1327.

14. Alcantara GA, Borges LL, Paula JR (2012) Seasonal variation in the content of phenolic compounds in barks of *Myrcia rostrata* DC. by influence of environmental factors. J Pharm Res 5: 1306-1309.

15. Borges LL, Alves SF, Bara MTF, Conceição EC, Ferri PH, et al. (2013) Influence of environmental factors on the composition of essential oils from leaves of *Myrcia tomentosa* (Aubl.) DC. BLACPMA 12: 572-580.

16. Rezende WP, Borges LL, Alves NM, Ferri PH, Paula JR (2013) Chemical variability in the essential oils from leaves of *Syzygium jambos*. Rev Bras Farmacogn 23: 433-440.

17. Hagerman AE, Butler LG (1978) Protein precipitation method for the quantitative determination of tannins. J Agr Food Chem 26: 809-812.

18. Waterman PG, Mole S (1987) A critical analysis of techniques for measuring tannins in ecological studies I: Techniques for chemically defining tannins. Oecologia 72: 137-147.

19. Waterman PG, Mole S (1987) A critical analysis of techniques for measuring tannins in ecological studies II: Techniques for biochemically defining tannins. Oecologia 72: 148-156.

20. Rolim A, Maciel CPM, Kaneko TM, Consiglieri VO, Salgado-Santos IMN, et al. (2005) Validation assay for total flavonoids, as rutin equivalents, from *Trichilia catigua* Adr. Juss (Meliaceae) and *Ptychopetalum olacoides* Bentham (Olacaceae) Commercial Extract. J AOAC Int 88: 1015-1019.

21. Brazil (2011) Instituto Nacional de Pesquisas Espaciais.

22. Silva SC (2009) Manual de análises químicas de solos, plantas e fertilizantes. Embrapa Informação Tecnológica.

23. Draper NR, Smith H (1981) Applied regression analysis. (2ndedn) New York: John Wiley.

24. Ward JH (1963) Hierarchical grouping to optimize an objective function. J Am Stat Assoc 58: 236-244.

25. Piaw CY (2005) Asas statistik penyelidikan. McGraw-Hill, Kuala Lumpur, Malaysia.

26. Yamada T (2004) Resistência de plantas às pragas e doenças: pode ser afetada pelo manejo da cultura? Inf Agronom 108: 1-7.

27. Bowerman BL, O'Connell RT, Koehler AB (2005) Forecasting, time series and regression. Brooks/Cole Thomson Learning Inc, USA.

28. Ghani IMM, Ahmad S (2010) Stepwise Multiple Regression Method to Forecast Fish Landing. Procedia Soc Behav Sci 8: 549-554.

29. Padda MS, Picha DH (2008) Effect of low temperature storage on phenolic composition and antioxidant activity of sweet potatoes. Postharvest Biol Tec 47: 176-180.

30. Albert A, Sareedenchai V, Heller W, Seidlitz HK, Zidorn C (2009) Temperature is the key to altitudinal variation of phenolics in *Arnica montana* L. cv. ARBO. Oecologia 160: 1-8.

31. Nguyen TBT, Ketsa S, Doorn WG (2003) Relationship between browning and the activities of polyphenol oxidase and phenylalanine ammonia lyase in banana peel during low temperature storage. Postharvest Biol Tec 30: 187-193.

32. Rivero RM, Ruiz JM, García PC, López-Lefebre LR, Sánchez E, et al. (2001) Resistance to cold and heat stress: accumulation of phenolic compounds in tomato and watermelon plants. Plant Sci 160: 315-321.

33. Nicholas KA, Matthews MA, Lobell DB, Willits NH, Field CB (2011) Effect of vineyard-scale climate variability on *Pinot noir* phenolic composition. Agric For Meteorol 151: 1556-1567.

34. Kovácik J, Klejdus B (2008) Dynamics of phenolic acids and lignin accumulation in metal-treated *Matricaria chamomilla* roots. Plant Cell Rep 27: 605-615.

35. Jin CW, You GY, He YF, Tang C, Wu P, et al. (2007) Iron deficiency-induced secretion of phenolics facilitates the reutilization of root apoplastic iron in red clover. Plant Physiol 144: 278-285.

10

Basic Concepts of using Solid Phase Synthesis to Build Small Organic Molecules using 2-Chlorotrityl Chloride Resin

Blake Bonkowski, Jason Wieczorek, Mimansa Patel, Chelsea Craig, Alison Gravelin and Tracey Boncher*

Department of Pharmaceutical Sciences, College of Pharmacy, Ferris State University, 220 Ferris Drive Big Rapids MI 49307, USA

Abstract

Solid Phase Synthesis (SPS) is a chemical strategy that was developed and refined by Bruce Merrifield in the early 1960s and later led to a Nobel Prize in 1984. This discovery paved the way for chemists to be able to construct proteins with high yields, less time-consuming purification and much faster synthetic routes. The strategy utilizes an insoluble solid polystyrene cross-linked support resin, 2-chlorotrityl-chloride (2-CCR), to form an ester linkage with an acid so proteins or small molecules can be built from the N terminal one amino acid at a time. This same chemistry can be employed to construct peptidomimetics and small non-peptide molecules. This chemistry is especially useful for building molecules that require temporary protection of a carboxylic acid during the synthetic route. This article will provide the basic concepts and considerations when practicing SPS for small non-peptide molecule construction. Significant considerations in employing this chemistry include: resin selection, swelling of resin, coupling agents, solvents, mechanism, loading of resin, nucleophilic substitution, and cleavage from resin support, amine protecting groups, general reaction techniques as well as purification of final product.

Keywords: 2-Chlorotrityl chloride resin; Solid phase synthesis; Small molecules and solid phase; Solid phase techniques

Introduction

Peptide synthesis before the work of Bruce Merrifield was performed in solution and required time-consuming purification after each intermediate step. Attaching an Fmoc protected amino acid to an insoluble support, 2-CCR, allows for the temporary protection of the carboxylic acid while the molecule is coupled with additional building blocks. After the reaction is completed, the resin must be washed with excess solvents, allowing un-reacted building blocks and reaction by-products to be rinsed away. Synthesizing small non-peptide molecules using SPS is an efficient synthetic method with relatively short reaction and completion times in comparison to solution phase reactions [1-3]. This review is focused on using solid phase chemistry to build small peptidomimetic organic molecules. The techniques used in small molecule construction are different from the techniques used when synthesizing proteins (Scheme 1).

Helpful Techniques and Recommendations

A vortex apparatus is required to achieve the desired vibrational agitation necessary for complete coupling [1]. It is vortexed from 6-24 hours to drive reactions to completion. 2-CCR is a commonly used resin in solid phase synthesis for carboxylic acids. This resin is extremely moisture sensitive so it is highly recommended to store in desiccators under nitrogen gas. It is imperative that the resin acclimates to room temperature before opening and weighing desired resin. If this is not done then it will affect the ability of the resin to swell properly when solvent is added. When resin is removed from storage at low temperature it may take hours to reach room temperature so proper planning is required.

Glass and disposable peptide reaction vessels are commercially available. Glass peptide vessels are preferred when building small molecules because some of the solvents are corrosive to most disposable peptide vessels. Selecting the proper frit porosity for the vessel is critical because it is imperative to choose one that is smaller than the size of the resin bead to avoid getting trapped in the frit. For example, a 200-400 mesh was used with a C or D frit size to minimize any 2-CCR obstructing the frit. If after several reactions solvents are having difficulty filtering through the frit it is a sign that the frit needs cleaning [4]. The recommended procedure for cleaning a frit is to wash with trifluoroacetic acid (TFA), dichloromethane (DCM) and then double distilled water before being heated to 260°C (500°F) for 4 hours.

After the reaction vessel is removed from the oven, allow it to cool to room temperature before a final flushing of contaminants.For small molecule construction it is important that the glassware apparatus is dry and that only anhydrous solvents are used for reactions to ensure increased yields.

Resin Selection

Most commonly used resins are insoluble polystyrene x-linked divinylbenzene (DVB) beads. Generally smaller beads have faster reactions because of increased surface area to volume ratios [1]. Certain resins have a linker attached to the bead that helps with reactivity and selectivity for specific functional groups. The degree of cross-linking affects the swelling of the respective bead. A 1% DVB cross-linked resin swells 2-4 times the original volume in DCM while 2% DVB cross-linked resin swells 4-6 times the original volume in DCM. An increase in loading capacity corresponds to an increase in swelling and smaller resin beads [1,5]. When resin is swollen, active linker sites hidden inside resin are exposed increasing the amount of reagent that can be added to the resin [1]. A 1% cross-linked resin is adequate for optimal swelling [6]. The number of cross linking is important with respect to the number of reaction sites available for loading. Based on the cross-linking, some beads swell more than others but the more DVB cross-linking the greater the reaction sites. For protein construction it is desired to have fewer cross-links because the protein is being synthesized inside the bead and steric hindrance may become an issue. For small molecule construction, it is less important because sterics are not likely to be a concern due to the overall size of the molecules [7].

The selection of a particular insoluble resin is important as each resin has specific stereochemistry and affinities for different attachable functional groups [4]. Specific resins are cleaved in a variety of ways but

*Corresponding author: Tracey Boncher, Associate Professor, Department of Pharmaceutical Sciences, College of Pharmacy, Ferris State University, 220 Ferris Drive Big Rapids MI 49307, USA, E-mail: TraceyBoncher@ferris.edu

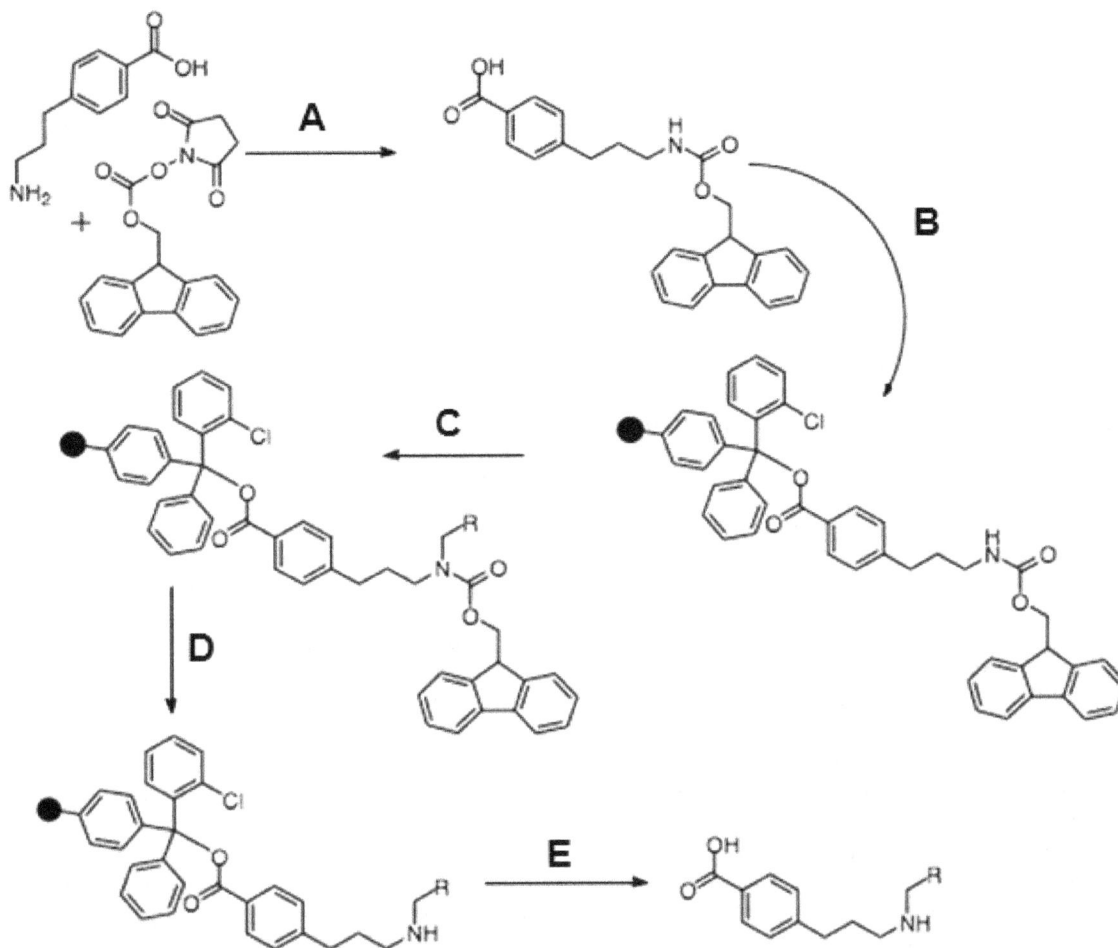

Scheme 1: Reagents and conditions: (A) 2eq NaHCO$_3$/H$_2$O, 0°C, 1.5eq Fmoc-OSu/5°C Dioxane, 24 hr and warm to R.T. after 1hr (B) 1eq 2-Chlorotrityl Chloride Resin, 1eq Fmoc Acid, DCM, 3 Eq DIPEA, 2eq TEA, 2-6hr, R.T. (C) 1.5eq TEA/DIPEA, DMF, NMP, R-Br (D) 20% Piperadine, DCM, R.T. 1Hr then repeat 15 min (E) 95% TFA, 2-3% TIPS, 2% H$_2$O, 1-2 Hr, R.T.

if the chemistry requires use of bases for coupling then an acid labile resin is desirable [8-10]. This way it is assured the resin continues to protect the carboxylic acid during the construction of the molecule without fear that the resin will be removed from the acid in the process. 2-CCR is useful when the desired molecule requires temporary protection of a carboxylic acid [11].

A consideration when selecting which mesh size of 2-CCR to use is the porosity of the frit in the reaction vessel. Mesh size is inversely related to the percent DVB cross-links which allows for the number of reaction sites. For the 2-CCR there is only a 100-200 or a 200-400 mesh selection commercially available. The 100-200-mesh resin will be larger in size but will also have fewer reaction sites available. This is recommended for peptide synthesis but will require more resin because there are fewer reaction sites. The 200-400 mesh is smaller, has greater reaction sites and is recommended for small molecule construction. A factor mesh size plays is in reaction time. For the 200-400 mesh the mmol per gram for loading is larger so this allows for greater loading and less resin will be required. A higher mesh size corresponds to a faster reaction time because of increased surface area [6]. Resins swell in solvents of low to medium polarity such as dimethylformamide (DMF) or DCM, but do not swell well in protic solvents such as alcohol or water [6]. If the resin does not swell properly the linker sites buried

in the resin will not be exposed decreasing the amount of carboxylic acid that can be loaded. To maximize the number of carboxylic acid substrate loaded to resin, the 2-CCR must be swollen 30 minutes in DCM prior to initiating coupling of the first amino acid. Prior to coupling the building block may require addition of DMF or N-Methyl-2-pyrrolidone(NMP) to help with solubility because it may be insoluble in DCM.

There are different techniques for loading an amino acid to the resin. The following is a method used for 2-CCR. First, dissolve and add one equivalent of Fmoc protected amino acid in DCM, (10 ml/g of resin), 3 equivalents of N,N-Diisopropylethylamine (DIPEA) and 2 equivalents of triethylamine (TEA) to the reaction vessel and let react for 5 minutes. After 5 minutes add an additional 1.5 equivalent of DIPEA. Next, add this mixture collectively to the pre-swollen resin and vortex for 1-2 hours [12]. After the amino acid is loaded to the resin the un-reacted sites must be end-capped with HPLC grade methanol (MeOH) (0.8mL per gram of resin) immediately to ensure that future reactions do not react at those unloaded sites. If methanol capping is not done the next amino acid will have the ability to add to a free reaction site on the resin and this will cause a mixture of final products. The process of methanol capping is fairly simple and involves a mixture of DCM, MeOH, and DIPEA in a 80:15:5 ratio (10 mL/g resin) added to the resin [1]. DCM is added because MeOH will not properly swell the resin and will not expose all of the linker sites [1,13]. After the MeOH is used to end-cap the resin (15 minutes), the resin must be washed to remove any excess MeOH and DIPEA before being re-swollen and coupled with the next building block.

Amine Protection

Amino acid building blocks that have both a carboxylic acid as well as a primary amine require the amine group be protected prior to loading to the resin. This is important because addition of asterically-hindered protecting group, such as Fmoc, to the nitrogen will decrease the likelihood of the amine attaching to the resin. The two most commonly used amine-protecting groups are 9-fluorenylmethyloxy carbonyl (Fmoc) or t-butyloxycarbonyl (Boc). The N-(9-fluorenylmethoxy carbonyloxy)succinimide (Fmoc-OSu) is preferred over the fluoroenylmethyloxycarbonyl chloride (Fmoc-Cl) as the yields increase due to the enhanced stability of a leaving group in Fmoc-OSu.

A common procedure for attaching an Fmoc to an amine is to first dissolve the amino acid in water before the addition of 2 equivalents of sodium bicarbonate and then stir in an ice bath at 0°C (32°F). In a separate beaker dissolve 1.5 equivalents of Fmoc-OSu or Fmoc-Cl in cooled p-dioxane before slowly adding to the deprotonated amino acid solution over one hour. The reaction should be vigorously stirred and warmed to room temperature overnight. If using Fmoc-OSu, the addition of some acetone may be needed, due to lower solubility of the Fmoc in p-dioxane [14]. To extract the Fmoc-amino acid, double distilled purified water is added to the beaker and then extracted twice with ethyl acetate (EtOAc). Treat the organic layer with a saturated sodium bicarbonate solution twice. Next, combine the aqueous layers and then acidify with concentrated HCl to a pH of 1-2 and then extract the organic layer with EtOAc three times. Combine all organic layers and dry over magnesium sulfate. Using chromatography the Fmoc-amino acid can then be purified with silica gel chromatography [5,15]. The Fmoc is base-sensitive so this group can easily be cleaved using 20% piperidine in DMF (10 mL/g resin) and gently vortex for 30 minutes [16]. This step is usually repeated twice with the second duration only lasting 15 minutes to ensure full Fmoc cleavage. After cleavage the resin is washed with DCM, DMF and NMP twice before a final wash with DCM and MeOH. Fmoc removal can be monitored via UV spectroscopy to ensure completion [5,15].

The Boc amine protection group is acid labile and there may be times where the Boc serves as a more appropriateamine-protecting group in comparison to Fmoc [17]. Specifically, when the addition of Boc is used to protect an amine attached to a base labile resin. First dissolve the primary amine in enough MeOH to make it soluble and then add 2.5 eq of triethylamine (TEA). The reaction should be allowed to stir for 10 minutes before heating to 55°C for 30 minutes. Slowly 1.6 eq of Di-tert-butyl dicarbonate (Boc₂O) was added and the reaction was allowed to go for 16 hours. Boc is acid labile and vulnerable to acid cleavage. A common procedure used to cleave the Boc off the amine while on a solid phase resin is to add a solution of 1:1 TFA in DCM (10 mL/1g resin) and vortex at room temperature for one hour [8,9]. The amine will form a TFA salt and must be washed with 3 sequential steps of 1:1 DCM and DIPEA (10 mL/1 g resin) to prepare the deprotected amine product for the next addition step. If using 2-chlorotrityl-chloride resin this will also cleave it form the resin as well as the amine. It is recommended to use a base labile resin if one needs to remove the Boc for further alkylation prior to deprotection from a resin.

Basic Techniques

Solvent washing

In SPS the resin is washed and filtered from reaction byproducts and is considered the purification step. Excess solvents are filtered through the resin in an ordered sequence of DCM (10 mL/g resin), DMF (5 mL/g resin), NMP (1 mL/g resin) and MeOH (1 mL/g resin) three times [18]. It may help to use a steel scoopula to agitate the mixture as the vacuum is applied to wash the resin. After the wash is completed the resin is washed one final time with DCM (10 mL per gram resin) and MeOH (1 mL/g resin) and dried under vacuum. MeOH helps to decrease the size of the resin bead and aids with evaporation to dry the resin prior to the next step. The resin will need to be re-swollen in DCM before any further coupling reactions proceed [1].

Nucleophilic Substitution

Since solid phase is not being used in this case to create peptides but rather peptidomimetic organic molecules, different methods are practiced when using this chemistry. Once the carboxylic acid molecule is attached to the resin and the amine is N-protected then S_N2 displacement can help assist with building the molecule. It is imperative that all glassware and solvents are dry and anhydrous to ensure greater yields. Bases such as DIPEA, TEA and sodium hydride (NaH) have been used withdeprotonation for various S_N2 displacement reactions. Due to the strength of sodium hydride base it is important to note that it has a potential to deprotect the Fmoc group. Sometimes this is done with the intention of generating a primary amine before reductive animation with an aldehyde while other times it does not cause any harm to the molecule because low concentrations of NaH are used.

Cleavage from the Resin

The cleavage cocktail used to remove the resin from the final compound may vary slightly but a common mixture uses 95% trifluoroacetic acid (TFA) with 2-3% triisopropylsilane (TIPS) and 2% water. One important consideration when selecting the cleavage cocktail is whether to add a scavenger, such as TIPS [17]. The reason scavengers are often added is because during the course of cleavage, highly reactive cationic species can be generated that can cause damage to the structure. The purpose of a scavenger is to quench any reactive species that may be generated during exposure from TFA cleavage. Other scavengers that can be used besides TIPS are thioanisole or ethanedithiol (EDT) [18].

Detectors

Ninhydrin is a useful chemical to detect primary or secondary amines. Primary amines are the most sensitive to ninhydrin and a strong blue color will immediately result if present [18]. The secondary amine is less sensitive to ninhydrin so a less intense brown-red color will be displayed indicatingpresence of a secondary amine [19]. To facilitate this test with solid phase, place a small sample of resin on TLC paper and then spray with ninhydrin before gently heating with a heat gun. The ninhydrin test is useful in solid phase synthesis to confirm successful addition of the amino acid to the N-terminal. However, bromocresol and malachite green are other chemical detectors that can be used to detect carboxylic acids. Both of these detectors will transition from green to yellow if a carboxylic acid is present [20].

Purification

Purification of final product should first be initiated by lyophilization prior to Reverse Phase Chromatography (RPC). Compounds that have greater differences in their R_f values will be easier to purify but C18 functionalized silica is required for optimal purification. RPC takes advantage of hydrophobic interactions between the stationary phase and the mobile phase for purification. Polar impurities elute faster than the hydrophobic ones. Therefore, reverse phase chromatography, combined with appropriate mobile phase polar to non-polar ratio, is most useful to desalt the compound and helppurify enantiomers. Hydrophobic molecules will elute from the column slower than hydrophilic ones so appropriate buffer solutions may need to be adjusted to extract those compounds from the functionalized silica column. A greater organic to polar solvent buffer solution may need to be used. The ability of RPC to tolerate both isocratic solvent systems and gradient solvent systems is important in purification of compounds synthesized via SPS due to presence of small amount of mostly polar impurities [21,22]. Typical solvents used in reverse phase purification processes are aqueous acetonitrile or methanol solutions.

Conclusion

Solid phase chemistry has advantages with protein construction, but it also is a valuable tool for the construction of small non-peptide molecules or peptidomimetics. This type of chemistry allows the chemist to accomplish their synthesis much more quickly compared to solution phase chemistry, which requires long purification methods after each intermediate step. Understanding the basic techniques discussed in this review should allowchemists to synthesize molecules with this method. Once the reaction is completed, reaction grade or ACS solvents can be used for washing the resin. The most important steps in the synthesis are the loading as well as cleaving from the resin. If poor yields are attained it is likely due to an error in either of those two steps so they should be investigated first. Numerous chemicals, apparatus and analytical techniques are used to confirm products between reaction steps. At any intermediate step throughout the synthetic procedure a small sample of resin may be cleaved from insoluble support so that NMR analysis can be conducted. Commonly used practices use chemical detectors, spectrophotometric analysis, IR, NMR and mass increases to help further clarify if reactions were successful. Once the molecule is removed from the resin and purified via HPLC the molecule will be fully analyzed by NMR to confirm the exact structure. Solid phase chemistry can be used as another tool to enhance small molecule construction.

Reference

1. Athanassopoulos P, Barlos K, Gatos D, Hatzi O, Tzavar C (1995) Application of 2-chlorotrityl chloride in convergent peptide synthesis. Tetrahedron Lett 36: 5645-5648.

2. Vaino AR, Janda KD (2000) Solid-phase organic synthesis: a critical understanding of the resin. J Comb Chem 2: 579-596.

3. Garcia-Martin F, Bayo-Puxan N, Cruz LJ, Bohling JC, Albericio F (2007) Chlorotrityl Chloride (CTC) Resin as a Reusable Carboxyl Protecting Group. QSAR Comb Sci 26: 1027-1035.

4. Myers AG, Gleason JL (1999) Asymmetric synthesis of α-amino acids by the alkylation of pseudoephedrine glycinamide. Prepatation of L-allylglycine and N-Boc L-allylgycine. Org Synth 76: 57-76.

5. Chandrudu S, Simerska P, Toth I (2013) Chemical methods for peptide and protein production. Molecules 18: 4373-4388.

6. Hoekstra WJ (2001) The 2-chlorotrityl resin: a worthy addition to the medicinal chemist's toolbox. Curr Med Chem 8: 715-719.

7. Greene TW, Wuts PGM (2007) Greene's protective groups in organic synthesis. John Wiley & Sons Inc., New Jersey, USA.

8. Kates S, Albericio F (2000) Solid-phase synthesis: A practicle guide. Taylor & Francis, USA.

9. Ermolat'ev DS, Babaev EV (2005) Solid-phase synthesis of N-(pyrimidin-2-yl) amino acid amides. ARKIVOC 4: 172-178.

10. Kim SJ, McAlpine SR (2013) Solid phase versus solution phase synthesis of heterocyclic macrocycles. Molecules 18: 1111-1121.

11. Toy PH, Lam Y (2012) Solid-phase organic synthesis: Concepts, strategies, and applications. John Wiley & Sons Inc., New Jersey, USA.

12. Mendonca AJ, Xiao XY (1999) Optimization of solid supports for combinatorial chemical synthesis. Med Res Rev 19: 451-462.

13. Snyder LR, Kirkland JJ, Dolan JW (2010) Introduction to modern liquid chromatography. John Wiley & Sons, New Jersey, USA.

14. Chiva C, Vilaseca M, Giralt E, Albericio F (1999) An HPLC-ESMS study on the solid-phase assembly of C-terminal proline peptides. J Pept Sci 5: 131-140.

15. Sigma-Aldrich (2003) Resins for Solid-phase peptide synthesis. Chemfiles 3: 1-30.

16. EMD Millipore corp. (2013) Novabiochem catalog 2012-2013. Darmstadt, Germany.

17. Lloyd-Williams P, Albericio F, Giralt E (1997) Chemical approaches to the synthesis of peptides and proteins. Taylor & Francis, USA.

18. Kaiser E, Colescott RL, Bossinger CD, Cook PI (1970) Color test for detection of free terminal amino groups in the solid-phase synthesis of peptides. Anal Biochem 34: 595-598.

19. Myers AG, Gleason JL, Yoon T, Kung DW (1997) Highly practical methodology for the synthesis of D- and L-α-amino acids, N-protected α-amino acids, and N-methyl-α-amino acids. J Am Chem Soc 119: 656-673.

20. Thompson LA, Ellman JA (1996) Synthesis and Applications of Small Molecule Libraries. Chem Rev 96: 555-600.

21. Merrifield RB (1963) Solid phase peptide synthesis. I. The sysnthesis of a tetrapeptide. J Am Chem Soc 85: 2149-2154.

22. Diamond D, Lau KT, Brady S, Cleary J (2008) Integration of analytical measurements and wireless communications--current issues and future strategies. Talanta 75: 606-612.

11

Treatment of Acidic Petroleum Crude Oil Utilizing Catalytic Neutralization Technique of Magnesium Oxide Catalyst

Norshahidatul Akmar Mohd Shohaimi, Wan Azelee Wan Abu Bakar*, Jafariah Jaafar and Nurasmat Mohd Shukri

Department of Chemistry, Universiti Teknologi Malaysia, Malaysia

Abstract

The presence of naphthenic acids in crude oils has caused a major corrosion problem to the production equipment, storage and transport facilities in the petroleum industry. The level of acidity of crude oil was determined by the value of Total Acid Number (TAN) in the oil samples. Two types of crude: Petronas Penapisan Melaka Heavy Crude and Petronas Penapisan Melaka Light Crude were studied. Various parameters studied were the amount of chemical dosing, type of catalyst, different catalyst calcination temperatures, and catalyst ratio of basic metal and dopant. The basic chemical used was ammonia solution in ethylene glycol (NH_3-EG) with a concentration range of 100-1000 mg/L. The best experimental condition for the possible TAN for the two samples is 1000 mg/L of NH_3-EG, and the catalyst reaction must be in the range of 35-40°C. Cu/Mg (10:90)/Al_2O_3 catalyst successfully reduced TAN in Heavy Crude for about 84.8% while for Light Crude, TAN was reduced 66.7% with the aids of Ni/Mg (10:90)/Al_2O_3 catalyst. Increase concentration of basic chemical, reduced the total acid number value of both crude oil.

Keywords: Naphthenic acid; Crude oil; Catalyst; Total acid number

Introduction

Naphthenic acid consist of saturated acyclic, monocyclic and polycyclic carboxylic acids with the general chemical formula of $C_nH_{2n+Z}O_2$; where n is the carbon number and Z represents the hydrogen atoms lost as the structures form rings [1]. Naphthenic acid problem had become a great concern to petroleum industry nowadays because of the corrosive properties of naphthenic acid tends to cause aggravated equipment corrosion, especially at high temperature (230-400°C), and this leads to high maintenance costs, may pose environmental disposal problems and low quality of the crude oils produced. There is no clear consensus on what constitutes a dangerous concentration of naphthenic acids, but corrosion will occur if the neutralization number is above 0.5 mg KOH /g of crude [2,3]. Therefore, any reduction in naphthenic acids content would alleviate corrosion related problems.

Many efforts had been done in order to remove or lowered down the naphthenic acid value in crude oil such as by catalytic decarboxylation [4] addition of caustic, blending, neutralization by caustic addition [5], blending petroleum crude oils with ionic liquid of ethanol [6], by adding solid acid catalyst and metal oxide catalyst [7]. But, these methods have their own drawbacks that created many problems and lead to high maintenance cost. Zhang et al. had study on the removal of naphthenic acid by ammonia solution in ethylene glycol, but that method required longer extraction time and the technique was a bit complicated [8]. Thus, the concern of this research is to formulate a simple catalytic neutralization technique utilizing a chemical base solution together with the aid of an alumina supported alkaline earth metal catalyst, in order to neutralized and remove the naphthenic acids to reduce the TAN lower than a value of one.

Problem statement

Naphthenic acid currently become the main issue in the refineries unit as it is one of the factors that contribute to the corrosion problem. High value of TAN lead to highly acidic crude oils that affected the quality of oils produced and contributed to the low performance of operations unit. This leads to high maintenance costs, and may create environmental disposal problems. So far, no research has been done to fabricate a specific chemical base solution and an effective catalyst to remove the naphthenic acids in crude oils. Hence, this study was executed to remove the naphthenic acid in crude oils using ammonia solution of ethylene glycol with the aid of alkaline earth metal catalyst.

Experimental

Preparation of catalysts

Magnesium based catalyst with dopant of Nickel and Copper will be studied. The ratio of based-dopant used were 10(dopant):90(based) and 30(dopant):70(based). Various calcination temperatures used were 400°C, 700°C and 1000°C. The metal precursors used in this research are nitrate salt. Generally, all the samples in this research were prepared by aqueous incipient wetness impregnation method [9]. Each of metal salts was weighed in a beaker according to the desired ratio and dissolved it in small amount of distilled water. The magnesium loading used was 90 wt%. Then, the solutions were mixed together and stirred continuously by magnetic bar for 30 minutes at room temperature to homogenize the mixture. In Mg based catalyst, Al_2O_3 is the most widely used support material. Thus, alumina beads with diameter of 4 mm to 5 mm were used as support material in this study. The support was immersed into the catalysts solution for 1 hour and transferred the supported catalysts onto evaporating dish with glass wool on it. It was then aging inside an oven at 80-90°C for 24 hours to remove water and allow good coating of the metal on the surface of the supported catalysts. It was then followed by calcination in the furnace at 400°C for 5 hours using a ramp rate of 5°C/min to eliminate all the metal precursor and excess of water or impurities. Similar procedure was repeated for the other ratio and calcination temperature of catalysts.

Catalysts characterization

Potential catalyst was characterized by several techniques to study its physical properties. The information obtained is highly useful in order to understand the relationship between the properties and its catalytic performance towards the neutralization activity. In this research, the

---

*Corresponding author: Wan Azelee Wan Abu Bakar, Professor, Department of Chemistry, Universiti Teknologi Malaysia, Malaysia
E-mail: wazelee@kimia.fs.utm.my

characterization techniques used were X-Ray Diffraction Spectroscopy (XRD), Scanning Electron Microscopy (SEM), and Energy Dispersive X-Ray (EDX).

Feedstock and basic chemical

The feedstock used in this study was light and heavy crude obtain from Petronas Penapisan Melaka. Ammonia solution in ethylene glycol was prepared to be used as the base chemical in the acid base neutralization process. The concentrations of ammonia solution in ethylene glycol use are 100 mg/L, 500 mg/L and 1000 mg/L.

Preparation of ammonia solution in ethylene glycol

400 µL of ammonia solution was added into 99.6 mL of ethylene glycol and stirred for about 15 min. The solution is then stir vigorously for 1 hour and stored in the dark place to avoid sunlight's penetration. The solution is then ready to be blend with the crude oil samples.

Total Acid Number (TAN) determination

The petroleum crude oil samples were titrated with potassium hydroxide solution (0.01 mol/L) with the addition of the developed base chemical to remove the organic acid present in crude oil. Prepared catalyst was also added in the solution to enhance the neutralization process. The indicator used was phenolphthalein solution where the stable red color was observed. The titration method was performed on two different types of crude oil samples which are PETRONAS Penapisan Melaka Light Crude and PETRONAS Penapisan Melaka Heavy Crude. In order to express the results, total acid number (TAN) of the sample was calculated in milligrams of potassium hydroxide per gram of sample (mg KOH/g) by using this equation:

$$TAN = 56.1 \times c \times \frac{(V_{KOH} - V_B)}{m}$$

Where,

c = concentration, in moles per liter, of standard volumetric potassium hydroxide solution.

V_{KOH} = volume, in milliliters, of titrant required for the determination.

V_B = volume, in milliliters, of titrant required for the blank test.

m = mass, in grams, of the test portion.

Results and Discussion

Characterization of the potential catalyst

The most potential catalyst undergone characterization process utilizing XRD and SEM-EDX analysis. The most effective catalyst that reduced the total acid number in the both crude oils are Cu/Mg (10:90)/ Al_2O_3 and Ni/Mg (10:90)/ Al_2O_3 which both calcined at 1000°C.

Characterization process using X-ray diffraction

Figures 1 and 2 show the XRD diffractograms of Cu/Mg (10:90)/ Al_2O_3 and Ni/Mg (10:90)/ Al_2O_3 catalysts calcined at 400, 700, and 1000°C for 5 hours. All diffractograms illustrates the catalysts Cu/Mg (10:90)/ Al_2O_3 and Ni/Mg (10:90)/ Al_2O_3 catalyst are highly amorphous, lack of periodicity and it have short range order. With increase of calcinations temperature also increase the intensity of the characteristic lines. From the diffraction patterns of both catalysts calcined at 400, 700 and those calcined at 1000°C, it can be suggested that the phase is dominated by Al_2O_3 cubic phase at 2θ values of 37.682°, 45.845° and 66.845° with d spacing values of 2.385, 1.978 and 1.398Å was observed as seen in figures 1 and 2 all of the peaks assignment of Al_2O_3 in the

diffractograms remained almost unchanged. Based on figure 1 Cubic CuO was revealed at 2θ values of 42.559° with d spacing values of 2.123 Å. Besides that, MgO cubic phase was obtained at 2θ values of 42.95° with d spacing value of 2.1064 Å. Rhombohedral phase Al_2O_3 was found at 2θ values of 57.497° with d spacing value of 1.602 Å. Figure 2 was only dominated by Al_2O_3 cubic phase at 2θ values of 37.682°, 45.845° and 66.845° with d spacing values of 2.385, 1.978 and 1.398 Å. The amorphous structure of figure 2 catalyst has made the MgO and NiO phase undetectable. The disappearance of MgO and NiO peak in XRD pattern was probably due to the small size MgO and NiO particles which was beyond the detection limit of XRD (detection limit of XRD = 4 nm) or incorporation of Mg and Ni species into the bulk matrixes (Figures 1 and 2).

Scanning Electron Microscope (SEM) Analysis

Figure 3 shows the SEM micrographs of Cu/Mg (10:90)/ Al_2O_3 catalysts, and figure 4 show SEM images for Ni/Mg (10:90)/ Al_2O_3 catalysts calcined at 400, 700 and 1000°C for 5 hours with magnification of 1000x. Both catalysts showed rough surface morphology with inhomogeneous spherical shape and comes with a mixture of small and large particles sizes. The micrograph shows that the particle is not well dispersed. The smaller particles size plays an important role to exhibit the higher catalytic activity.

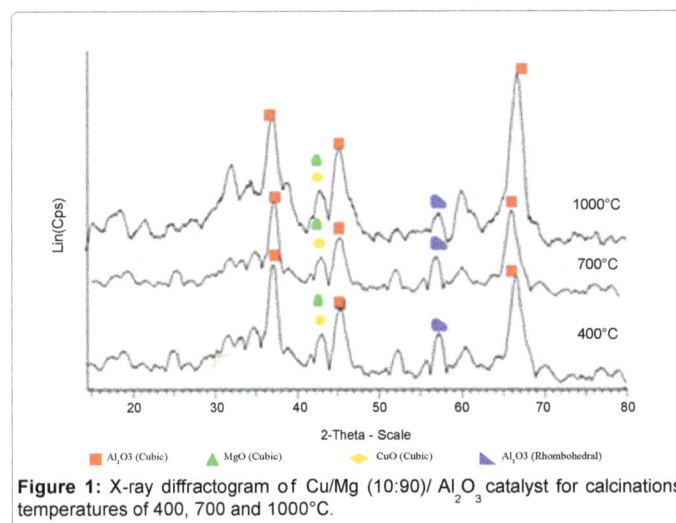

Figure 1: X-ray diffractogram of Cu/Mg (10:90)/ Al_2O_3 catalyst for calcinations temperatures of 400, 700 and 1000°C.

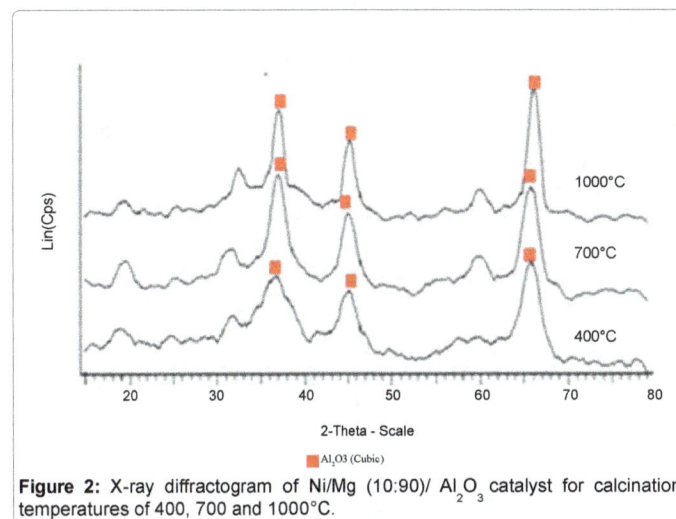

Figure 2: X-ray diffractogram of Ni/Mg (10:90)/ Al_2O_3 catalyst for calcination temperatures of 400, 700 and 1000°C.

This result is relevant to the results of XRD analysis which appeared with very broad peaks denoting an amorphous state observed in the diffractogram calcined at 400°C caused by the very small of particle size. The smaller particle size of the catalyst will lead to a higher dispersion of the catalyst and large surface area of supported magnesium oxide based catalyst. It is little difficult to find out the particle size and morphology of the sample from this image (Figures 3 and 4).

Energy Dispersive X-Ray analysis (EDX)

EDX analysis provides useful information on the percentage of the elements present in the sample, as well as the percentage of catalyst over the support in order to see how much of the catalyst that has successfully coated the support. Table 1 illustrates that at calcination temperature of 1000°C, the weight composition of Mg in the Cu/Mg (10:90)/ Al_2O_3 catalyst is 9.32, the highest compared to the catalysts calcined at temperature 400 and 700°C.

Meanwhile, the table 2 shows that the amount of Mg in the Ni/Mg (10:90)/ Al_2O_3 catalyst at calcination temperature of 1000°C was 10.28 higher than other in the different temperature. The elemental analysis performed by EDX confirmed the presence of Mg, Cu, Ni, Al, and O in the potential catalyst. Basically, the amount of support material have greater amount than metal as the support material play a role as to control, homogenize and to stabilize the metal oxide phases in the catalyst. From the elementally analysis, the results obtained indicated that element are in homogeneously distributed on the surface of the supported catalyst (Tables 1 and 2).

Formulated chemical base solution consists of ammonia solution of ethylene glycol

The ammonia solution of ethylene glycol with various concentrations

Calcination Temperature (°C)	Element	Weight Composition (%)
400	Al	57.87
	Mg	4.33
	Cu	1.28
	O	36.52
700	Al	50.32
	Mg	7.20
	Cu	6.46
	O	36.02
1000	Al	49.88
	Mg	9.32
	Cu	4.88
	O	35.92

Table 1: Elemental composition from EDX analysis for Cu/Mg (10:90)/ Al_2O_3 catalysts calcined at 400, 700, and 1000°C.

Calcination Temperature (°C)	Element	Weight Composition (%)
400	Al	56.53
	Mg	4.12
	Ni	1.33
	O	38.02
700	Al	53.77
	Mg	6.88
	Ni	4.67
	O	34.68
1000	Al	50.73
	Mg	10.28
	Ni	4.92
	O	34.07

Table 2: Elemental composition from EDX analysis for Ni/Mg(10:90)/ Al_2O_3 catalysts calcined at 400, 700, and 1000°C.

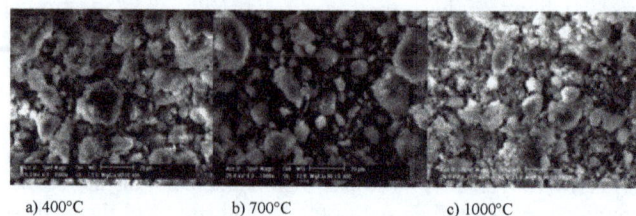

Figure 3: SEM micrographs of Cu/Mg(10:90)/ Al_2O_3 catalysts, calcined at 400,700 and 1000°C for 5 hours with magnification of 1000x.

a) 400°C   b) 700°C   c) 1000°C

a) 400°C   b) 700°C   c) 1000°C

Figure 4: SEM images for Ni/Mg(10:90)/ Al_2O_3 catalysts, calcined at 400,700 and 1000°C for 5 hours with magnification of 1000x.

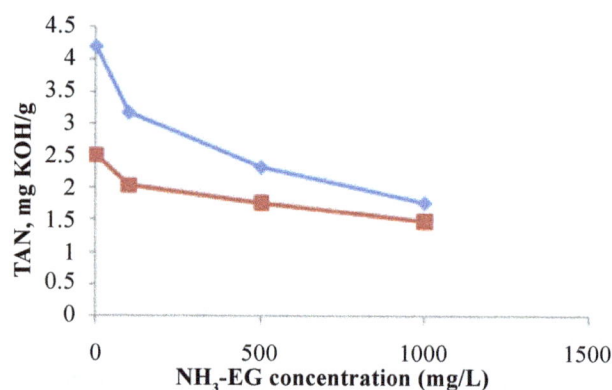

Figure 5: Effect of NH_3-EG concentration to the TAN reduction for (◆) CrudeA and (■) Crude B.

of 100, 500 and 1000 mg/L were synthesized. Figure 5, shows that only small amount of naphthenic acids in crude oil was reduced when an ammonia solution of ethylene glycol was added at concentration of 100, 500 and 1000 mg/L.

Crude A is a heavy crude oil, the total acid number (TAN) in the crude oil was 4.21. But, after treatment with 100 mg/L of ammonia solution, the number was reduced to 3.18. The number was further decreased to 2.33 and 1.78 after added with 500 and 1000 mg/L of ammonia solution respectively.

Initially, without addition of ammonia solution of ethylene glycol, the total acid number (TAN) in the crude B which is light crude oil was 2.52. But, with addition of 100 mg/L of ammonia solution, the TAN was lowered to 2.05. TAN was decrease as the amount of ammonia solution is increase; this has been shown when the number of acid in crude B was decreased to 1.77 and 1.49 for 500 and 1000 mg/L of ammonia solution.

It can be concluded that only 57.72 percent of acid number in the crude A and 40.87 percent of crude B were reduced when treated

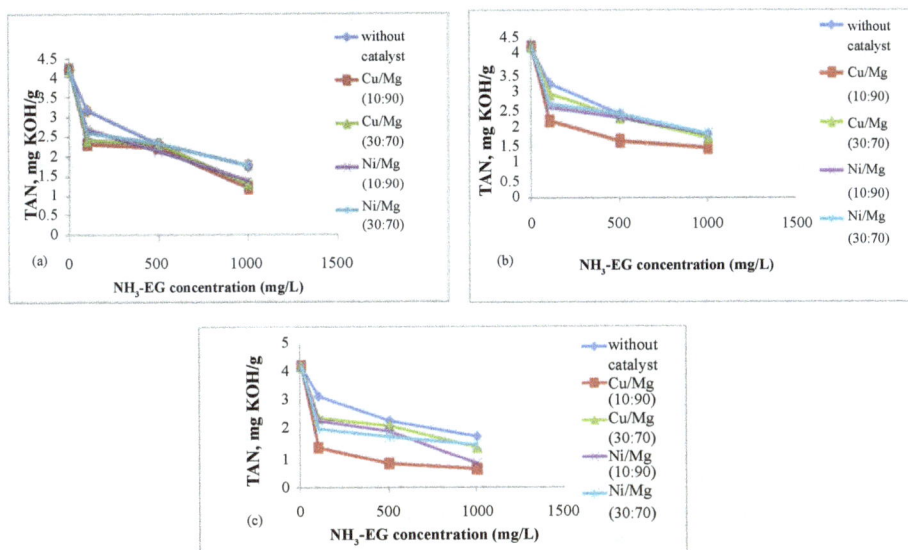

Figure 6: Effect of catalysts on the reduction of TAN for crude oil A with different calcination temperatures of (a) 400°C, (b) 700°C and (c) 1000°C.

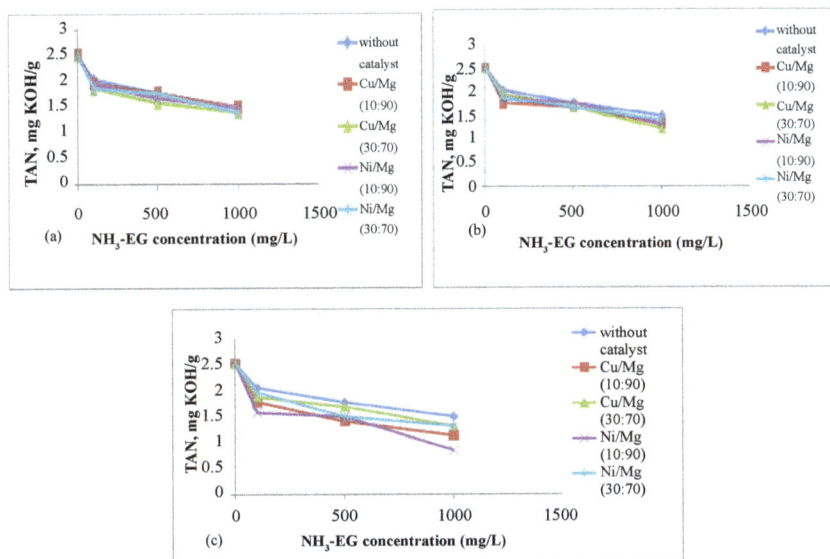

Figure 7: Effect of catalysts on the reduction of TAN for crude oil B with different calcination temperatures of (a) 400, (b) 700 and (c) 1000°C.

with ammonia solution of ethylene glycol due to insufficient amount concentration added. Thus, in order to completely reduce the number of acid in both crude oil, the concentration of the ammonia solution must be greater than 1100 mg/L or by addition of catalysts.

Catalytic activity of alkaline earth metal oxide catalyst

The metal oxide catalysts had been used in this research study was a heterogeneous catalyst. The prepared catalyst with respect to each ratio and with different calcination temperature were brought to conduct the test involve of removal of naphthenic acid in the both crude oils utilizing ASTM D664 standard titration method. Figures 6a-6c illustrates the total acid number in Crude A (heavy crude) by using ammonia solution of ethylene glycol as a base chemical with the addition of a catalyst. As shown in this figure, the aim of the addition

of the catalyst is to lower the total acid number through the catalytic decarboxylation reaction compared to the neutralization process without the addition of a catalyst.

From the plotted graphs, it's obviously shown that the most effective catalyst in the lowered down the acid number is Cu/Mg (10:90) /Al$_2$O$_3$ calcined at a temperature of 1000°C since the number of total acid number is lowered than one which is 0.64. Followed by Ni/Mg (10:90)/ Al$_2$O$_3$, 1000°C and the least effective catalyst was Ni/Mg (30:70)/ Al$_2$O$_3$, at 700°C. This proves that the higher the temperature, the activity of the catalyst also higher and the more favored the catalyst to convert the carboxyl compound to the carbon dioxide. Thus, enhance the catalytic decarboxylation process to occur (Figure 6).

Based on the figures 7a-7c, the highly potential catalyst that can reduce the total acid number in Crude oil B to a TAN less than one is Ni/Mg (10:90) /Al$_2$O$_3$, which calcined at 1000°C. This catalyst reduced the TAN of Crude oil B from 2.52 to 0.84. Next, is Cu/Mg (30:70)/ Al$_2$O$_3$, calcined at 700°C which decrease the 51.98 percent of the number of acid in the crude oil B. Thus, to make this possible for this catalyst to reduce TAN, the dose of ammonia solution of ethylene glycol must be higher. The ineffective catalyst for crude oils B is Cu/Mg (10:90)/ Al$_2$O$_3$, due to the catalyst only activate at the high temperature of calcined and this catalyst still have impurities that cannot be remove at low temperature of calcination.

As per conclusion, the most preferable catalyst for lowered TAN in the crude oils A is Cu/Mg (10:90)/ Al$_2$O$_3$. Meanwhile, for crude B is Ni/Mg (10:90)/ Al$_2$O$_3$. These two catalysts were characterized to know its properties, which will be discussed further (Figure 7).

Summary and Conclusion

The research study has proven that the formulated chemical base solution consisting of ammonia solution of ethylene glycol and addition of alkaline earth metal oxides as catalyst had been successfully applied to neutralize and remove the acidic compound in the crude oil, thus reducing the total acid number (TAN) below than one. The TAN value decreased as the concentration of ammonia solution of ethylene glycol increased with the aid of a catalyst. In this study, it has been found that semi micro color titration method is one of the effective methods to determine the total acid number (TAN) in crude oil.

Catalytic decarboxylation was proven as an effective technique in the removal of naphthenic acids in the crude oil samples. In this study, the effective catalysts for both crude oil have been identified. For crude A, the most effective catalyst is Cu/Mg (10:90) Al$_2$O$_3$ and for crude B, is Ni/Mg (10:90) Al$_2$O$_3$. The calcination temperature for both catalysts was at 1000°C.

The characterization of the potential catalyst has been achieved using XRD and SEM-EDX. From XRD diffractograms, it showed that both catalysts were highly amorphous while SEM images showed surface morphology with inhomogeneous spherical shape and appeared with a mixture of small and large particles sizes.

Acknowledgements

The authors would like to thank the Malaysian Ministry of Higher Education, Universiti Teknologi Malaysia for their financial funding through FRGS grant 78468 and research university grant 01H58.

References

1. Vaz de Campos MC, Oliveira EC, Filho PJ, Piatnicki CM, Caramao EB (2006) Analysis of tert-butyldimethylsilyl derivatives in heavy gas oil from Brazilian naphthenic acids by gas chromatography coupled to mass spectrometry with electron impact ionization. J Chromatogr A 1105: 95-105.

2. Huang M, Zhao S, Li P, Huisingh D (2006) Removal of naphthenic acid by microwave. Journal of Cleaner Production 14: 736-739.

3. Wang Y, Chu Z, Qiu B, Liu C, Zhang Y (2006) Removal of naphthenic acids from a vacuum fraction oil with an ammonia solution of ethylene glycol. Fuel 85: 2489-2493.

4. Zhang A, Ma Q, Wang K, Liu X, Shuler P, et al. (2006) Naphthenic acid removal from crude oil through catalytic decarboxylation on magnesium oxide. Applied Catalysis A: General 303: 103-109.

5. Ding L, Rahimi P, Hawkins R, Bhatt S, Shi Y (2009) Naphthenic acid removal from heavy oils on alkaline earth-metal oxides and ZnO catalysts. Applied Catalysis A: General 371: 121-130.

6. Shi LJ, Shen BX, Wang GQ (2008) Removal of naphthenic acids from Beijiang crude oil by forming ionic liquids. Energy Fuels 22: 4177-4181.

7. Fu X, Dai Z, Tian S, Long J, Hou S, et al. (2008) Catalytic decarboxylation of petroleum acids from high acid crude oils over solid acid catalysts. Energy Fuels 22: 1923-1929.

8. Wang Y, Chu Z, Qiu B, Liu C, Zhang Y (2006) Removal of naphthenic acids from vacuum fraction oil with an ammonia solution of ethylene glycol. Fuel 85: 2489-2493.

9. Wan Abu Bakar WA, Ali R, Toemen S (2011) Catalytic methanation reaction over supported nickel-rhodium oxide for purification of simulated natural gas. Journal of Natural Gas Chemistry 20: 585-594.

12

Spectral, Electrochemical and Molecular Orbital Studies on a New Solvatochromic Binuclear Mixed Ligand Copper(II) Complexes

Taha A and Ahmed HM*

Faculty of Education, Ain Shams University, Roxy, Cairo, Egypt

Abstract

A new series of solvatochromic binuclear mixed ligand complexes with the general formula: Cu2(DMCHD)(Am)2X3 (where, DMCHD: 5.5-Dimethyl cyclohexanate 1,3-dione, Am=N,N,N'-trimethylethylenediamine (Me3en), N,N,N',N'-tetramethylethylenediamine (Me4en), or N,N,N',N',N''-penta-methyldiethylenetriamine (Me5dien) and X=ClO4- or Cl- have been synthesized and characterized by the analytical, spectral methods, magnetic and molar conductance as well as electrochemical measurements. The formation constant values for HDMCHD ligand with various metal ions are much lower than expected for similar β-diketones revealing a monobasic unidentate nature of this ligand. The d-d absorption bands of the prepared complexes in weak donor solvents suggest square-planar, distorted octahedral and/or distorted trigonal bipyramid geometries for the perchlorate and chloride of diamine in addition to triamine complexes, respectively. However, an octahedral structure was identified for the complexes in strong donor solvents. Perchlorate of diamine complexes show a remarkable color change from violet to green as the Lewis basicity of the donor solvent or anions increases, whereas chloride complex is mainly affected by the Lewis acidity of the acceptor solvent. Specific and non-specific interactions of solvent molecules with the complexes have been investigated using the unified solvation model. Band's oscillator strength of the d-d transition has been calculated and discussed. Cyclic voltammetric measurements on the prepared complexes in different solvents showed a quasi-reversible or irreversible and mainly diffusion controlled reduction process. Such behavior has been explained according to the EECE mechanism. A linear correlation has been found between the Cu(II) reduction potential and the spectral data. Structural parameters of the free ligands and their Cu(II)- complexes have been calculated on the basis of semiemperical PM3 level and correlated with the experimental data.

Keywords: Formation constant; pH-metric; Solvatochromic; Red shift; Oscillator strength; Cyclic voltammetry; Semiemperical; Molecular modeling; Biological activity

Introduction

Solvatochromism is widely used in many fields of chemical and biological research to study bulk and local polarity in macro-systems (membranes, etc.), even though its wide use, solvatochromism (a color change as a result of solvent effect) still remains a largely unknown phenomenon due to the enormously complex coupling of many different interactions and dynamical processes which describe it. Metal- chelates with O_2N_2 as Schiff-base or mixed ligands have been studied as solvatochromic indicators [1-5]. In view of the limited information available regarding the spectroscopic properties of mixed ligand metal complexes containing 5.5-dimethyl cyclohexane1,3-dione (HDMCHD) the present study which is a continuation of our studies on chromotropic of the metal chelates with mixed ligands containing O-O and N-N in various solvents [6-10]. In the preceding communication detailed structural, molecular orbital and optical characterizations of solvatochromic mixed ligand copper(II) complex of 5,5-dimethyl cyclohexanate 1,3-dione and N,N,N',N',N''-penta-methyldiethylenetriamine, Cu(DMCHD)(Me5dien)NO3, the structure elucidated as mononuclear and HDMCHD ligand acted as mono-basic monodentate [11].

Thus, the purpose of the current work was to achieve a microscopic understanding of the intermolecular effects which govern the absorption spectral properties of solvated complex probes, such as solubility, Lewis-acid base, and specific and non-specific interactions. Thus, the present investigation includes: (a) effect of solvent anion and secondary ligand (polyamine) on energetic shifts of d-d bands and change the color of solution, (b) solvation effect on the structure and stability of the resulting complexes. For this intention a new series of copper(II) complexes abbreviated as Cu2(DMCHD)(Am)2X3 (where DMCHD=5.5-dimethyl cyclohexanate1.3-dione, Am=N,N,N'-trimethylethylenediamine (Me3en), N,N,N',N'-tetramethylethylenediamine (Me4en), or

N,N,N',N'N''-pentamethyldiethylenetriamine (Me5dien) and X=ClO4- or Cl- have been synthesized and characterized by spectral, conductivity and magnetic as well as electrochemical techniques. The main goal of this study is to examine the applicability of the current complexes as a Lewis acid-base color indicator. As further support to our experimental work, molecular orbital calculations have been accomplished using Hyperchem 7.52 for the ligands and their complexes, the structural parameters data are correlated with the experimental results.

Experimental

Materials

All chemicals used were of the analytical reagent grade and obtained from either Merck or Aldrich and were used without further purification. Solvents used, nitromethane (CH_3NO_2), acetonitrile (MeCN), acetone (Me_2CO), methanol (MeOH), pyridine (py), N,N-dimethylformamide (DMF) and dimethylsulfoxide (DMSO), propylenedicarbonate (PDC), formamide (Fa) were "Spectro-grade" and further purified using standard methods [12].

Physical measurements

The infrared spectra in KBr (400-4000 cm^{-1}) were recorded using Shimadzu FTIR 8101 spectrometer. Electronic spectra were obtained on UV-2101 pc w/full spectrophotometer using 10 mm quartz cells

*Corresponding author: Ahmed HM, Faculty of Education, Ain Shams University, Roxy, Cairo, Egypt, E-mail: hany_magdy2013@yahoo.com

thermostated at 25°C. Magnetic moments were obtained using a MSB-AUTO magnetic susceptibility balance by the Gouy method. The molar conductance was measured with Metrohm 660 conductor in DMF solutions at 25°C. The potentiometric titrations were carried out by means of a digital 520 WTW pH-meter with a conventional pH-electrode. Titrations and calculations of the proton-dissociation and stepwise formation constants were performed as described earlier [12]. Cyclic voltammetric measurements were performed as reported elsewhere [8,13].

Syntheses of $Cu_n(DMCHD)(Am)_nX_m$ complexes

These complexes were prepared by adding a mixture of 10 mmol of the HDMCHD ligand in 25 mL absolute EtOH and solid anhydrous Na_2CO_3 (20 mmol) to an ethanolic solution of 20 mmol appropriate copper(II) salt. The mixture was continuously stirred for about 30 minutes resulting in a green solution, which was then filtered. Then a solution of 20 mmol diamine or triamine derivative (Am) in 20 mL EtOH was added dropwise to the filtrate with a continuous stirring for an additional 45 min. After that, the resulting solution was filtered off and left to stand overnight. The complexes obtained were re-crystallized from dichloromethane and stored over silica gel.

Results and Discussion

The analytical and physical data of the mixed ligands solid complexes with Cu(II) ion are given in Table 1. It is well known that most of β-diketones are mixtures of keto-enol forms. Their keto-enol tautomerism is responsible for the formation of inner complexes with metal ions; so they behave in general as monobasic bidentate nature [14]. This statement cannot apply to characterize the current mixed ligands complexes of copper(II) with 5,5-dimethylcyclohexane-1,3dione (HDMCHD) and polyamine ligands (Am). This motivated us to study the complexation equilibrium of HDMCHD ligand with various metal ions in solution to focus the light on the dentate nature of HDMCHD ligand. Thus, poteniometric titration studies were carried out for HDMCHD ligand with various metal ions in 50% (v/v) aqueous solution.

Potentiometric studies

Potentiometric titration curves of free HDMCHD ligand (2.5×10^{-3} M) in the absence and presence of different metal ions (5×10^{-4} M) in a ratio 5:1(L:M) revealed: (i) divalent metal ions showed a distinct inflection point at m=2, (m=number of moles of base added per mole of metal ion), indicating the formation of ML_2, (ii) however an inflection at m=3 for trivalent metal ions corresponding to the formation of ML_3 or ML_2OH complexes.

The titration curve of the free ligand against KOH solution shows one ionizable proton as expected with the value of pK^H=6.29 and 6.16, in 50% (v/v) dioxane-water and isopropanol-water, respectively which agrees with the value obtained in pure water (pK^H=5.22) [15]. The trend of deprotonation constant is in the opposite of increasing dielectric constant of the medium. The lower basicity of HDMCHD compared to acetylacetone arises from the existence of 95.3% of HDMCHD in enol form [15,16].

Formation constants values given in Table 2, evident that $logK_1$ and K_2 values are too low compared to the basicity of HDMCHD ligand [17]. Moreover allied complexes derived from 2-acylcyclohexanone have sufficiently large formation constant values supporting its bidentate nature. These facts in addition to the small ratio of the two successive formation constants would lead to suggest monodentate nature of HDMCHD. Furthermore, the half-chair boat [18] conformation of the enol form of HDMCHD clearly depicts that due to imposition of the planar configuration of the cyclohexane ring bearing the groupings -C(OH)=C-C=O the proximity of the C-OH and C=O groups (the positions of which are non-flexible) are so wide that intramolecular hydrogen bonding is precluded in HDMCHD. Thus chelation is not possible, rather complex formation would take place typically by the replacement of the hydrogen ion of the enolic-OH group [15]. Thus, HDMCHD ligand behaves as monobasic mono-dentate or binuclear bidentate ligand.

The values in Table 2 reveal that, the formation constants increase in the following order: Cd(II)<Mn(II)<Co(II) ~ Ni(II)<Cu(II)<<Fe(III). This order except Fe(III) (owing to the higher oxidation state) is largely reflects the change in the heat of complex formation across the series, which arises from a combination of the influence of the polarizing ability of metal ion and crystal field stabilization energies [12]. The formation constant values of some lanthanides (Ln^{3+}) -DMCHD complexes increase in the order of increasing 1/r (r=ionic radii of Ln(III) ions): Pr(III)<Sm(III)<Yb(III).

Infrared spectra

The main characteristic IR absorption frequencies of the free ligand (HDMCHD) and its Cu(II)-mixed ligand complexes are given in Table 3. The observed bands may classified into those originating from the ligands, those emanating from the counter-balancing anions and those arise from the bonds formed between copper(II) and coordinating sites. The bands are assigned in comparison with similar Cu(II)-complexes [7-11,19]. The broad band observed in the range 3225-3520 cm^{-1} for complex (1) assigned to v_{NH} of the unsymmetrical alkylated diamine Me_3en- ligand. The blue shift of the observed IR absorption bands at

No.	Complex/ Emperical Formula	Formula weight	Color	Anal, Found (Calc)				M.P	Λ/S.cm^2 Mol^{-1}	µeff. (B.M)
				M%	C%	H%	N%			
1	$Cu_2(DMCHD)(Me_3en)_2(ClO_4)_3.2C_2H_5OH$ $C_{22}H_{51}N_4O_{16}Cl_3Cu_2$	861	*Bluish violet*	15.25 (14.75)	31.42 (30.69)	5.98 (5.97)	6.46 (6.51)	196	199	1.65
2	$Cu_2(DMCHD)(Me_4en)_2(ClO_4)_3 2C_2H_5OH$ $C_{24}H_{53}N_4O_{16}Cl_3Cu_2$	889	Reddish-violet	17.80 (14.30)	32.5 (32.39)	5.90 (5.96)	6.21 (6.29)	154	245	1.74
3	$Cu_2(DMCHD)(Me_4en)_2Cl_3.2C_2H_5OH$ $C_{24}H_{55}N_4O_4Cl_3Cu_2$	697	Green	19.02 (18.22)	41.70 (41.35)	7.68 (7.95)	7.43 (8.04)	110	90	2.67
4	$Cu_2(DMCHD)(Me_5dien)_2(ClO_4)_3$ $2C_2H_5OH$ $C_{30}H_{69}N_6O_{16}Cl_3Cu_2$	1003	Reddish-violet	12.01 11.61	37.25 (37.22)	8.19 (7.66)	7.78 (7.66)	157	155	1.62

Table 1: Physical and analytical data of $Cu_2(DMCHD)(am)_2X_{m3}$ complexes.

| Cation | HDMCHD | | | | | | Hacac | | |
| | Dioxane-water | | | Isopropanol-water | | | | | |
	LogK1	LogK2	logβ	LogK1	LogK2	logβ	LogK1	LogK2	logβ
H+	6.29	-	-	6.16	-	-	11.78	-	7.27
Mn+2	3.45	3.11	6.56	-	-	-	4.18	3.07	-
Co+2	3.64	3.22	6.86	3.52	3.13	6.65	5.40	4.40	-
Ni+2	3.62	3.21	6.83	3.62	3.21	6.83	6.69	6.73	-
Cu+2	3.62	3.33	6.95	3.49	3.15	6.64	8.22	8.23	-
Cd+2	3.43	3.07	6.50	3.51	3.11	6.62	3.82	2.76	-
Fe+3	6.12	5.87	11.99	5.62	5.41	11.03	-	-	-
Pr+3	3.58	3.20	6.78	3.39	3.07	6.46	-	-	2.64
Sm+3	3.69	3.28	6.97	3.48	3.13	6.61	5.59	4.46	2.86
Yb+3	3.74	3.34	7.08	3.65	3.26	6.91	5.18	4.86	-

Table 2: Stability constants of metal ccomplexes of 5,5-dimethyl cyclohexane 1,3-dione (HDMCHD) 25°C (□=0.10 in 50 (v/v)% solvent –water)

1613 and1582 cm^{-1} in the spectrum of free HDMCHD ligand which assigned to v(C=O) to 1624-1674 and 1598-1618 cm^{-1}, respectively suggest the involvement of the carbonyl group in the coordination sphere of Cu(II) [19].

The coordination modes of perchlorate anion have been implied from the IR data. Two intense stretching vibrational bands at 1140 and 625 cm^{-1} are observed for the perchlorate complexes 1, 2 and 4. the weak splitting observed for the strong broad band of $v_{(ClO4^-)}$ at 1140 cm^{-1} indicates a mixture modes for perchlorate anions, properly one being partially contributing to the coordination sphere of Cu(II) as bidentate, and the other being free [7-11,19].

Conductance measurements

The molar conductance values, in DMF (10^{-3} mol/L), of all complexes are given in Table 1. These values indicate that all complexes in DMF showed an electrolyte behavior, which might arises to the Lewis basicity of DMF that drive out the coordinating anions, especially when the Lewis basicity of anion (such as ClO_4) is much weaker than DMF solvent [20]. Complexes 1, 2 and 4 have molar conductance values 199, 245 and 155 ohm^{-1}cm^2 mol^{-1}, respectively indicating 1:3 electrolyte [21]. Whereas, complex 3 has molar conductivity value 90 ohm^{-1} cm^2 mol^{-1}, suggesting the ionic nature of complex 3 (1:1 electrolytes).

Electronic spectra

The prepared complexes are freely soluble in most organic solvents and their UV/Vis spectra were measured at room temperature in a variety of solvents, selected to give a wide spread of donor and acceptor strengths as possible; the data obtained are collected in Table 4. The positions of the d-d absorption band of the $Cu_2(DMCHD)(Am)_2X_3$, where X=perchlorate or chloride anions in weak donor solvent (such as CH_3NO_2) are observed in the range 19.05×10^3 and 14.47-13.97×10^3 cm^{-1}. These values could be assigned to the following d-d transitions: dxy dx^2-y^2 ($^2B_{1g} \rightarrow {}^2B_{2g}$), dx^2-y$^2 \rightarrow$ dz^2 ($^2E_g \rightarrow {}^2T_{2g}$), and/or dz$^2 \rightarrow$ dx^2-y^2 [22,23]. These transitions suggest a square planar for diamine perchlorate complexes 1 and 2; distorted octahedral for triamine perchlorate complex 4 and/or square based pyramidal distorted trigonal bipyramide (SBRDTBP) geometries for chloride complex 3 [24,25]. This could be attributed to the polyamine, which varies between diamine for complexes 1-3 and triamine in complex 4, and coordination abilities of anions towards Cu(II). This interpretation further confirmed by the strong effect of anions on the d-d bands of perchlorate complexes 1 and 2 in weak donor solvents such as MeNO$_2$ (Table 5). The red shift of v_{max} versus the donor strength of these anions ($DN_{X, MeNO_2}$) [20] yield the following linear correlations: v_{max}/cm^{-1}=20340-168.44 $DN_{X, MeNO_2}$ r=0.99, and v_{max}/cm^{-1}=18640-

125.28, DN_{X, CH_3NO_2} r=0.95, for $Cu_2(Me_3en)_2(DMCHD)(ClO_4)_3$ and $Cu_2(Me_4en)_2(DMCHD)(ClO_4)_3$, respectively.

The position of the d-d transition band of diamine perchlorate complexes 1 and 2 exhibits a red shift forming a wide range of color solutions (violet → blue → green) as the Lewis basicity of solvent increases revealing a positive solvatochromic [9]. Figure 1 shows the visible absorption spectra of the $Cu_2(DMCHD)(Me_4en)_2(ClO_4)_3$ complex 2, in different organic solvents. There is only one broad band observed in the visible region that assigned to the promotion of an electron from dz^2 orbital of the Cu(II) ion to the hole in dx^2-y^2 orbital of the Cu(II) ion (d^9) forming different distorted tetragonal depending on the Lewis basicity (DN) of the donor solvent or anion [23]. This red shift could be attributed to the strong repulsion of the electrons in dz^2 orbital by the lone pair of the incoming Lewis base such as solvent molecules or anion species, those are axially coordinated to the central copper (II) ion, consequently, a less energy will be required to transfer the electron to dx^2-y^2. In view of the fact that all the d$_{xy}$, d$_{yz}$ and d$_{xz}$ orbitals of the Cu(II) ion are raised up by its interaction with the lone pair of electron on the solvent molecules approaching from above and below of the molecular plane, the broad d-d transition band of complexes moves gradually to the red with the increase of the Lewis basicity (DN) of the solvent. This originates in variation of Lewis acid-base interaction between the complex ion and the respective solvent molecules or anion.

Some remarks on the solute-solvent interaction resolved from the data in Table 4 are of immediate notes with discussing., (a) the $λ_{max}$

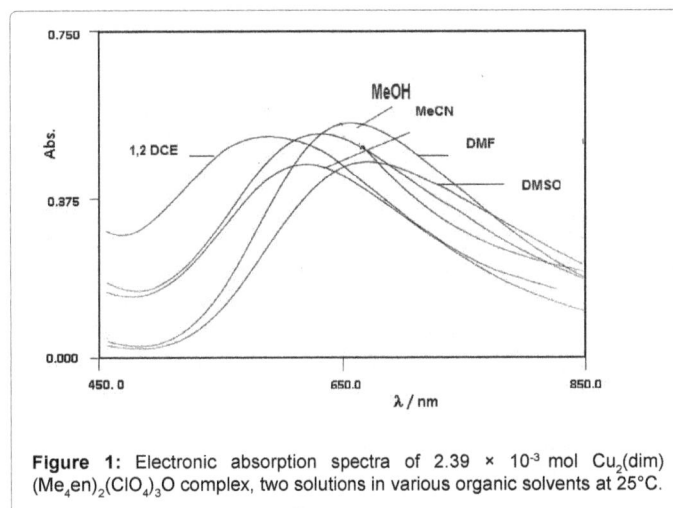

Figure 1: Electronic absorption spectra of 2.39 × 10^{-3} mol Cu$_2$(dim) (Me$_4$en)$_2$(ClO$_4$)$_3$O complex, two solutions in various organic solvents at 25°C.

of the present complexes in 1,2-dichloroethane (DCE) occur at longer wavelength than in nitro methane (CH_3NO_2), although the donor number of DCE (DN=0) is lower than that of CH_3NO_2 (DN=2.7). This anomaly was ascribed to the formation of ion pairs by mean of an axial coordination of ClO_4 anion in DCE. As the relative dielectric constant of nitromethane 28.5 versus 8.9 for DCE at room temperature; facilitating the dissociation of cationic complex and minimizes the axial interactions of ClO_4^- with the central copper(II) ion [26]. (b) Finally, although donor strength (DN), of acetone is higher than that of acetonitrile, the solutions of the present perchlorate of diamine-complexes 1 and 2 in acetone show a slightly lower $_{max}$ values than acetonitrile solution. This discrepancy was recognized to the effect of steric and π-bonding in acetone and acetonitrile, since they are acting in opposite directions [27].

The extraction of chemical information from the data in Table 4 can be carried out by statistical method of Multiple Linear Regression Analysis (MLRA). In this method, a dependent variables Y (v_{max}) is described in terms of a series of explanatory variables X. In this respect, the well known Gutmann's and Mayer's donor-acceptor numbers (DN and AN) [28] and the unified solvation model of Drago [29-31] were used, according to Equations 1 and 2; respectively.

$$v_{max}/10^3 \text{ cm}^{-1}=v^0+a \text{ (DN)}+b\text{(AN)}. \qquad (1)$$

$$v_{max}/10^3 \text{ cm}^{-1}=W+P(S)+E_BE_A+C_BC_A \qquad (2)$$

It is assumed that all the explanatory variables are independent of each other and truly additive as well as relevant to the problem under study [30,31] where DN and AN refer to the solvent Lewis basicity and acidity; respectively [28]. S is the solvent bipolarity term; P is a measure of the susceptibility of the complex to solvation. E_B, C_B and E_A, C_A quantify the electrostatic and covalent contributions to the Lewis basicity and acidity; respectively [29-31]. For acceptor probes, (complexes 1 and 2), E_B and C_B solvent parameters are used in Equation 2. Nevertheless, E_A and C_A reflect the physicochemical acceptor parameters of the complex under investigation. However E_A and C_A solvent parameters were used as an alternative in case of complex 3.

The following discussion examines the data of the current complexes. The overall picture which emerges from the MLRA based on Guttmann's donor-acceptor concept is considered followed by the unified solvent model proposed by Drago. The frequencies of d-d absorption transition band (v_{max}) of the current complexes in various solvents (Table 4) are fitted in Equations 1 and 2. The Regression

coefficients and constant values are shown in Table 6.

Fitting spectral data of the current complexes (Table 4) in Equation 1 indicates dependence of the d-d absorption transitions on the DN and AN parameters. The relative percentage of the influences of DN and AN parameters on the v_{max} values were calculated from the coefficients a and b, and found in the ranges 16.23-96.7% and 3.23-83.77%; respectively. The data in Table 6 suggests that, the DN parameter of the solvent has the dominant contribution (61.30-96.77%) in the shift of d-d absorption band of the perchlorate complexes 1, 2 and 4. However, the AN parameter (83.77%) is more pronounced than the DN parameter (16.23%) in case of chloride complex 3. The negative signs of the coefficients a and b for the current complexes indicate a red shift as the Lewis acid- base interactions increases. Complexes 1 and 2 are highly sensitive towards the Lewis basicity of solvent than complex 4 as indicated from their contribution percentage of DN (89.7-96.8) for the former and 61.3% for the later.

Fitting the v_{max} data (Table 4) in Equation 2 indicates dependence of the observed spectral shifts on the dipolarity/polarizability parameter P(S) and both components of the Lewis acid-base quantification, E and C parameters. The regression coefficients reported in Table 6 indicate that, the shift of the d-d absorption transitions of the chloride complex 3 is mainly influenced by the specific term of solute-solvent interaction (contribution percentages, 96%); in which the contribution of the covalency part (C) is more pronounced than the electrostatic part (E). However the contributions of E (67.65 and 56.00%) are more pronounced than C (11.7 and 3.3) for the perchlorate complexes 1 and 4, respectively; whereas in case of perchlorate complex 2, E smaller than C (22.40 and 33.51%). In the same way the non-specific term plays a significance role on the d-d transition of complexes 2 and 4 than complex 1.

The d-d transition band in the visible region of the electronic absorption spectrum of the investigated complexes is completely resolved, which allows an accurate determination of the band's oscillator strength (f) as described earlier [8]. The oscillator strength,f, is a dimensionless quantity that is used to express the electronic transition probability [23]. Table 7 collects f-values of the present complexes, in several solvents.

Once more, the data in Table 7 reveals the dependence of the d-d transitions on the Lewis acid base interactions of the current Cu(II)-complexes with the solvent molecules as indicated from the

No	Complex	v(C=O)		C=C	v(ClO$_4^-$) or			v(Cu-O)	v(Cu-N)
	HDMCHD	1613	1582	1474	-	-	-	-	-
1	Cu$_2$(DMCHD)(Me$_3$en)$_2$(ClO$_4$)$_3$.2C$_2$H$_5$OH	1624 (m)	-	1468 (w)	1141	1085	626	515	458 (w)
2	Cu$_2$(DMCHD)(Me$_4$en)$_2$(ClO$_4$)$_3$.2C$_2$H$_5$OH	1638	1570	1410 (s)	1144	1090	627	525	478
3	[Cu$_2$(DMCHD)(Me$_4$en)$_2$Cl$_2$] Cl.2C$_2$H$_5$OH	1674	1602 (s)	1528 (m)	-	-	-	568	500
4	Cu$_2$(DMCHD)(Me$_5$dien)$_2$(ClO$_4$)$_3$.2C$_2$H$_5$OH	1637	1618	1572	1143	1089	-	574	478

Table 3: Characteristic infrared frequencies (cm^{-1}) and their tentative assignments of free HDMCHD ligand and its mixed ligand complexes

No	Complex	Py	DMSO	DMF	MeOH	Me$_2$CO	MeCN	1.2DCE	MeNO$_2$
1	Cu$_2$(DMCHD)(Me$_3$en)$_2$(ClO$_4$)$_3$.2C$_2$H$_5$OH	670	646	633	605	571	596	590	525
2	Cu$_2$(DMCHD)(Me$_4$en)$_2$(ClO$_4$)$_3$.2C$_2$H$_5$OH	664	646	632	614	576	588	595	525
3	[Cu$_2$(DMCHD)(Me$_4$en)$_2$Cl$_2$] Cl.2C$_2$H$_5$OH	749	709	696	661	678	685	620	691
4	Cu$_2$(DMCHD)(Me5dien)$_2$(ClO$_4$)$_3$.2C$_2$H$_5$OH	695	692	675	655	645	625	640	630

Table 4: Absorption maxima bands, λ_{max}/nm for the Cu$_2$(DMCHD)(am)$_2$X$_3$ complexes solutions in various solvents at 25°C.

Chelate	ClO_4^-	I^-	Br^-	Cl^-	N_3^-	$CF_3SO_3^-$	CO_3^{-2}
DN_X(anion)/$MeNO_2$	8.44	28.9	33.7	36.2	34.3	16.9	13.3
$Cu_2(DMCHD)(Me_3en)_2(ClO_4)_3.2C_2H_5OH$	525	670	677	690	684	-	-
$Cu_2(DMCHD)(Me_4en)_2(ClO_4)_3 2C_2H_5OH$	567	647	707	705	608	635	573

Table 5: Absorption maxima bands, λ_{max}/nm for the $Cu_2(DMCHD)(diam)_2(ClO_4)_3$ complexes with various anions in $MeNO_2$ solution at 25°C.

Complex	MLRA using Gutmann parameter, Eqn. 2				MLRA using Drago parameter, Eqn.					Relative contrib.%		
	v_s^0 /10^3	a	b	r	W	P	E	C	r	Non-specific	Specific	
$Cu_2(DMCHD)(Me_3en)_2(ClO_4)_3.2C_2H_5OH$	19.86	-3.50 (94.59)*	-0.20 (5.41)*	0.97	18.40	0.468	-1.533	-0.265	0.99	20.65	67.65	11.70
$Cu_2(DMCHD)(Me_4en)_2(ClO_4)_3.2C_2H_5OH$	19.72	-3.69 (95.35)*	-0.18 (4.65)*	0.99 0.99	14.39	0.745	-0.378	0.566	0.97	44.1	22.4	33.51
$[Cu_2(DMCHD)(Me_4en)_2Cl_2]$ $Cl.2C_2H_5OH$	15.72	-0.62 (16.23)*	-3.20 (83.77)*	0.998	15.32	-0.23	2.18	-3.10	0.99	4.0	39.50	56.30
$Cu_2(DMCHD)(Me5dien)_2(ClO_4)_3.2C_2H_5OH$	16.47	-2.075 (61.30)*	-1.031 (38.70)*	0.98	13.78	0.50	0.69	-0.041	0.97	40.70	56.00	3.30

*Relative contribution percentage.
Table 6: Solvatochromic parameters of the $Cu_2(DMCHD)(am)_2X_3$ complexes, using Gutmann's and Drag's models (Equations 1and 2).

No.	Solvent/Complex	F × 103							F=f₀+aDN+bAN					el.con	
		PY	DMSO	DMF	MeOH	Me2CO	MeCN	CH2Cl2	f₀	a	b	r	n	DN	AN
1	$Cu_2(DMCHD)$ $(Me_3en)2(ClO4)_3.2C_2H_5OH$	24.827	24.523	25.027	25.531	27.745	25.916	28.194	28.16	-0.112	-	0.93	6	100	0
2	$Cu_2(DMCHD)$ $(Me_4en)_2(ClO_4)_3.2C_2H_5OH$	24.634	25.321	22.00	25.308	28.398	30.600	29.553	32.4	-7.37	-4.64	0.99	5	61.4	38.6
3	$[Cu_2(DMCHD)(Me_4en)_2Cl2]$ $Cl.2C_2H_5OH$	14.126	14.018	14.280	15.521	14.659	15.445	16.548	16.2	-3.24	1.33	0.92	7	70.9	29.1
4	$Cu_2(DMCHD)$ $(Me5dien)_2(ClO_4)_3.2C_2H_5OH$	11.3	9.581	8.706	9.586	9.734	9.242	8.633	8.3	3.25	-0.68	0.99	5	84.3	15.7

Table 7: Oscillator strength (f) of d-d transition absorption bands of the $Cu_2(DMCHD)(am)_2X_3$ complexes.

linear plots of the f-values versus the Lewis basicity (DN) and Lewis acidity (AN) of solvent: $f=f_0+a$ DN+b AN, the multi-parametric linear regression data were collected in Table 7. The intercept (f_0) represent the extrapolated oscillator strength, which was found to be higher for square planar diamine complexes 1 and 2 than the five coordinated, chloro of diamine and the perchlorate of triamine, complexes 3 and 4. For the same complex cation, $[Cu_2(DMCHD)(Me_4en)_2]^{3+}$, as the coordination ability of Lewis base increases f-values decreases.

Electrochemical studies

Cyclic voltammetric studies were performed for some copper(II) complexes 2 and 3 using a platinum microelectrode, of 1.0 mM L^{-1} complex solutions in different solvents and 0.1M L^{-1} Bu_4ClO_4 as supporting electrolyte at scanning rate of 100 mV s^{-1}, ferrocine/ ferrocinium (Fc/Fc$^+$) and bis(biphenyl)chromium (I)/(0) were used as internal standards for assignment of the potential values, the experimental data are given in Table 8. Since the ligands used in this work are not reversibly oxidized or reduced at the applied potential range values, the redox processes are assigned to the metal centers only. The reduction process of the investigated complexes in all solvents exhibited a quasi-reversible or quasi-irreversible, redox associated with a two-electron reduction, and mainly diffusion controlled as indicated from the linear dependence of the current peak on the square root of the scan rate [32-34].

Taking into account the cyclic voltammetric behavior of $[Cu_2(DMCHD)(Me_4en)_2Cl_2]Cl$ complex 3, in DMF solution (Figure 2) one can propose the following mechanism consisting of electrode and chemical reactions:

(a) the cathodic peak can be assigned to the cathodic reduction

Solvent	$Cu_2(DMCHD)(Me_4en)_2(ClO_4)_3$				$Cu_2(DMCHD)(Me_4en)_2(Cl_3)$			
	E_{pc1}	E_{pc2}	E_{pa1}	E_{pa2}	E_{pc1}	E_{pc2}	E_{pa1}	E_{pa2}
MeNO$_2$	0.83	–	0.52	–	0.80	0.87	0.015	–
MeCN	0.48	–	0.149	–	0.49	0.58	0.12	–
Fa	0.62	0.91	0.225	0.62	0.61	–	0.62	–
DMF	0.914	0.742	0.127	-	0.632	–	0.567	0.262
DMSO	0.933	0.723	0.713	0.328	0.728	0.908	0.377	0.327

Table 8: Electrochemical data for the redox processes of $Cu_2(DMCHD)(diam)_2X_3$ in various solvents potentials E/V vs. bis(biphenyl)-chromium(1) at (pt) ref electrode

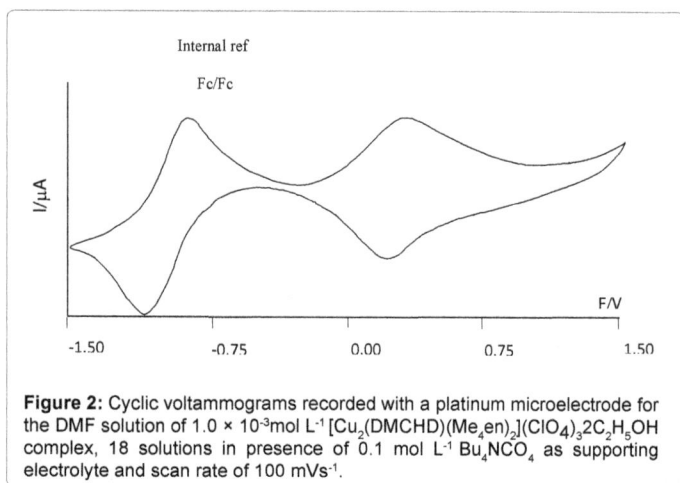

Figure 2: Cyclic voltammograms recorded with a platinum microelectrode for the DMF solution of 1.0×10^{-3} mol L^{-1} [Cu$_2$(DMCHD)(Me$_4$en)$_2$](ClO$_4$)$_3$.2C$_2$H$_5$OH complex, 18 solutions in presence of 0.1 mol L^{-1} Bu$_4$NCO$_4$ as supporting electrolyte and scan rate of 100 mVs^{-1}.

of the dicopper (II) complex to the dicopper (I) complex in a quasi-irreversible mechanism.

$$Cu^{II}Cu^{II} \rightleftharpoons Cu^{I}Cu^{I}$$

or

(b) Dissociation of the dicopper (II) into a mononuclear complex in solution first, followed by a reduction process for the product; according to the following sequences:

The anodic peak can be assigned to the anodic oxidation of the complex Cu(0), if this one could exist for very short time, to complex Cu(I):

$$Cu(I)\text{-complex} + e^- \rightarrow Cu(0)\text{-complex}$$

The decomplexation reaction is a chemical reaction occurring because the Cu(0) atom is unable to keep the ligands in a coordination arrangement:

$$Cu(0)\text{-complex} \rightarrow Cu(0) + ligands$$

This chemical reaction is fast enough in order to decrease the height of the corresponding anodic peak in which complex Cu (0) is involved.

The very prominent anodic peak, probably due to the anodic oxidation of the complex Cu(I) to complex Cu(II).

$$Cu(I)\text{-complex} \rightarrow Cu(II)\text{-complex} + e^-$$

So that one may conclude that the mechanism of the single cyclic voltammetry behaviour of the complex Cu(II) is described by an EECE or EEC mechanism. (C=chemical complication following the charge transfer E), rather than a simple EE mechanism. It is more

likely that, according to the Marcus theory, the activation barrier to electron transfer is increased, slowing down the rate of heterogeneous charge transfer, i.e., causing distinct deviation from the pure reversible character of the charge transfer [35].

The potential values of the redox peaks (E_{pc} and E_{pa}) of the investigated complexes in different solvents are given in Table 8. The E_{pc} potentials for the binuclear perchlorate and chloride Cu(II) complexes depend again on the Lewis acid-base solvent's properties. The trend of data also agrees well with the visible absorption spectral results in various solvents where, complex 2: $E_{pc_1}/v = 6.562 - 3.63 \times 10^{-4}$ v/cm^{-1}, r=0.94 (4 points, except MeNO$_2$); and $E_{pa1}/v = 1.716 + 1.16 \times 10^{-4}$ v/cm^{-1}, r=0.93 (4 points, except DMSO)

However complex 3 $Cu_2(DMCHD)(Me_4en)_2Cl_3$: $E_{pc1}/v = 7.7.78 - 4.78 \times 10^{-4}$ v/cm^{-1}, r=0.99 (3 points, except MeNO$_2$ and Fa); and $E_{pa_1}/V = 8.26 + 5.73 \times 10^{-4}$ v/cm^{-1}, r=0.99 (3 points, except DMF and DMSO)

T he combination of electrochemical and spectroscopic studies enables one to investigate the axial ligation of the solvatochrmic complexes in detail that can utilized to achieve a microscopic understanding of the specific and non-specific Lewis–acid base interactions of the current Cu(II)-complexes with solvent molecules.

Molecular orbital calculations

Table 9 shows the structural parameters data of the free ligands and their Cu(II)-complexes as calculated by means of a semi-empirical molecular orbital calculations at the PM3 level provides by the hyperchem 7.52 program. The calculated energies of the Frontier orbitals, lowest unoccupied (E_{LUMO}) and highest occupied molecular orbitals (E_{HOMO}) and E_{gap} ($E_{gap} = E_{Lumo} - E_{HOMO}$) of the ligands and their complexes are correlated with the current experimental data (Tables 3 and 4): $E_{gap}/eV = 9.597 - 0.0157 \Delta(_{C=O-C=C})/cm^{-1}$, r=0.92, and $E_{HOMO}/eV = 8.558 - 0.0131 \Delta v(_{C=O-C=C})/cm^{-1}$, r=0.92 except complex 1. The negative slope of these linear relationships reveal increasing of both Egap and E_{HOMO} accompanied by decreasing the separation in frequencies between the carbonyl group, $v_{C=O}$ and adjacent $v_{C=C}$. This finding further emphasized by the negative slope of the linear relationship of the calculated E_{HOMO} (Table 9) versus the d-d transition v_{max}: $E_{HOMO}/eV = 14.127 - 4.54 \times 10^{-4} v_{max}/cm^{-1}$, r=0.92, except complex 1. The negative slope recommends again increasing of E_{HOMO} was accompanied by red shift of d-d band. Figure 3 shows the optimized structure of the current complex 1 as an example.

Biological activity studies: Table 10, summarized the biological activity of $Cu_2(DMCHD)(Me_5dien)_2(ClO_4)_2$ complex. Investigation was carried out against the sensitive organisms *Staphylococcus aureus* (ATCC 25923) and *Bacillus subtilis* (ATCC 6635) as Gram-positive bacteria, *Escherichia coli* (ATCC 25922) and *Salmonella typhimurium* (ATCC 14028) as Gram-negative bacteria, yeast; *Candida albicans* (ATCC 10231) and fungus; *Aspergillus fumigates* as described elsewere [36]. Inspection of the data reveals that the metal complex exhibited intermediate activity towards the Gram- positive bacteria *Bacillus subtilis* (ATCC 6635), also intermediate activity towards the Gram-negative bacteria *Salmonella typhimurium* (ATCC 14028). However, low antimicrobial activity was observed against Yeasts and Fungi; similar to that observed earlier for Cu(DMCHD)(Me₅dien)(NO₃) [11].

Conclusion

A newly compounds have been synthesized and characterized on the basis of spectral, conductance, magnetic and electrochemical studies on the complexes were carried out in solution and in solid state; UV-visible spectra on the solution of the complexes in various solvents are found to be depending upon the type of solvent. The absorption coefficient spectrum in the UV-Vis region shows absorption bands and generally interpreted in terms of π-π^* excitation. The metal complex $Cu_2(DMCHD)(Me_5dien)_2(ClO_4)_2$ showed intermediate activity toward the Gram- positive and Gram-negative bacteria, and low antimicrobial activity against yeasts and fungi.

References

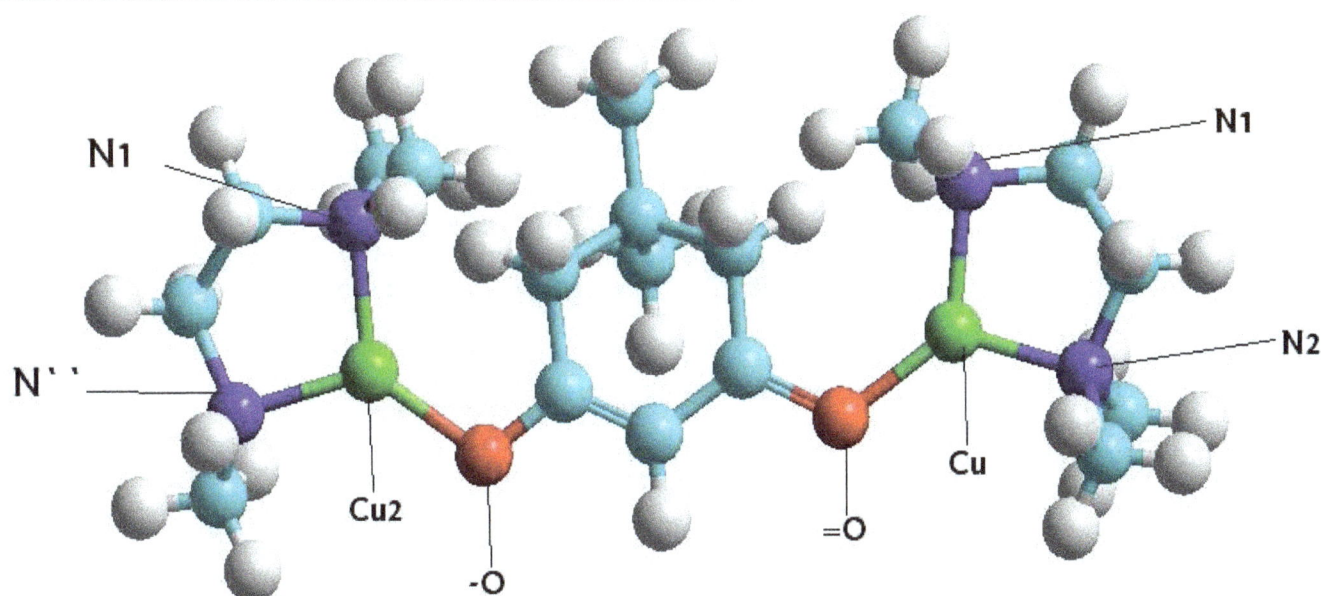

Figure 3: Optimized structure of complex 1 {Cu₂(DMCHD)(Me₃en)₂}(ClO₄)₃.2C₂H₅OH.

No	Organic ligand and complex	E_{tot} Kcal/mol	Dipole moment	ΔHf Kcal/mol	Electr E Kcal/mol	Nuclear E Kcal/mol	Binding E Kcal/mol	E_{homo} ev	E_{lumo} ev	ΔEgap
A	Me₄en	-29581.82	1.20	-13.30	-158075.95	128494.13	-2098.27	-8.981	2.388	11.369
B	Me₃en	-26136.84	1.74	-11.45	-128743.73	102606.88	-1821.32	-9.092	2.422	11.514
C	Dimedone free (HDMCHD)	-39685.20	4.94	-90.39	-204106.53	164421.32	-2201.85	-9.712	-0.014	9.698
D	Me₅dien	-44015.22	2.40	-16.52	-284075.25	240059	-30092.00	-8.850	2.270	11.12
1	[Cu₂(DMCHD)(Me₃en)₂ClO₄	-146288.39	6..90	-219.03	-1130492.39	984204.5	-6059.55	-0.661	1.971	2.632
2	Cu₂(DMCHD)(Me₄en)₂(ClO4)₃	-153186..66	3.715	-231.38	-1264009.33	1110822.65	-6622.08	-5.314	1.026	6.34
3	[Cu₂(DMCHD)(Me₄en)₂Cl₂]Cl	-175811.37	15.85	-255.50	-1576159.37	1400348.00	-6837.38	-7.190	-0.5448	6.646
4	Cu₂(DMCHD)(Me₅dien)₂(ClO₄)₃	-110636.36	5.871	-107.24	-943728.00	833091.93	-5322.66	-7.439	1.467	8.906

Table 9: Structural parameters of the organic ligands and their metal complexes using hyperchem 7.5 at pm3 level.

Organism	Gram-positive bacteria				Gram-negative bacteria				Yeasts and Fungi**			
	Staphylococcus aureus (ATCC 25923)		Bacillus subtilis (ATCC 6635)		Salmonella typhimurium (ATCC 14028)		Escherichia coli (ATCC 25922)		Candida albicans (ATCC 10231)		Aspergillus fumigatus	
Concentration / Sample	1 mg/ml	0.5 mg/ml	1 mg/ml	0.5 mg/ml	1 mg/ml	0.5 mg/ml	1 mg/ml	0.5 mg/ml	1 mg/ml	0.5 mg/ml	1 mg/ml	0.5 mg/ml
Cu(DMCHD)(Me₅dien)(NO₃)	10 L	7 L	13 I	9 I	12 L	8 L	15 I	11 I	9 L	7 L	10 L	6 L
Control#	35	26	35	25	36	28	38	27	35	28	37	26

**Identified on the basis of routine cultural, morphological and microscopical characteristics.

L: Low activity=Mean of zone diameter ≤1/3 of mean zone diameter of control.

I: Intermediate activity=Mean of zone diameter ≤2/3 of mean zone diameter of control.

H: High activity=Mean of zone diameter >2/3 of mean zone diameter of control.

#Chloramphenicol in the case of Gram-positive bacteria, cephalothin in the case of Gram-negative bacteria and cycloheximide in the case of fungi.

Table 10: Antimicrobial activity of $Cu_2(DMCHD)(Me_5dien)_2(ClO_4)_3 \cdot 2C_2H_5OH$

1. Linert W, Fukuda Y, Camard A (2001) Chromotropism of coordination compounds and its applications in solution. Coord Chem Rev 218: 113-152.

2. Soliman AA, Taha A, Linert W (2006) Spectral and thermal study on the adduct formation between square planar nickel(II) chelates and some bidentate ligands. Spectrochim Acta A Mol Biomol Spectrosc 64: 1058-1064.

3. Kuzniarska-Biernacka I, Bartecki A, Kurzak K (2003) UV–Vis–NIR spectroscopy and colour of bis(N-phenylsalicylaldiminato)cobalt(II) in a variety of solvents. Polyhedron 22: 997-1007.

4. Golchoubian H, Moayyedi G, Bruno G, Rudbari HA (2011) Syntheses and characterization of mixed-chelate copper(II) complexes containing different counter ions; spectroscopic studies on solvatochromic properties. Polyhedron 30: 1027-1034.

5. Ashraf AM, Mohammed JM, Ismail Y (2012) Synthesis, Characterization and Biological Studies of 2-(4-Nitrophenylaminocarbonyl)Benzoic Acid and its Complexes With Cr(III), Co(II), Ni(II), Cu(II) and Zn(II). Iran J Chem and Chem Eng 31: 9-14.

6. Linert W, Taha A (1993) Spectroscopic, Thermodynamic and Quantum Mechanical Studies on Solvatochromic Mixed Ligand Copper(II)-Chelates. J Coord Chem 29: 265-276.

7. Taha A (2001) Spectroscopic and Molecular Orbital Studies of Chromotropic Ternary Complexes of Copper(II) with Thenoyltrifluoroacetone and Diamine Derivatives. Synth React Inorg Metal Org 31: 227-238.

8. Taha A (2001) Spectroscopic and electrochemical studies on the chromotropism of ternary copper(II) complexes. New J Chem 25: 853-858.

9. Taha A (2003) Spectral, electrochemical and molecular orbital studies on solvatochromic mixed ligand copper(II) complexes of malonate and diamine derivatives. Spectrochim Acta A Mol Biomol Spectrosc 59: 1373-1386.

10. Taha A (2003) Spectroscopic studies on chromotropic mixed-ligand copper(II) complexes containing o-hydroxy benzoyl derivatives and dinitrogen bases. Spectrochim Acta A Mol Biomol Spectrosc 59: 1611-1620.

11. Taha A, Farag AA, Ammar AH, Ahmed HM (2014) Structural, molecular orbital and optical characterizations of solvatochromic mixed ligand copper(II) complex of 5,5-Dimethyl cyclohexanate 1,3-dione and N,N,N′,N″-pentamethyldiethylenetriamine. Spectrochim Acta A Mol Biomol Spectrosc 122: 512-520.

12. Taha A (2001) Metal Complexes of Triazine Schiff Bases: Synthetic, Thermodynamic, Spectroscopic, and Electrochemical Studies on Complexes of some Divalent and Trivalent Metal Ions of 3-(A-Benzoylbenzylidenhydrazino)-5,6-Diphenyl-1,2,4-Triazine. Synth React Inorg Met-Org Chem 31: 205-218.

13. Linert W, Taha A, Jameson RF (1992) The Electrochemical Behaviour of Mixed Nickel(II)-Chelates in Nonaqueous Solvents. J Coord Chem 25: 29-41.

14. Kucharsky J, Safarik L (1965) Titrations in Nonaqueous Solvents. Elsevier, Amsterdam, Netherlands.

15. Dutt NK, Sarma UUM (1975) Chemistry of lanthanous-XL. Solution stabilities of lanthanide ions with dimedone (5,5-dimethyl 1,3-cyclohexanedione). J Inorg Nucl Chem 37: 606-607.

16. Frankel LS, Stengle TR, Langford CH (1968) Preferential solvation of Co2+ and Ni2+ ions in mixed solvents: n.m.r. methods. Can J Chem 46: 3183-3187.

17. Dutt NK, Sanyal S, Sarma UUM (1972) Chemistry of lanthanons—XXXV: Formation constants of rare-earth elements with a few oxygen donating ligands. J Inorg Nucl Chem 34: 2261-2264.

18. Nakamoto K (1986) Infrared and Raman Spectra of Inorganic and Coordination. (4thedn) Wiley-Interscience New York, USA.

19. Linert W, Jameson RF, Taha A (1993) Donor numbers of anions in solution: the use of solvatochromic Lewis acid–base indicators. J Chem Soc Dalton Trans 1993: 3181-3186.

20. Geary WJ (1971) The use of conductivity measurements in organic solvents for the characterization of coordination compounds. Coord Chem Rev 7: 81-122.

21. Huheey JE, Keiter EA, Keiter RL (1993) Inorganic Chemistry: Principles of Structures and Reactivity. (4thedn) Pearson Education Inc., USA.

22. Lever ABP (1984) Inorganic Electronic Spectroscopy. (2ndedn) Elsevier, Amsterdam, Netherlands.

23. Fukuda Y, Sone K (1987) Inorganic Thermochromism. Inorganic Chemistry Concepts. Springer-Verlag Berlin, Heidelberg, Germany.

24. O'Sullivan C, Murphy G, Murphy B, Hathaway B (1999) Crystal structures, electronic properties and structural pathways of thirty [Cu(bipy)₂X][Y] complexes, where X = Cl⁻, Br⁻ or I⁻. J Chem Soc Dalton Trans 2: 1835-1844.

25. Soukup RW, Sone K (1987) (Acetylacetonato)(N,N,N′,N′-tetramethylethylenediamine)copper(II) Tetraphenylborate as a Solvent Basicity Indicator. Bull Chem Soc Jpn 60: 2286-2288.

26. Taha A, Gutmann V, Linert W (1991) Spectroscopic and thermodynamic studies on solvatochromic nickel(II) complexes. Monatsh Chem 122: 327-339.

27. Gutmann V (1978) The Donor-Acceptor Approach to Molecular Interactions. Plenum Press, New York, USA.

28. Drago RS (1980) The interpretation of reactivity in chemical and biological systems with the E and C model. Coord Chem Rev 33: 251-277.

29. Drago RS, Ferris DC, Wong N (1990) A method for the analysis and prediction of gas-phase ion-molecule enthalpies. J Am Chem Soc 112: 8953-8961.

30. Drago RS (1992) A unified scale for understanding and predicting non-specific solvent polarity: a dynamic cavity model. J Chem Soc Perkin Trans 2: 1827-1838.

31. Gritzner G, Kuta J (1982) Recommendations on reporting electrode potentials in nonaqueous solvents. J Appl Chem 54: 1527-1532.

32. Gritzner G, Kuta J (1984) Recommendations on reporting electrode potentials in nonaqueous solvents (Recommendations 1983). J Appl Chem 56: 461-466.

33. Gritzner G (1986) Solvent effects on half-wave potentials. J Phy Chem 90: 5478-5485.

34. Mabbott GA (1983) An introduction to cyclic voltammetry. J Chem Edu 60: 697.

35. Nakamura M, Okawa H, Kida S (1982) Binuclear metal complexes. XLVI[1]. Electronic and electrochemical properties of copper(II)M(II) binuclear complexes of N,N′-bis(5-t-butylsalicylidene)alkanediamines. Inorg Chim Acta 62: 201-205.

36. Vogel AI (1978) Vogel's textbook of quantitative inorganic analysis, including elementary instrumental analysis. (4thedn) Longman, London.

13

Electrical Properties of Pristine and Electron Irradiated Carbon Nanotube Yarns at Small Length Scales

Francisco Solá*

NASA Glenn Research Center, Materials and Structures Division, Cleveland, OH 44135, USA

Abstract

In this report the effect of e-beam irradiation on Carbon Nanotube (CNT) yarns electrical resistivity was studied by employing electron beam irradiation on a Transmission Electron Microscope (TEM) follow by two probe resistivity method in a Scanning Electron Microscope (SEM). Both local crosslinking and amorphous regions within the CNT yarn were observed with increased electron dosage, as revealed by High Resolution TEM (HRTEM). The resistivity lower bound value was obtained at the maximum dosage used, which was below the resistivity of the pristine yarn. The resistivity data is explained by a proposed model that takes into account the microstructural changes.

Keywords: CNTs; Crosslinking; Defects; E-beam irradiation; Electrical resistivity; FIB; SEM; TEM

Introduction

NASA is currently exploring routes to potentially replace conventional carbon fiber composites with CNT based composite materials. This could translate to approximately a one-third reduction of unfueled weight of space vehicles and structures, and if successful, will diminish considerably vehicle launch costs. To achieve this, commercially available CNT based materials must have at least two times the strength of conventional carbon fibers. CNT yarns are currently one of the best commercially available CNT based materials in terms of mechanical properties. However, their tensile strength is about half of conventional carbon fibers. This is related to the weak shear interactions between carbon shells and bundles within a yarn [1]. Therefore, current efforts are focused in developing protocols to improve the mechanical properties of these materials.

One potential route to achieve the mechanical improvement is the cross-linking method induced by electron beam irradiation. The weak shear interactions between adjacent C-shells/CNT can be improved by the formation of sp^3 C-C bonds induced by e-beam irradiation [1]. This can occur at both the interwall sites of individual multiwall CNT (MWCNT) and between CNTs neighbors and both can potentially increase the mechanical response of CNT yarns [2,3]. E-beam energies greater than 80 keV are needed to displace C atoms and to induce complex kinetics and recombination of lattice defects within the hexagonal carbon network, which eventually leads to cross-linking [4]. For one isolated MWCNT (and small bundles), 100-200 keV are effective energies to crosslink [5,6]. Being CNT yarns fibers composed of several MWCNTs, the question arises as to what extend energies in this range will still promote crosslinking effectively.

The study of the electrical response of CNT yarns as a function of electron dose can be a complementary route to monitor possible cross-linking events, and is important to establish multifunctional properties of CNT yarns. Potential applications of CNT yarns that possess both good electrical and mechanical properties include antennas and lightning strike protection of aircraft [7]. Considerable efforts have been focused until now in e-beam irradiation methods that lead to mechanical improvement. Although Mikó and coworkers reported the effects of e-beam irradiation on the electrical resistivity of Single-Walled CNT (SWCNT) fiber systems [8,9], work on the e-beam irradiation effects on the electrical properties of CNT yarns is lacking. In this brief report, small segments of CNT yarns are exposed to e-beam irradiation in a TEM operated at 200 keV and at different doses. The electrical resistivity as a function of e-beam irradiation is studied by the two probe method, using micromanipulators inside an SEM.

Experimental

CNT yarns were obtained from Nanocomp Technologies, Inc. As-received samples from Batch 5279 were used in this work. E-beam irradiation experiments were performed in a TEM Philips CM200 microscope [10] operated with a beam energy of 200 keV, and e-beam flux of ~ 5×10^{12} e/cm^{2}s.

Small segments of CNT yarns were mounted on special TEM sample mounts from Norcada, Inc. These TEM windows consist of 5 mm × 5 mm, 200 μm Si frames, decorated from both sides with a silicon nitride film of ~100 nm. The silicon nitride films serve as electrical isolation material. In the center of the TEM sample mount there is an open (without silicon nitride material) 0.5 mm×0.5 mm area that provides a compatible window for both e-beam irradiation experiments and TEM imaging investigation. A schematic of the e-beam irradiation is depicted in Figure 1. Possible microstructural changes on the CNT yarns were monitored by high resolution HRTEM using a Gatan Imaging Filter (GIF) camera.

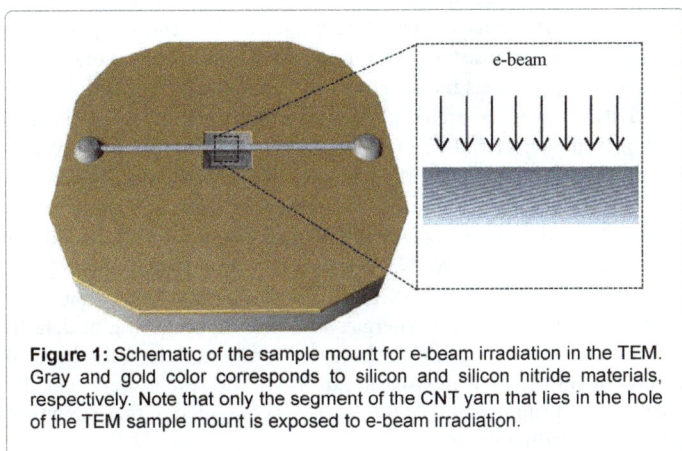

Figure 1: Schematic of the sample mount for e-beam irradiation in the TEM. Gray and gold color corresponds to silicon and silicon nitride materials, respectively. Note that only the segment of the CNT yarn that lies in the hole of the TEM sample mount is exposed to e-beam irradiation.

*Corresponding author: Francisco Solá, NASA Glenn Research Center, Materials and Structures Division, Cleveland, OH 44135, USA
E-mail: francisco.sola-lopez@nasa.gov

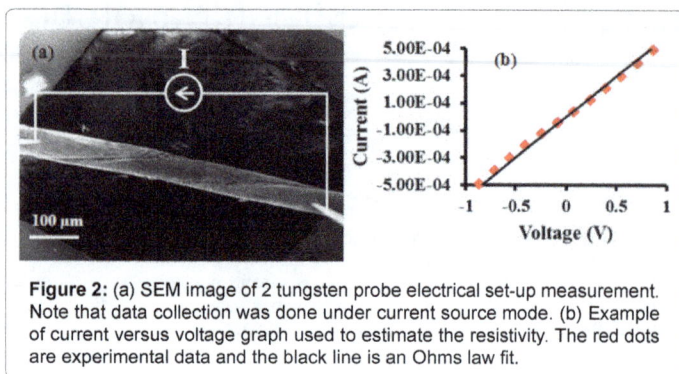

Figure 2: (a) SEM image of 2 tungsten probe electrical set-up measurement. Note that data collection was done under current source mode. (b) Example of current versus voltage graph used to estimate the resistivity. The red dots are experimental data and the black line is an Ohms law fit.

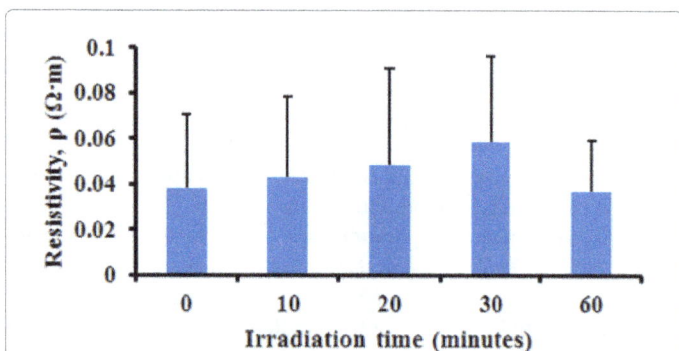

Figure 3: Effect of e-beam irradiation on CNT yarn resistivity. Four samples were tested for each case.

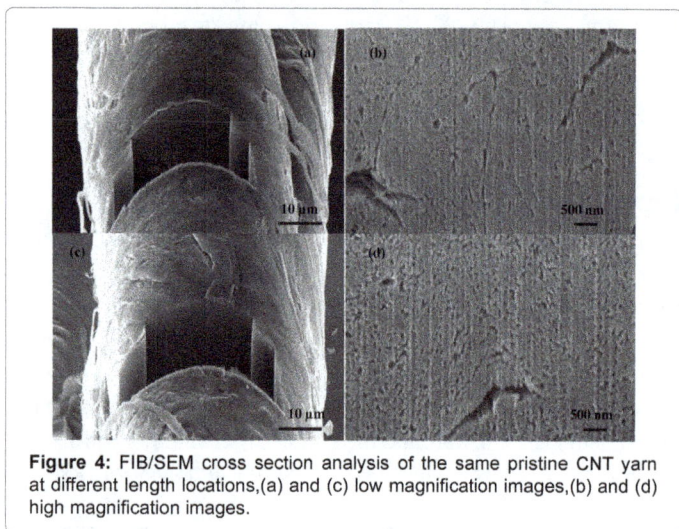

Figure 4: FIB/SEM cross section analysis of the same pristine CNT yarn at different length locations,(a) and (c) low magnification images,(b) and (d) high magnification images.

Resistivity studies were done inside (in vacuum) an Auriga Focused Ion Beam (FIB) microscope from Carl Zeiss, with a specimen vacuum of ~ 10^{-6}Torr [11]. The Auriga instrument is a crossbeam FIB instrument, but only the electron beam column was used for image formation, which means that it was used as a standard SEM. All SEM images were acquired using an In-lens Secondary Electron (SE) detector [12,13], and a beam energy of 5 keV. Two electrochemically etched tungsten tips were connected to two Kleindiek micromanipulators equipped with low current plug-ins [14]. Electrical experiments were controlled with a Keithley 2400 source meter in current source mode, and a LabView program. The SEM beam was blanked before collecting the data. The vacuum chamber, including the micromanipulators and sample stage holder, were cleaned with a plasma cleaner to remove contamination,

prior to sample insertion. The sample itself and the TEM sample mounts were not exposed to plasma cleaning, as this creates defects on the yarns and damage the silicon nitride films.

Cross section imaging of the CNT yarns was conducted in the Auriga microscope. Milling procedures involved the formation of a trench in the CNT yarns using an ion beam current of 2nA at 30 kV, followed by low current cleaning steps of the imaged surface [15].

Results and Discussion

Figure 1 depicts the schematic of the sample mount for e-beam irradiation in the TEM. Only the segment of the CNT yarn that lies in the hole of the TEM sample mount is exposed to e-beam irradiation. Figure 2a is a SEM image which illustrates the two probe electrical characterization of CNT yarns. Typically, the measurements were conducted to probe CNT yarns length of ~ 450 μm, and diameters of ~ 77-84 μm. An example of current versus voltage dependence is plotted in Figure 2b. Note that the relationship is linear and therefore the use of Ohm's law (V = IR) to estimate system resistance (R) is justified. To estimate the resistivity of CNT yarns, the experimental set-up is modeled by a series resistor, where each component resistor is additive, $R_{total} = R_{yarn} + \Sigma R_{other}$ (internal electrical source resistance, cables and W probes). Furthermore, contact resistance is assumed to be negligible here. This is a reasonable assumption when a considerably pressure force between probe and sample is established [16], as in this work. Before establishing contact with the CNT yarns, the two end-tips of the tungsten probes were contacted and conditioned for electrical measurements [17]. By doing so, contact resistance is further reduced, and from the linear current voltage dependence of the conditioned probes, ΣR_{other} can be estimated and extracted from the total resistance to obtain R_{yarn}. Then, the yarns resistivity (ρ) is obtained using equation 1

$$\rho = \pi R_{yarn}^2 \frac{D^2}{4L} \tag{1}$$

where D and L are the diameter and length of the yarn tested, respectively [18], and both quantities can be extracted directly from SEM images.

The electrical resistivity as a function of e-beam irradiation is presented in Figure 3. Irradiation times range from 10-60 min, which correspond to dosages of ~3×10^{15}-2×10^{16} e/cm². The average values of the resistivity increased with irradiation time for up to 30 min and decrease with further irradiation, though with the presence of scattering in the data. The scattering of the data is attributed to the local variation of the microstructure (diameter, internal porosity as supported in the SEM and focused ion beam microscopy results) of the CNT yarns in Figure 4. Similar local changes in resistivity for larger length scale CNT yarns due to variations in microstructure have been reported [19]. Note that the maximum resistivity at 30 min of irradiation corresponds to an increase of ~ 1.5 times the resistivity of the pristine yarn, and that the resistivity at 60 min is just a slight decrease of the pristine resistivity. For comparison purposes, the resistivity of CNT yarns of the current study are about 10^2 times smaller than aerosol-like (non-twisted) yarns, but are ~10^3 times larger than other twisted yarns (although these values were estimated with yarn segment lengths of 50 mm) [19], and about 10 times higher than SWCNT fibers of 3-5 mm tested length [8,9]. Interestingly, our group at NASA likewise noticed that larger scale CNT yarns irradiated with MeV energies also present scattering of data in regards to tensile strength, something that can be explained also by local variations on the yarn microstructure [20]. The results were that the maximum tensile strength of the yarns was obtained at the maximum e-beam irradiation dosage.

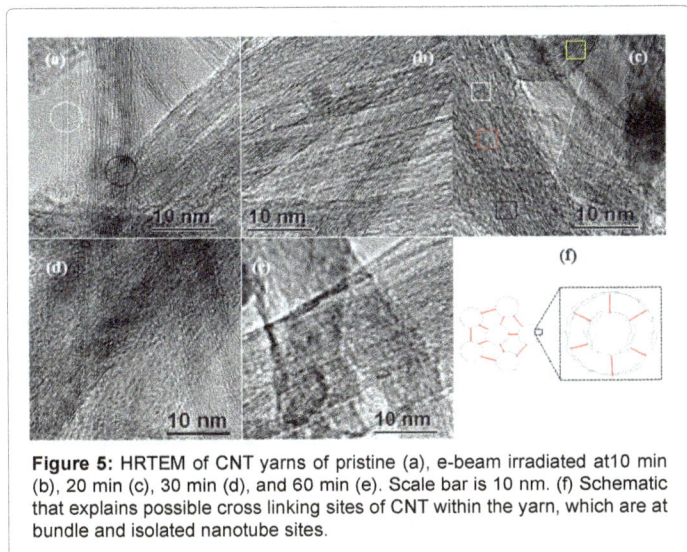

Figure 5: HRTEM of CNT yarns of pristine (a), e-beam irradiated at 10 min (b), 20 min (c), 30 min (d), and 60 min (e). Scale bar is 10 nm. (f) Schematic that explains possible cross linking sites of CNT within the yarn, which are at bundle and isolated nanotube sites.

To obtain possible explanations to the e-beam irradiation effects on resistivity, the author conducted HRTEM analysis on the microstructure of CNT yarns. HRTEM of CNT yarns of pristine (a), e-beam irradiated with 10 min (b), 20 min (c), 30 min (d), and 60 min (e), are presented in Figure 5. Figure 5f is a schematic that summarizes possible crosslinking sites (marked with red lines) of CNT constituents within the yarn at two different scales. The images were taken in thin areas located at the edges of the yarns. From these images, the different planes of CNTs oriented in a given twisted direction can be observed.

For the purpose of the following discussion, only areas of the images that are in focus are described. The area enclosed by a white circle in Figure 5a shows that the CNTs of yarns are double walled, while the area enclosed by a black circle is consistent with a CNT bundle structure. The alternating black/white fringes correspond to the lattice planes of CNTs in the bundle, with each lattice plane built from a row of CNTs [21,22]. Crosslinking sites on CNTs can be monitored by these HRTEM images and typically correspond to areas where the fringes are less coherent but do not lose completely their structure to form an amorphous carbon (a-C) structure [2]. Cross-linking events can be observed at 10 min of irradiation (Figure 5b). Several types of microstructural changes of CNTs within the yarn are evident at 20 min of irradiation (Figure 5c). These include cross-linked sites (area enclosed by a black box), pristine-non-cross-linked sites (yellow box), a-C structure sites (red box), and sites with a mixture of a-C and cross-linked sites (white box). With further irradiation it can be seen that both a-C and cross-linked sites grow, however, in overall, the crystallinity (fringes structure) of the CNTs is preserved (in the sense that it is not totally lost).

In order to understand these microstructural changes and how to correlate them to the corresponding resistivity results, a brief explanation on e-beam irradiation on CNT defect formation is needed. E-beam electrons can displace C atoms located at the hexagonal lattice network of CNT, only when a critical minimum energy is used known as the displacement threshold. However, the displacement threshold depends on the local arrangement of carbon atoms relative to the electron beam and type of CNT. This is due to the direction of momentum transfer to C atoms distributed in the hexagonal lattice. For instance, displacement threshold energies of 82 keV have been reported for small CNTs oriented perpendicular to the e-beam, and up to 240 keV for relative bulky CNTs oriented tangential to the e-beam [4]. For MWCNTs, displacement threshold energies correspond to 100 keV [4].

Once C atoms are displaced, lattice defect formation in the form of interstitials and vacancies will take place. Based on quantum mechanics calculations, defects on the form of di-vacancies, interstitials and Frenkel pair (interstitial-vacancy pair) defects, were shown to crosslink graphitic layers [23]. However, at the same time e-beam irradiation can lead to unwanted loss of lattice coherence, a process known as amorphitization. This is related to the kinetics of defects (production rate, dynamics) on specific sites of the C lattice and agglomeration of point defects that leads to larger defects. The dynamics of defects depend on temperature. E-beam irradiation at room temperature (as in this work) leads to the formation of vacancies and interstitials (for energies above the threshold energy), however both remain relatively localized (immobile) at specific lattice sites, and if they do not recombine to form cross-linking sites, it can lead to a high concentration and agglomeration of defects as e-beam time increases, which eventually causes the lattice to lose its crystallinity at those sites. This model is consistent with the amorphous regions encountered in Figures 5c-5e.

In terms of electrical results, our data can be explained in terms of a competitive process between crosslinking and amorphitization. Crosslinking sites reduce the resistivity by reducing the CNT to CNT distance, while amorphitization increases resistivity. Note that the reduction of resistivity by crosslinking events is not due to the conduction of electrons through sp^3 C-C bonds. The increase in resistivity at 10 min of e-beam irradiation can be explained by the formation of defects in the lattice that have not produced enough cross-linking sites that enhance conductivity. At irradiation times of 20-30 minutes, although crosslinking events are increasing, the resistivity is dominated by amorphitization events. The reduction of resistivity at 60 min can only be explained by a significant increase in crosslinking population that dominates the overall electrical conduction of electrons in the yarn. This is consistent with the microstructural data presented in Figure 5.

Conclusion

In summary, this is the first report on the e-beam irradiation effects on the electrical resistivity of CNT yarns. The author conducted systematic e-beam irradiation experiments on a TEM, follow by two probe analysis in an SEM. Both crosslinking and amorphous regions within the CNT yarn were observed by HRTEM, and both appear to increase with irradiation. Resistivity data were explained taking into account microstructural changes. The resistivity lower bound value corresponds to the maximum dosage used, and was below the resistivity of the pristine yarn. Taking into consideration (from other reports) that e-beam induced crosslinking can improve the mechanical properties of the yarn; our data suggests that e-beam processing can be a suitable route to achieve multifunctional CNT yarns. Note that we expect more significant improvement in regard to mechanical improvements of CNT yarns using the e-beam protocol of the current study. As pointed out by Cornwell and Welch (using molecular dynamics simulations), short fibers with just few cross-link density number can considerably improve fiber strength [24]. In contrast to the results presented by Mikó et al., CNTs within the yarn structure of the current study do not completely transform to a-C during irradiation.

Although it was established that 200 keV energies can induce cross-linked MWCNTs within the yarn, aspects about crosslinking uniformity across the entire cross section of the yarn needs further investigation. This is more critical in regards to the mechanical properties, because regions that are weaker due to less crosslinking densities can potentially start to fail faster than areas with high crosslinking densities, and those relative weak sites may further grow in size causing failure of the yarn at lower strength values than expected. Note that the penetration of electrons at 200 keV ($\sim$234 µm for carbon material) is enough to penetrate the whole CNT yarn, however, the corresponding mean free path of electrons is just below 1.5µm [25]. This means that the electrons will suffer energy loss beyond a depth of $\sim$1.5 µm. How much and how dependent is on depth is unknown at the moment, but if the energy loss is below the threshold energy displacement of C atoms, crosslinking will not take place.

Acknowledgement

This work was supported by the NASA Space Technology MissionDirectorate's Game Changing-Carbon Nanotube Materials Development Project.

References

1. Filleter T, Espinosa HD (2013) Multi-scale mechanical improvement produced in carbon nanotube fibers by irradiation cross-linking. Carbon 56: 1-11.

2. Kis A, Csányi G, Salvetat JP, Lee TN, Couteau E, et al. (2004) Reinforcement of single-walled carbon nanotube bundles by intertube bridging. Nat Mater 3: 153-157.

3. Xia ZH, Guduru P, Curtin WA (2007) Enhancing mechanical properties of multi-wall carbon nanotubes via sp^3 interwall bridging. Phys Rev Lett 98: 245501.

4. Krasheninnikov AV, Nordlund K (2010) Ion and electron irradiation-induced effects in nanostructured materials. J Appl Phys 107: 071301.

5. Peng B, Locascio M, Zapol P, Li S, Mielke SL, et al. (2008) Measurements of near-ultimate strength for multiwalled carbon nanotubes and irradiation-induced crosslinking improvements. Nat Nanotechnol 3: 626-631.

6. Filleter T, Bernal R, Li S, Espinosa HD (2011) Ultrahigh strength and stiffness in cross-linked hierarchical carbon nanotube bundles. Adv Mater 23: 2855-2860.

7. Lebrón-Colón M, Meador MA, Gaier JR, Solá F, Scheiman DA, et al. (2010) Reinforced thermoplastic polyimide with dispersed functionalized single wall carbon nanotubes. ACS Appl Mater Interfaces 2: 669-676.

8. Mikó C, Milas M, Seo JW, Couteau E, Barišić N, et al. (2003) Effect of electron irradiation on the electrical properties of fibers of aligned single-walled carbon nanotubes. Appl Phys Lett 83: 4622-4624.

9. Mikó C, Seo JW, Gaál R, Kulik A, Forró L (2006) Effect of electron and ultraviolet irradiation on aligned carbon nanotube fibers. Phys Status Solidi b 243: 3351-3354.

10. Solá F, Xia Z, Lebrón-Colón M, Meador MA (2012) Transmission electron microscopy of single wall carbon nanotube/polymer nanocomposites: A first-principles study. Phys Status Solidi RRL 6: 349-351.

11. Solá F, Niu J, Xia Z (2013) Heating induced microstructural changes in graphene/Cu nanocomposites. J Phys D: Appl Phys 46: 065309.

12. Steigerwald MDG, Arnold R, Bihr J, Drexel V, Jaksch H, et al. (2004) New detection system for GEMINI. Microsc Microanal 10: 1372.

13. Solá F, Hurwitz F, Yang J (2011) A new scanning electron microscopy approach to image aerogels at the nanoscale. Nanotechnology 22: 175704.

14. Solá F, Biaggi-Labiosa A, Fonseca LF, Resto O, Lebrón-Colón M, et al. (2009) Field Emission and Radial Distribution Function Studies of Fractal-like Amorphous Carbon Nanotips. Nanoscale Res Lett 4: 431-436.

15. Sears K, Skourtis C, Atkinson K, Finn N, Humphries (2010) Focused ion beam milling of carbon nanotube yarns to study the relationship between structure and strength. Carbon 48: 4450-4456.

16. Weeden O (2003) Probe card tutorial. Keithley Instruments, Inc., 24370303: 1-40.

17. Chen Q, Wang S, Peng LM (2006) Establishing Ohmic contacts for in situ current-voltage characteristic measurements on a carbon nanotube inside the scanning electron microscope. Nanotechnology 17: 1087-1098.

18. Rebouillat S, Lyons MEG (2011) Measuring the electrical conductivity of single fibres. Int J Electrochem Sci 6: 5731-5740.

19. Miao M (2011) Electrical conductivity of pure carbon nanotube yarns. Carbon 49: 3755-3761.

20. Miller SG, Williams TS, Baker JS, Solá F, Lebrón-Colón M, et al. Increased tensile strength of carbon nanotube yarns and sheets through chemical modification and electron beam irradiation, under review.

21. Lambin Ph, Loiseau A, Culot C, Biró LP (2002) Structure of carbon nanotubes probed by local and global probes. Carbon 40: 1635-1648.

22. Solá F, Lebrón-Colón M, Ferreira PJ, Fonseca LF, Meador MA, et al. (2010) In-situ TEM-STM observations of SWCNT ropes/tubular transformations. Mater Res Soc Symp Proc 1204: 1204-K10-26.

23. Telling RH, Ewels CP, El-Barbary AA, Heggie MI (2003) Wigner defects bridge the graphite gap. Nat Mater 2: 333-337.

24. Cornwell CF, Welch CR (2011) Very-high-strength (60-GPa) carbon nanotube fiber design based on molecular dynamics simulations. J Chem Phys 134: 204708.

25. Yao N, Wang ZL (2005) Handbook of microscopy for nanotechnology. 362-363 pages..

14

Tumor Differentiation Factor (TDF) and its Receptor (TDF-R): Is TDF-R an Inducible Complex with Multiple Docking Sites?

Urmi Roy, Izabela Sokolowska, Alisa G Woods and Costel C Darie*

Biochemistry and Proteomics Group, Department of Chemistry and Biomolecular Science, Clarkson University, USA

Abstract

Tumor Differentiation Factor (TDF) is a protein produced by the pituitary and secreted into the blood stream. TDF targets breast and prostate and induces cell differentiation. However, the mechanism of cell differentiation, the TDF receptor and the TDF pathway have not been adequately investigated. Here, we provide some insights about the possible composition of the TDF-R. TDF-R may be a protein complex, composed of GRP78, HSP70 and HSP90 proteins, and all three protein subunits have a docking site for TDF-P1. The question of whether the TDF-R complex is a stable or transient/inducible complex is currently being investigated.

Keywords: Tumor differentiation factor; Protein complex; Protein-protein interactions

Introduction

Virtually, all expressed proteins in a given cell are arranged into multi-protein complexes [1-9]. Identification of individual components of those complexes is extremely important for their functional characterization [5,8,10-20]. One of the most powerful methods in identifying proteins is mass spectrometry, in particular, Liquid Chromatography-Tandem Mass Spectrometry (LC-MS/MS) [5,8,15-19]. Combination of LC-MS/MS with a biochemical purification or fractionation strategy makes LC-MS/MS even more powerful, as the protein fractionation allows the LC-MS/MS to increase the number of proteins identified from a particular sample. Affinity Purification-Mass Spectrometry (AP-MS), a combination of Affinity Purification (AP) and Liquid Chromatography-Tandem Mass Spectrometry (LC-MS/MS), allows for screening of multiple-protein complexes, and for accurate identification of their components [6,7,9,13,14,21]. Therefore, a large number of the available Protein-Protein Interactions (PPIs) data, both stable and transient, have been discovered using AP-MS [2,22-26]. Using one protein as bait in AP-MS experiments will usually lead to identification of several potential interactors, and will help to organize them into functional interacting units. Tumor Differentiation Factor (TDF) is a protein produced by the pituitary and secreted into the blood stream. TDF targets breast and prostate and induces cell differentiation. However, the mechanism of cell differentiation, TDF receptor and TDF pathway, have not been thoroughly enough investigated. Here, we provide some insights about the possible composition of the TDF-R, as well as a review of research to date.

Methods

All biochemical purification and proteomics identification of the TDF-R candidates were performed, as described in [22,23]. All STRING PPIs were performed as in [4,27-31]. All structural biology experiments were performed as in [3,22,23].

Results and Discussion

Tumor Differentiation Factor (TDF) is a protein produced by the pituitary and secreted into the blood stream [32-35]. The target organs as breast and prostate, and the final effect is cell differentiation [32-34]. Work in our lab also identified TDF in the brain, specifically in neurons, but not in the astrocytes. Additional work in our lab also focuses on identification of the mechanism of TDF-induced cell differentiation. Therefore, some of the questions that we initially asked were 1) what are the potential TDF receptor (TDF-R) candidates? 2) How does TDF-R transduce the differentiation effect across the cell membrane, 3) Is TDF a hormone? To answer to one of these questions, we used TDF-P1, a 20 amino acid peptide from the open reading frame of TDF protein, cross-linked to agarose beads to purify potential TDF-R candidates. In our experiments using DU145 prostate cancer cells and MCF7 breast cancer cells, but not in experiments using HeLa, fibroblasts or other cells, we identified several proteins from the 70 kDa and 90 kDa family of Heat Shock Proteins (HSPs) as TDF-R candidates, with glucose-regulated protein/HSPA5/GRP78, HSP70 and HSP90 being the most likely TDF-R candidates [7,22,23,34,36]. Examples of MS/MS spectra that led to the identification of these proteins as TDF-R candidates are shown in Figure 1. The results from our AP-MS experiments could potentially expand the interactome map for those proteins and lead to better understanding of their function in breast and prostate cancer.

To further investigate GRP78, HSP70 and HSP90 proteins as potential TDF-R candidates, and whether these proteins interact with each other and possibly form a protein complex, we have used String database to predict Protein-Protein Interactions (PPIs) and the protein's functional relationships with its partner proteins [28,29,31]. We took dnaK (chaperone HSP70, co-chaperone with DnaJ; *Escherichia coli* strain K-12 substr. MG1655), 78 kDa glucose-regulated protein (heat shock 70kDa protein 5 or HSPA5) and HSP90 (heat shock 90kDa alpha) as examples to study their interaction and relation to their functional partners. Network architecture of protein-protein interactions and their functional relatives can be identified and estimated using String. String network (direct and indirect relations) uses several active prediction methods that include "co-expression", "experiments" and "text mining". (Figures 2A and 2B, 3A and 3B, and 4A and 4B) display possible network of multiple interacting partner proteins (nodes) of dnaK, HSPA5 and HSP90, respectively. A node is the representative of a protein and an edge is the interaction or linkage between two protein partners. Figures 2C, 3C and 4C are the graphic representation of the observed connectivity between dnaK/GRP78/HSP90 protein and their ten predicted partners. All these views are in

*Corresponding author: Costel C Darie, Biochemistry and Proteomics Group, Department of Chemistry and Biomolecular Science, Clarkson University, 8 Clarkson Avenue, Potsdam, NY, 13699-5810, USA
E-mail: cdarie@clarkson.edu

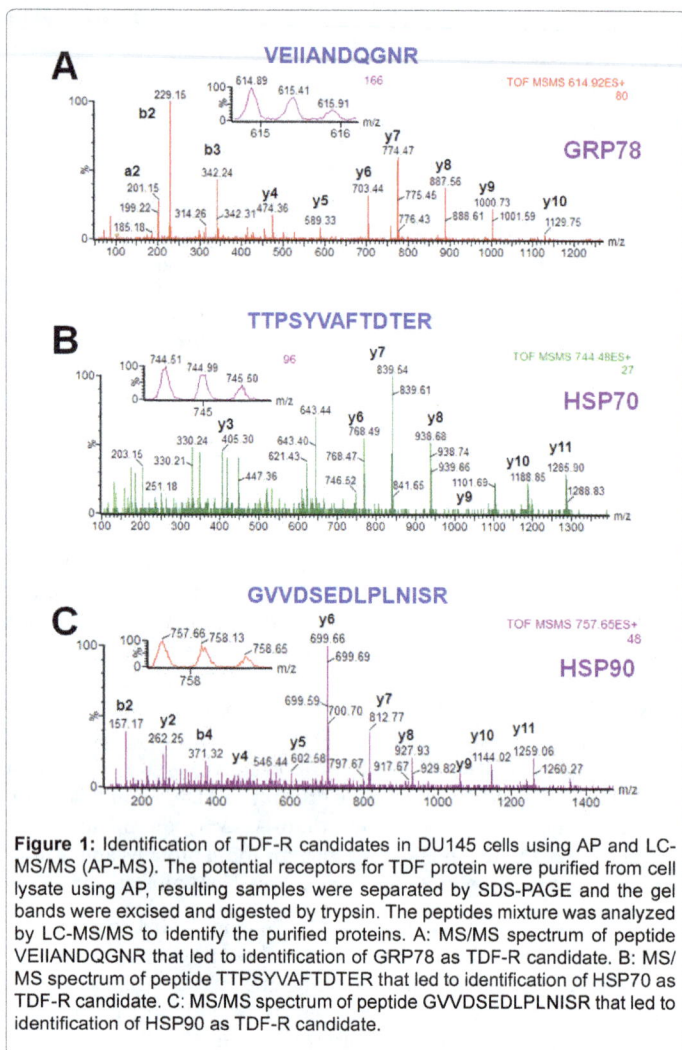

Figure 1: Identification of TDF-R candidates in DU145 cells using AP and LC-MS/MS (AP-MS). The potential receptors for TDF protein were purified from cell lysate using AP, resulting samples were separated by SDS-PAGE and the gel bands were excised and digested by trypsin. The peptides mixture was analyzed by LC-MS/MS to identify the purified proteins. A: MS/MS spectrum of peptide VEIIANDQGNR that led to identification of GRP78 as TDF-R candidate. B: MS/MS spectrum of peptide TTPSYVAFTDTER that led to identification of HSP70 as TDF-R candidate. C: MS/MS spectrum of peptide GVVDSEDLPLNISR that led to identification of HSP90 as TDF-R candidate.

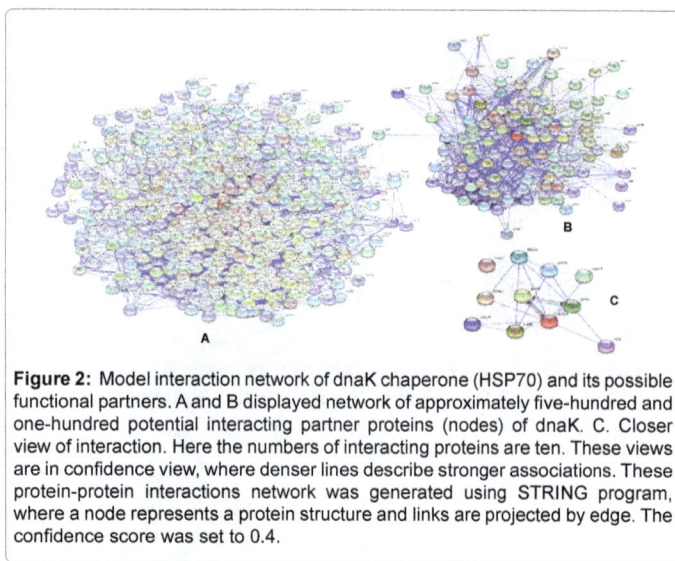

Figure 2: Model interaction network of dnaK chaperone (HSP70) and its possible functional partners. A and B displayed network of approximately five-hundred and one-hundred potential interacting partner proteins (nodes) of dnaK. C. Closer view of interaction. Here the numbers of interacting proteins are ten. These views are in confidence view, where denser lines describe stronger associations. These protein-protein interactions network was generated using STRING program, where a node represents a protein structure and links are projected by edge. The confidence score was set to 0.4.

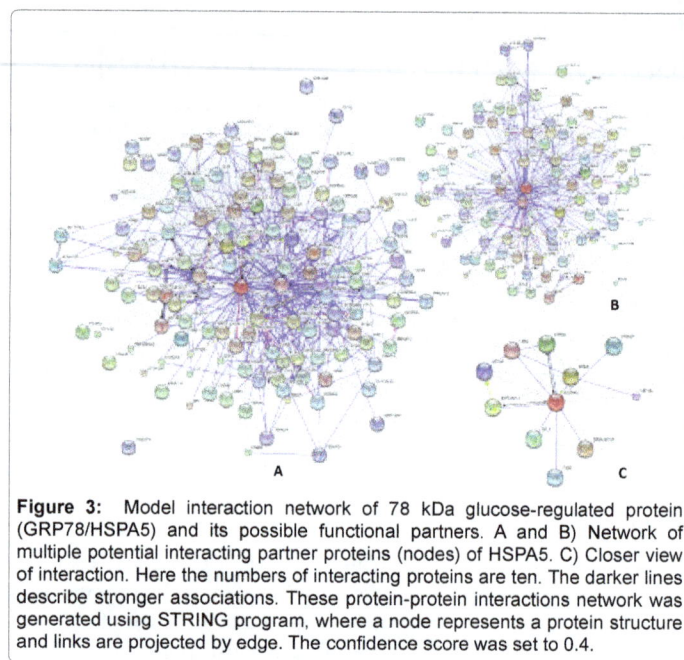

Figure 3: Model interaction network of 78 kDa glucose-regulated protein (GRP78/HSPA5) and its possible functional partners. A and B) Network of multiple potential interacting partner proteins (nodes) of HSPA5. C) Closer view of interaction. Here the numbers of interacting proteins are ten. The darker lines describe stronger associations. These protein-protein interactions network was generated using STRING program, where a node represents a protein structure and links are projected by edge. The confidence score was set to 0.4.

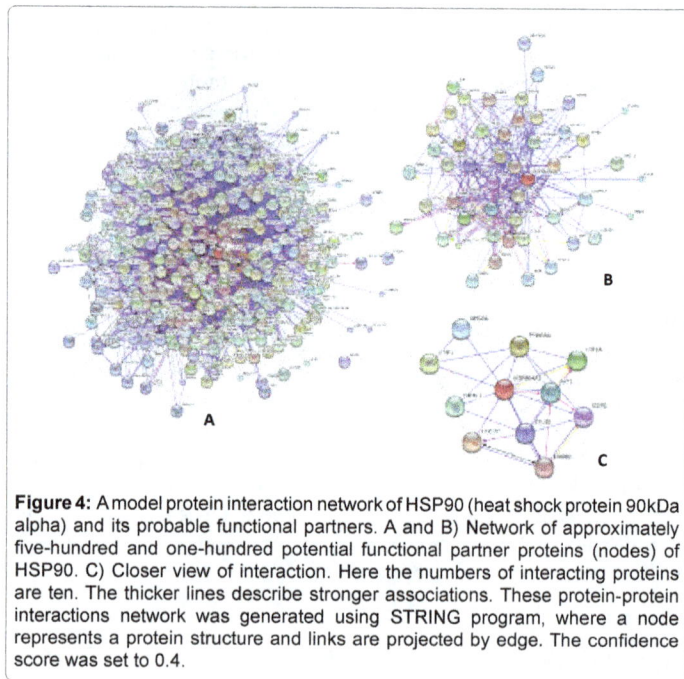

Figure 4: A model protein interaction network of HSP90 (heat shock protein 90kDa alpha) and its probable functional partners. A and B) Network of approximately five-hundred and one-hundred potential functional partner proteins (nodes) of HSP90. C) Closer view of interaction. Here the numbers of interacting proteins are ten. The thicker lines describe stronger associations. These protein-protein interactions network was generated using STRING program, where a node represents a protein structure and links are projected by edge. The confidence score was set to 0.4.

action view, where dark lines describe stronger associations. Based on the published results and the String PPI network, it looks indeed as if these three proteins do interact with each other and possibly form a protein complex.

Treatment of MCF7 human breast cancer cells and DU145 prostate cancer cells with TDF-P1 leads to differentiation of these cells; this effect is not observed on other non-breast, non-prostate cancer or normal cells [32,33]. TDF-P1 is a peptide from the N-terminal part of the TDF that has demonstrated differentiation activity on breast and prostate cancer cells as the full length protein [32,33]. Therefore, to interact with TDF-P1 and transduce a differentiation signal, the three TDF-R candidates (GRP78, HSP70 and HSP90) must be present at the cell surface. However, it is still not clear to us whether these proteins form a stable protein complex or is a transient, inducible protein complex. This question is still being investigated in our laboratory. Also, not known is whether the knock down of GRP78, HSP70 and HSP90 will prevent binding of TDF and TDF-P1 to its receptor and will promote cell differentiation. This question is currently investigated in our laboratory.

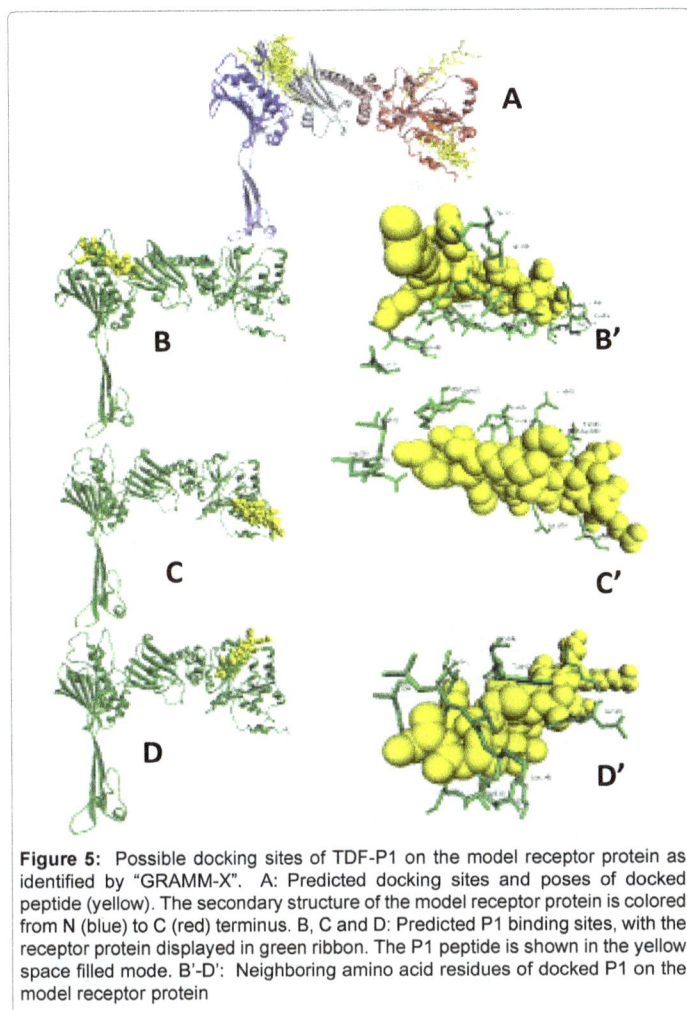

Figure 5: Possible docking sites of TDF-P1 on the model receptor protein as identified by "GRAMM-X". A: Predicted docking sites and poses of docked peptide (yellow). The secondary structure of the model receptor protein is colored from N (blue) to C (red) terminus. B, C and D: Predicted P1 binding sites, with the receptor protein displayed in green ribbon. The P1 peptide is shown in the yellow space filled mode. B'-D': Neighboring amino acid residues of docked P1 on the model receptor protein

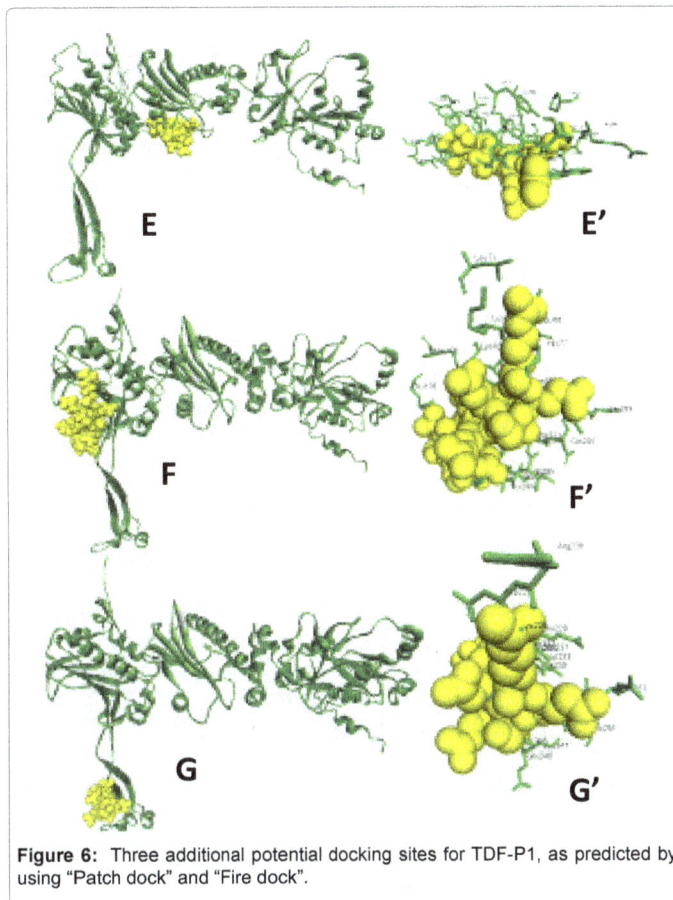

Figure 6: Three additional potential docking sites for TDF-P1, as predicted by using "Patch dock" and "Fire dock".

The next question that we asked was whether HSP90, in addition to GRP78 and HSP70, have docking sites for TDF-P1. We already knew that both GRP78 and HSP70 have several docking sites for TDF-P1. Therefore, we investigated HSP90 for possible TDF-P1 docking sites. The crystal structure 2CG9 (chain B, heat shock protein 90-alpha) was used as a template to set up a homology model of HSP 90 [37]. HSP 90 proteins are composed of N terminal, middle and C terminal domains. Figure 5A presents the homology model of 2CG9B starting from the N (colored blue) to C (colored red) terminals, and the model receptor was established using the SWISS-MODEL server [38,39]. The α-carbon Root-Mean-Square Deviation (RMSD) of 2CG9B crystal structure and homology model is 4.13 Å [40]. The docking site of P1 peptide onto the receptor model was identified using the GRAMM-X Protein-Protein Docking Web Server v.1.2.0, as used in our published work [22,23,35,36,41,42]. A second run for this identification was carried out using the "Patch dock" and "Fire dock" servers [43-46]. Detailed descriptions of these docking experiments are described in our previous papers [22,23,35,36], and Discovery Studio Visualizer 3.5 was used to plot the tentatively identified binding pockets [47].

We then used structural biology to investigate the possibility that the members of the HSP90 family of proteins are a docking place for TDF-P1. Among the first 10 highest ranked structures developed by "GRAMM-X" web server, P1 was docked onto three regions of the model receptor (Figure 5A). These three potential docking sites and neighboring amino acid residues of P1 peptide are shown in Figures

5B-5D. "Patch dock" and "Fire dock" simulation servers identified three additional potential docking sites for P1 peptide. These three additional potential docking sites and neighbor residues of docked peptide on the receptor model are shown in Figure 6. Therefore, based on these investigations, our molecular modeling experiments indeed found possible docking sites within HSP90 for TDF-P1.

Conclusions

Overall, the data allowed us to conclude that the TDF-R may indeed be a protein complex, composed of GRP78, HSP70 and HSP90 proteins, and all three protein subunits have a docking site for TDF-P1. The question of whether the TDF-R complex is a stable or transient/inducible complex is currently being investigated. Current investigations in our laboratory will also allow us to clarify whether there is only one subunit as the main TDF-R, or there is more than one natural docking site for TDF.

Acknowledgement

Part of this work was started in the Protein Core Facility, Columbia University, New York NY (Dr. Mary Ann Gawinowicz), USA. This work was supported in part by the Keep A Breast Foundation (KEABF-375-35054) and by support from the U.S. Army research office through the Defense University Research Instrumentation Program (DURIP grant #W911NF-11-1-0304). This work was also supported in part by Mary Joyce, Kenneth Sandler, Robert Matloff, and by the SciFund challenge contributors.

References

1. Ngounou Wetie AG, Sokolowska I, Woods AG, Roy U, Deinhardt K, et al. (2013) Protein-protein interactions: switch from classical methods to proteomics and bioinformatics-based approaches. Cell Mol Life Sci.

2. Ngounou Wetie AG, Sokolowska I, Woods AG, Roy U, Loo JA, et al. (2013)

Investigation of stable and transient protein-protein interactions: Past, present, and future. Proteomics 13: 538-557.

3. Ngounou Wetie AG, Sokolowska I, Wormwood K, Beglinger K, Michel TM, et al. (2013) Mass spectrometry for the detection of potential psychiatric biomarkers. J Mol Psychiat 1: 8.

4. Sokolowska I, Dorobantu C, Woods AG, Macovei A, Branza-Nichita N, et al. (2012) Proteomic analysis of plasma membranes isolated from undifferentiated and differentiated HepaRG cells. Proteome Sci 10: 47.

5. Sokolowska I, Gawinowicz MA, Ngounou Wetie AG, Darie CC (2012) Disulfide proteomics for identification of extracellular or secreted proteins. Electrophoresis 33: 2527-2536.

6. Guerrera IC, Kleiner O (2005) Application of mass spectrometry in proteomics. Biosci Rep 25: 71-93.

7. Sokolowska I, Woods AG, Wagner J, Dorler J, Wormwood K, et al. (2011) Mass spectrometry for proteomics-based investigation of oxidative stress and heat shock proteins. Oxidative Stress: Diagnostics, Prevention, and Therapy 13: 369-411.

8. Woods AG, Ngounou Wetie AG, Sokolowska I, Russell S, Ryan JP, et al. (2013) Mass spectrometry as a tool for studying autism spectrum disorder. J Mol Psychiat 1: 6.

9. Woods AG, Sokolowska I, Yakubu R, Butkiewicz M, LaFleur EM, et al. (2011) Blue native page and mass spectrometry as an approach for the investigation of stable and transient protein-protein interactions. Oxidative Stress: Diagnostics, Prevention, and Therapy 12: 341-367.

10. Darie CC (2013) Mass spectrometry and proteomics: Principle, workflow, challenges and perspectives. Mod Chem Appl 1: e105.

11. Darie CC (2013) Investigation of protein-protein interactions by Blue Native-PAGE and mass spectrometry. Mod Chem Appl 1: e111.

12. Darie CC (2013) Mass spectrometry and its application in life sciences. Aust J Chem 66: 719-720.

13. Darie CC, Litscher ES, Wassarman PM (2008) Structure, processing, and polymerization of rainbow trout egg vitelline envelope proteins. Applications of Mass Spectrometry in Life Safety 23-36.

14. Darie CC, Shetty V, Spellman DS, Zhang G, Xu C, et al. (2008) Blue Native PAGE and mass spectrometry analysis of the ephrin stimulation-dependent protein-protein interactions in NG108-EphB2 cells. Applications of Mass Spectrometry in Life Safety 3-22.

15. Aitken A (2005) Identification of post-translational modifications by mass spectrometry. The Proteomics Protocols Handbook 431-437.

16. Ngounou Wetie AG, Sokolowska I, Woods AG, Wormwood KL, Dao S, et al. (2013) Automated mass spectrometry-based functional assay for the routine analysis of the secretome. J Lab Autom 18: 19-29.

17. Sokolowska I, Ngounou Wetie AG, Roy U, Woods AG, Darie CC (2013) Mass spectrometry investigation of glycosylation on the NXS/T sites in recombinant glycoproteins. Biochim Biophys Acta 1834: 1474-1483.

18. Woods AG, Sokolowska I, Darie CC (2012) Identification of consistent alkylation of cysteine-less peptides in a proteomics experiment. Biochem Biophys Res Commun 419: 305-308.

19. Woods AG, Sokolowska I, Taurines R, Gerlach M, Dudley E, et al. (2012) Potential biomarkers in psychiatry: Focus on the cholesterol system. J Cell Mol Med 16: 1184-1195.

20. Thome J, Coogan AN, Woods AG, Darie CC, Häßler F (2011) CLOCK Genes and Circadian Rhythmicity in Alzheimer Disease. J Aging Res 2011: 383091.

21. Darie CC, Deinhardt K, Zhang G, Cardasis HS, Chao MV, et al. (2011) Identifying transient protein-protein interactions in EphB2 signaling by blue native PAGE and mass spectrometry. Proteomics 11: 4514-4528.

22. Sokolowska I, Woods AG, Gawinowicz MA, Roy U, Darie CC (2012) Identification of potential tumor differentiation factor (TDF) receptor from steroid-responsive and steroid-resistant breast cancer cells. J Biol Chem 287: 1719-1733.

23. Sokolowska I, Woods AG, Gawinowicz MA, Roy U, Darie CC (2012) Identification of a potential tumor differentiation factor receptor candidate in prostate cancer cells. FEBS J 279: 2579-2594.

24. Chautard E, Fatoux-Ardore M, Ballut L, Thierry-Mieg N, Ricard-Blum S (2011) MatrixDB, the extracellular matrix interaction database. Nucleic Acids Res 39: D235-D240.

25. Blagoev B, Kratchmarova I, Ong SE, Nielsen M, Foster LJ, et al. (2003) A proteomics strategy to elucidate functional protein-protein interactions applied to EGF signaling. Nat Biotechnol 21: 315-318.

26. Koh GC, Porras P, Aranda B, Hermjakob H, Orchard SE (2012) Analyzing protein-protein interaction networks. J Proteome Res 11: 2014-2031.

27. Franceschini A, Szklarczyk D, Frankild S, Kuhn M, Simonovic M, et al. (2013) STRING v9.1: protein-protein interaction networks, with increased coverage and integration. Nucleic Acids Res 41: D808-D815.

28. Jensen LJ, Kuhn M, Stark M, Chaffron S, Creevey C, et al. (2009) STRING 8-A global view on proteins and their functional interactions in 630 organisms. Nucleic Acids Res 37: D412-D416.

29. http://string.embl.de/

30. von Mering C, Jensen LJ, Kuhn M, Chaffron S, Doerks T, et al. (2007) STRING 7--recent developments in the integration and prediction of protein interactions. Nucleic Acids Res 35: D358-D362.

31. von Mering C, Jensen LJ, Snel B, Hooper SD, Krupp M, et al. (2005) STRING: Known and predicted protein-protein associations, integrated and transferred across organisms. Nucleic Acids Res 33: D433-D437.

32. Platica M, Chen HZ, Ciurea D, Gil J, Mandeli J, et al. (1992) Pituitary extract causes aggregation and differentiation of rat mammary tumor MTW9/PI cells. Endocrinology 131: 2573-2580.

33. Platica M, Ivan E, Holland JF, Ionescu A, Chen S, et al. (2004) A pituitary gene encodes a protein that produces differentiation of breast and prostate cancer cells. Proc Natl Acad Sci U S A 101: 1560-1565.

34. Woods AG, Sokolowska I, Deinhardt K, Sandu C, Darie CC (2013) Identification of tumor differentiation factor (TDF) in select CNS neurons. Brain Struct Funct.

35. Roy U, Sokolowska I, Woods AG, Darie CC (2012) Structural investigation of tumor differentiation factor. Biotechnol Appl Biochem 59: 445-450.

36. Sokolowska I, Woods AG, Gawinowicz MA, Roy U, Darie CC (2013) Characterization of tumor differentiation factor (TDF) and its receptor (TDF-R). Cell Mol Life Sci 70: 2835-2848.

37. Ali MM, Roe SM, Vaughan CK, Meyer P, Panaretou B, et al. (2006) Crystal structure of an Hsp90-nucleotide-p23/Sba1 closed chaperone complex. Nature 440: 1013-1017.

38. Arnold K, Bordoli L, Kopp J, Schwede T (2006) The SWISS-MODEL workspace: a web-based environment for protein structure homology modelling. Bioinformatics 22: 195-201.

39. Schwede T, Kopp J, Guex N, Peitsch MC (2003) SWISS-MODEL: An automated protein homology-modeling server. Nucleic Acids Res 31: 3381-3385.

40. Maiti R, Van Domselaar GH, Zhang H, Wishart DS (2004) SuperPose: A simple server for sophisticated structural superposition. Nucleic Acids Res 32: W590-594.

41. Tovchigrechko A, Vakser IA (2005) Development and testing of an automated approach to protein docking. Proteins 60: 296-301.

42. Tovchigrechko A, Vakser IA (2006) GRAMM-X public web server for protein-protein docking. Nucleic Acids Res 34: W310-314.

43. Andrusier N, Nussinov R, Wolfson HJ (2007) FireDock: Fast interaction refinement in molecular docking. Proteins 69: 139-159.

44. Duhovny D, Nussinov R, Wolfson HJ (2002) Efficient unbound docking of rigid molecules. Algorithms in Bioinformatics 2452: 185-200.

45. Mashiach E, Schneidman-Duhovny D, Andrusier N, Nussinov R, Wolfson HJ (2008) FireDock: A web server for fast interaction refinement in molecular docking. Nucleic Acids Res 36: W229-232.

46. Schneidman-Duhovny D, Inbar Y, Nussinov R, Wolfson HJ (2005) PatchDock and SymmDock: Servers for rigid and symmetric docking. Nucleic Acids Res 33: W363-W367.

47. Accelrys Software Inc., Discovery Studio Modeling Environment, Release 3.5, San Diego: Accelrys Software Inc., 2012.

15

Novel Hydrido-Rhodium (III) Complexes with Some Schiff Bases Derived from Substituted Pyridines and Aryl Amines

Abdulhamid Alsaygh[1], Jehan Al-Humaidi[2] and Ibrahim Al-Najjar[1]*

[1]Petrochemicals Research Institute, King Abdulaziz city for Science and Technology, P.O.Box 6086, Riyadh, 11442, Kingdom of Saudi Arabia
[2]Chemistry Department, College of Science, Princes Nora Bent Abdulrahman University, Riyadh, Saudi Arabia

Abstract

A Series of rhodium (III) cyclometallated complexes of the type (RhHCl(NC$_5$H$_2$C=N Ar (PPh$_3$)$_2$} (Ar=Substituted aryl), have been synthesized and characterized. Schiff bases derived from a substituted benzaldehyde and 2-amino pyridine substituents were allowed to react with [RhCl(PPh$_3$)$_3$] or [Rh(μ-Cl)(COD)]$_2$ in the presence of 4 equivalents of PPh$_3$ (or Ph$_2$BzP) to give Rh(III) Cyclometallated complexes, in which the imine C-H bond was added oxidatively to the rhodium metal to give (H-M-C). The complexes were characterized using IR and NMR spectroscopy confirmed by elemental micro-analysis. The absorption of the hydride ligand was inferred as trans to N-donar ligand.

Keywords: Rhodium; Schiff-bases; Phosphine complexes; Hydrido complexes; Oxidative-addition; Ligand substitution

Introduction

Although the Cyclometallation of aromatic and to a lesser extent aliphatic C-H groups is widely recognized [1,2], these are relatively little known concerning with the cyclometallation of aldehydes [3] and imine functions [4-6]. We have shown that Schiff bases of 2-substituted benzylideneaminothiazoles [5], and 2-(benzylideneamino) pyridines [6], can be form cyclometallated complexes at the imine carbon by using Rh (I) complex. A number of studies have exploited ligands such as quindine-8-carbaldehyde [3,7] and 2-(benzylideneamino) pyridines [8]. Complexation of the metal with aromatic nitrogen gives a favorable geometry for the insertion of the metal into the neighboring C-H or C-C bond [4,7,9,10]. In most recent application for ruthenium, rhodium and iridium complexes have been used as therapeutic agents and a number of kinetically inert ruthenium(II), iridium(III) and rhodium(III) complexes have been reported as inhibitors of protein kinases [11-15]. Chung-Hang Leung and Dik-Lung Ma group [14] has also actively pursued the development of kinetically inert metal complexes as inhibitors of various bimolecular targets, including DNA, enzymes and protein–protein interactions [13]. The synthesis and characterization of a variety of new rhodium (III) complexes of {N-benzylideneamino} pyridines, in which the imine C-H bond has undergone oxidative addition to the metal, are reported here.

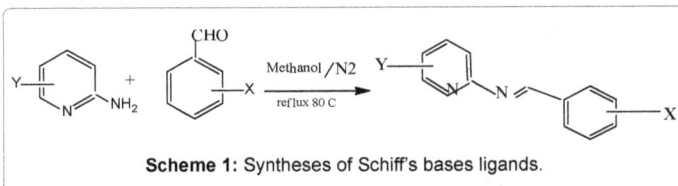

Scheme 1: Syntheses of Schiff's bases ligands.

Complex No.	Y(pyridine)	X(aryl)	Complex No.	Y(pyridine)	X(aryl)
1.	2-OH	H	8.	4-NO$_2$	4-Me
2.	H	3-Me	9.	4-Br	4-Me
3.	2-OH	3-Me	10.	H	5-Cl
4.	4-NO$_2$	3-Me	11.	2-OH	5-Cl
5.	4-Br	3-Me	12.	4-NO$_2$	5-Cl
6.	H	4-Me	13.	4-Br	5-Cl
7.	2-OH	4-Me			

Table 1: The prepared Schiff bases (free ligands).

Experimental

Materials and reagents

All chemicals used such as pyridine substituent's, benzaldehyde substituent's, RhCl$_3$xH$_2$O, phenyl phosphine (PPh$_3$) cyclo-1,5-octadiene (COD), tetrahydrofuran (THF), were obtained from Winlab, Aldrich Chemicals and Strem chemicals, respectively and were used without further purification.

Instruments

Open capillaries were used to determine melting points and were uncorrected using Gallenkamp Melting Points Apparatus. Elemental microanalysis of the separated solid chelates for C, H, N, were performed at Perkin Elmer 2400 CHN. The analyses were repeated twice to check the accuracy of the results obtained. Infrared spectra were recorded on a Nexus 470-670-760 spectrometer and FT-IR Spectrometer, Spectrum 8400s. The ^{1}H, ^{13}CNMR and ^{31}P NMR spectra were recorded using 400 MHz Joel Spectrometer.

Synthesis of ligands

All experiments were carried out under an atmosphere of nitrogen by Schlenk techniques. The Schiff bases were prepared by mixing equivalent amount of substituted benzaldehydes and 2-amino pyridine derivatives in methanol solution. This mixture was boiled under reflux with stirring for 8h, at 80°C in an oil bath, and then the mixture was concentrated by rotary evaporation to give yellow precipitate. Which was filtered off, dried, yields are 70%-80% (Scheme 1,Table 1). The results of UV, IR, ^{1}H and ^{13}C, Spectroscopy and elemental analyses for Schiff's bases were published elsewhere [16].

Rhodium compounds of {RhCl(COD)}$_2$ and {RhCl(PPh$_3$)$_3$} were prepared by literature procedures [17,18]. In this work rhodium cyclometallated complexes, were prepared by the reaction of the Schiff

*Corresponding author: Ibrahim Al-Najjar, Petrochemicals Research Institute, King Abdulaziz city for Science and Technology, P.O.Box 6086, Riyadh 11442, Kingdom of Saudi Arabia, E-mail: alnajjar@kacst.edu.sa

Complex No.	X	Y	L	Complex No.	X	Y	L
14.	2-OH	H	BzPh₂P	21.	2-OH	4-Me	PPh₃
15.	2-OH	H	PPh₃	22.	4-NO₂	4-Me	PPh₃
16.	H	3-Me	PPh₃	23.	4-Br	4-Me	PPh₃
17.	2-OH	3-Me	PPh₃	24.	H	5-Cl	PPh₃
18.	4-NO₂	3-Me	PPh₃	25.	2-OH	5-Cl	PPh₃
19.	4-Br	3-Me	PPh₃	26.	4-NO₂	5-Cl	PPh₃
20.	H	4-Me	PPh₃	27.	4-Br	5-Cl	PPh₃

Table 2: The prepared rhodium complexes (14-27).

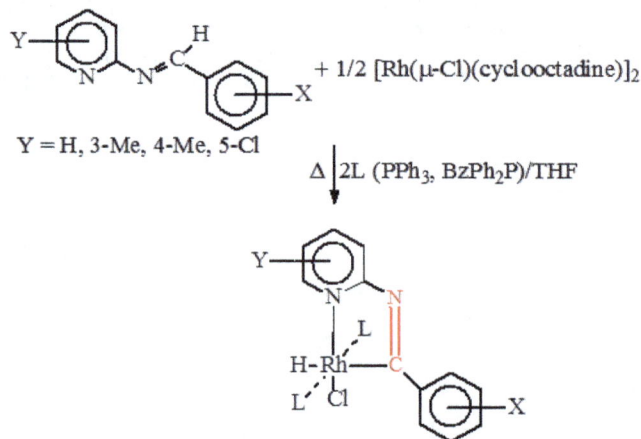

Scheme 2: Synthesis of rhodium complexes.

No.	L	X	M.P. (°C)	M.F.	Calculated (%)			Found (%)		
					C	H	N	C	H	N
14.	Ph₂BzP	2-OH	137	RhC₅₀H₄₄P₂N₂OCl	67.53	5.34	3.15	67.66	5.23	3.34
15.	PPh₃	2-OH	225	RhC₄₈H₄₀P₂N₂OCl	66.94	4.68	3.25	67.30	4.89	3.21
16.	PPh₃	H	84	RhC₄₉H₄₂P₂N₂Cl	68.49	4.92	3.26	68.28	4.71	3.47
17.	PPh₃	2-OH	100	RhC₄₉H₄₂P₂N₂OCl	67.24	4.83	3.20	67.63	4.82	3.09
18.	PPh₃	4-NO₂	140	RhC₄₉H₄₁P₂N₃O2Cl	73.46	5.15	5.24	73.35	4.98	5.34
19.	PPh₃	4-Br	98	RhC₄₉H₄₁P₂N₂BrCl	70.46	4.94	3.35	69.93	4.93	3.45
20.	PPh₃	H	98	RhC₄₉H₄₂P₂N₂Cl	68.49	4.92	3.26	68.28	4.81	3.14
21.	PPh₃	2-OH	139	RhC₄₉H₄₂P₂N₂OCl	67.24	4.83	3.2	66.93	4.86	3.34
22.	PPh₃	4-NO₂	155	RhC₄₉H₄₁P₂N₃O2Cl	73.44	5.15	5.24	73.35	5.01	5.35
23.	PPh₃	4-Br	97	RhC₄₉H₄₁P₂N₂BrCl	70.46	4.94	3.35	70.33	4.53	3.13
24.	PPh₃	H	216	RhC₄₈H₃₉P₂N₂Cl2	64.40	4.46	3.18	64.53	4.43	3.28
25.	PPh₃	2-OH	349	RhC₄₈H₃₉P₂N₂OCl2	64.37	4.38	3.12	64.77	4.48	3.99
26.	PPh₃	4-NO₂	192	RhC₄₈H₃₈P₂N₃O2Cl2	62.35	4.14	4.54	62.34	4.38	4.58
27.	PPh₃	4-Br	195	RhC₄₈H₃₈P₂N₂BrCl₂	60.14	3.99	2.92	60.07	3.96	2.85

Table 3: Physicochemical Properties of the rhodium complexes (14-27).

base with either {RhCl(PPh₃)} or with {Rh(μ-Cl)(COD)}₂. Two typical examples are described here.

1. A solution containing {RhCl(PPh₃)₃} (300 mg, 0.325 mmol) and an equivalent amount of Schiff base (in ca. 20 ml, of dry THF was boiled under reflux for 1 hr under nitrogen atmosphere. After cooling, addition of n-hexane led to precipitation of the product as a yellow powder which was filtered off (the product recrystallized twice from CH₂Cl₂/hexane, yield 40%-50% (Table 2 and Scheme 2).

2. A solution of {Rh (μ-Cl)(COD)}₂ (200 mg, 0.28 mmol) Schiff base (0.56 mmol) and PPh₃ (293 mg, 1.12 mmol) in ca. 20 ml of dry THF was boiled under reflux for 1 hr addition of n-hexane induced precipitation of the product, which was filtered off (the product could be recrystallized form CH₂Cl₂/hexane (Table 2, Scheme 2).

Results and Discussion

The physical, analytical data and UV, IR, ¹H, ¹³C-NMR Spectroscopy for Schiff bases were published elsewhere [16]. The corresponding Rh-complexes of different Schiff base ligand are investigated also by analytical, physical and different spectroscopy methods (Tables 3-5).

Characterization of Rh-Complexes

Infrared Spectra: Infrared spectra of the complexes were recorded to confirm their structure. The vibration frequencies and their tentative assignments for imines ligand (Scheme 1) and their Rh-complexes were assigned by comparison with the vibrational frequencies of the free ligand and their related complexes. The main futures in the infrared of the complexes is the shift of the stretching frequencies of the azomethine (-C=N-) group of the transition metal complexes to lower frequencies

Complex No.	X	Y	L	δ ¹H Hydride (ppm)	δ ³¹P{¹H} (ppm)	²J(³¹P-¹H) (Hz)	¹J(¹⁰³Rh-¹H) (Hz)	¹J(¹⁰³Rh-³¹P) (Hz)
14.	2-OH	H	BzPh₂P	-11.78	25.6	11.00	14.3	105.0
15.	2-OH	H	PPh₃	-11.43	30.2	11.00	13.2	112.0
16.	H	3-Me	PPh₃	-11.20	30.30	12.40	12.3	111
17.	2-OH	3-Me	PPh₃	-11.35	3069	12.42	12.4	112.5
18.	4-NO₂	3-Me	PPh₃	-11.21	30.63	12.45	13.7	114.5
19.	4-Br	3-Me	PPh₃	-11.27	30.65	12.44	13.3	114.3
20.	H	4-Me	PPh₃	-11.19	33.36	11.60	13.44	114.6
21.	2-OH	4-Me	PPh₃	-11.78	33.7	11.00	13.90	112.0
22.	4-OH₂	4-Me	PPh₃	-11.29	31.86	12..24	13.44	114.6
23.	4-Br	4-Me	PPh₃	-11.32	32.69	11.00	12.20	118.0
24.	H	5-Cl	PPh₃	11.19 -11.19	34.67 18.79	11.20	13.41	121.5 98.7
25.	2-OH	5-Cl	PPh₃	-11.41	n	11.23	14.10	n
26.	4-NO₂	5-Cl	PPh₃	-11.31	n	11.0	14.42	n
27.	4-Br	5-Cl	PPh₃	-11.32	20.10 26.82	11.30	14.52	114.3 104.3

n=not measured

Table 4: ¹H and ³¹P NMR (δ ppm) and coupling constants (Hz) of the rhodium complexes (14-27).

Complex No.	X	Y	δ C(7) (ppm)
17.	2-OH	3-Me	235.56
18.	2-NO₂	3-Me	225.16
19.	4-Br	3-Me	236.24
20.	H	4-Me	237.60

Table 5: ¹³C-NMR for iminoyl carbon (C-7) (δ ppm) in the rhodium complexes (17-20).

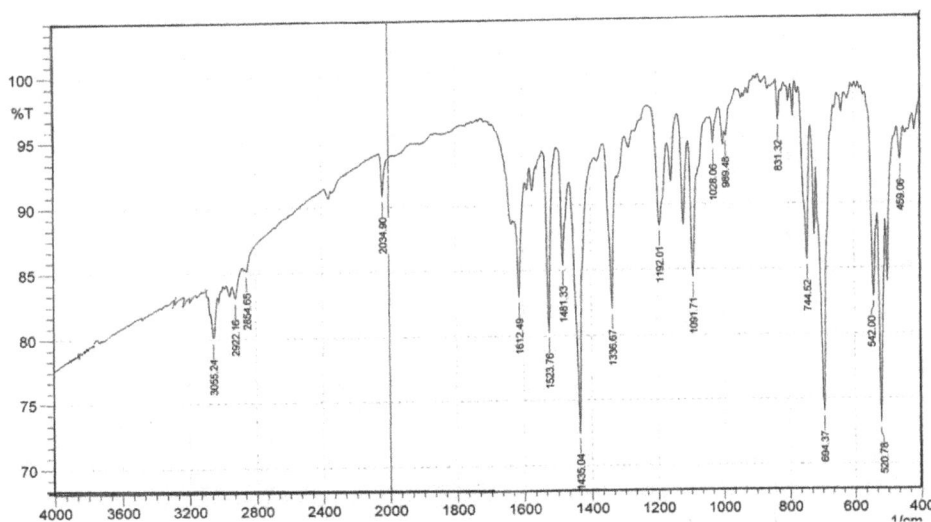

Figure 1: IR Spectra for complex (22).

in the range, 1600-1576 cm⁻¹, compared with free imine ligand, v(1690-1620 cm⁻¹) due to the coordination of the azomethine moiety, v(C=N) to the metal [19]. Further evidence of the bonding is given by the observation of new bands in the spectra of the metal complexes of medium or week intensity at the region 467-435 cm⁻¹ due to v(M-N) stretching vibration supporting the involvement of the nitrogen atom of the azomethine group via coordination [20,21] (Figure 1), complex (22). Further evidence come from the spectra of ¹H, ¹³C and ³¹P NMR (Tables 4 and 5).

¹H, ¹³C and ³¹P NMR Spectra: The ¹H, ¹³C and ³¹P NMR spectra of the rhodium complexes have been studied in CDCl₃. The ¹H NMR spectrum of each of the new rhodium complexes in CDCl₃, shows a hydride resonance between δ11.19-11.78 ppm (Table 4). The imines C-H signals for the starting free imines appear at δ 9.01-9.44 ppm and after complexation these signals are absent, providing evidence for insertion of Rh metal into the C-H bond of the imines. Strong confirmation evidence comes from appearance of the resonance of the hydride signal in each complex at high field [22,23], ca. (average) δ -11.29 ppm. The hydride signals in the complexes are split by compiling to two equivalent ³¹P nuclei of the rhodium complex. As both of these spin-spin couplings are ca. 11.00-14.52Hz, frequently. ¹J (¹⁰³Rh-¹H). Hz, and 2J (³¹P-¹H), ca. 11.00 - 12.45 Hz (Table 4). The hydride multiple often appears as a pseudo quartet, but at higher resolution studies usually reveal the expected doublet of triplets (Figure 2 and Figure 3), complexes (23 and 24) .The phosphine (PPh₃) rhodium complexes

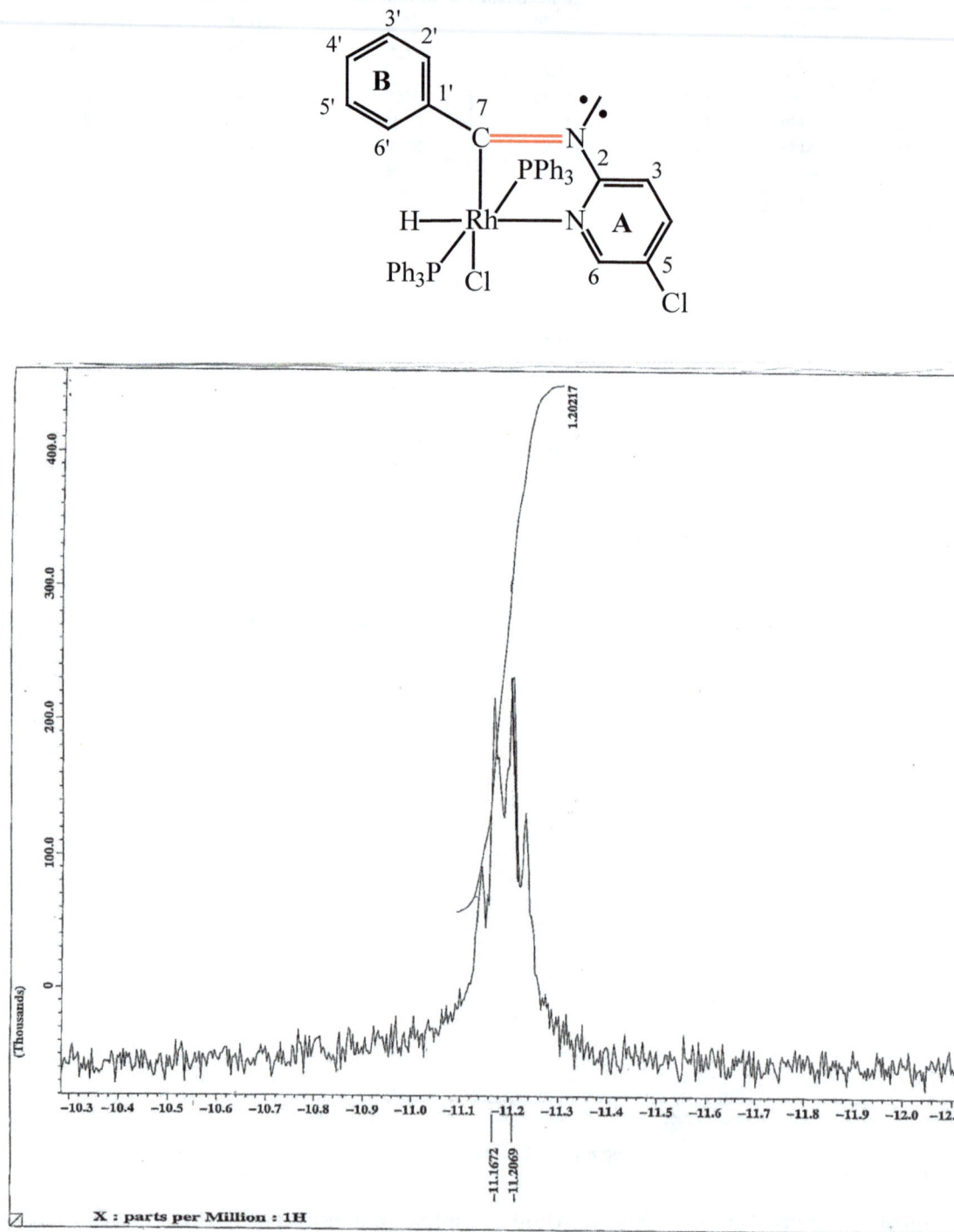

Figure 2: ^{1}H NMR (hydride) spectra for complex (23).

(Figure 4 and Figure 5), complexes (24 and 27), show a ^{31}P signal at ca. 18.79-34.67 Hz, (Table 4), with ^{1}J(^{103}Rh-^{31}P) 98.7-118.0 Hz as a doublet in keeping with previous report [3,10,16], depending on the type of the substituent group on pyridine ring (Table 4). The majority of the rhodium imine hydride complexes are only moderately soluble in most organic solvents. The signal of ^{13}C=N of the imino group is observed at ca. δ 225.06-237.60ppm (Table 5). The ^{13}C [^{1}H] NMR spectrum, in particular the signal from the metal-bonded carbon atom, is consistent with the presence of the cyclometallated ring [22,23]. The

signal from the metal-bonded carbon, C(7) (iminoyl carbon), appear as a doublet or triplets owing to coupling of two equivalent ^{31}P nuclei and the ^{103}Rh nucleus, whereas the corresponding signal from the uncomplexed imines is found at ca. δ146.24-164.97 ppm [22]. This low-field position for C(7) has been observed in other cases in what a chelating atom is incorporated in a five member-ring [24], and is not unusual for a cyclometallated sp^2 carbon [25], similar to carbene-carbon. The remaining ^{1}H and ^{13}C data are as expected. Steric effects are extremely important to structures, spectroscopic properties, and

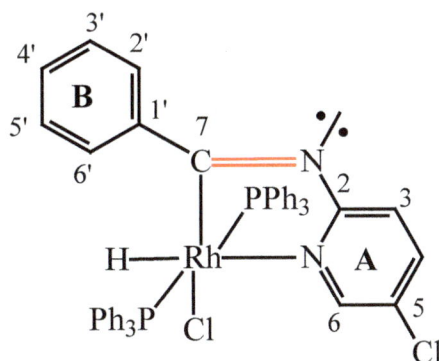

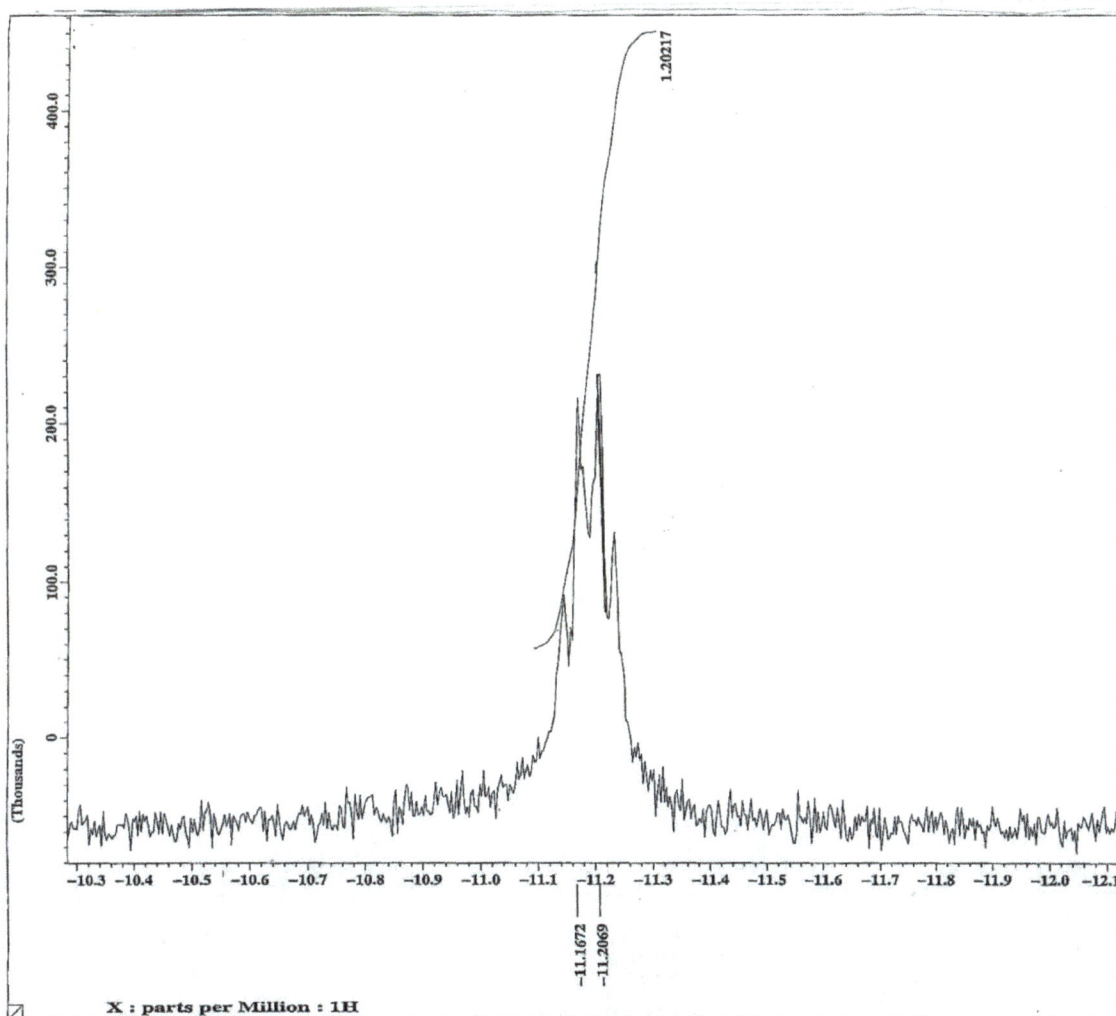

Figure 3: ^{1}H NMR (hydride) spectra for complex [24].

chemical behavior of phosphorus ligands and their complexes [26]. In this study two types of phosphorus ligands (PPh$_3$ and PBzPh$_2$) were used with different steric and electronic effects. The cone-angle data of Tolman [27] allows some comparisons of relative ligand steric effects to be made and demonstrates phosphine ligands such as PBzPh$_2$ (ca. 153°) and PPh$_3$ (ca. 145°). Increasing the size of the substituents on phosphorus will tend to reduce the s character in the phosphorus long

pair, thus decreasing [1]J(M-P)[21]. Data from Table 5, shows the δ [31]P [^{1}H] at 25.60ppm, with [1]J ([103]Rh-^{31}P), 112.0Hz when ligand BzPh2P and δ [31]P [^{1}H] at 105.0Hz with [1]J ([103]Rh-^{31}P), 105.0Hz when ligand PPh$_3$ [27,28].

The position of the ligand signals in both IR(v Rh-H, 2034.9cm^{-1}) for complex 22 (Figure 1) and ^{1}H-NMR (δ-11.29ppm) Spectra, are as expected for a Rh-H bond trans to N-donor ligand. Furthermore,

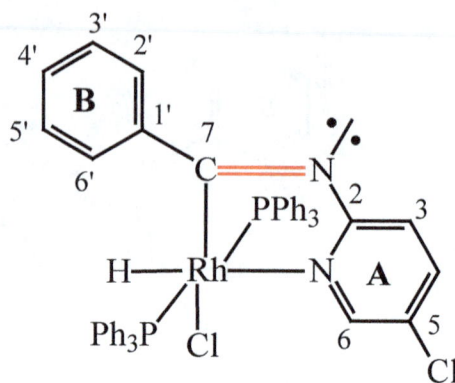

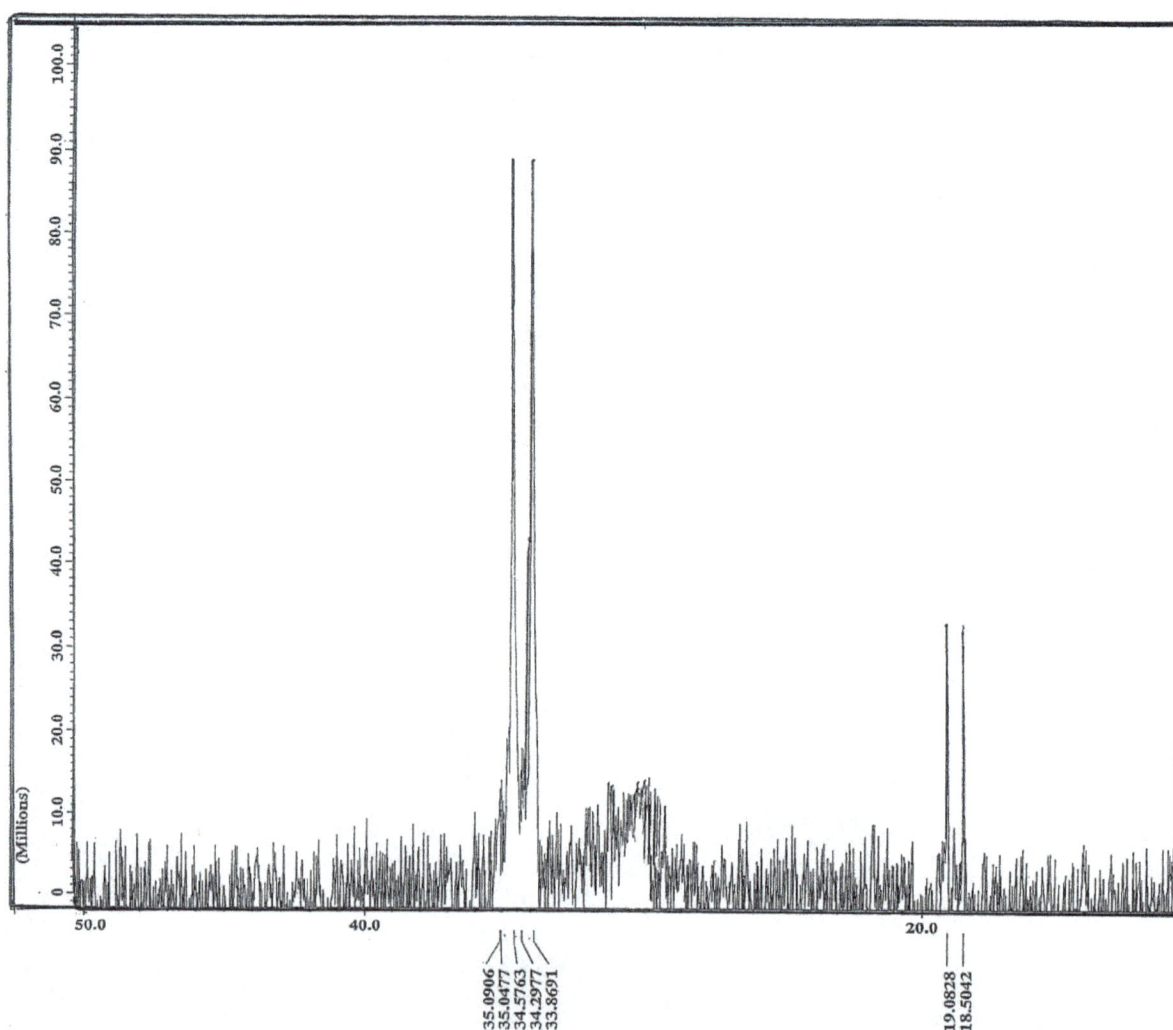

Figure 4: ^{31}P NMR spectra for complex (24).

the $^1J(^{31}P\text{-}1H)$ value is consistent with a hydride located cis to two magnetically equivalent PPh$_3$ groups [29], which in turn are mutually trans, as inferred from ^{31}P [^{1}H]NMR spectrum (Table 4).

Interestingly, the hydride and ^{31}P NMR spectrum of complexes 24 and 27, the ^{31}P-NMR presented in two types of spectrum, for ^{31}P-NMR-spectra , which δ -observed at 34.67 and 18.79 ppm (for complex 24), and at δ 20.10 and 26.82 ppm (for complex 27), with ^2J (^{31}P-^{1}H) 11.20Hz

and 11.30 Hz , (Figure 4 and Figure 5) respectively, and with $^1J(^{103}$Rh-^{31}P), of 121.50 Hz , 98.70 Hz and 121.50, 104.2 Hz respectively (Table 4).

This result may be due to complex instability. The similarity of present of Cl- atom at C5 results of two or three ^{31}P absorption spectrum. By substitution of Br-atom at C-4 of aryl ring (Figure 5) a significant change in signal of ^{31}P was recorded in Figures 4,5 and Table 4. It was also observed that the signal for C-7 (iminoyl carbon ^{13}C=N) is

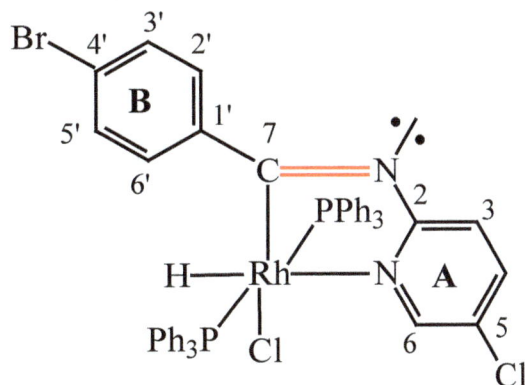

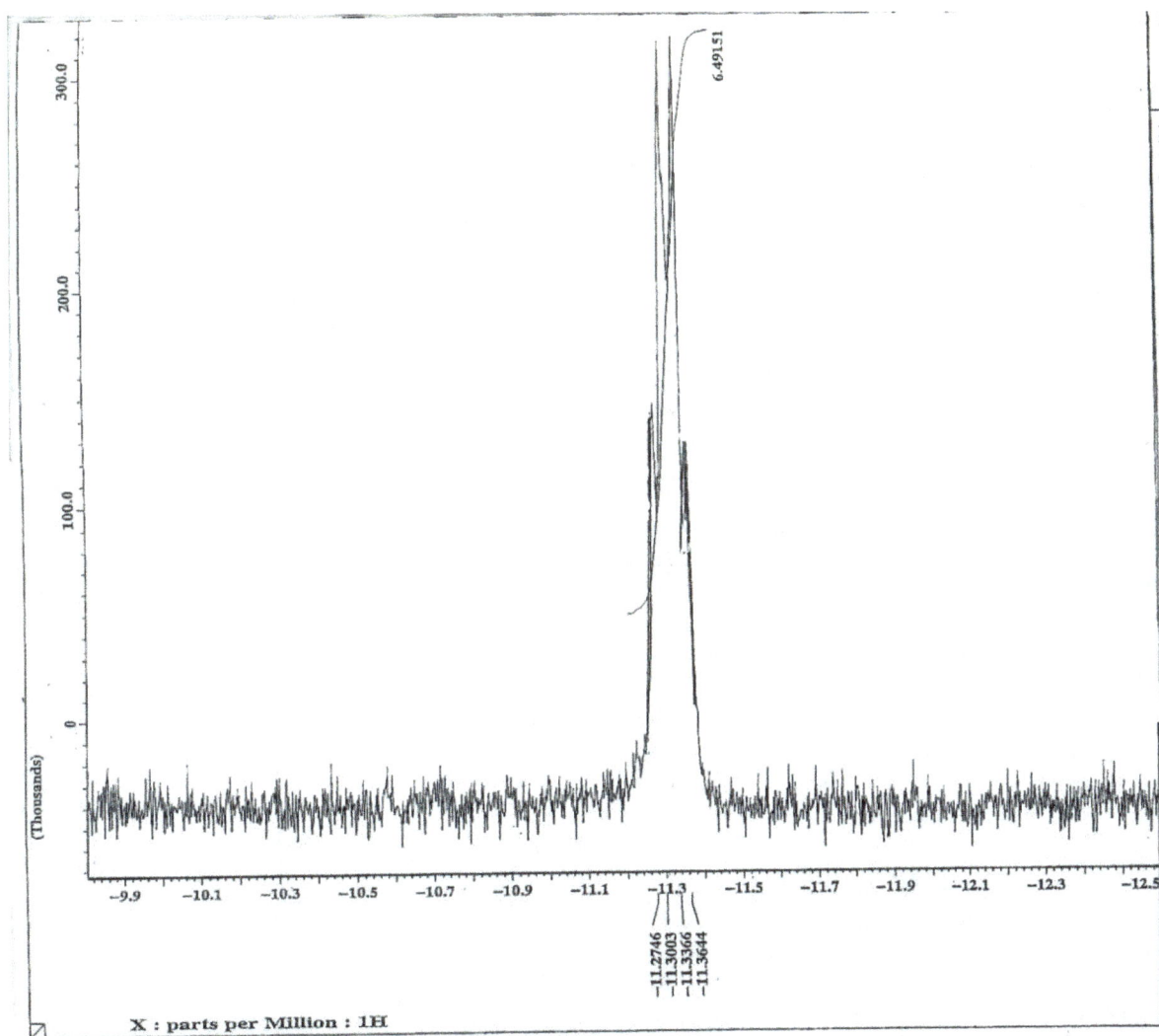

Figure 5: ¹H NMR spectra for complex (27).

at low magnetic field, at δ225.16-237.60ppm with ¹J (¹⁰³Rh-¹³C), 32-33 Hz and ²J (³¹P-¹³C), 8-9Hz (Table 5).

The rhodium complexes are only moderately soluble in organic solvents, and so we have not obtained many ¹³C spectra, however, some ¹³C (7) data for few complexes are shown in Table 5. The signal for C-7 is all at 225.16-237.60 ppm, whereas the uncomplexed imines C-7 signal is found at δ159.39-164.97 ppm. This low field position is suggestive of carbine-like properties; however, the δ ¹³C=N for complex (24) is observed at low magnetic field at δ 237.67ppm (Table 5 and Figure 6).

Unfortunately, treatment of some of imines prepared in this work

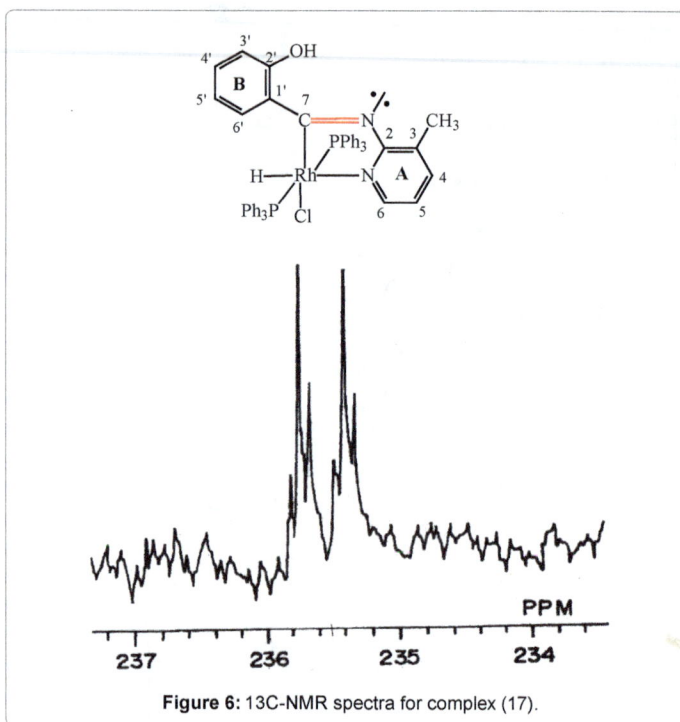

Figure 6: 13C-NMR spectra for complex (17).

with 1,5-hexadiene in toluene at 110°C for 6 h under [RhCl (PPh$_3$)$_3$] in screw-capped vial, gives only imonoacyl rhodium(III) complex. The chromatographic results show no indication of forming hex-5'-enylketimine. These results indicated that the bond between rhodium and hydrogen is not active enough, very stable and can't go for further reactions.

Conclusion

The new cyclometallated rhodium complexes have been characterized by elemental analysis, UV, IR, ^{1}H, ^{31}P (occasionally) and ^{13}C-NMR-spectroscopy. Interestingly the hydride ligand signal in IR (v 2034.9 cm^{-1} and ^{1}H-NMR (δ -11.29 ppm), complex (22). The result obtained from the spectra was expected for Rh-H group trans position to the N-donor ligand.

However, the ^{31}P-NMR for some cyclometallated complexes shows signal at δ 31.86ppm, complex (22). Furthermore, the ^{2}J (^{31}P-1H) value account for H cis to two magnetically equivalent PPh$_3$-groups, which in turn are mutually trans, as inferred from ^{31}P(1H) NMR spectrum. This result is supported from ^{1}H and ^{13}C NMR spectra.

Interestingly, the ^{13}C-NMR of the iminoyl carbon (^{13}C=N) signal in Rh(III) (δ 225.16-237.60 ppm). This low-field position for cyclometallated complexes is suggestive of carbene-like properties. The result from the study indicated that the bond between rhodium and hydrogen is not active enough, very stable and can't go for further reactions.

Acknowledgement

The author would like to thank the Research Centre, College of Science, Princess Nora Bent Abdulrahman University and King Abdulaziz city for Science and Technology for the financial support to this research project (AT-17-171).

References

1. Constable EC (1984) Cyclometallated complexes incorporating a heterocyclic donor atom; the interface of coordination chemistry and organometallic chemistry. Polyhedron 3: 1037-1057.

2. Bruce MI (1977) Cyclometalation Reactions.Angewandte Chemie International Edition in English 16: 73-86.

3. Albinati A, Anklin CG, Ganazzoli F, Ruegg H, Pregosin PS (1987) Preparative and proton NMR spectroscopic studies on palladium(II) and platinum(II) quinoline-8-carbaldehyde (1) complexes. X-ray structures of the cyclometalated acyl complex PdCl(C(O)C9H6N)PPh3).cntdot.PPh3 and trans-PtCl2(Q)(PEt3). Inorg Chem 26: 503-508.

4. Albinati A, Arz C, Pregosin PS (1987) Synthesis, structure and NMR spectroscopy of some rhodium(III) cyclometallated Schiff's base complexes derived from 2-benzylidene-3-methylpyridines. Crystal structure of [RhHI{2-(3-nitrobenzylidene)-3-methylpyridine}; (PPh$_3$)$_2$]. Journal of Organometallic Chemistry 335: 379-394.

5. El-Baih FEM, Abu-Loha FM, Gomma Z, Al-Najjar IM (1994) Synthesis and characterization of some rhodium(III) cyclometallated complexes of 2-substituted benzylideneamino thiazoles.Transition Metal Chemistry 19: 325-328.

6. Amin HB (1997) Synthesis and Characterization of Some Rhodium (III) CyclometallatedSchiff's Base Complexes Derived from 2-Benzylidene Amino substituted Pyridines.J King Saud University, Science 9: 65-75.

7. Suggs JW, Wovkulich MJ, Cox SD (1985) Synthesis, structure, and ligand-promoted reductive elimination in an acylrhodium ethyl complex. Organometallics 4: 1101-1107.

8. Suggs JW (1979) Activation of aldehyde carbon-hydrogen bonds to oxidative addition via formation of 3-methyl-2-aminopyridyl aldimines and related compounds: rhodium based catalytic hydroacylation. J Am Chem Soc 101: 489-493.

9. Suggs JW, Jun CH (1985) Metal-catalysed alkyl ketone to ethyl ketone conversions in chelating ketones via carbon–carbon bond cleavage. J Chem Soc Chem Commun 92-93.

10. Meiswinkel A, Werner H (2004) Five- and six-coordinate hydridorhodium(III) complexes containing metalated Schiff-bases as ligands.InorganicaChimica Acta 357: 2855-2862.

11. Leung CH, He HZ, Liu LJ, Wang M, Chan DSH, et al. (2013) Metal complexes as inhibitors of transcription factor activity.Coordination Chemistry Reviews 257: 3139-3151.

12. Zhong HJ, Leung KH, Liu LJ, Lu L, Chan DS, et al. (2014) Antagonism of mTOR Activity by a Kinetically Inert Rhodium(III) Complex. ChemPlusChem79: 508-511.

13. Liu LJ, Lin S, Chan DS, Vong CT, Hoi PM, et al. (2014) A rhodium(III) complex inhibits LPS-induced nitric oxide production and angiogenic activity in cellulo. J Inorg Biochem 140C: 23-28.

14. Ma DL, Liu LJ, Leung KH, Chen YT, Zhong HJ, et al. (2014) Antagonizing STAT3 Dimerization with a Rhodium(III) Complex.Angew Chem Int Ed Engl 53: 9178-9182.

15. Leung CH, Yang H, Ma VPY, Chan DSH, Zhong HJ, et al. (2012) Inhibition of Janus kinase 2 by cyclometalated rhodium complexes.Med Chem Commun 3: 696-698.

16. Alsaygh A, Al-Humaidi J, Al-Najjar I (2014) Synthesis of Some New Pyridine-2-yl-Benzylidene-IminesInternational Journal of Organic Chemistry 4: 116-121.

17. Colquhoun HM, Holton J, Thompson DJ, Twigg MV (1984) New Pathways for Organic Synthesis: Practical Applications of Transition Metals. Springer Press Science.

18. Osborn JA, Wilkinson G (1967) Tris (triphenylphosphine) halorhodium(I). Inorganic Syntheses 10: 67- 71.

19. Nakamoto K(1997) Infrared and Raman Spectra at Inorganic and Coordination Components (5th edition). John Wiley and Sons, New York.

20. Chohan HZ, Naseer MM (2007) Organometallic based biologically active compounds: synthesis of mono- and di-ethanolamine derived ferrocenes with antibacterial, antifungal and cytotoxic properties.Applied Organometallic Chemistry 21: 1005-1012.

21. El-Shiekh SM, Abd-Elzaher MM, Eweis M (2006) Synthesis, characterization and biocidal studies of new ferrocenylthiadiazolo-triazinone complexes.Applied Organometallic Chemistry 20: 505-511.

22. Dowerah D, Radonovich LJ, Woolsey JF, Heeg MJ (1990) Reaction of 2-((. alpha.-R-benzylidene)amino)pyridines [R=CH$_3$, 4-(CH$_3$O)C$_6$H$_4$] with RhCl(L)3 or Rh$_2$Cl$_2$(CO)$_4$: formation and structure of a rhodium(II) dimer. Organometallics 9: 614-620.

23. Suggs JW, Chul-Ho J (1984) Directed cleavage of carbon-carbon bonds by transition metals: the α-bonds of ketones. J Am Chem Soc 106: 3054-3056.

24. Giordano G, Crabtree RH (1979) Preparation of (1,5-cyclooctadiene) chlororhodium (I) dimer; rhodium complex. Inorganic Syntheses 19: 218-220.

25. Garrou PE (1981) DELTA.R-ring contributions to phosphorus-31 NMR parameters of transition-metal-phosphorus chelate complexes. Chem Rev 81: 229-266.

26. Foot RG, Heaton BT (1973) Metallation of 2-vinylpyridine by rhodium (III). J Chem Soc ChemCommun838-839.

27. Tolman CA (1977) Steric effects of phosphorus ligands in organometallic chemistry and homogeneous catalysis. Chem Rev 77: 313-348.

28. Al-Najjar IM (1987) ^{31}P and ^{195}Pt NMR characteristics of new binuclear complexes of [Pt$_2$X$_4$](PR$_3$)$_2$] cis/trans isomers and of mononuclear analogs. InorganicaChimica Acta 128: 93-104.

29. Kaesz HD, Sailant RB (1972) Hydride complexes of the transition metals. Chem Rev 72: 231-281.

16

The Use of Fourier Transform Infrared (FTIR) Spectroscopy and Artificial Neural Networks (ANNs) to Assess Wine Quality

Snezana Agatonovic-Kustrin[1]*, David W. Morton[1] and Ahmad Pauzi Md. Yusof[2]

[1]*School of Pharmacy and Applied Science, La Trobe University, Australia*
[2]*Physiology Department, Medical School, Universiti Teknologi Mara, Selangor, Malaysia*

Abstract

The aim of this study was to develop a simple method to assess wine quality from its Fourier Transform Infrared Spectroscopy (FTIR) spectrum with minimal or no sample preparation. FTIR spectral data of selected wine samples, grape variety, wine barrel type, wine type and production year were correlated with total phenolic content, total and volatile acidity and alcohol content using Artificial Neural Networks (ANNs). A total of 20 (2 whites and 18 reds) different wines used in this study came from three different states across Australia; New South Wales, Victoria and South Australia.

FTIR spectroscopy proved to be a promising technique that provides a rapid and accurate method in the quality assessment of wine. A plot of the values predicted by the validated ANN models showed excellent correlation with the experimentally measured values for acetic acid concentration, alcohol content, total phenols, and total acidity (r=0.898-0.942).

Keywords: Fourier transform infrared spectroscopy; Artificial neural network; Wine quality assessment; Polyphenolic content

Introduction

It is known that moderate consumption of red wine and fruit juices can reduce the risk for cardiovascular disease [1,2]. The protective effects of wine have been attributed to polyphenols that are efficient scavengers of free radicals and breakers of lipid peroxidative chain reactions [3]. One of the most fascinating observation is the 'French paradox', a term first coined by Dr. Serge Renauld, in the 1980s [4,5] that refers to the fact that French population have an incredible low coronary heart disease death rates despite high intake of dietary cholesterol and saturated fat [5,6]. Numerous studies have documented the health benefits of red wine consumption, including anti-oxidative, anti-carcinogenic, anti-inflammatory, anti-cardiovascular and antibacterial properties [7,8]. Grapes and wines are rich in a large number of polyphenolic compounds, belonging to non-flavonoids, flavonoids and phenolic-protein-polysaccharide complexes, which possess high antioxidant activity and are believed to reduce and prevent oxidative stress related diseases [9,10], as well as several organic acids, such as tartaric and malic acids, which have antimicrobial effects, especially at the low pH of wine [11,12]. The amount of these potentially beneficial compounds present in red wine usually varies depending on the variety of grapes and the vinification process used [13]. The major polyphenolic compounds found in red wine contribute significantly to the major organoleptic qualities such as mouth feel, taste and colour, and therefore play an important role in the overall quality of the wine. It is believed that wine quality variations are related to both the origin (structure related) and quantity (concentration factor) of the polyphenols present in red wine [14]. Polyphenolics in wine are responsible for varietal and flavor characteristics of red wines such as color and tannin characteristics [15]. Phenolics also affect the sensory characteristics of wines, contributing to bitterness and astringency. Astringency and bitterness are produced primarily by flavonoids that are extracted from the skins and seeds of grapes. The types and amounts of different phenolic compounds present have been used as broad indicators of wine quality and good correlation has been obtained between several aspects of the phenolic content and assessed quality of red wine [16]. However, it would be useful to be able to perform a single measurement that is representative of a wine's composition universally correlates with perceived wine quality [17].

The development of more effective and efficient methods to assess grape and wine quality is important to the wine industry. It is desirable that these methods require minimal sample preparation and are able to produce rapid results, preferably providing information on multiple parameters simultaneously. The aim of this study was to develop a simple and rapid method based on Fourier Transform Infrared Spectroscopy (FTIR) spectroscopy combined with Artificial Neural Network data modeling to assess multiple quality indicators of wine samples. Using non-destructive FTIR spectroscopy with a horizontal Attenuated Total Reflectance (ATR) accessory enables wine spectra to be obtained using relatively small volumes (around 0.5 to 1 mL) of wine with minimal sample preparation and reagent consumption. Mid-IR spectrometry has been previously used in the analysis of a number of foods, including wine. The typical analytes that have been measured using this technique are ethanol, pH, organic acids, sugars and glycerol [17,18]. However, only a few studies have applied FTIR to analyse selected polyphenolic compounds such as tannins [19] and anthocyanins [20] and total antioxidant capacity [21]. The use of Artificial Neural Networks (ANNs), as a non-linear statistical data modelling tool was chosen to correlate FTIR spectra of selected wine samples, grape variety, wine barrel (oak type), wine type (red or white) and production year with total phenolic content, total and volatile acidity, and alcohol content. The aim of this modelling was to: (1) successfully predict the overall polyphenolic content of a wine sample from its FTIR spectra; and (2) to determine how parameters such as alcohol content, pH etc. affect the overall polyphenolic content in red wine. This type of information should enable winemakers to better optimise the concentration of polyphenolic compounds during the winemaking process.

*Corresponding author:** Snezana Agatonovic- Kustrin, Associate Professor of Pharmacy, School of Pharmacy and Applied Science, Faculty of Science, Technology and Engineering, La Trobe University, Bendigo, Post Box 199, Bendigo 3552, Australia, E-mail: s.kustrin@ latrobe.edu.au

Wine sample	State/ Region	Type of Wine	Varietal Composition	Year	Alcohol Content	Type of Oak
1	VIC	Red	Cabernet	2008	14.5%	French
2	VIC	Red	Cabernet	2006	14%	French
3	VIC	Red	Cab. Sauvignon 31% Merlot 30% Cab. Franc 24% Malbec 15%	2008	15.6%	
4	VIC	Red	Shiraz Cabernet	2007	15%	French
5	VIC	Red	Shiraz	2008	15%	French
6	NSW	Red	Shiraz	2008	13.0%	Unknown
7	VIC	White	Semillon	2009	14%	French
8	VIC	Red	Cab. Sauvignon 60% Cab. Franc 20% Merlot 20%	2008	13.3%	French
9	VIC	White	Riesling	2008	13%	Tank
10	VIC	Red	Shiraz	2008	14%	American
11	SA	Red	Shiraz	2009	14%	French American
12	SA	Red	Shiraz 96% Petit Verdot 4%	2008	14%	American
13	VIC	Red	Shiraz	2010	13.5%	French
14	VIC/ SA	Red	Shiraz Merlot	2009	13.5%	Unknown
15	VIC	Red	Cabernet Merlot	2009		
16	VIC	Red	Shiraz	2010	15%	American
17	VIC	Red	Shiraz	2001	15%	American
18	VIC	Red	Shiraz	2008	15.5%	American
19	VIC	Red	Shiraz 87% Malbec 7% Cab. Sauvignon 6%	2008	14%	American
20	VIC	Red	Cabernet Sauvignon	2008	15%	American

Table 1: Wine samples collected from different wineries with the information found on the label.

Materials and Methods

Wine samples

A total of 20 different wines (2 whites and 18 reds) used in this study came from three different states across Australia; New South Wales, Victoria and South Australia. The wines selected varied considerably in regards to growing region, varietal composition, year, alcohol content and type of oak used (Table 1).

Total acidity

Total acidity (free acid) or titratable acidity of the wine samples, expressed the concentration of tartaric acid (g/L), was determined titrimetrically using a standard 0.05M NaOH solution with phenolphthalein indicator. The color of the end point was a stable gray for red wines and a faint pink color for white wines. In order to overcome the problem associated of the intense color of the red wines, samples were diluted by 1 in 75 using distilled water prior to the titration. All pH and total acidity determinations were replicated twice.

Volatile acidity

The volatile acidity, expressed as the amount of acetic acid (g/L) was determined spectrophotometrically using an enzymatic analysis kit (Vint essential Laboratories, Dromana, Australia) with absorbance measurements made at a wavelength of 340 nm.

Total phenolic content

The total phenolic concentration was determined according to the Folin-Ciocalteu colorimetric method. Samples of wine (1.00 mL) were first diluted with distilled water (4.00 mL). An aliquot (0.200 mL) of the diluted wine sample and 1.00 mL of Folin-Ciocalteu reagent (Sigma Chemicals Co, St. Louis, MO, USA), were added to a 20.00 mL volumetric flask. Exactly after 1 minute, 4.00 mL of sodium carbonate (20% w/v) (Merck, Vic, Australia) was added and the solution was made to a total volume of 20.00 mL using distilled water. Finally, the mixture was allowed to stand at room temperature in the dark for 30 min after which time the absorbance of the solution was measured at 750 nm. The total polyphenolic concentration was calculated from a calibration curve using Gallic acid (Aldrich Chemical Co (WI, USA) as a standard (1-6 mg/L). Polyphenolic concentration was expressed in grams of Gallic acid equivalents per Litre (GAE g/L) and was an average of three measurements.

Apparatus

UV-Vis spectra were collected using a UV-Vis single beam spectrometer (Mini 1240, Shimadzu). The IR spectra of the wine samples were examined over the range of 400-4000 cm^{-1} using a Bruker Equinox 55 FT-IR spectrometer equipped with a horizontal Attenuated Total Reflectance (ATR) device with a diamond crystal. Spectra were recorded using OPUS software (Bruker Optik, Germany) by averaging 100 scans for each spectrum with resolutions of 2 cm^{-1}. Background spectra were obtained and subtracted from each sample IR spectra. Small quantities of untreated wine samples were smeared directly onto the ATR diamond crystal using a disposable pipette. In order to avoid the strong interference due to the presence of water and alcohol in each wine sample, a 12% (v/v) ethanol spectrum was subtracted from each wine spectrum. Statistical Neural Networks 4.0 F (Stat Soft Inc., Tulsa,

OK, USA) was used to model the spectral data and develop predictive ANN models.

Pre-processing the FTIR data

The FTIR spectra were sampled between 400-4000 cm^{-1} and obtained spectra were smoothed to reduce the noise and improve signal-to noise ratio using the Moving Average method by counting the average of 20 data sets in every tenth wavelength record [22]. Significant signal noise in the spectra would degrade the signal-to-noise ratio and resolution of the spectra and hence reduce the accuracy and precision of a model. The resulting 185 spectral intensities for each wine sample, together with variety of grape, wine barrel (oak type), and production year were used as ANN inputs and experimentally measured total phenolic content, total acidity, volatile acidity (acetic acid) and pH, were used as ANN outputs.

Artificial Neural networks (ANNs)

ANNs are biologically inspired computational model designed to simulate the way in which the human brain processes information [23]. ANNs are composed of many individual processing units or artificial neurons which are extensively inter-connected with connection weights to form a network. They collect their knowledge by detecting patterns and relationships in inputs and outputs and learn from experience from previously seen data, rather than using a pre-designed equation in a model. Artificial neurons are typically organized into an input layer, one or more hidden layers, and an output (prediction) layer. They function by linking the input neurons (i.e. spectral intensity/peak) to output neurons (i.e. measured wine quality indicator), through a set of connections with adjustable strengths (weights). The standard supervised network architectures (multilayer perceptrons and radial basis functions) are models in which connection weights and the number of hidden neurons are adjustable parameters that are optimized during the learning phase. This is performed using the training and validation sets of compounds. Training is performed iteratively, such that as it progresses, the ANN will generate a more accurate output and establish a (linear or non-linear) relationship between input and output data. Following this, true predictive ability of the model can be tested using an independent set of compounds, validation set.

Network training and design

Averaged spectral intensities, variety of grape (Merlot, Cabernet, Shiraz, Semillon, Sauvignon, Malbec), wine barrel (French oak, American oak, mixed, tank, unknown), and production year were used as inputs and calculated concentrations of total phenols expressed as Gallic Acid Equivalents (GAE g/L), acetic acid concentration (g/L), total acidity (g/L), alcohol content (%), and measured pH were used as outputs in the ANN. The most straightforward approach was used to build the ANN model. Input/output data sets were were automatically randomised into training (60%), testing (20%) and validation (20%) subsets. The training set was used for learning and to fit network parameters (weights) while the testing set was used to optimise the network topology and avoid over fitting. An external validation subset was used to assess the predictive performance (generalization) of a developed neural network. The extent of training was monitored internally by the ANN program and was stopped when the Root Mean Square (RMS) error failed to improve during training cycles and the testing RMS error started to increase. A number of networks with different topologies were trained and tested to determine the optimum topology for the dopamine receptor data and back-propagation Multi Layer Perceptron (MLP) with one hidden layer selected due to its

superiority in network performance. In contrast to linear statistical techniques, there is no known method for the automatic determination of an optimal network structure to fit a specific dataset [24]. As a result, training algorithms are run a number of times through automated network searches so that the best networks could be selected.

Selection of input variables

Sensitivity analysis was used for feature selection to identify the most important inputs that are directly correlated to total phenolic content, total and volatile acidity, pH and alcohol content by determining how sensitive a model is to changes in the input values. Sensitivity is defined as the ratio of error of a retrained optimum model, which does not contain the information of a specific molecular descriptor, to the error of the optimum model that includes the information from the molecular descriptor [25]. Since ANNs compute the output as a sum of nonlinear transformations of linear combinations of the inputs, sensitivity shows the percentage contribution of a corresponding input to the output value and reveals the effect that a change in that particular input has on output. Hence, inputs with low sensitivity are considered to only have a minimal contribution to the model being analysed so were eliminated from subsequent models.

Molecular descriptors with sensitivities less than one were sequentially removed after each automated network run until models that contained only molecular descriptors of relative importance for assessed wine quality parameters were developed.

Results and Discussion

Fourier Transform Infrared (FT-IR) spectroscopy is a non-destructive analytical technique that provides structural information on molecular features of a large range of compounds.

In general, no two wine samples will show exactly the same IR absorption pattern, thus a fingerprint IR spectrum can be produced, unique to each wine. However, the common constituents of wine (i.e. water and ethanol) produce IR absorption bands, which may disguise the characteristic IR vibrations of phenols. This is due to the fact that water, ethanol and organic acids absorb in the same MIR region as phenols. To eliminate this interference, a 12% (v/v) ethanol in water spectrum was subtracted from the wine spectra used in this work.

All wine samples including those from the Bendigo region, NSW and wines ranging over different vintages gave rise to similar spectra patterns (Figure 1). Several absorption bands were identified including those within the region between 800-1750 cm^{-1} which are categorized as C=C-C aromatic ring stretches (1580-1615 cm^{-1}; 1450-1510 cm^{-1}) while IR bands in the area from 820 to 760 cm^{-1} can be attributed

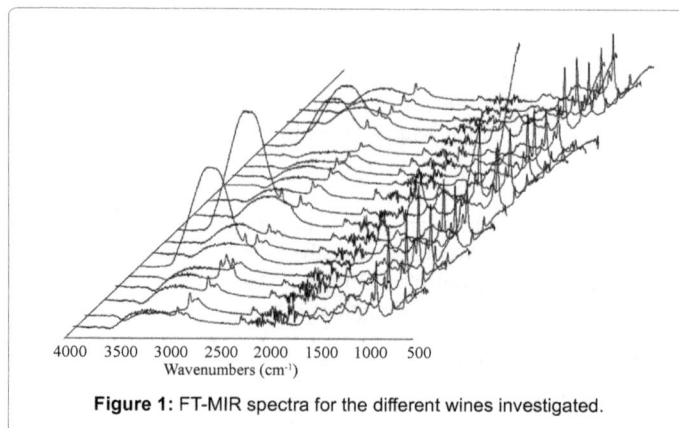

Figure 1: FT-MIR spectra for the different wines investigated.

Wine sample	Total acidity (g/L)	pH (±0.3)	Calculated acetic acid concentration (g/L)	Total phenols (GAE* g/L)
1	6.75	3.51	0.13	2.65
2	5.24	3.74	0.27	2.31
3	6.73	3.38	0.37	2.53
4	4.49	3.67	0.72	1.83
5	6.18	3.66	1.65	1.70
6	5.99	3.53	1.12	1.21
7	6.83	3.54	0.24	0.17
8	7.02	3.25	0.31	2.18
9	6.56	3.45	0.15	0.36
10	5.68	3.55	0.52	1.52
11	5.83	3.46	0.15	1.65
12	5.49	3.50	1.12	1.75
13	4.45	3.88	1.17	1.44
14	6.31	3.41	0.21	1.65
15	6.18	3.38	0.27	1.52
16	5.2	3.77	1.54	2.42
17	7.04	3.69	1.44	2.46
18	5.93	3.38	1.3	2.15
19	6.37	3.42	0.02	1.98
20	4.63	3.48	0.52	3.06

*GAE=Gallic acid equivalent

Table 2: Total acidity, pH, acetic acid concentration, and total phenolic concentration for the wine samples.

	Correlation				ANN architecture*
	Training	Testing	Validation	Average	
Acetic acid concentration (g/L)	0.960	0.998	0.924	0.898	25-1-1
Alcohol content	0.914	0.990	0.789	0.942	147-1-1
Total polyphe-nols (g/L GAE)	0.999	0.904	0.925	0.918	186-5-1
Total acidity (g/L)	0.954	0.999	0.801	0.898	143-1-1

*Number of inputs-*hidden neurones*-number of outputs

Table 3: Correlation data for the four developed ANN models.

to ring vibrations [26]. Furthermore, peaks between 670-900 cm⁻¹ can be attributed to aromatic C-H out of plane (750-1000 cm⁻¹) and in plane bending (950-1225 cm⁻¹) [27]. The IR regions of significant importance to this study were from 1542 to 965 cm⁻¹, usually referred to as the "fingerprint" region and various IR bands, including those corresponding to the vibration of the C-O, C-C, C-H and C-N bonds, occurs in this region [1,28]. This area provides important information regarding organic compounds such as sugars, alcohols and organic acids present in the sample. The distinct absorbance peaks in the wave number regions 3626-2970 cm⁻¹ and 1716-1543 cm⁻¹, are the result of the absorbance of water [29]. Other absorption bands of interest involved those at 1044 and 1085 cm⁻¹, which are indicative of an alcohol functional group. The 1382 cm⁻¹ absorption band attributes to the O-H in plane deformation in polyphenols [3]. The cyclic nature of the ether was reflected by the peaks located at 1283-1247cm⁻¹ range and as well at 1158 cm⁻¹, which was produced by the aromatic C-O bond stretching. The 1739 cm⁻¹ absorption band may attributed to the carbonyl group, C=O of the galloyl unit on epicatechin gallate. The deformation vibration of the carbon–carbon bonds in the phenolic groups adsorb in the region of 1500-1400 cm⁻¹ [2].

IR absorption due to the presence of sugar functional groups

are within the range of 1200 and 950 cm⁻¹, more specifically the peaks observed at 1157, 1107, 1065, 1014 cm⁻¹. The stretch vibration of the C=O group is around 1700 cm⁻¹. Peaks at 1618 and 1407 cm⁻¹ corresponds to symmetrical and asymmetrical stretching vibration for the carboxyl ion (COO-) indicating the existence of carboxylic acid, ester, or carbonyl groups [29]. Peaks located in the region of 1450 to 1410 cm⁻¹ originate from symmetric stretching vibration of C-O, and those around 1500 cm⁻¹ can be assigned to C-C stretching in rings [1].

In the same way, the IR peaks that are observed at approximately 1300 and 1150 cm⁻¹ can be assigned to S=O stretch present in a sulfate group Peaks at 1350 and 1175 cm⁻¹ for sulfonates and peaks in the area between 1000-750 cm⁻¹ can be assigned to S-O [30].

Liquid sulfur dioxide shows IR bands at 530, 1142 and 1330 cm⁻¹. For the sulphuryl compounds such as sulphuric acid, the SO2 bend is in the range 500-600 cm⁻¹ [31]. The IR absorbance at 1000 cm⁻¹ may indicates the existence of PO_4^{3-} [32]. The presence of polyphenolic compounds is indicated by characteristic bands for gallic acid at 669, 763, 1025, 1100 and 1654 cm⁻¹, tannic acid at 669, 860, 1172, 1511 and 1627 cm⁻¹ [4] (Figure 1).

Four separate nonlinear ANN models were developed to correlate the recorded IR spectra, variety of grape, wine barrel type and production year, with the experimentally measured total polyphenolic content, total and volatile acidity, pH and alcohol content in the selected wine samples (Table 2). The strength of the correlation between experimental and predicted data was assessed by the accuracy of the predicted values for the validation data sets (Table 3).

The developed ANN models confirmed that the characteristic IR bands, grape variety, year, wine type and wine barrel used for aging are directly related to the experimentally determined amounts of phenols, total acidity, acetic acid content, pH, and alcohol content. Good linearity (Table 3) of the ATR-FTIR method indicates that this technique can be used to assess the quality of wine accurately and specifically.

Total polyphenols

Not surprisingly the ANN model for the polyphenolic content (186-5-1) has the wine barrel as the most important input contribution in addition to the peaks at 1870,3740 2200 and 530-550 due to alcohol, carbonyl, alkyne, nitriles and vibrations in the fingerprint region. The composition of untreated wine samples consists of a complex mixture of various phenolic compounds and hence classification of each specific phenol by their spectral peak profile can be difficult. Never the less, the distinctive absorption bands within the finger print region of the IR spectrum can be assigned to particular functional groups present in several phenolic compounds. Moreover, wine consists of several components including water (80 to 90 percent), alcohol, sugar, carboxylic acids, tannins, polyphenols, amino acids, vitamin C, inorganic components and numerous fragrance ingredients [3]. Therefore, acknowledging their characteristic IR functional group absorption peak(s) can assist in more accurate determination of the phenolic compounds present.

Red wines are rich in polyphenolic substances, mainly flavanols and anthocyanins, which contribute to the sensory properties of wines, such as colour, taste, astringency and flavour [15,33]. Furthermore, the total phenol content directly correlates with the antioxidant activity of wines [34]. The antioxidant activity of wine polyphenols has been demonstrated in many studies and they act via various chemical pathways (i.e. as free radical terminators, singlet oxygen quenchers, and chelators of metal ions) [35,36]. Inhibition of low-density lipoprotein oxidation, inhibition of platelet aggregation, and anti-

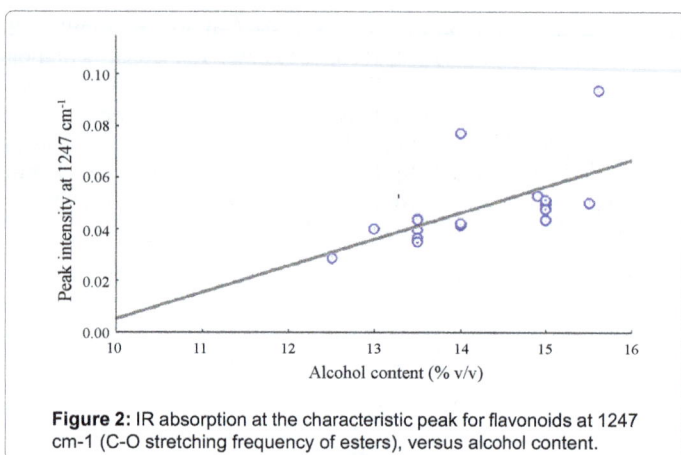

Figure 2: IR absorption at the characteristic peak for flavonoids at 1247 cm-1 (C-O stretching frequency of esters), versus alcohol content.

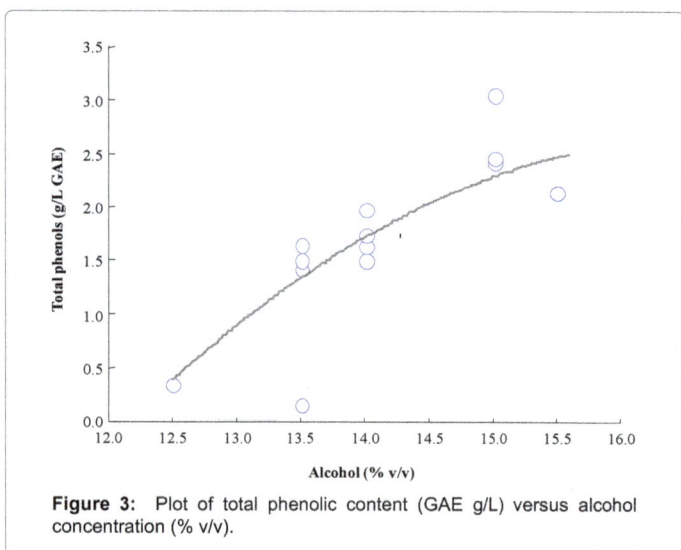

Figure 3: Plot of total phenolic content (GAE g/L) versus alcohol concentration (% v/v).

Furthermore, the total phenolic content increases as the alcohol content of the wine increases (Figure 3). Optimal antioxidant activity is achieved at an alcohol concentration of approximately 15% v/v. Above this range, however, antioxidant activity declines. Results show that the high alcohol content wines in the group (wine samples 4, 5, 16, 17, and 20), also have the highest overall antioxidant concentrations. However, wine sample 8, with a much lower alcohol content of 13.2% v/v had the highest antioxidant concentration of all of the wines in the group. This result however, may not be reliable as it was an outlier from the general trend and may have been subject to experimental error. Further testing of a wider range of wine samples with varying alcohol content is needed in order to confirm the reliability of this correlation. Moreover, the true antioxidant activity of a wine sample in vivo cannot be accurately determined unless physiological testing is also performed (Figure 3).

The phenolic composition of red wine depends on many factors, such as grape variety, wine maturation and ageing. During wine maturation and aging the polyphenolic components extracted from the grapes undergo various reactions, the most important being condensation reactions between anthocyanins and flavonoids, and reactions involving other wine components, such as acetaldehyde, pyruvate, hydroxycinnamic acids and vinyl phenols, which give rise to more stable oligomeric or polymeric pigments [42,43]. The rate and extent of these reactions is influenced by various parameters, including the initial concentration of reactant species, pH of the medium, temperature, ageing conditions, oxygen availability and the concentration of antioxidant agents such as sulphites [44-46]. It is also known that the effect of the type of wine barrel used is a contributing factor to the total phenolic content of red wine (Figure 4).

The aging of wines in wine barrels can potentially provide a modest contribution to the phenolic content of wine. The oak barrels contain gallotanins and ellagitannins which leach into the wine from the barrels resulting in an increase in the total tannin levels found in the wine [47]. For this reason, we need to consider the type of oak barrel used in wine storage as an important factor in determining the overall level of total phenolic compounds found in wine. Results indicate that variations in wine barrel oak type (i.e. French or American oak) will affect the amount of tannins extracted into the wine. (Figure 4) indicates that French oak based wines achieved a small yet higher content of total polyphenols than American oak based wines. According to studies, French oak contains an overall higher concentration of tannins than American oak [48]. For this reason several wineries opt for more

inflammatory properties by red wine phenolics [9,37] may be a major factor contributing to the health benefits of red wine and the decreased risk of cardiovascular diseases despite a high-fat diet in certain French populations [38]. The main component in red wine that is believed to provide cardiovascular protection is resveratrol (3,4,5-trihydroxy-trans-stilbene) and together with polyphenols are produced exclusively in the vine leaf epidermis and grape skin [39]. The resveratrol content of wine is related to the length of time the grape skins are present during the fermentation process. It was found that the synergistic contribution due to the presence of both phenolic and polyphenolic compounds, might be more important for antioxidant activity than any specific phenolic concentration [40]. Polyphenolic compounds are found in the skin and seeds of grapes. Red wines have a considerably higher amount of polyphenolic compounds than white wines, because the skins are removed earlier during white-wine production, lessening the amount that is extracted [6]. When wine is made, the alcohol produced by the fermentation process dissolves the polyphenols contained in the skin and seeds.

Based on Raman measurements of pure flavones and related compounds, the intense band at 1247 cm^{-1} can be attributed to flavonoid-type compounds [41]. The intense band at 1247 cm^{-1}, that could be assigned to bending of the OH groups coupled to C-H in plane bend in flavonoids, correlates well with the alcohol content found in a wine sample (Figure 2), but does not relate with total polyphenolic content, acidity and pH [16].

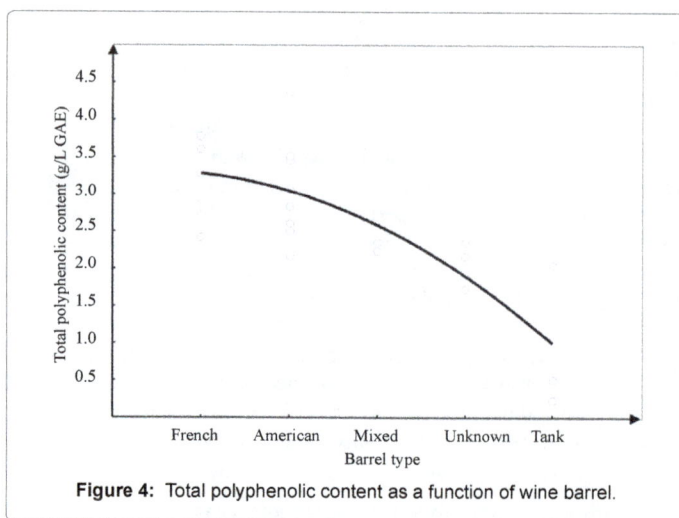

Figure 4: Total polyphenolic content as a function of wine barrel.

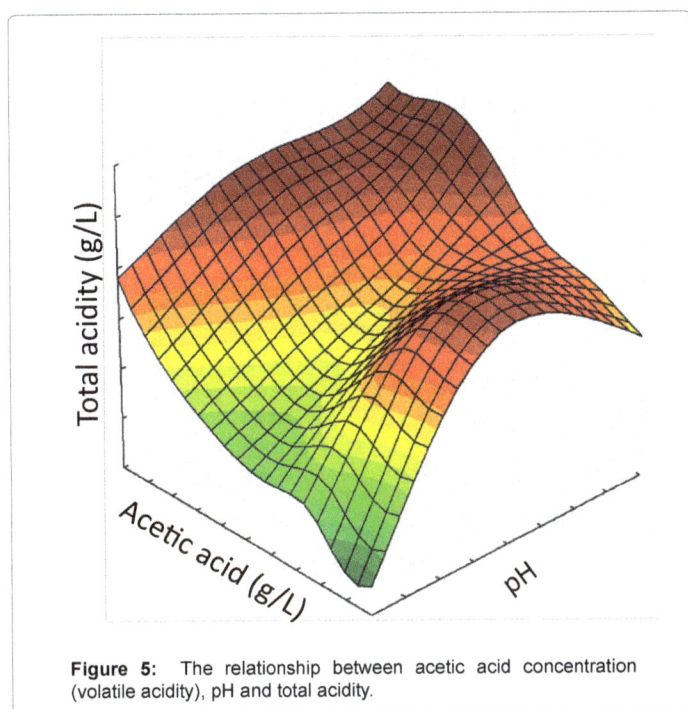

Figure 5: The relationship between acetic acid concentration (volatile acidity), pH and total acidity.

expensive French oak in order to help release phenols and tannins into the wine in order to improve flavor and structure. Wines aged in French and American oak both showed higher absorbance readings than wines that have been stored in fermenting tanks.

As expected, our results show that the total phenolic content was higher in the red wine samples. Furthermore, despite a spread in the amounts of the phenolic content present in the different wine varieties and vineyards, cabernet sauvignon wines had higher levels of phenolics compared to merlot wines.

Acetic acid content (Volatile acidity)

The ANN model correlating IR spectral bends with acetic acid content was the simplest with only 25 input descriptors, mostly related to carbonyl stretching vibration between 1650-1750 cm^{-1} and peaks in the fingerprint region (500-550 cm^{-1}) in addition to wine variety (red wine versus white wine) and type of the wine barrel used for aging. Volatile acids consist of a number of low molecular weight fatty acids [49] with acetic acid as the predominant acid. High levels of acetic acid are indicative of spoilage, mostly from *Acetobacter*. A small amount of acetic acid is normal, but should not exceed 0.3 g/L. Higher levels of acetic acid; especially 0.6-0.9 g/L is enough to provide a sensory indication of spoilage. At the level of 1.5 g/L the wine has essentially turned into vinegar. It can be seen from the 3D surface plot (Figure 1), that there is a correlation between pH, total acidity and acetic acid content, presenting optimal aspects for wine quality. As discussed earlier, in order to have a better quality wine, ideally the wine would have a higher content of total acidity and a lower acetic acid concentration. This can be seen at a pH of approximately 3.6 where spoilage is at a low (acetic acid < 0.75 g/L) yet total acidity is not affected. Therefore according to (Figure 1), a pH of 3.6 and a total acidity of 7.0 g/L would optimise wine quality in relation to these parameters. It is important to note the wines with a pH and total acidity in this region also have higher a phenolic content (and antioxidant capacity) than the other wines investigated. At higher pH levels, acetic acid content becomes more significant thereby affecting wine quality.

Titratable or total acidity

The most important descriptors that were included in the model for the total acidity were bends attributed the broad stretch of carboxylic acid O-H (2500-3000 cm^{-1}), carbonyl group stretching vibrations (1670-1820 cm^{-1}), mono substituted benzene ring bend (strong band at 690 cm^{-1}), nitrile (2210-2260 cm^{-1}), wine type and wine barrel (oak type). The total acidity of the wine is a measure of all types of acids present, i.e. inorganic acids, such as phosphoric acid, and organic acids where tartaric acid and malic acid are the predominate components, contributing to around 90% of the overall total acidity of the wine. There are also amino acids whose contribution to titratable acidity is not very well known. Often the total acidity is quantified simply as the measure of the total amount of tartaric acid present [49].

Organic acids (i.e. tartaric acid) make major contributions to the taste, feel and colour of the wine. More importantly their preservative properties also enhance wines microbiological and physicochemical stability. Red wines are stable at lower acidity, due to the presence of phenols which enhance acidity, balance the sweet taste of the alcohols and help to maintain stability throughout the aging process. Acidity also greatly influences the taste and colour of wine, and is important when assessing the general quality to the wine [50]. Grapes grown in warmer climates tend to have higher sugar content and lower acidity, compared with colder climate grapes. In general the total acidity of wine varies from 0.4% v/w to 1.0% w/v (4-10 g/L), but usually in red wines the acidity is from 0.6% w/v to 0.8% w/v (6–8 g/L). The measured total acidity of the wine samples in this work is within an acceptable range of 4.45 to 7.04 g/L. A total of 8 different wines had acetic acid levels greater than 0.5 g/L, with 2 of them above 1.5 g/L. A high level of spoilage is due to the concentration of acetic acid being anything above 1.5 g/L, but for good quality wine, the level should be less than 0.5 g/L [51].

Wine pH

As a wine quality parameter, pH is of equal or greater significance than the titratable total acidity. It affects wine colour, taste, oxidation-reduction potential, ratio of free to bound sulphur dioxide, and the extent of iron phosphate cloudiness present in the wine [52]. The pH and total acidity of a wine follows an inverse proportional relationship, where an increase in pH correlates to a lower total acidity. However, while there is a general relationship between total acidity and pH, these parameters are not directly related to each other. The association is regarded as an empirical relationship between the pH and the ratio of potassium bitartrate to total tartaric acid. Note that pH only measures the free hydrogen ions in solution, while titratable acidity measures the concentration of all of the available hydrogen ions, both those free in solution and those bound to undissociated acid molecules. Most of the wines samples in this study came from the central Victorian region, which is classified as a warm climate wine growing area. These wines are expected to have lower levels of acids and higher pH values, and are known for deep colour, low tannins and high levels of acid (which is unusual for a warm climate red grape). As the pH shifts to the lower end of the scale, the colour of the wine shifts to a more intense red form. With the taste of wine, pH values around 4.0 results in a flat taste while a pH of less than 3.0 results in a tart and sour taste. On average, the pH of red wine should not exceed 3.6 [53], with the optimum pH range for wine between 3.2-3.6. Although according to literature, there is no direct correlation proven between pH, volatile acidity and total acidity, there is a more general non-linear relationship between the total acidity, pH and volatile acidity, which can be seen in Figure 5 [54].

Conclusion

Parameters used to assess grape and wine quality such as total and volatile acidity, pH, alcohol, and antioxidant content were determined for each wine sample using conventional analytical techniques. Each of these parameters affects the overall quality of the wine, either contributing to taste, colour, protective and preservative properties, and potential health benefits.

The total acidity of the wines ranged from 4.45 to 7.04 g/L, pH from 3.25 to 3.88, acetic acid concentration from 0.02 to 1.54 g/L, and polyphenolic content from 0.17 to 3.06 g/L. Within the FTIR fingerprint region obtained for each wine sample, several absorption bands were identified as functional groups representative of those present in phenolic compounds. A plot of the values predicted by the validated ANN models showed excellent correlation with the experimentally measured values (r=0.898-0.942) for acetic acid concentration, alcohol content, total phenols, and total acidity. Experimental data collected from the wine samples showed that the phenolic content of the wines may be influenced by the type of oak used for aging the wines. Also there appeared to be a direct correlation between the alcohol content and the total phenolic content of the wines. FTIR spectroscopy proved to be a promising technique that provides a rapid method in the quality assessment of wine, for determining acetic acid concentration, alcohol content, total polyphenols, and total acidity. Moreover, this method uses small sample volumes (a few milliliters) with almost no sample preparation required.

References

1. Movasaghi Z, Rehman S, Rehman IU (2007) Raman spectroscopy of biological tissues. Applied Spectroscopy Reviews 42: 493-541.

2. Nakano Y, Takeshita K, Tsutsumi T (2001) Adsorption mechanism of hexavalent chromium by redox within condensed-tannin gel. Water Res 35: 496-500.

3. Ozacar M, Soykan C, Sengil IA (2006) Studies on synthesis, characterization, and metal adsorption of mimosa and valonia tannin resins. Journal of Applied Polymer Science 102: 786-797.

4. Kannan RRR, Rajasekaran A, Perumal A (2011) Fourier transform infrared spectroscopy analysis of sea grass polyphenols. Current Bioactive Compounds 7: 118-125.

5. Ferrieres J (2004) The French paradox: lessons for other countries. Heart 90: 107-111.

6. Kopp P (1998) Resveratrol, a phytoestrogen found in red wine. A possible explanation for the conundrum of the 'French paradox'? Eur J Endocrinol 138: 619-620.

7. Goldberg DM, Hahn SE, Parkes JG (1995) Beyond alcohol: beverage consumption and cardiovascular mortality. Clin Chim Acta 237: 155-187.

8. Just JR, Daeschel MA (2003) Antimicrobial effects of wine on Escherichia coli O157:H7 and Salmonella typhimurium in a model stomach system. Journal of Food Science 68: 285-290.

9. Frankel EN, Waterhouse AL, Teissedre LP (1995) Principal phenolic phytochemicals in selected California wines and their antioxidant activity in inhibiting oxidation of human low-density lipoproteins. J Agric Food Chem 43: 890-894.

10. Soleas GJ, Diamandis EP, Goldberg DM (1997) Wine as a biological fluid: history, production, and role in disease prevention. J Clin Lab Anal 11: 287-313.

11. Ikigai H, Nakae T, Hara Y, Shimamura T (1993) Bactericidal catechins damage the lipid bilayer. Biochim Biophys Acta 1147: 132-136.

12. Jayaprakasha GK, Selvi T, Sakariah KK (2003) Antibacterial and antioxidant activities of grape (Vitis vinifera) seed extracts. Food Research International 36: 117-122.

13. Vaquero MJR, Alberto MR, de Nadra MCM (2007) Influence of phenolic compounds from wines on the growth of Listeria monocytogenes. Food Control 18: 587-593.

14. Kennedy JA, Matthews MA, Waterhouse AL (2000) Changes in grape seed polyphenols during fruit ripening. Phytochemistry 55: 77-85.

15. Robichaud JL, Noble AC (1990) Astringency and bitterness of selected phenolics in wine. Journal of the Science of Food and Agriculture 53: 343-353.

16. Canamares MV, Lombardi JR, Leona M (2009) Raman and surface enhanced raman spectra of 7-hydroxyflavone and 3',4'-dihydroxyflavone. e-Preservation Science 6: 81-88.

17. Bauer R, Nieuwoudt H, Bauer FF, Kossman J, Koch KR, et al. (2008) FTIR spectroscopy for grape and wine analysis. Anal Chem 80: 1371-1379.

18. Urtubia A, Perez-Correa JR, Pizarro P, Agosin E (2008) Exploring the applicability of MIR spectroscopy to detect early indications of wine fermentation problems. Food Control 19: 382-388.

19. Jensen JS, Egebo M, Meyer AS (2008) Identification of spectral regions for the quantification of red wine tannins with fourier transform mid-infrared spectroscopy. J Agric Food Chem 56: 3493-3499.

20. Versari A, Boulton RB, Parpinello GP (2006) Effect of spectral pre-processing methods on the evaluation of the color components of red wines using Fourier-transform infrared spectroscopy. Italian Journal of Food Science 4: 423-432.

21. Versari A, Parpinello GP, Scazzina F, Del Rio D (2010) Prediction of total antioxidant capacity of red wine by Fourier transform infrared spectroscopy. Food Control 21: 786-789.

22. Fuller WA (1996) Introduction to statistical time series. John Wiley & Sons, Inc., New York, USA.

23. Agatonovic-Kustrin S, Beresford R (2000) Basic concepts of artificial neural network (ANN) modeling and its application in pharmaceutical research. J Pharm Biomed Anal 22: 717-727.

24. Agatonovic-Kustrin S, Turner JV, Glass BD (2008) Molecular structural characteristics as determinants of estrogen receptor selectivity. J Pharm Biomed Anal 48: 369-375.

25. Turner JV, Maddalena DJ, Cutler DJ, Agatonovic-Kustrin S (2003) Multiple pharmacokinetic parameter prediction for a series of cephalosporins. J Pharm Sci 92: 552-559.

26. Lin-Vien D, Colthup NB, Fateley WG, Grasselli JG (1991) The handbook of Infrared and Raman characteristic frequencies of organic molecules. Academic Press, New York, USA.

27. Yadav BS, Tyagi SK, Seema (2006) Study of vibrational spectra of 4-methyl-3-nitrobenzaldehyde. Indian Journal of Pure & Applied Physics 44: 644-648.

28. Smith BC (1999) Infrared spectral interpretation: A systematic approach. 1st edition, CRC Press LLC, Florida, USA.

29. Zhang Y, Chen J, Lei Y, Zhou Q, Sun S, et al. (2010) Discrimination of different red wine by Fourier-transform infrared and two-dimensional infrared correlation spectroscopy. Journal of Molecular Structure 974: 144-150.

30. Flor S, Tripodi V, Contin M, Dobrecky C, Lucangioli S (2012) Spectroscopic approach of the association of heparin and its contaminant and related polysaccharides with polymers used in electrokinetic chromatography. Journal of Chemical and Pharmaceutical Research 4: 972-979.

31. Anderson A, Savoie R (1965) Raman spectrum of crystalline and liquid SO2. Canadian Journal of Chemistry 43: 2271-2278.

32. Yu D, Wu C, Kong Y, Xue N, Guo X, et al. (2007) Structural and catalytic investigation of mesoporous iron phosphate. J Phys Chem C 111: 14394-14399.

33. Eiro MJ, Heinonen M (2002) Anthocyanin color behavior and stability during storage: effect of intermolecular copigmentation. J Agric Food Chem 50: 7461-7466.

34. De Beer D, Harbertson JF, Kilmartin PA, Roginsky V, Barsukova T, et al. (2004) Phenolics: A comparison of diverse analytical methods. Am J Enol Vitic 55: 389-400.

35. Kinsella JE, Frankel EN, German B, Kanner J (1993) Possible mechanisms for the protective role of antioxidants in wine and plant foods. Food technology 47: 85-89.

36. Packer L, Hiramatsu M, Yoshikawa T (1999) Antioxidant food supplements in human health. Academic Press, London, UK.

37. Saito M, Hosoyama H, Artiga T, Kataoka S, Yamaji N (1998) Antiulcer activity of grape seed extract and procyanidins. J Agric Food Chem 46: 1460-1464.

38. Renaud S, de Lorgeril M (1992) Wine, alcohol, platelets, and the French paradox for coronary heart disease. Lancet 339: 1523-1526.

39. Liu BL, Zhang X, Zhang W, Zhen HN (2007) New enlightenment of French Paradox: resveratrol's potential for cancer chemoprevention and anti-cancer therapy. Cancer Biol Ther 6: 1833-1836.

40. Lin YT, Vattem D, Labbe RG, Shetty K (2005) Enhancement of antioxidant activity and inhibition of Helicobacter pylori by phenolic phytochemical-enriched alcoholic beverages. Process Biochemistry 40: 2059-2065.

41. Yamauchi S, Shibutani S, Doi S (2003) Characteristic raman bands for artocarpus heterophyllus heartwood. Journal of Wood Science 49: 466-468.

42. Bakker J, Timberlake CF (1997) Isolation, identification, and characterization of new color-stable anthocyanins occurring in some red wines. J Agric Food Chem 45: 35-43.

43. Monagas M, Nunez V, Bartolome B, Gomez-Cordoves C (2003) Anthocyanin-derived pigments in graciano, tempranillo, and cabernet sauvignon wines produced in Spain. Am J Enol Vitic 54: 163-169.

44. Schwarz M, Wabnitz TC, Winterhalter P (2003) Pathway leading to the formation of anthocyanin-vinylphenol adducts and related pigments in red wines. J Agric Food Chem 51: 3682-3687.

45. Romero C, Bakker J (2000) Anthocyanin and colour evolution during maturation of four port wines: Effect of pyruvic acid addition. Journal of the Science of Food and Agriculture 81: 252-260.

46. Romero C, Bakker J (2000) Effect of storage temperature and pyruvate on kinetics of anthocyanin degradation, vitisin A derivative formation, and color characteristics of model solutions. J Agric Food Chem 48: 2135-2141.

47. Waterhouse AL (2002) Wine phenolics. Ann N Y Acad Sci 957: 21-36.

48. Fernandez K, Agosin E (2007) Quantitative analysis of red wine tannins using Fourier-transform mid-infrared spectrometry. J Agric Food Chem 55: 7294-7300.

49. Ough CS, Amerine MA (1988) Methods of analysis of musts and wines. 2nd edition, John Wiley & Sons, New York, USA.

50. Mataix E, de Castro MDL (1999) Sequential determination of total and volatile acidity in wines based on a low injection-pervaporation approach. Analytica Chimica Acta 381: 23-28.

51. Zoecklein BW, Fugelsang KC, Gump BH, Nury FS (1999) Wine analysis and production. Chapman & Hall, New York, USA.

52. Peynaud E, Blouin J (1996) The taste of wine: The art science of wine appreciation. John Wiley and Sons, New York, USA.

53. Jackson RS (2000) Wine science: Principles and Applications. 3rd edition, Academic Press, San Diego, USA.

54. Boulton R (1980) The relationships between total acidity, titratable acidity and ph in wine. American Journal of Enology and Viticulture 31: 76-80.

17

Snow Chemistry at Mukteshwar in Central Himalayan Region of India

Bablu Kumar[1], Gupta GP[1], Sudha Singh[1], Lone FA[2] and Kulshrestha UC[1*]

[1]School of Environmental Sciences, Jawaharlal Nehru University, New Delhi 110067, India

[2]Centre for Climate Change, Mountain and Agriculture, SKUAST, Shalimar, Srinagar, 191123, Jammu & Kashmir, India

Abstract

The present study reports snow chemistry and source apportionment at Mukteshwar in central Himalayan region of India during winter 2012-13. In this study, fresh snowfall samples were collected at Mukteshwar during winter season of 2012-13. The results showed that the pH of the snowmelt samples ranged from 5.47 to 7.95 with an average of 6.37 indicating alkaline nature of precipitation which is similar to the range reported. The concentration of ions followed the following order- Ca^{2+} > Cl^- > Na^+ > SO_4^{2-} > HCO_3^- > NH_4^+ > NO_3^- > Mg^{2+} > K^+ > F^-. Very high concentration of Ca^{2+} indicated the dominance of crustal sources. Source fraction calculations revealed that crustal, marine and anthropogenic sources contributed 40%, 38% and 22% ionic components in snowmelt, respectively. Since, Mukteshwar is remote site as compared to Delhi, values of NO_3^- were compared with the NO_3^- reported in the precipitation (rain water) of Delhi as NO_3^- is an indicator of vehicular pollution in urban areas. Such comparison of NO_3^- values suggested that though Mukteshwar precipitation had 1/3 of NO_3^- in precipitation as compared to Delhi, but considering it as a small town, precipitation at Mukteshwar is significantly influenced vehicular sources possibly due to Long Range Transport (LRT) of pollution.

Keywords: Himalaya; Precipitation; Vehicular pollution; NO_3^-; Crustal sources

Introduction

Air pollution is considered as one of the major environmental challenges. Huge amount of air pollutants is injected into the atmosphere due to rapid urban and industrial growth during past few decades. South East and South Asian are the major air pollution emitters due to rapidly increasing energy demand for their fast growing economy. Among south Asian countries, Indian emissions are significant. After China, India is the second biggest emitter of SO_2 in Asia [1]. However, the levels of gas like SO_2 are quenched by atmospheric dust in India resulting in very low ambient SO_2 [2,3]. Nevertheless, emissions of atmospheric aerosols and gaseous pollutants have caused the problems of acid deposition, ozone depletion, climate change and monsoon modification etc. [4-6]. Once emitted, most of these pollutants are scavenged by removal processes. Wet deposition is one of the most effective deposition processes of airborne pollutants.

Long term study of wet deposition can be used to notice the changes in oxides of N & S content thereby relating to coal & petroleum energy consumption patterns. Hence, rain and snow chemistry become very important to know the sources of pollution and their possible effects on ecosystems and environment. Precipitation chemistry in general, provides information about the deposition fluxes of various air pollutants to different ecosystems. Due to its significance, snowfall chemistry has been studied extensively throughout the world [7-14]. Fresh snow chemistry in Himalaya ranges had shown relatively low concentration of air borne pollutants are representative of remote site [9,11,15]. The ionic content in Himalayan snow is highly affected by the impact of long range transport of anthropogenic sources of air pollutants [16]. Snow chemistry at Shanghai in China suggested that fossil fuel combustion and biomass burning is the major source of air pollution [14].

Acid deposition, a serious threat to terrestrial, aquatic and marine ecosystem has been studied comprehensively at various places in North America, Europe, Japan and other countries of the world [2,17-19]. Acidic precipitation is determined primarily by the interaction of acidic and basic species in the atmosphere. Acidic species are dominated by SO_2 and NO_x which is precursor of H_2SO_4 and HNO_3 respectively which are mainly emitted by fossil fuel combustion. On the other

hand, alkaline species (Ca^{2+}, Mg^{2+} and NH_4^+) help in enhancing pH of precipitation. The main source of these species is atmospheric dust which is made up of carbonates and bicarbonates of Ca and Mg [2,3,20]. Unlike North America and Europe, it is interesting that higher pH of precipitation is reported in India even at higher SO_4^{2-} concentration which might be due to huge amount of dust in the atmosphere [3].

In India, several studies have been reported on rain chemistry mostly in urban areas [2,21-26] with few studies from rural sites [27-30]. The detailed and updated studies on fresh snowfall chemistry are even rare in rural areas of Himalayan region of India [16,31,32]. Hence, this study was carried out to fill this knowledge gap about the chemical characteristics of snow in central Himalaya by selecting Mukteshwar as a rural representative site. The present study also focuses on quantification of relative contributions of marine and non-marine sources during winter season. Further, non-marine fractions have been quantified into crustal associated and anthropogenic sources. An attempt has been also made to highlight the extent of influence of vehicular pollution in urban area like Delhi vs. remote area like Mukteshwar by considering NO_3^- concentration as an indicator of vehicular pollution.

Materials and Methods

Sampling site

Mukteshwar is located at 29.47°N 79.64°E in Nainital district of Uttarakhand state of India (Figure 1). It is situated in the Kumaon Hills of central Himalaya at an altitude of 2286 meters. It lies approx. 51 km NE of Nainital city, 51 km from Nainital, 72 km from Haldwani and 395 km from Delhi city. Mukteshwar is rich in scenic beauty,

*Corresponding author: Kulshrestha UC, School of Environmental Sciences, Jawaharlal Nehru University, New Delhi 110067, India
E-mail: umeshkulshrestha@gmail.com

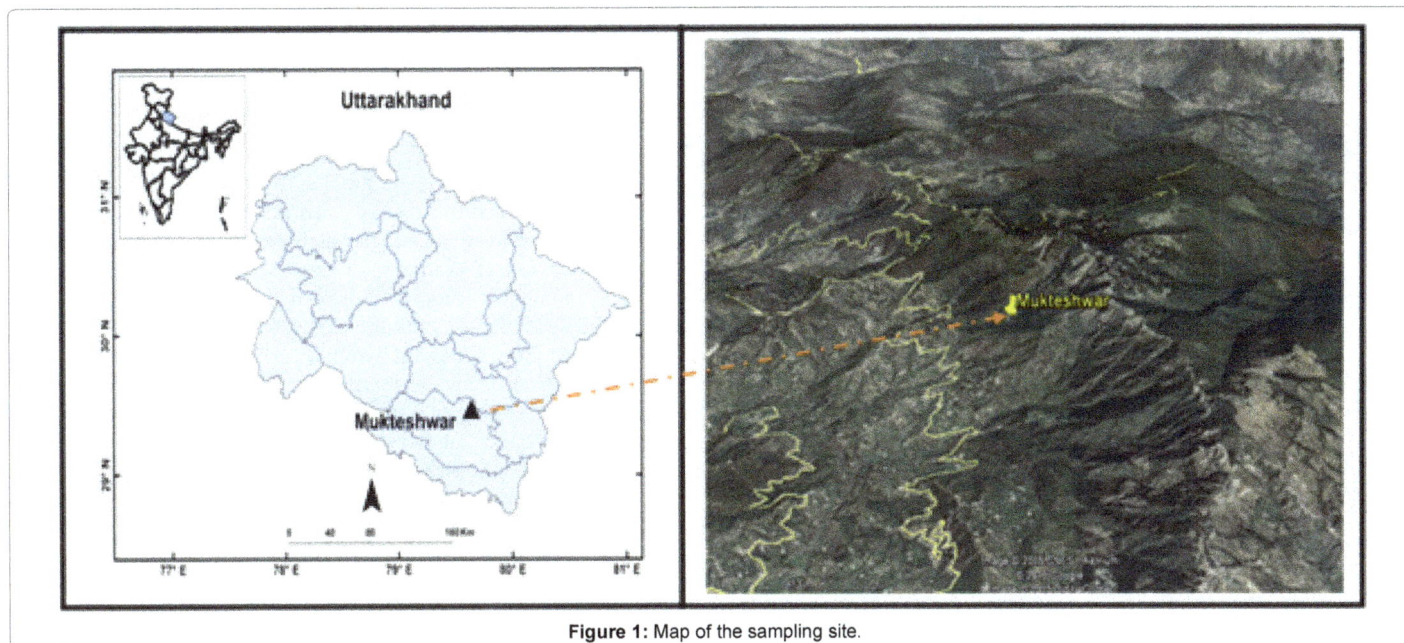

Figure 1: Map of the sampling site.

with magnificent views of the central Himalayas. It is one of the most famous tourist spots in north India which receive domestic as well as foreign tourists every year. The mean minimum temperature in this village ranges from -5°C to 0°C during January whereas the mean maximum temperature ranges from 25°C to 30°C during the month of June. There are no any major industrial units around this site. Most of the people use biomass as a source of energy for domestic heating and cooking purposes. The major sources of air pollution at this site include vehicular pollution used for tourist activities and emissions from agricultural activities.

Collection of snowfall samples

Fresh snowfall samples were collected with the help of plastic trays (30 cm diameter) on event basis during winter season of 2012-2013. Generally, winter season spreads between November and February but the site receives maximum snowfall during December and January. The tray was washed properly with high quality deionized water and dried before collection of snowfall. Sampler was placed at ~ 2 m height above the ground level. The collector tray was kept outside just before the start of snowfall and was removed immediately after snowfall to avoid contamination. In order to cover entire Mukteshwar area, the collection was done at five points at same time on the day of snowfall event. These points are located at around 100 m distance between each point. Fresh snowfall samples deposited on tray were transferred into pre-cleaned polypropylene bottles using polyethylene gloves. These samples were stored in the refrigerator at the site which was later transferred to the laboratory within 15 days from the collection period. Samples were later processed for analysis of pH, Electrical Conductivity (EC) and major ions.

Analysis of samples

The collected samples were brought to the laboratory and analysed for major anions, cations, EC and pH. Determination of major cations (Na^+, NH_4^+, K^+, Ca^{2+} and Mg^{2+}) and major anions (F^-, Cl^-, NO_3^- and SO_4^{2-}) were performed by using ion chromatography (Metrohm 883 Basic IC Plus). Metrosep A SUPP 4, 250/4.0 column and an eluent of 1.8 mmol/L Na_2CO_3 and 1.7 mmol/L $NaHCO_3$ at a flow rate of 1.0 with Metrohm suppressor technique were used for determination

of anions. While Metrosep C4-100/4.0 column and an eluent of 1.7 mmol/L Nitric acid and 0.7 mmol/L Dipicolinic acid at a flow rate of 0.9 without suppressor was used for determination of cations. HCO_3^- was determined by using 0.0025 N H_2SO_4 [33].

Quality analysis

Quality control and Quality Assurance (QA/QC) of chemical analysis was performed by checking ion balance and conductivity balance [24,34,35]. A significant correlation (R^2 =0.81) was found between sum of anions (F^-, Cl^-, NO_3^-, SO_4^{2-} and HCO_3^-) and sum of cations (Na^+, NH_4^+, K^+, Ca^{2+} and Mg^{2+}) which indicated good ion balance for samples. A very good correlation between measured Electrical Conductivity (EC) and calculated conductivity (R^2=0.87) further confirmed good quality of dataset.

Results and Discussion

pH variation in snowmelt samples

The pH of the snowmelt samples ranged from 5.47 to 7.95 with an average of 6.37. Approximately, 95% samples had pH more than 5.6. The pH of precipitation in a clean atmosphere is generally 5.6 due to its equilibration with atmospheric CO_2 [36]. The snowmelt samples with pH more than 5.6 indicating inputs of alkaline components at this sampling site, which has been discussed in subsequent sections. Alkaline precipitation is a typical feature of Indian region due to suspended atmospheric dusts rich in calcium carbonate [2]. Similar range of pH distribution has been reported by many workers in global precipitation (Table 1). At this site, only 5% acidic precipitation has been observed. Results of this site were compared with other sites in Indian region. Satyanarayana et al. [30] have reported 11% acidic precipitation at Hudegade in ecologically sensitive region of Western Ghats, India. The frequency distribution of pH of snow samples showed that approx. 17% samples were acidic at Kothi in north western Himalayan region of India [16]. Very high frequency of acid rain occurrence has been reported over Indian Ocean during INDOEX due to high concentration of non-sea salt sulphate [37].

Ionic composition of snowmelt samples

Table 2 gives statistical parameters of major ions of snowfall. The ion concentration followed the following sequence- $Ca^{2+} > Cl^- > Na^+ > SO_4^{2-} > HCO_3^- > NH_4^+ > NO_3^- > Mg^{2+} > K^+ > F^-$ (Figure 2). Among cations, the percent concentration followed the sequence – $Ca^{2+} > Na^+ > NH_4^+ > Mg^{2+} > K^+$ while anions followed the sequence- $Cl^- > SO_4^{2-} > HCO_3^- > NO_3^- > F^-$.

The most abundant ion was Ca^{2+}, with concentrations ranging from 43 to 111 µeq/L. The average concentration of Ca^{2+} was 87 µeq/L accounts for approximately 24% of all ions and 44% among all cations. We compared snowfall concentration of Ca^{2+} at this site with global reports. It was observed that the average concentration of Ca^{2+} at Mukteshwar was higher than Khumbu-Himal [9]. The highest concentration of Ca^{2+} indicating the dominance of crustal and marine sources which has been discussed in source contribution section. High levels of Ca^{2+} due to local as well as transported dust have been reported in precipitation [30,38]. The suspended soil dust might be significant local source for Ca^{2+} since the the soil in this region are loosely bound in the earth crust. Since, the site is very fascinating place for tourists attracting its scenic beauty. The ongoing construction activities

especially building resorts to accomodate the maximum number of tourists. Besides this, road dust also contributes Ca^{2+} in precipitation. These activities are considered as significant sources of Ca^{2+} [16,20]. Very high concentration of Ca^{2+} due to suspended dust has been reported in indian precipitation [24].

After Ca^{2+} and Na^+, NH_4^+ ion has been most abundant. NH_4^+ concentrations ranged from 4 to 52 µeq/L. The average concentration of NH_4^+ was recorded as 24 µeq/L which accounted for approximately 7% of all ions and 12% among all cations. Very high concentration of NH_4^+ among all cations indicated a significant influence of anthropogenic sources at Mukteshwar. The most important sources of NH_4^+ are agriculture activities, livestock, excreta of human and animal etc. [19,39]. Local people in this region are dependent upon mainly agriculture activities for their livelihood. Due to this region, agriculture activities and livestock might be a significant source of NH_4^+ at this site. Apart from these sources, local people as well as tourists reaching here also might be significant source for NH_4^+ in this region. Singh and co-workers [39] reported that humans are also a significant source of NH_3. Since many local people are forced to go in the open field for excretion since they don't have proper toilet facility. Open excreta of human and animal also might be a good source of NH_4^+ at Mukteshwar. Apart from above mentioned sources, some contribution of long range transport can't be ruled out at this site which has not been discussed in this paper.

SO_4^{2-} also had relatively higher concentrations ranging from 2.11 to 99.95 µeq/L. The average concentration was observed to be 34.74 µeq/L which accounted for approximately 9% of all ions and 21% among all anions. A comparision with other global data, we found that average concentration of SO_4^{2-} at this site was lower than Yulong snow [40] but higher than Mt. Everest [11] and Mt. Logan Massif [41]. Since,

Sampling site	Country	pH	Reference
Mukteshwar, Uttarakhand	India	6.37	Present study
Gulmarg, Jammu & Kashmir	India	6.70	[31]
Kothi, Himachal Pradesh	India	5.69	[16]
Central and Southern Californis	USA	5.20	[59]
Scottish Catchment	Europe	4.20	[60]
Larsemann Hills	Antartica	5.71	[58]

Table 1: Geographical comparison of pH of snowmelt samples.

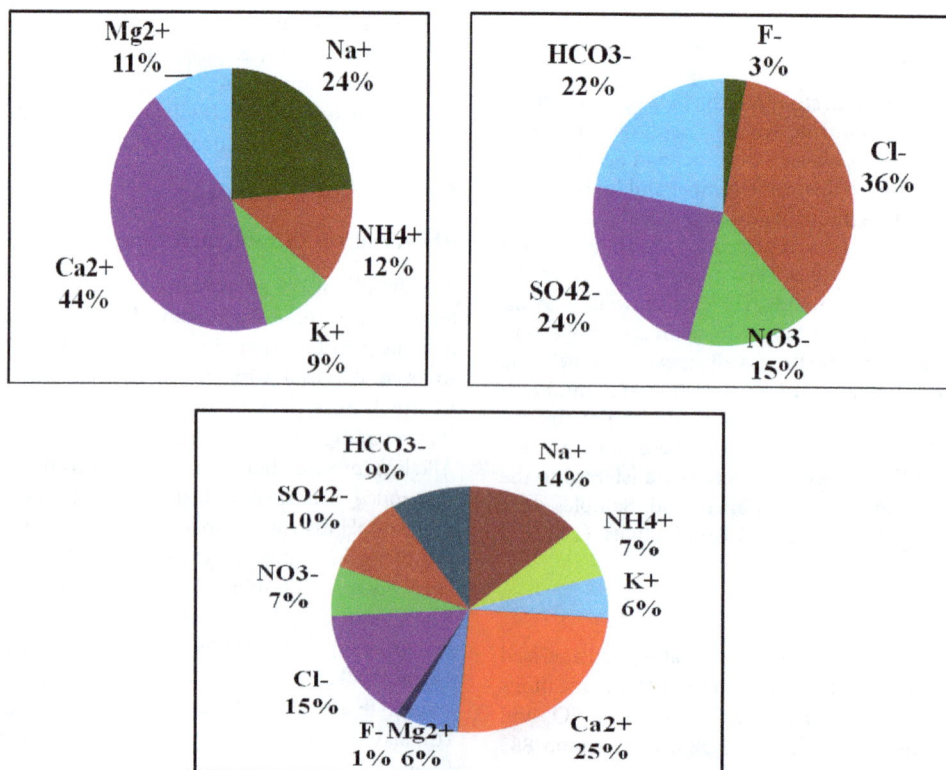

Figure 2: Percent contribution of ions (a) Cations (b) Anions (c) Total ions.

there is no any industrial activities at this site. Due to this reason, the maximum contribution of SO_4^{2-} might be due to transported SO_4^{2-} from other site in India or transboundary and long range transport SO_4^{2-}. Very similar range of SO_4^{2-} concentration has been reported worldwide [2,26,30,38,42,43]. High concentration of SO_4^{2-} indicated a significant influence of anthropogenic sources emitting SO_2. Among local sources, diesel driven vehicular traffic might be the possible source of SO_4^{2-}. Vehicular flow due to increased tourist activities during winter season might be considered as a good source of SO_2 at Mukteshwar.

NO_3^- was found to have concentrations ranging from 2.8 to 48.8 µeq/L. The average concentration was observed to be 22.6 µeq/L accounts approximately 6% of total ions and 14% of total anions. NO_3^- in snowfall samples might be due to emission of NO_x mainly from vehicular sources which is transformed into the atmosphere which give rise to NO_3^- in precipitation samples [44,45]. Another local source might be biomass burning during winter season to produce heat by local people cannot be ruled out [46]. Apart from local emissions, very high concentration of NO_3^- might be due to long range transport of NO_x / NO_3^- from various air masses approaching at Mukteshwar.

Figure 3 compares NO_3^- in precipitation samples of Mukteshwar (snowmelt) with urban site, Delhi (rain water) [47]. On an average, it is very clear from the Figure that urban sites had 3 times high NO_3^- in precipitation. As compared to the size of town and local activities NO_3^- values in snowmelt at Mukteshwar are significantly high when we compare the size and the amount of local activities of Delhi city. This indicates that even though Mukteshwar is a remote hilly site, it is significantly affected by long range transport of pollution while most of NO_3^- in the precipitation of Delhi can be considered as local vehicular contributions. The varying composition of precipitation samples at different sites might be due to differences in local emission, meteorological conditions, methods of sampling, regional and global scale transport of pollutants in relation to air masses, type of sampling site, elevation from sea level etc. [9,24]. In order to understand the possible reason behind the variation of NO_3^- concentration in precipitation samples, we are comparing our study site i.e. Mukteshwar (rural characteristics) with Delhi city (urban characteristics). Mukteshwar is a small village with no major vehicular pollution except those by tourist activities. Apart from local vehicular emission, significant contribution of NO_3^- in precipitation at Mukteshwar might be due to transported NO_x/ NO_3^- from other places within India as well as long range transport of pollutants. Raatikainen and co-workers also observed the transport of pollution from Indo-Gangetic plain to Mukteshwar in Himalayan region of India. Delhi had very high NO_3^- in precipitation which might be due to a vast increase in vehicular population. In 1994, Delhi had 24,32,295 mobile vehicles which reached to 69,32,706 in 2010 [48]. This has increased the consumption of petrol by 44.7%. In addition, NO_3^- might also be contributed by increased number of industrial units. According to Delhi statistics, the number of industries is increased by 39.54% in Delhi region from 1994-2010 [48].

Source contribution

In order to find out the contribution of various sources categories, marine and non-marine fractions have been calculated. Further non-marine fraction has been calculated as crustal associated and anthropogenic fractions following the approach as reported by Kulshrestha et al. [49].

Marine contribution: The maximum fraction of sea salt is composed of Na^+, Cl^- and Mg^{2+} with some fraction of Ca^{2+}, SO_4^{2-} and K^+. In order to estimate the marine contribution in snowmelt samples, sea salt ratios were calculated by assuming Na^+ as a sea salt tracer

with marine origin only [50]. The sea salt ratios deviated considerably for all components from standard sea salt ratios (Table 3) indicating significant influence of non-marine contribution. The results showed that Cl^-/Na^+ ratio is very close to sea water ratios indicating its major contribution from marine sources (Table 3). Mg^{2+}/Na^+ ratios are slightly higher than the standard sea water ratio reflecting contribution of marine sources with some contribution from non-marine sources. Sodium ratios of SO_4^{2-}, K^+ and Ca^{2+} are higher than standard sea water ratios indicating their likely contribution from soil or anthropogenic sources. Elevated ratios of SO_4^{2-}/Na^+, K^+/Na^+ and Ca^{2+}/Na^+ have been reported at various other sites in Indian region [2,26,29].

In order to estimate sea salt fraction (ssf) and non-sea salt fraction (nssf) the following formula were used:

$$\%ssf = \frac{100}{EF_{sea water}} \qquad (1)$$

$$\text{Here } EF_{sea water} = \left[\frac{X}{Na^+}\right]_{snowfall} \bigg/ \left[\frac{X}{Na^+}\right]_{sea water} \qquad (2)$$

Where, [X] is the concentration of desired ionic species in ueq/l.

$\%nssf = 100 - \%ssf$

The percent contribution of sea salt fraction and non-sea salt fraction of Cl^-, SO_4^{2-}, K^+, Ca^{2+} and Mg^{2+} in snowfall samples has been given in Table 3. It is very clear from the table that the maximum fraction of SO_4^{2-}, K^+ and Ca^{2+} originated from non-marine sources. Similar to our results, large fraction of $nssSO_4^{2-}$ and $nssCa^{2+}$ have been

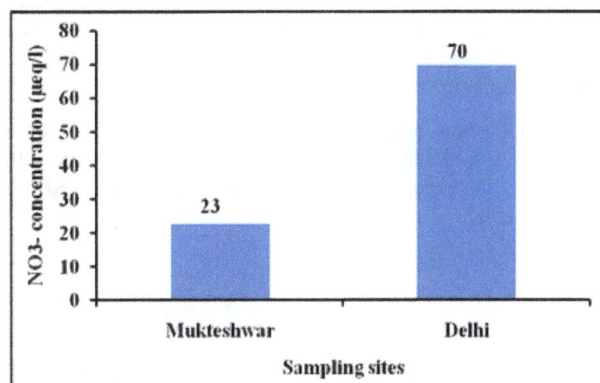

Figure 3: Comparison of NO_3^- in precipitation at Mukteshwar (nowmelt) and Delhi (rain water). NO_3^- value of Delhi precipitation has been taken from Singh et al. [50].

	Average	Min	Max
pH	5.47	7.95	6.37
EC	27.6	20.9	42.5
Na⁺	47	22	82
NH₄⁺	24	4	52
K⁺	19	5	42
Ca²⁺	87	43	111
Mg²⁺	22	6	56
F⁻	4	1	9
Cl⁻	52	26	85
NO₃⁻	23	10	38
SO₄²⁻	35	13	49
HCO₃⁻	32	5	78

Table 2: pH, EC and ionic composition (µeq/L).

	Cl⁻	SO₄²⁻	K⁺	Ca²⁺	Mg²⁺
Sea water ratios (Keene et al. 1986)	1.16	0.125	0.022	0.044	0.227
Snowmelt ratios	1.12	0.74	0.40	1.86	0.46
EF (sea water)	0.97	5.93	18.36	42.30	2.03
%ssf	100	17	5	2	49
%nssf	0	83	95	98	51

Table 3: Sodium ratios of snowmelt and percent sea salt fraction and non sea salt fraction of major ions at Mukteshwar

Ions	Marine	CF	Anthropo
Cl⁻	100	0	0
NO₃⁻	0	12.3	87.7
SO₄²⁻	16.7	22.4	60.9
K⁺	5.4	94.6	0
Ca²⁺	2.3	97.7	0
Mg²⁺	48.9	51.1	0
Na⁺	100	0.0	0
NH₄⁺	0	4.2	95.8
F⁻	0	1.0	99.0

Table 4: Percent contribution of marine, crustal (CF) and anthropogenic fraction (Anthropo) of all ions.

reported at various sites in snow samples [12,31,40,51].

Non-marine contribution: Further, non- marine fraction was differentiated into two categories- i) crustal associated and ii) anthropogenic sources. It is well known that anthropogenic sources are mainly responsible for lowering the pH of precipitation i.e. enhancing acidity while crustal sources i.e. alkaline species increases the pH and decrease the acidity of precipitation [3,52,53]. The contribution of crustal associated fraction (CF) and anthropogenic fraction (Anthro) of ionic species was calculated by the same approach as it was used by Kulshrestha et al. [49] which is an appropriate approach to explain pH value in Indian precipitation.

The nss Ca²⁺ was considered as reference element in snowfall samples in order to determine the crustal associated fraction (CF) of major ions [2]. The percent contribution of CF was calculated by the following formula:

$$\%CF_X = \frac{100}{EF_{soil}}$$

$$\text{Where, } EF_{soil} = \frac{\left[\dfrac{y}{nss\,Ca^{2+}}\right]_{snowfall}}{\left[\dfrac{y}{nss\,Ca^{2+}}\right]_{soil}} \quad (3)$$

y= (nss SO₄²⁻, nss Ca²⁺, nss K⁺, nss Mg²⁺, NO₃⁻, NH₄⁺ and F⁻) in µeq/L.

The percent contribution of Anthro fractions of each ion was calculated by the following formula-

$$\% Anthro_x = \%nss_x - \%CF_x \quad (4)$$

Using the above mentioned formula, the source characterization for different ions of snowfall has been calculated as given in Table 4. It is very clear from the table that the maximum contribution of Ca²⁺ is associated with crustal fraction. The percent contribution of SO₄²⁻, NH₄⁺ and NO₃⁻ was the highest in Anthro in comparison to CF. The highest contribution of crustal Ca²⁺ might be due to contribution from local sources such as local soil, construction activities, road dust etc. as well as long range transport of fine dusts. The significant contribution of long range transport has been reported in snow samples globally

[7,54]. Very high percent contribution of Anthro SO₄²⁻ at this site might be due to transported SO₂/ SO₄²⁻ mainly since there is no any significant source of SO₂ nearby the sampling site. The crustal SO₄²⁻ at this site might be due to reaction of SO₂ with CaCO₃ rich dust particles forming calcium sulphate [2,3] which is very common in Indian region. The highest contribution of NH₄⁺ by anthropogenic sources might be due to significant contribution from local sources like agricultural activities, animal and human excreta, biomass burning, biogenic sources etc. with some contribution due to transported NH₄⁺ carried out by various air masses at this site.

Very high fraction of Anthro NO₃⁻ might be due to the emissions of NOₓ from fossil fuel combustion and biomass burning. Among local sources, vehicles used by tourist might be the major source of NOₓ. The significant contribution of NO₃⁻ has been reported in air masses due to long range transport of NOx/ NO₃⁻ in precipitation samples [16,30].

Average contribution of different sources: The average contribution of different sources viz. marine, crustal and anthropogenic was calculated by the following formula:

$$Average\ marine\% = \frac{\sum X_{ssf}}{\sum X_{Total}} \times 100$$

Here Σ X_{ssf}=Sum of concentration of all ssf values in each air mass cluster in µeq/L.

Σ X_{Total}=Sum of concentration of all components in each air mass cluster in µeq/L

$$Average\ crustal\% = \frac{\sum X_{CF}}{\sum X_{Total}} \times 100$$

Here Σ X_{CF}=Sum of concentration of all CF values in each air mass cluster in µeq/L.

Average Anthro%=100-Average marine%-Average crustal%

Using the above approach, the average percent contribution of marine, crustal and anthropogenic sources was calculated and given in Figure 4. The order of dominance followed the sequence- Crustal (40%) >Marine (38%) >Anthropogenic (22%). The significant contribution of crustal fraction has been observed in snow samples which might be due to local contribution with long range transport [15,16]. After crustal contribution, it was observed that the marine contribution dominated over anthropogenic sources. A very similar kind of observation has been reported in precipitation samples in Indian region [38]. The pH of snowmelt samples at this site suggesting significant dominance of alkaline species (crustal components) over acidic species (anthropogenic components) since sea salt don't play any important role in deciding pH of precipitation samples.

Neutralization factor

It is well known that NO₃⁻ and SO₄²⁻ are the major acidic species while nss Ca²⁺, nss Mg²⁺ and NH₄⁺ are the major alkaline species in precipiation samples. Due to this fact, nss Ca²⁺, nss Mg²⁺ and NH₄⁺ are mainly considered for calculation of neutralization factor.

The neutralization factor of these species were calculated by the following formula-

$$NF_{Xi} = \frac{[X_i]}{[NO_3^- + nss\,SO_4^{2-}]}$$

Where [X_i] = The concentration of desired ionic species i.e. nss Ca²⁺, nss Mg²⁺ and NH₄⁺ in µeq/l.

Using the above mentioned formula, the neutralization factor of nss Ca²⁺, nss Mg²⁺ and NH₄⁺ was calculated and are given in Figure 5.

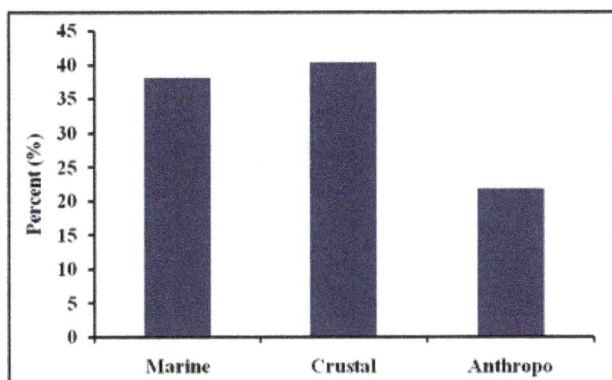

Figure 4: Average contribution of marine, crustal and anthropogenic sources.

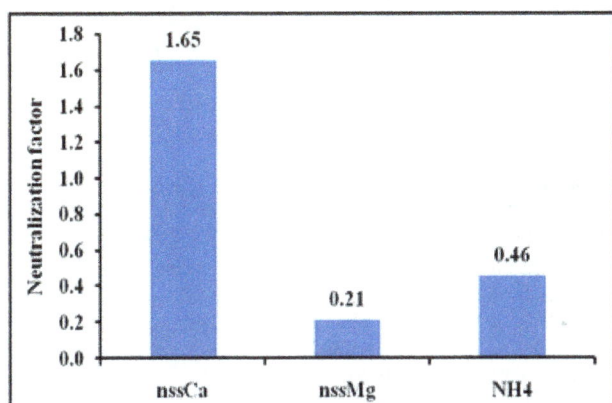

Figure 5: Neutralization factor of nss Ca^{2+}, nss Mg^{2+}, NH$_4^+$.

The neutralization factor for these species followed the order- nss Ca^{2+} > NH$_4^+$ > nss Mg^{2+}.

The maximum neutralization of snowfall acidity occurs by nssCa^{2+} which might be due to suspended soil dust rich in CaCO$_3$ in Indian atmosphere [29]. These soil dusts might be contributed significantly by long range transport. The importance of transported dust in the neutralization of acidic components in precipitation samples has been observed globally [30,38]. Very high value of neutralization factor for nssCa^{2+} and NH$_4^+$ has been reported in precipitation samples globally [2,30,53,55-60].

Conclusion

The pH of fresh snowfall collected at Mukteshwar was noticed in the range of 5.47-7.95 with an average of 6.37 which is similar to the pH of precipitation reported in other studies in this region. This is mainly due to the interferences of crustal sources. Among all ionic species, Ca^{2+} dominated. CaCO$_3$ rich aerosols are contributed by the suspension of soil, road dusts and construction activities. Scavenging of such aerosols gives rise to high pH of precipitation due to buffering action. Even at high SO$_4^{2-}$ levels, the pH is relatively higher as the acidity created by sulphuric acid is buffered by the presence of Ca^{2+}. Relative source contribution to the snowmelt was calculated as crustal 40%, marine 38% and anthropogenic 22% at the site. Comparison of NO$_3^-$ values revealed that precipitation at Delhi is highly influenced by vehicular sources showing very high NO$_3^-$ as compared to Mukteshwar. Interestingly, in spite of small town having lesser vehicular activities, precipitation at Mukteshwar has relatively high NO$_3^-$ values which

might be due to long range transport of NO$_x$/ NO$_3^-$.

Acknowledgement

Financial support received from JNU as CBF, LRE and DST-PURSE grants is gratefully acknowledged. Author Bablu Kumar, Gyan Prakash Gupta and Sudha Singh acknowledge the award of fellowships from UGC. Our sincere thanks to Mr. Kaushal Ji for his help in collection of snowfall samples.

References

1. Klimont Z, Smith SJ, Cofala J (2013) The last decade of global anthropogenic sulfur dioxide: 2000-2011 emissions Environ Res Lett 8: 014003.

2. Kulshrestha UC, Kulshrestha MJ, Sekar R, Sastry GSR, Vairamani M (2003) Chemical characteristics of rainwater at an urban site of south-central India. Atmos Environ 37: 3019-3026.

3. Kulshrestha UC (2013) Acid rain. In: Encyclopedia of Environmental Management. Jørgensen SE (editor) Taylor and Francis, Manila Typesetting Company.

4. Andreae MO, Crutzen PJ (1997) Atmospheric aerosols: biogeochemical sources and role in atmospheric chemistry. Science 276: 1052-1058.

5. Ramanathan V, Crutzen PJ, Kiehl JT, Rosenfeld D (2001) Aerosols, climate, and the hydrological cycle. Science 294: 2119-2124.

6. IPCC (2014) Climate Change 2014: Synthesis report. Fifth assessment report (AR5) of the Intergovernmental Panel on Climate Change. Cambridge University Press, Cambridge, United Kingdom and New York, NY, USA.

7. Mayewski PA, Lyons WB, Ahmad N (1983) Chemical composition of a high altitude fresh snowfall in the Ladakh, Himalayas. Geophys Res Lett 10: 105-108.

8. Wang C, Zhu W, Wang Z, Guicherit R (2000) Rare Earth elements and other metals in atmospheric particulate matter in the western part of the Netherlands. Water Air Soil Pollut 12: 109-118.

9. Marinoni A, Polesello S, Smirraglia C, Valssecchi S (2001) Chemical composition of fresh snow samples from the southern slope of Mt. Everest region (Khumbu-Himal region, Nepal). Atmos Environ 35: 3183-3190.

10. Kang SC, Qin DH, Mayewski PA, Sneed SR, Yao TD (2002) Chemical composition of fresh snow on Xixabangma peak, Central Himalaya, during the summer monsoon season. J Glaciol 48: 337-339.

11. Kang SC, Mayewski PA, Qin DH, Sneed SR, Ren JW, et al. (2004) Seasonal differences in snow chemistry from the vicinity of Mt. Everest, central Himalayas. Atmos Environ 38: 2819-2829.

12. Niu H, He Y, Zhu G, Xin H, Du J, et al. (2013) Environmental implications of the snow chemistry from Mt. Yulong, southeastern Tibetan Plateau. Quat Int 313-314: 168-178.

13. Xu J, Zhang Q, Li X, Ge X, Xiao C, et al. (2013) Dissolved Organic Matter and Inorganic Ions in a Central Himalayan Glacier: Insights into Chemical Composition and Atmospheric Sources. Environ Sci Technol 47: 6181-6188.

14. Zhang Y, Xiu G, Wu X, Moore CW, Wang J, et al. (2013) Characterization of mercury concentrations in snow and potential sources, Shanghai, China. Sci Total Environ 449: 434-442.

15. Shrestha AB, Wake CP, Dibb JE, Whitlow SI (2002) Aerosol and Precipitation Chemistry at a Remote Himalayan Site in Nepal. Aerosol Sci Technol 36: 441-456.

16. Kulshrestha UC, Kumar B (2014) Airmass Trajectories and Long Range Transport of Pollutants: Review of Wet Deposition Scenario in South Asia. Adv Meteor 2014: 596041.

17. Rodhe H, Granat L (1984) An evaluation of sulfate in European precipitation 1955-1982. Atmos Environ 18: 2627-2639.

18. Hara H, Akimoto H (1993) National level variations in precipitation chemistry in Japan. In Proceedings of the International Conference on Regional Environment and Climate Changes in East Asia, Taipei. November 30-December 3, 1993.

19. Galloway JN (1995) Acid deposition: perspectives in time and space. Water Air Soil Poll 85: 15-24.

20. Rahn KA (1976) The chemical composition of the atmospheric aerosol technical report. Graduate School of Oceanography, University of Rhode Island, Kingston.

21. Khemani LT (1989) Physical and chemical characteristics of atmospheric

aerosols. Cheremisionoff PN (editor) Air pollution control Volume 2, Gulf Publishing, USA.

22. Kumar N, Kulshrestha UC, Saxena A, Khare P, Kumari KM, et al. (1993) Effect of anthropogenic formate and acetate level in precipitation at four sites in Agra. Atmos Environ 27B: 87-91.

23. Kulshrestha UC, Sarkar AK, Srivastava SS, Parashar DC (1996) Investigation into atmospheric deposition through precipitation studies at New Delhi (India). Atmos Environ 30: 4149-4154.

24. Kulshrestha UC, Granat L, Engardt M, Rodhe H (2005) Review of precipitation monitoring studies in India a search for regional patterns. Atmos Environ 39: 7403-7419.

25. Safai PD, Rao PSP, Momin GA, Ali K, Chate DM, et al. (2004) Chemical composition of precipitation during 1984-2002 at Pune, India. Atmos Environ 38: 1705-1714.

26. Tiwari S, Kulshrestha UC, Padmanabhamurty B (2007) Monsoon rain chemistry and source apportionment using receptor modeling in and around National Capital Region (NCR) of Delhi, India. Atmos Environ 4: 5595-5604.

27. Mahadeven TN, Negi BS, Meenakshy V (1989) Measurement of element composition of aerosol matter and precipitation from a remote continental site in India. Atmos Environ 23: 869-874.

28. Rao PSP, Momin GA, Naik MS, Safai PD, Pillai AG, et al. (1990) Impact of Ca and SO4 on pH of rain water in rural environment in India. Indian J Environ Pollut 10: 941-943.

29. Jain M, Kulshrestha UC, Sarkar AK Parashar DC (2000) Influence of crustal aerosols on wet deposition at urban and rural sites in India. Atmos Environ 34: 5129-5137.

30. Satyanarayana J, Reddy Kumar AL, Kulshrestha JM, Kulshrestha CU (2010) Chemical composition of rain water and influence of airmass trajectories at a rural site in an ecological sensitive area of Western Ghats (India). J Atmos Chem 66: 101-116.

31. Naik MS, Khemani LT, Momin GA, Rao PS, Safai PD, et al. (1995) Chemical composition of fresh snow from Gulmarg, North India. Environ Pollut 87: 167-171.

32. Lone FA, Khan MA, Qureshi N, Kirmani NA, Sidiquee SH, et al. (2010) Environmental chemistry of a rare muddy snowfall occurrence on Alpine zone glaciers of Gulmarg, Kashmir Himalaya, India. Curr World Environ 5: 271-278.

33. American Public Health Association (1998) Standard Methods for the Examination of Water and Wastewater. American Water Works Association, Water Environment Federation.

34. Ayers GP (1995) Some practical aspects of acid deposition measurement. Presentation to the Third Expert Meeting on Acid Deposition Monitoring Network in East Asia. 14-16 November 1995, Niigata Prefecture, Japan, pp. 1-20.

35. WMO (1994) Report of the workshop on precipitation chemistry laboratory techniques. In: Mohnen, V., Santroch, J., Vet, R. (Eds.), Hradec Kralove, Czech Republic 17-21 October. WMO Report no 102.

36. Charlson RJ, Rodhe H (1982) Factors controlling the acidity of natural rainwater. Nature 295: 683-695.

37. Kulshrestha UC, Jain M, Mandal TK, Gupta PK, Sarkar AK, et al. (1999) Measurements of acid rain over Indian Ocean and surface measurements of atmospheric aerosols at New Delhi during INDOEX pre-campaigns. Curr Sci 76: 968-972.

38. Budhavant KB, Rao PSP, Safai, PD Ali K (2011) Influence of local sources on rainwater chemistry over Pune region, India. Atmos Res 100: 121-131.

39. Singh S, Kulshrestha UC (2012) Abundance and distribution of gaseous ammonia and particulate ammonium at Delhi, India. Biogeosciences 9: 191-207.

40. Zhu GF, Pu T, He YQ, Shi PJ, Zhang T (2012) Seasonal variations of major ions in fresh snow at Baishui Glacier No. , Yulong Mountain, China. Environ Earth Sci 69: 1-10.

41. Yalcin K, Wake CP, Dibb JE, Whitlow SI (2006) Relationships between aerosol and snow chemistry at King Col, Mt. Logan Massif, Yukon, Canada. Atmos Environ 40: 7152-7163.

42. Avila A, Alacron M (1999) Relationship between precipitation chemistry and meteorological situations at a rural site in NE Spain. Atmos Environ 33: 1663-1677.

43. Granat L, Norman M, Leck C, Kulshrestha UC, Rodhe H (2002) Wet scavenging of sulfur compound and other constituents during the Indian Ocean Experiment (INDOEX). J Geophys Res 107: 8025.

44. Finlayson-Pitts JB, Pitts JN (1986) Atmospheric Chemistry: Fundamentals and experimental techniques. John Willey & Sons, USA.

45. Seinfeld JH, Pandis SN (2006) Atmospheric Chemistry and Physics: From Air Pollution to Climate Change (2nd Edition), John Wiley & Sons, USA.

46. Singh S, Gupta GP, Kumar B, Kulshrestha UC (2014) Comparative study of indoor air pollution using traditional and improved cooking stoves in rural households of Northern India. Energy Sustain Dev 19: 1-6.

47. Singh S, Kumar B, Gupta GP, Kulshrestha, UC (2015) Signatures of increasing energy demand of past two decades as captured in rain water composition and airmass trajectory analysis at Delhi (India). J Energy Environ and Carbon credits 4.

48. Delhi Statistical Hand Book (2011) Directorate of Economics and Statistics Government of National Capital Territory of Delhi.

49. Kulshrestha UC, Jain M, Saxena AK, Kumar A, Parashar DC (1997) Contribution of sulphate aerosol to the rain water at urban site in India. In proceedings of IGAC international symposium on atmospheric chemistry and future global environment, Nagoya congress center, Nagoya, Japan, 11-13 November, 1997.

50. Keene WC, Pszenny AP, Galloway JN, Hawley ME (1986) Sea-salt corrections and interpretations of constituent ratios in marine precipitation. J Geophys Res 91: 6647-6658.

51. Qin DH, Hou SG, Zhang DQ, Ren JW, Kang SC, et al. (2002) Preliminary results from the chemical records of an 80.4 m ice core from East Rongbuk Glacier, Qomolangma (Everest), Himalaya. Annals of Glaciology 35: 278-284.

52. Rodhe H, Dentener F, Schulz M (2002) The global distribution of acidifying wet deposition. Environ Sci Technol 36: 4382-4388.

53. Zhang M, Wang S, Wu F, Yuan X, Zhang Y (2007) Chemical composition of wet precipitation and anthropogenic influence at a developing urban site in Southeastern China. Atmos Res 84: 311-322.

54. Liu Y, Geng Z, Hou S (2010) Spatial and seasonal variation of major ions in Himalayan snow and ice: a source consideration. J Asian Earth Sci 37: 195-205.

55. Das R, Das SN, Mishra VN (2005) Chemical composition of rainwater and dustfall at Bhubaneswar in the east coast of India. Atmos Environ 39: 5908-5916.

56. Calvo AI, Olmo FJ, Lyamani H, Alados-Arboledas L, Castro A, et al. (2010) Chemical Composition of Wet Precipitation at the Background EMEP Station in Viznar (Granada, Spain) (2002-2006). Atmos Res 96: 408-420.

57. Saxena A, Sharma S, Kulshrestha UC, Srivastava SS (1991) Factors affecting alkaline nature of rain water in Agra (India). Environ Pollut 74: 129-138.

58. Budhavant KB, Rao PSP, Safai PD (2014) Chemical Composition of Snow-Water and Scavenging Ratios over Costal Antarctica. Aerosol Air Qual Res 14: 666-676.

59. Gunj DW, Hoffmann MR (1990) Field investigation on the snow chemistry in central and southern California-I. Inorganic ions and hydrogen peroxide. Atmos Env 24A: 1661-171.

60. Tranter M, Daveis TD, Abrahams PW, Blackwood I, Brimblecombe P, et al. (1987) Spatial variability in the chemical composition of snowcover in a small, remote, scottish catchment. Atmos Environ 2: 853-862.

18

Adsorption Isotherm of Dibenzyl Toluene and its Partially Hydrogenated Forms Over Phenyl Hexyl Silica

Rabya Aslam[1,2]* and Karsten Muller[1]

[1]*Institute of Separation Science and Technology, Friedrich-Alexander-Universität Erlangen-Nürnberg, Germany*
[2]*Institute of Chemical Engineering and Technology, University of the Punjab, Lahore, Pakistan*

Abstract

Liquid organic hydrogen carriers (LOHC) are an interesting option for chemical energy storage and hydrogen transportation. Dibenzyltoluene (H0-DBT), heat transfer oil, is capable of reversibly storing hydrogen emerged as a feasible LOHC system. However, it is not available as a pure compound but consists of an isomeric mixture of 6 to 8 compounds. During the hydrogen storage process a high number of stable intermediate species is formed. These compounds can be categorized into four main classes according to their degree of hydrogenation. To implement H0-DBT as a LOHC system, thermophysical data of these intermediate compounds are required. In our previous work, a reversed phase HPLC method was developed using phenylhexyl silica stationary phase and acetone/water as eluent to separate these partially hydrogenated fractions with a purity >98%. For further designing a batch or continuous HPLC process, adsorption isotherm data are required. In this work, adsorption isotherms for dibenzyltoluene and its partial and fully hydrogenated forms namely hexahydro-dibenzyltoluene, dodecahydro-dibenzyltoluene, and octadecahydro-dibenzyltoluene are measured over phenylhexyl silica in acetone/water solvent using the static method. Sip's equation (Combined Langmuir-Freundlich isotherm) fits the data better as compared to simple Freundlich, Langmuir or competitive Langmuir adsorption isotherms.

Keywords: Adsorption isotherms; LOHC; RP-HPLC

Introduction

Liquid organic hydrogen carriers (LOHC) are a promising approach for hydrogen storage and transportation. This concept is based on the reversible cycle of hydrogenation and dehydrogenation for uptake and release of hydrogen [1-5]. The LOHC system consists of a pair of dehydrogenated and hydrogenated compounds representing the hydrogen lean and rich forms, respectively. Due to the chemical bonding of the hydrogen in the hydrogenated LOHC compound, the latter can store H_2 at ambient conditions over longer periods. Recent studies shows the potential of dibenzyltoluene (H0-DBT) as a competitive LOHC system due to its reasonable hydrogen storage capacity and thermal stability [5,6]. Each H0-DBT molecule can store 9 molecules of hydrogen in the form of octadecahydro-dibenzyltoluene (H18-DBT). However, to implement it as LOHC system, its thermophysical data are required. Since it is not available as pure compound and partial hydrogenation further produces a number of stable intermediate species. More than 24 compounds are observed in a partially hydrogenated reaction mixture which can be classified into four main fractions on the basis of their degree of hydrogenation namely dibenzyltoluene (H_0-DBT), hexahydro-dibenzyltoluene (H_6-DBT), dodecahydro-dibenzyltoluene (H_{12}-DBT), and octadecahydro-dibenzyltoluene (H_{18}-DBT). In our previous work, a reversed phase high pressure liquid chromatography method was developed and successfully scaled up to semi preparative scale for the separation of these partially hydrogenated derivatives of dibenzyltoluene with phenylhexyl silica as stationary phase and a acetone water mixture as mobile phase (96/4, *v/v*%) [7]. In order to simulate and design a linear or non-linear chromatographic processes, adsorption isotherm data over mobile and stationary phase are required which describe the distribution behavior of the DBT fractions in the chromatographic system [8-10]. Adsorption isotherms can experimentally be determined using static (shake-flask and adsorption-desorption methods) or dynamic methods (frontal analysis method, elution curve methods and minor perturbation methods) [8,11,12]. Moreover, for a mixture, competitive isotherms can be predicted using single solute adsorption isotherm but it often comes with limited applicability range and accuracy [8]. In this work, adsorption isotherm data of single isomers of dibenzyltoluene derivatives such as dibenzyltoluene, hexahydro-dibenzyltoluene, dodecahydro-dibenzyltoluene, and octadecahydro-dibenzyltoluene are determined using static method over phenylhexyl silica from a acetone/water solution (96/4, v/v%) within a concentration range of 0-25 mg/ml at 22 ± 1°C.

Materials and Methods

Materials

The stationary phase phenylhexyl silica (Luna 15 μm, 100°A) was purchased from Phenomenex, Germany. The specifications of the adsorbent provided by the supplier are summarized in Table 1. The single isomers of dibenzyltoluene, hexahydro-dibenzyltoluene, dodecahydro-dibenzyltoluene and octadecahydro-dibenzyltoluene were separated and purified via vacuum distillation and reversed phase high pressure liquid chromatography with >99% purity. The details of the separation process are described in section 2.2.1. The HPLC grade acetone was purchased from VWR Germany and 18 MΩ water produced by a Milli Q integral-3 system (Merck Millipore) was used in the experimental work. The details of the chemicals used are given in Table 2.

Methods

Separation of dibenzyltoluene derivatives: Single isomers of dibenzyltoluene derivatives representing each fraction of dibenzyltoluene were separated in two steps. Firstly, vacuum distillation was used to separate a partially hydrogenated reaction mixture. The

---

*Corresponding author: Rabya Aslam, Institute of Separation Science and Technology, Friedrich-Alexander-Universitat Erlangen- Nurnberg, Institute of Chemical Engineering and Technology, University of the Punjab, Lahore, Pakistan, E mail: rabya.icet@pu.edu.pk

Chemicals	Purity (mass%)	Degree of hydrogenation (%)	Analysis method
Dibenzyltoulene (H0-DBT)[+]	>99	0	GC-MS[a]
Hexahydro-dibenzyltoluene (H$_6$-DBT)[+]	>99	33	GC-MS[a]
Dodecahydro-dibenzyltoluene (H$_{12}$-DBT)[+]	>99	67	GC-MS[a]
Octadecahydro-dibenzyltoluene (H$_{18}$-DBT)[+]	>99	100	GC-MS[a]
Acetone	>99	-	-

Table 1: Specifications of phenyl hexyl Silica provided by manufacturer.

Parameter	Value
Particle size	11-15 µm
Particle size distribution	1.8 (dp 90/10)
Pore diameter	100°A
Surface diameter	390 m²/g
Bulk density	0.58 g/cm³
Total carbon	16%
Metal contents	11 ppm

Table 2: Purity information of chemicals used.

distillation was performed in packed distillation column (Normag GmbH, Germany) equipped with Sulzer structured packing (L=2 m, DXP, SS 1.4404). Due to the overlapping range of boiling points, pure compounds were not obtained via distillation but fractions consisting of compounds with similar boiling points were obtained. The batch plot for vacuum distillation experiment and boiling point range of DBT-fractions is provided in the supporting information. Then in the second step, using those fractions as feed, single isomers representing each fraction of dibenzyltoluene were separated on semi-preparative scale column (250 mm (L) × 50 mm (I.D.)) using reversed phase liquid chromatography. The detailed procedure can be found somewhere else [7]. The GC analysis and NMR analysis of each isomer used in this work is given in the supporting information.

Measurement of adsorption isotherm: The adsorption isotherms were measured using a static method (shake-flask method) [13]. Solutions in acetone/water (96.4%, v/v) containing equal concentration of dibenzyltoluene (H0-DBT), hexahydro-dibenzyltoluene (H6-DBT), dodecahydro-dibenzyltoluene (H12-DBT), and octadecahydro-dibenzyltoluene (H18-DBT) were prepared in 10 ml bottles for the concentration range of 0.1-25 mg/ml. The upper threshold of concentration was determined on the basis of the least soluble compound, i.e., H18-DBT in acetone/water solvent. The uncertainty of the initial concentration was ± 0.01 mg/ml. Then a fixed amount (0.5 ±0.003 g) of phenylhexyl silica was added to each bottle and the tightly closed bottles were kept for 24 hours under stirring. Preliminary experimental studies indicate that the adsorption equilibrium was reached after 18 hours for all components. The same procedure was repeated for the determination of the isotherms of dibenzyltoluene and octadecahydro-dibenzyltoluene fractions. It should be noted that fraction represents the isomeric mixture of components with same degree of hydrogenation. The initial and final concentrations were measured using Gas Chromatography (Agilent technology 7890 A system). The amount of solutes adsorbed on phenylhexyl silica were determined with initial and equilibrium concentrations using eq. (1).

$$q_e = \frac{(c_i - c_e) \cdot v}{m} \tag{1}$$

where Ci and Ce are initial and equilibrium concentrations of solutes, respectively; m is the mass of the phenylhexyl silica phase in gram and v is the volume of solution. Each measurement was repeated three times and the standard deviation was found to be less than 5%.

Results

The experimental data for the adsorption isotherm for fraction of dibenzyltoluene and octadecahydro-dibenzyltoluene and single components of dibenzyltoluene, hexahydro-dibenzyltoluene, dodecahydro-dibenzyltoluene, and octadecahydro-dibenzyltoluene over phenylhexyl silica in acetone/water solvent (96/4, v/v%) at 22 ± 1°C are presented in Table 3 and 4. It was observed that adsorption to stationary phase increased with degree of hydrogenation of compounds i.e., phenylhexyl silica has three times more adsorption capacity for fully hydrogenated octadecahydro-dibenzyltoluene in comparison to fully dehydrogenated dibenzyltoluene at the same concentration. Moreover, the results show that within the studied concentration range (0 to 25 mg/ml), the adsorption isotherm for H0-DBT and H18-DBT were not influenced by other species as single component data were comparable with the competitive adsorption isotherms within deviation of only 1% as shown in Figure 1.

Correlation of experimental data

Due to the negligible difference between the single and competitive adsorption isotherms (Figure 1), it was assumed that components do not influence each other during adsorption within concentration range studied. Several equations for correlating single component adsorption data such as Freundlich, Langmuir and combined Langmuir-Freundlich (also known as Sip's equation) were evaluated for the correlation of the experimental data. The Freundlich isotherm with two parameters is given by eq (2) [14,15].

$$q_e = q_{sat} \cdot \frac{b.c_i}{1 + b.c_i} \tag{2}$$

Where q_e is the amount of solute adsorbed on the stationary phase, k and n are Freundlich parameters. The Langmuir adsorption isotherm with two parameters is described by eq (3) [14,15].

$$q_e = q_{sat} \cdot \frac{b.c_i}{1 + b.c_i} \tag{3}$$

Where q_{sat} is the saturation capacity of monolayer and b is second Langmuir parameter which quantifies the adsorption energy H_{ads} using eq (4) [14].

$$q_e = q_{sat} \frac{(b.c_i)^n}{1 + (b.c_i)^n} \tag{4}$$

The parameters for the Freundlich equation, Langmuir equation, and Sip's equation are presented in Table 4. The Freundlich correlation did not fit the experimental data satisfactorily (with 13 data points for each compound), especially at low concentrations. The average relative error for the Freundlich model was 11.6% with a maximum deviation of 18.4% observed at low concentrations. For the Langmuir model, the fitting results were better in comparison to the Freundlich

H$_0$-DBT		H$_6$-DBT		H$_{12}$-DBT		H$_{18}$-DBT	
C$_e$	q$_e$	C$_e$	q$_e$	C$_e$	q$_e$	C$_e$	q$_e$
mg·ml^{-1}	mg·g^{-1}	mg·ml^{-1}	mg·g^{-1}	mg·ml^{-1}	mg·g^{-1}	mg·ml^{-1}	mg·g^{-1}
0.25	17.9	0.25	36.5	0.24	41.8	0.24	47.9
0.38	27.2	0.39	59.7	0.37	69.7	0.38	84.9
0.56	39.7	0.57	92.2	0.54	109.1	0.57	143.1
1.00	70.5	0.98	169.6	0.93	206.9	0.98	283.2
1.92	133.3	1.92	357.9	1.84	450.3	1.92	656.0
2.87	195.8	2.89	553.6	2.77	704.8	2.86	1055.4
3.64	244.2	4.61	928.0	4.86	1264.8	4.52	1760.9
4.98	325.1	5.27	1025.8	6.88	1754.4	4.92	1927.4
8.81	533.4	7.52	1435.4	9.24	2260.5	7.04	2749.7
10.61	620.5	10.08	1856.8	11.73	2721.7	9.55	3602.3
14.58	793.2	14.45	2470.3	14.40	2960.0	14.06	4814.8
26.55	1190.4	25.38	3564.0	23.42	4160.4	23.98	6484.8

Table 3: Experimental competitive adsorption isotherm data for dibenzyl toluene derivatives from acetone/water (96/4, v/v) solution over phenylhexyl silica.

H$_0$-DBT		H$_{18}$-DBT	
C$_e$	q$_e$	C$_e$	q$_e$
mg·ml^{-1}	mg.g^{-1}	mg·ml^{-1}	mg.g^{-1}
0.24	17.7	0.24	48.0
0.36	27.1	0.37	84.7
0.54	39.5	0.58	143.0
0.95	70.5	0.99	283.0
1.88	133.0	1.95	656.3
2.85	195.8	2.80	1054.0
3.6	244.2	4.50	1760.2
4.92	325.3	4.88	1925.4
8.9	533.4	7.10	2749.7
10.58	620.5	9.56	3602.3
14.55	793.3	14.10	4814.8
25.98	1190.2	24.05	6487.8

Table 4: Single component adsorption isotherm data for dibenzyltoluene and octadecahydro-dibenzyltoluene from acetone/water (96/4, v/v) solution over phenylhexyl silica.

correlations. However, the Langmuir model fits the H0-DBT and H6-DBT data satisfactorily but gives poor fitting for H12-DBT and H18-DBT data. The average relative error for the Langmuir isotherm was 6.2% with a maximum error of 13%. The relative error was high at low concentrations while the fit was observed to be better at high concentrations. A comparison of the Langmuir and Freundlich isotherms with experimental data is presented in Figure 2.

The Sip's equation with three parameters (eq (5)) correlates the experimental data within the working concentration range better compared to the other correlations.

$$q_e = q_{sat} \frac{(b.c_i)^n}{1 + (b.c_i)^n} \tag{5}$$

Although the number of parameters (three instead of two) increases, higher values of adj. R^2 (adjusted determination coefficient) and Q^2 (cross validated determination coefficient) show the better fit quality (compare Table 5). Additionally, the calculated average relative deviation was only 0.5% with a maximum value of 0.6%. The comparison of the Sip's equation data with the experimental data is presented in Figure 3.

Additionally, the bi-Langmuir adsorption isotherm (eq (6)) was used to correlate the experimental data by assuming the maximum loadability equal for all components in the system [14,15].

$$q_{i,e} = q_{i,sat} \cdot \frac{b_j.c_i}{1 + \sum_{j=1}^{n} b_j \cdot c_j} \tag{6}$$

As expected, the model based on this assumption did not fit the data well because the loadability for various hydrogenated fractions is fairly different for the different compounds. If maximum loadability is considered to be different for all compounds, 8 parameters would be needed to fit eq (6) to 13 experimental data points. Such a high number of parameters do not seem to be reasonable. Thus, within the studied concentration range, Sip's equation was found to be the best suited correlation for the adsorption isotherms of all partially hydrogenated derivatives of dibenzyltoluene.

Conclusion

In this work, adsorption isotherm data for single isomers of dibenzyltoluene, hexahydro-dibenzyltoluene, dodecahydro-dibenzyltoluene and octadecahydro-dibenzyltoluene are presented. The distribution of these components has been investigated between a acetone/water (96/4, v/v) liquid phase and a phenylhexyl silica (stationary phase) at 22 ± 1°C within the concentration range from 0 to 25 mg/ml. Langmuir and Freundlich isotherms with 2 parameters and Sip's equation (combined Langmuir-Freundlich equation) with 3 parameters and multicomponent isotherm based on the Langmuir

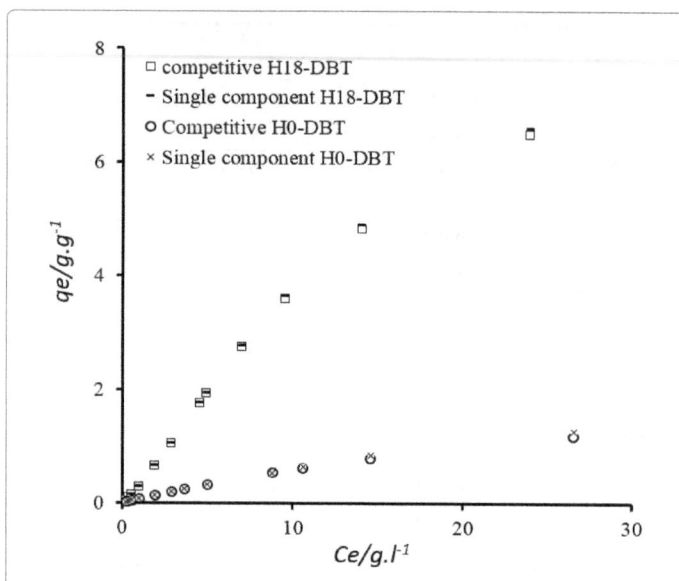

Figure 1: Comparison of single component isotherm with competitive isotherm for H0-DBT and H18-DBT.

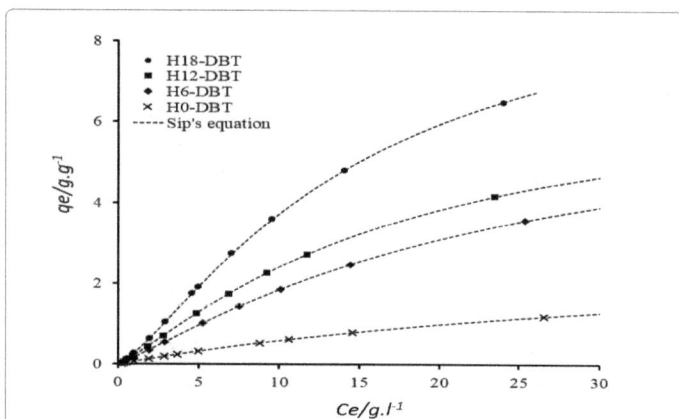

Figure 2: Comparison of experimental and modelled data using Langmuir and Freundlich isotherms.

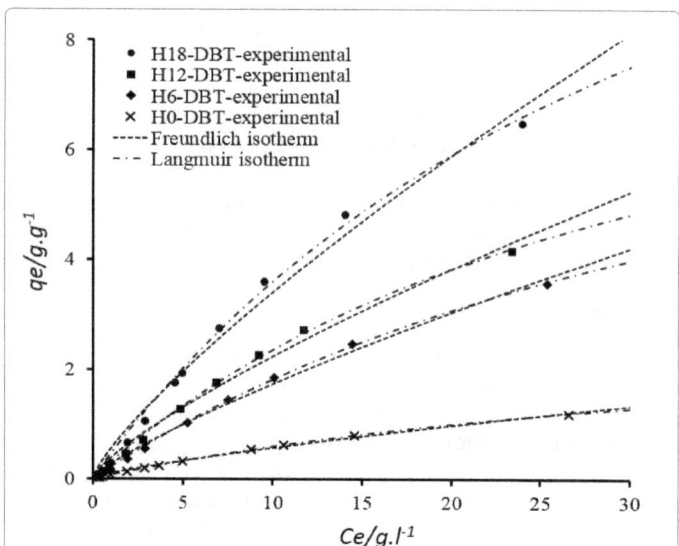

Figure 3: Comparison of experimental and modelled data using Sip's equation.

Parameters	H_{18}-DBT	H_{12}-DBT	H_6-DBT	H_0-DBT
Langmuir equation				
q_{sat}	14.215	10.207	9.883	3.068
b	0.034	0.029	0.023	0.024
Adj. R^2	0.996	0.998	0.999	0.999
Q^2	0.995	0.977	0.995	0.999
ARD/%	13.5	6.21	3.89	1.01
Freundlich equation				
k	0.55	0.37	0.293	0.099
n	0.79	0.78	0.779	0.752
Adj. R^2	0.979	0.988	0.961	0.996
Q^2	0.945	0.920	0.942	0.991
ARD/%	14.5	16.4	9.8	5.8
Sip's equation				
q_{sat}	9.92	7.042	6.915	2.991
b	0.068	0.058	0.042	0.025
n	1.32	1.2	1.15	1.01
Adj. R^2	0.999	0.999	0.999	0.999
Q^2	0.999	0.999	0.999	0.999
ARD/%	0.0023	0.0078	0.0003	0.0001

Table 5: Model parameters for adsorption isotherms with statistical parameters.

model (with 8 parameters) were used to correlate the experimental data. Good agreement with experimental data was found for Sip's equation as compared to the other models. Moreover, adsorption isotherms data for single components of dibenzyltoluene and octadecahydro-dibenzyltoluene were comparable with their competitive adsorption isotherms. Thus, influence of the individual components on each other were found to be negligible within the studied concentration range.

Acknowledgements

This work has been done within the framework of the Bavarian Hydrogen Center and has been funded by the state of Bavaria. The authors wish to thank Prof. Peter Wasserscheid, Dr. Andreas Bösmann, and Michael Müller for the provision of partially hydrogenated dibenzyltoluene mixture and Prof. Wolfgang Arlt as well as Prof. Malte Kaspereit for valuable discussion.

References

1. Muller K, Arlt W (2013) Status and Development in Hydrogen Transport and Storage for Energy Applications. Energy Technology 1: 501-511.

2. Muller K (2012) Amine Borane Based Hydrogen Carriers: An Evaluation. Energy & Fuels 26: 3691-3696.

3. Usman MR (2011) The Catalytic Dehydrogenation of Methylcyclohexane over Monometallic Catalysts for On-board Hydrogen Storage Production and Utilization. Energy Sources Part A: Recovery Utilization and Environmental Effects 33: 2231-2238.

4. Teichmann D, Arlt W, Wasserscheid P (2012) Liquid Organic Hydrogen Carriers as an efficient vector for the transport and storage of renewable energy. International Journal of Hydrogen Energy 37: 18118-18132.

5. Markiewicz M (2015) Environmental and health impact assessment of Liquid Organic Hydrogen Carrier (LOHC) systems - challenges and preliminary results. Energy & Environmental Science 8: 1035-1045.

6. Bruckner N (2014) Evaluation of Industrially Applied Heat-Transfer Fluids as Liquid Organic Hydrogen Carrier Systems. Chem Sus Chem 7: 229-235.

7. Aslam R (2016) Development of a liquid chromatographic method for the separation of a liquid organic hydrogen carrier mixture Separation and Purification Technology. Separation and Purification Technology 163: 140-144.

8. Lisec OP, Hugo, Seidel AM (2001) Frontal analysis method to determine competitive adsorption isotherms. Journal of Chromatography A 908: 19-34.

9. Schmidt TH (2005) Preparative Chromatography of Fine Chemicals and Pharmaceutical Agents, p: 485.

10. Ching CB, Chu KH, Ruthven DM (1990) A study of multicomponent adsorption equilibria by liquid chromatography. AICHE Journal 36: 275-281.

11. Knoerzer K (2011) Innovative Food Processing Technologies. Advances in Multiphysics Simulation.

12. Myers AL, Prausnitz JM (1965) Thermodynamics of mixed-gas adsorption. AICHE Journal 11: 121-127.

13. Bonomo R (2003) Multicomponent adsorption of whey proteins by ion exchanger. Braz J Food Technol 6: 323-326.

14. Hindarso H (2001) Adsorption of Benzene and Toluene from Aqueous Solution onto Granular Activated Carbon. Journal of Chemical & Engineering Data 46: 788-791.

15. Wan X, Lee J, Row K (2009) Nonlinear isotherm of benzene and its derivatives by frontal analysis. Korean Journal of Chemical Engineering 26: 182-188.

19

Behavior of Adhesion Forces of the Aqueous-Based Polyurethane Adhesive Magnetically Conditioned

Wanderley da Costa[1]*, Edith MM Souza[2], Leonardo GA Silva[3] and Helio Wiebeck[4]

[1]Department of Metallurgical Engineering and School of Polytechnic, USP, Sao Paulo, Brazil
[2]Department of Metallurgical and Materials Engineering of the Polytechnic School, USP, São Paulo, Brazil
[3]Centro Technology Radiation, IPEN, São Paulo, Brazil
[4]Departament of Metallurgical and Materials Engineering of the Polytechnic School, USP, São Paulo, Brazil

Abstract

This research presents a new proposal to water based adhesives manufacturing process, utilizing magnetic conditioning. Using this technique it is possible to increase the adhesion capacity between the adhesive and the substrate. The formulation proved to be efficient after the magnetic conditioning and without reactivation which also generated a significant alteration of the viscosity which was 350 mPa.s to 1100 mPa.s, without the necessity of the addition of any other product as thickener. If we compare the same adhesive formulation, with magnetic conditioning and without reactivation and an adhesive without magnetic conditioning and with reactivation, the increase of adhesion capacity was 42.11%, in the magnetically conditioning. When we verify the shear stresses in adhesive without reactivation, without magnetic conditioning and irradiated at 15 kGy we found an average of 3.29×10^5 Pa in the shear stresses, that is 2.16 times lower, compared with the average value found in the test specimens, magnetically conditioned, without reactivation, which was 7.11×10^5 Pa. If the non-ionizing radiation dose of 25 kGy passes, the shear stresses decrease considerably, regardless of the process, with or without magnetic conditioning has the consequence of breaking of the bonds of the polymeric matrix with the test specimens interface.

Keywords: Polyurethane; Ionizing radiation; Magnetic conditioning

Introduction

It is difficult to imagine a product in our home, in industry, transport or any other place that does not use any adhesive or sealant. Based on this fact, it can be affirmed that the adhesives and sealants are present in all segments of the world economy and for this reason have great significance. Since remote times there have been evidences of the use of substances with adhesive capacity, such as animal substances, asphalt material, resin trees, among others. With the development of organic chemistry, new materials and methods have arisen ushering in the production of synthetic adhesives. In order to improve productivity, new processing techniques and equipment have been contributing to a production performance in terms of more competitive operating costs. Currently, the choice of adhesive to promote the union between substrates, rather than mechanical joining methods have some advantages in the use of materials of various thicknesses; absorbing vibrations, sealing capacity and mechanical resistance, thus minimizing the problems of fatigue and decreasing the specific mass of the material, as well as reducing the production costs. Some challenges still exist concerning the use of adhesives, such as need to prepare the substrate, limited service temperature, lower mechanics forces and toughness in relation to metal joints, that do not allow dismounting when the joint is produced by electric welding. Adhesives and sealants are usually formulated based on the following substances: polymeric matrix, fillers, pigments, stabilizers, plasticizers and other additives that are required to confer specific characteristics to the final product. The processes used may involve simple mixtures, as well as sophisticated known as copolymerization processes [1-3].

With so many choices of materials to be adhered, it is necessary to list a few factors of choice for a particular type of adhesive. The adhesive should wet the surface where it will be applied; If the surface is difficult to wet, it should be prepared to receive the adhesive. The adhesives should be more rigid than the substrate, so there would be no stress concentration in adhesives. After the adhesive drying and hardening, the glue joint must resist environmental and mechanical conditions. Both the adhesive, and the method of application thereof, must have low-cost. Other factors that should be highlighted when considering the formulation and choice of adhesives, are the adhesion capacity and cohesion capacity. The polymeric matrixes and some additives should promote or increase these characteristics. Regarding the adhesion, it is understood that this is the result of forces between the adhesive and the substrate, whereas cohesion is the result of forces between the molecules within the adhesive, i.e., the interaction between the components of the formulation. The adhesive studied in this research, called polyurethane, has already been in the market in water base since 2000. Nowadays, the adhesive industry goes through a rereading, seeking high efficiency and substances in compliance with the environmental and market requirements. Studies on the replacement of organic solvents are being intensified and currently 58% of the adhesives already produced use water as solvent [4].

Analyzing this scenario, it can be said that the water-based polyurethane adhesive still requires better results for shear stresses and resistance to hydrolysis. Although there are several formulations in the market, some applications that use adhesives on solvent-based version, mainly due to its properties of open time and adhesion capacity, are used in specific applications in the automotive industry. Based on this fact, this research was held in order to innovate the concept of formulating water based polyurethane adhesives, based on a new technical proposal of the adhesive production technique using the advanced technologies such as magnetic conditioning [5-10] for the production that adhesive. This is designed to improve the properties of the shear stresses of the cured adhesive as much under ambient conditions as in exposure to non-ionizing irradiation in the above-mentioned processes [11].

*Corresponding author: Wanderley da Costa, Department of Metallurgical Engineering and School of Polytechnic, USP, Sao Paulo, Brazil
E-mail: wandher3@usp.br

Water-based systems

The water-based systems are dispersions of polymeric matrixes that may be polyurethane or any others, which, in turn, may be used practically on the same conditions of the solvent-based systems, representing a better alternative and ecologically perfectly viable. The long and positive experience of the industry with the solvent-based adhesives, arose several questions in the market regarding the advantages of the technology of water-based adhesive, which uses water as a solvent. It should be noted that the water based adhesives polyurethane are free from odor, environmentally friendly and safe. The demand of this market grows every year in relation to the solvent-based polyurethane adhesives [12].

Polyurethane adhesive

Polyurethane adhesives are derived from raw materials polyol and isocyanates, such as described in the general equation Figure 1, because of its molar mass, have low open time, which requires thermal reactivation step. These adhesives have very good initial adhesion, and are easy to apply. It is noted that the water-based polyurethane adhesive formulation provides a uniform and thin thickness application, resulting in an excellent final bonding. They can be used in wide variety of substrates, they can be welded on themselves upon application of pressure and have high cohesive efficiency. They are flexible, have a solid tenor between 15 to 20% when solvent-based is between 40 to 55% in water-based. They are widely used in the footwear, belts and handbags industries. This family of adhesives, denominated contact adhesives must be applied to both substrates. To carry out the union, it is mandatory to wait the solvent evaporate (water) or perform the thermal reactivation for bonding both substrates. Shortly after this contact a 110 psi pressure is performed on the glued sets [13-22].

In traditional formulations, both the solvent-based and the water-based adhesives, must be reactivated so that adhesive cohesive forces are intensified. The commercial water-based formulations have high adhesion capacity so they are able to resist the average shear stresses in 3.92×10^5 Pa. This value is lower than that observed in tests with solvent-based polyurethane adhesives, at the average value of reference which is 4.90×10^5 Pa. To solve the difference between the shear stresses, it is necessary to find innovative alternatives regarding the choice of raw materials used in adhesive preparation, as well as to introduce process stages of production, as it was done in this research work with the magnetic conditioning [1,2,15-17] (Figure 1).

Water based adhesive polyurethane

Traditional water-based adhesives are composed of polymeric matrixes, antioxidants, catalysts and other additives to grant ideal

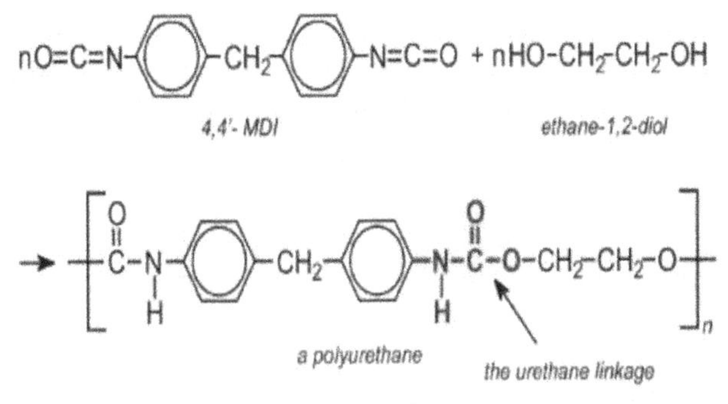

Figure 1: General chemical structure of polyurethane.

properties for their application. In the water-based formulation, the polymeric matrix, normally, is an emulsion in water. In this formulation, the water present in the emulsion must be evaporated to proceed to the bonding of materials. However, there are currently polymeric matrixes that can be used in conjunction with catalysts called blocked, which are substances that remain inactive at room temperature, and only in the presence of an adequate amount of heat are unlocked and from this point begins the curing process, resulting in the development of the mechanism of cross-reactions. It requires special care on the validity of this product, which is shorter compared to solvent-based, as its validity ranges from six months to one year depending on the formulation and its storage conditions that cannot be in places with temperatures above 50°C and below 5°C.

A relevant factor, to develop water-based adhesives resides in the possibility of not depending too much on the solvent-based adhesives, due to its advantages. Besides the fact of reducing the use of organic solvents in the formulations of solvent-based polyurethane adhesives. It can also reduce the use of other substances such as, for example, phthalates, flame retardants, pyrogenic silica and other additives, which has the purpose of improving the adhesive properties [13,14,18-20].

Production of water-based adhesive polyurethane

Normally to produce water based adhesives water dispersions of polyurethane and other substances which are conveniently mixed are used in the following proportions [17-20]. Main Polyurethane Dispersion: 90-98% in weight, Antioxidant or auxiliary additives 0.1 to 2% in weight.

Once formulated, the water based polyurethane adhesive can be magnetically conditioned as described in the following item.

Magnetic conditioning of aqueous solutions

The proposal of using the process of magnetic conditioning after the formulation of a water-based adhesive, lies in the fact that there are studies reporting evidence such as PH changes, electrical conductivity, surface stress, magnetic susceptibility and ionization constant. All this generates a reversal of the spins of electrons, affecting directly the hydrogen bonds and changes in salt and crystal structures of the solutions that were subjected to this process. In other words, in the water-based adhesive case, it is possible to obtain some important changes in the properties of magnetically conditioned adhesives [5-9,14]. As this is an unprecedented research work for polyurethanes aqueous based and there were no specific studies explaining how the action of magnetic fields can change the properties of the adhesives water-base polyurethane, especially how the shear stresses behave in interface of materials, until such research. Magnetic conditioning of water-based solutions as presented in the literature, organizes the molecular structures facilitating their ordination in a predefined direction, thus positively interfering in its viscosity. This improves the applicability of the product and increases the bonding strength on the surface of the substrates. With the increase of these properties, the adhesive improves its efficiency on the substrate surface and the formed film, not allowing the adhesive percolation into the substrate, thereby improving the efficiency of the adhesion conditions. Thus, the wettability or moistening can be defined as the measure of the contact angle between the adhesive and the adherent (substrate). The adhesion capacity is closely related to the spreading of the adhesive due to its surface tension at the interface of the substrate. When the liquid portion of the adhesive, solvent (water) is evaporated, the disposal of the solid part on the surface and its regularity in the formation of the film, will contribute to a better result in shear stresses. Adhesion is understood as a result of the action of forces opposing separation of adhesive molecules on the surface of the substrate, i.e., the ratio of attraction between different substances of the entire system resulting in an intermolecular force. This adhesion is often caused by molecular interactions between the adhesive and substrate and not necessarily by chemical bonds.

Cohesion is different of adhesion, because involves forces between the adhesive and the substrate and the cohesion, it is the attraction between the substances that makes the adhesive formulation, that is, it refers to the union of molecules of the same substance. This is defined as an action of forces that oppose the separation of molecules of a homogeneous body. This technique contributed to the realization of a research [10] on which it was reported the use of magnetic conditioning with water based polychloroprene adhesive, obtaining excellent results in the properties of shear stresses [16].

Application after conditioning magnetic

After the magnetic conditioning process the water-based polyurethane adhesive may be catalyzed in a proportion of 3 to 5%, with Desmoldur® DN (hexamethylene diisocyanate, HDI), according to the formula and applied by brush or spraying process. As already mentioned, it is necessary the insertion additives, for example, stabilizers and antioxidants in order to provide stability of the formulation during storage, to ensure its use, thereby protecting against aging and discoloration.

In another study in the adhesive area, non-ionizing radiation was used in adhesive sealant silicone links and butyl mastic. This process has improved the shear stresses in the butyl sealant formulations, whereas the silicon sealant has degraded the entire gluing system. This study was important to verify how the non-ionizing radiations interferes with links of the silicone sealant adhesion and butyl mastic, which according to the shear stresses tests, we were able to understand and to verify the behavior of such products in terms of the adherence [11-16].

Materials and Methods

The methodology of this research may be defined as a bibliographic, documentary, experimental and comparative research. For this purpose, 8 samples groups were organized: 4 groups containing 25 sets of substrates each one and 4 groups containing 6 sets of substrates each one. The sample groups were designate as follows:

Group 1: Samples in which was applied water based polyurethane adhesive, already formulated, without magnetic conditioning and with reactivation air at the temperature of 60°C.

Group 2: Samples in which was applied water based polyurethane adhesive, already formulated, with magnetic conditioning and with reactivation air at the temperature of 60°C.

Group 3: Samples in which was applied water based polyurethane adhesive, already formulated, without magnetic conditioning and without reactivation of heated air.

Group 4: Samples in which was applied water based polyurethane adhesive, already formulated, with magnetic conditioning and without reactivation by heated air.

Group 5: Samples in which was applied water based polyurethane adhesive, already formulated, without magnetic conditioning, without reactivation by hot air and submitted to non-ionizing irradiation dose of 15 kGy.

Group 6: Samples in which was applied water based polyurethane adhesive, already formulated, with magnetic conditioning, without

reactivation by hot air and submitted to non-ionizing irradiation dose of 15 kGy.

Group 7: Samples in which was applied water based polyurethane adhesive, already formulated, without magnetic conditioning, without reactivation by hot air and submitted to non-ionizing irradiation dose of 25 kGy.

Group 8: Samples in which was applied water based polyurethane adhesive, already formulated, with magnetic conditioning, without reactivation by hot air and submitted to non-ionizing irradiation dose of 25 kGy.

The methodology of this work consisted of the following steps: preparation of the formulation of the water base polyurethane adhesive; magnetic conditioning process; application of the adhesive in the substrates; conduction of tests to determine the shear stresses with and without non-ionizing radiation.

Materials and equipment's

For carrying out this research work several necessary equipment and materialswere provided by the companies. The list of all materials and equipments is

a) Main polymer matrix Dispercoll U® 54 (Anionic high-weight polyurethane dispersion) of Covestro® company, batch C41480060, validity October 16, 2015, Brookfield viscosity, LVR model (sp 3, 10 rpm) equal to 380 mPa.s, solids tenor (oven 180°C, 15 min) equal to 49%; with high crystallization speed, which will provide mechanic strength properties to the adhesive [21].

b) Catalyst Desmoldur DN® (Hexamethylene diisocyanate-HDI catalyst) of Covestro® company, batch LDC 1300029, validity December 20, 2015, Brookfield viscosity, LVR model (sp 3,10 rpm) equal to 1,400 mPa.s, solids tenor (oven 180°C, 15 min) equal to 99.8% used as cross-linking, completing the crosslink connections [21].

c) Antioxidant Rhenofit DDA®50 EN of Rhein Chemie Rheinau GmbH company, a substance derived from diphenylamine in water emulsion. Used as an antioxidant to latex compounds and water-base adhesives. It prevents damage caused by the action of heat and oxygen, in addition to being effective against dynamic crack formation (cracking) caused by oxygen. The concentration of diphenylamine derivative is 50%, batch 0700008972, Brookfield viscosity LVR model (sp 3,10 rpm) equal to 100 mPa.s and its specific mass is 1 g/cm^3 used as antioxidant.

d) PVC substrate, composed of 60% PVC resin (polyvinyl chloride) with 40% plasticizer DOP (dioctylphthalate).

e) Instron® dynamometer, model 3367, with load cell 30 kN.

f) Blower Thermal Steinel®, HL 1500 model, air flow 240/400 L/min, power 1400 W, 220 V.

g) Magnetic conditioner Hidrolink®, CME Model # 0075.

h) Temperature gauge laser MINIPA®, MT 350 model, -30 to 100°C.

i) Gaussmeter Globalmag®, model TLMP-HALL.

j) Soft bristle brush (white).

k) Submersible pump Atman®, AT-302 model, flow 450 L/h, 110 V.

l) Mechanical agitators high-torque microprocessor, Q250M1, 110 V model, 100 W, of Quimis® company, serial number 13040004 and 14120002.

m) Electron accelerator DC model 04/25/1500 JOB 188, Dynamitron˚ company.

n) Precision balance Mettler-Toledo, model ML 4002E with a maximum capacity of 4.2 kg.

o) silicone hoses with ¼ diameter.

p) 2 L Becker cups.

Preparation of adhesive

The water based adhesive was prepared with the polymeric matrix Despercoll U˚ 54 (batch C41480060) and the antioxidant Rhenofit DDA˚ 50 EN (batch 0700008972). Both were weighed out on precision balance Mettler-Toledo (ML model 4002E) as the formulation (Table 1), mixed in mechanical agitator high torque microprocessor for ten minutes at a speed of 50 rpm. The amount of formulated adhesive was 3500 grams.

Conditioning process

Magnetic conditioning process consists on passing the adhesive of two containers (Becker Cup) 2 L, with the help of silicone hoses and the submerged pumps ATMAN (Figure 2), during the time period of three hours under the effect of magnetic field of 0.02 T (Tesla) using a magnetic conditioner. The measurement of the magnetic field equipment was the Gaussmeter (TLMP-HALL model Globalmag), according to Figure 3.

Experimental procedure

For the formulation of the water-based polyurethane adhesive, the mixture was held at mechanical agitator of high torque microprocessor, after the weighing of raw materials, according to data presented in Table 1. After the formulation of the adhesive formulation, the resting time

Figure 2: Submersible pump.

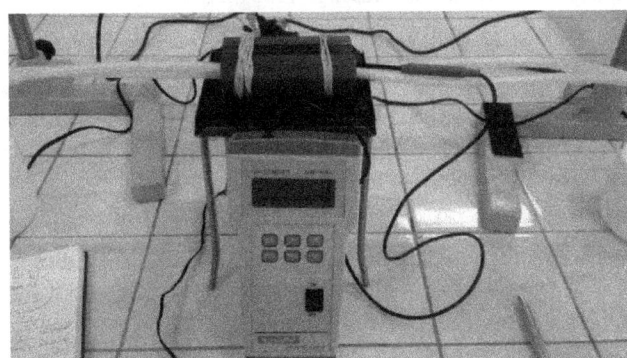

Figure 3: Magnetic conditioner.

is 10 minutes, a part of the adhesive was separated for the application, without subjected to the magnetic conditioning process. The other portion of the adhesive was subjected to the magnetic conditioning process.

The magnetic conditioning was carried out in a system called magnetic cell according Figure 4 composed of the following components: two containers where the adhesive passed; two submerged centrifugal pumps (whose flow rate was 450 L/h, 7 W power and 0.89 mwc of manometric cargo) silicone hoses to transfer the adhesive; magnetic conditioner (comprising a set of neodymium-iron-boron magnets). The intensity of the magnetic field presented a value of 2,120 Gauss (0.02120 T) measured by a Gauss meter and two micro processed agitators with a working speed of 50 rpm. The magnetic conditioning process was performed in 3 hours. Three hours is suggested time in studies on magnetic conditioning of water-based solutions. After such time, it was possible to proceed to the next step that it is the catalyzing of the adhesive. After that, the application of the adhesive in the substrates, with the union of them, denominated test specimens, was performed. Before application of the adhesive, the test specimens have been properly filed down and cleaned, as surface preparation procedure there. In the application of the adhesive, with and without the magnetic conditioning, a brush with soft bristles was used in the substrates.

The polymer matrix is responsible for the adhesion of the test specimens and their efficiency is measured by shear stresses. The catalyst used was Desmodur DN˚ (batch LDC 1300029), using 3% of the formulated adhesive (Table 1) to start the curing process with a time of workability (pot life) of three hours, intensifying the process of coagulation and increasing the formation of the internal forces of the adhesive cohesion. That increased the mechanical properties mainly in the shear stresses of the studied adhesive. An antioxidant was added to the formulation to protect the polymeric matrix from the degradation of the adhesive and to maintain its validity term for marketing.

Adhesive application in the substrates

The formulated quantity was 3500 grams of the water based polyurethane adhesive and submitted to the processes as described in Groups 1 to 8. The adhesives were catalyzed with 3% Desmoldur DN˚ (batch LDC 1300029) on the total mass of each according to their respective groups. All adhesives were properly prepared and two coats were applied on all substrates, with a soft bristle brush, to compose Groups of 1 to 8, as already described above.

The test specimens were prepared in the dimensions of 200 mm × 25 mm × 3 mm, with sticking area of 150 mm × 25 mm. It was analyzed an area of 25 mm × 25 mm, for each universe studied, according to Figures 5-7. All the substrates have had an open time by air drying in ambient air, by at least, twenty minutes before joining them.

All the PVC substrates were cleaned with the cosolvent COLABRAS ST (provided by Brascola company). After that, the formulated adhesive (Table 1) and following the process of distribution of the groups, already described above, and after about 20 minutes for the evaporation of water in ambient air, the adhesive was applied.

The application of water based polyurethane adhesive in the PVC substrates in which the test specimens were waiting for evaporation of the water fraction, in ambient air, during the drying process. At the end of water evaporation time in the PVC substrates, according to its groups, they were reactivated with the heated air using a thermal blower. Below the distribution of the applications process, according to the already mentioned groups.

Raw materials	Individual Quantity (%)
Dispercoll U® 54	98.0
Rhenofit DDA® 50 EM	2.0

Table 1: Water based polyurethane adhesive formulation.

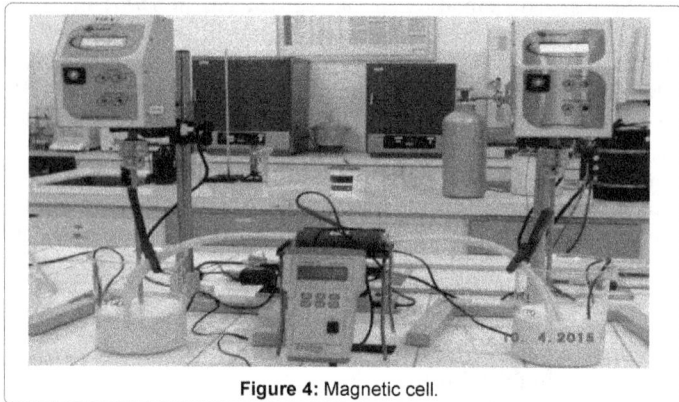

Figure 4: Magnetic cell.

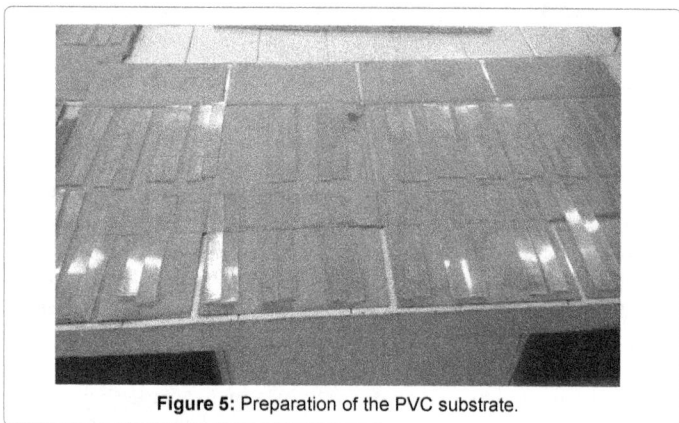

Figure 5: Preparation of the PVC substrate.

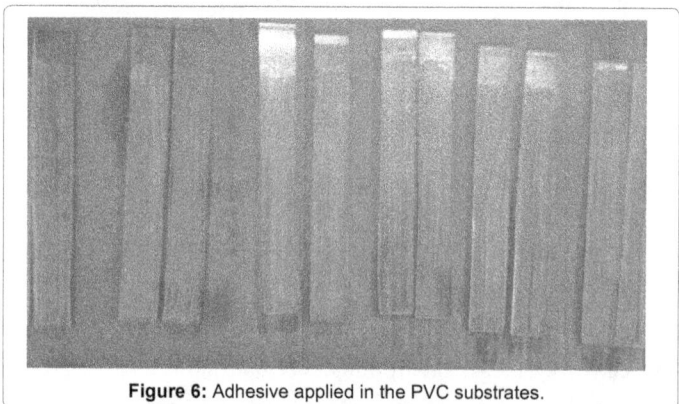

Figure 6: Adhesive applied in the PVC substrates.

After application of the adhesive in the groups 1 and 2, with soft bristle brush, we waited for the open time required of at least twenty minutes and the adhesive was reactivated with hot air at 60°C, thus promoting union of the substrates with the application of pressure of 110 psi for the bonding. They were cured for ten days and tested after 72 h. The test specimens were subjected to tensile tests for determining the shear stresses to verify the adhesion capacity of the adhesive with the magnetic conditioning process. This step was performed with the intention of obtaining data for subsequent comparison between the current manufacturing processes and the new proposal with a magnetic conditioning demonstrated here.

In Groups 3 and 4, the adhesive application procedure was with soft bristle brush, waiting for the open time of at least twenty minutes for the evaporation of water without reactivating it. Figure 6 shows the process of applying a quantity of hot air, at, at least 60°C, for the reactivation of the adhesive and subsequent bonding of the test specimens. The groups from 5 to 8 the test specimens were also prepared for being subjected to non-ionizing radiation, in order to verify the shear stresses behavior and consequently the possible bond breaking in the substrate adhesion links in the material interface. It is emphasized that the bonding process obeyed the same process described by Groups 3 and 4 which had not had reactivation by heated air. The groups from 5 to 8 were cured for 10 days. After this period, the test samples were subjected to non-ionizing irradiation and performed the shear stresses tests after a period of 72 h. This study used an electron accelerator model DC1500/25/4 JOB188 (Dynamitron®), with an energy of 0.5 MeV and 1.5 MeV, a beam current 25 mA, scanning from 60 to 120 cm with a maximum power of 37.5 kW beam. Though the dose rate can vary from 1.07 kGy/s to 161.67 kGy/s, the test specimens were irradiated with doses of 15 and 25 kGy. The experiment was limited to 25 kGy, because above such value the process will be economically impracticable. The test specimens were irradiated in a direction perpendicular to the substrate polyvinyl chloride (PVC) surface. The electron energy range was selected to allow maximum penetration of the electron beam through the PVC plates, because in this region of power PVC braking was small (1.779 MeV cm^2/g), and the electrons reach the adhesive cured in the substrates (PVC). The irradiation was performed in atmospheric air (atmosphere). In order to evaluate the behavior of the adhesive forces on the test specimens, the tests of shear stresses were conducted using a universal tensile testing machine (Instron, model 3367, with a 30 kN load cell), complimentary Brascola company. In the discussion of the following results, all the groups from 1 to 8 waited ten days to the curing process and more 72 h to start the shear stresses tests, with the purpose to make sure that the curing process would not interfere in the tests results.

After the adhesive total cure all the test specimens were submitted to the shear stresses test, including the irradiated samples. The standard used to test the shear stresses was ASTM D-1002 [Standard Test Method for Apparent Shear Strength of Single-Lap-Joint adhesively Bonded Metal Specimens by Tension Loading (Metal-to-Metal)], adapted for PVC substrates with PVC.

Results and Discussion

Viscosity

It is found that the initial viscosities of water based polyurethane adhesives as well as the formulation without magnetic conditioning that currently exist on market is 350 mPa.s. In this research with the magnetic conditioning process technique, the viscosity of the formulated adhesive was increased to 1100 mPa.s.

Therefore, a significant thickening of the water-based adhesive, leaving it with the viscosity closer to that of the commercial solvent based polyurethane adhesives that is about 2000 mPa.s. No type of thickener was added to the formulation of the adhesive to raise its viscosity. The only differential was the magnetic conditioning process.

Shear stresses

Regarding the adhesion capacity, the results of shear stresses tests for non-irradiated samples are presented in Table 2. Figures 8 and 9 present the results of the shear stresses of the substrates, the adhesive with and without magnetic conditioning; have shown different behaviors in the adhesion forces. These results were compared with the

Water-base				
Groups				
	1	2	3	4
Test Specimen	Without magnetic conditioning with reactivation (SCMR)	With magnetic conditioning with reactivation (CCMR)	Without magnetic conditioning without reactivation (SCM)	With magnetic conditioning without reactivation (CCM)
Reactivation	at 60°C	at 60°C	-	-
1	3.73	3.35	3.84	4.89
2	4.88	1.99	5.34	5.67
3	4.32	2.83	5.30	5.78
4	4.84	3.32	5.83	5.80
5	3.78	3.02	2.45	4.80
6	3.88	3.16	4.53	5.90
7	3.71	1.35	3.18	5.67
8	3.90	2.29	3.43	5.77
9	3.23	5.03	2.23	5.85
10	4.70	4.45	2.74	5.92
11	4.13	3.72	3.18	5.89
12	3.71	3.44	3.16	6.03
13	4.27	2.15	4.62	5.87
14	4.58	2.09	4.18	6.11
15	4.60	3.02	3.20	5.96
16	3.83	2.10	2.72	6.45
17	2.93	1.30	2.28	6.30
18	4.88	4.35	4.91	5.87
19	3.50	3.79	2.31	6.33
20	4.38	1.48	4.33	6.39
21	4.01	2.29	3.01	6.29
22	4.80	3.93	2.92	5.86
23	4.47	1.88	2.10	6.36
24	4.80	2.06	1.91	6.39
25	4.67	2.79	1.91	6.42
Average	4.18	2.85	3.42	5.94

Table 2: Results of traction Tests - Shear Tension in 10^5 Pa.

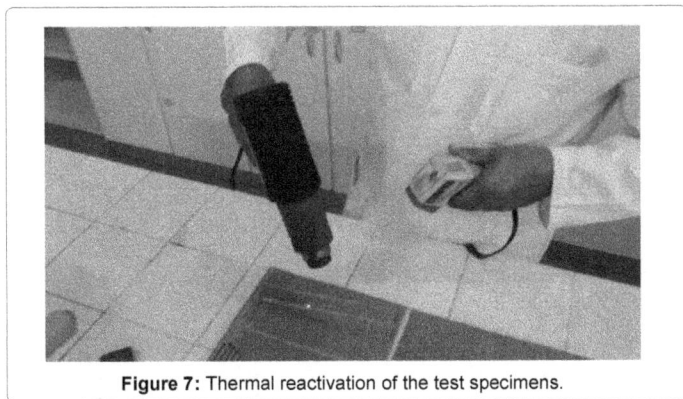

Figure 7: Thermal reactivation of the test specimens.

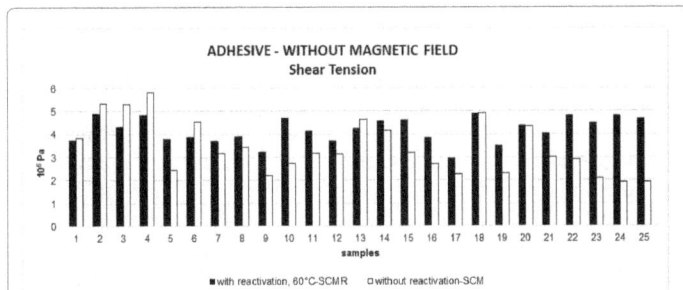

Figure 8: Tensile test results for Group 1, without magnetic conditioning and reactivated at 60°C and for Group 3, without the magnetic conditioning and no reactivation with heated air.

second step of this work, as shown in Table 3, Figures 10 and 11. The test specimens was submitted to non-ionizing irradiation where it was possible to evaluate the behavior of shear stresses of the formulated adhesive and also evaluate the effect of the magnetic field on it. To facilitate the visualization of the behavior of shear stresses supported by test specimens from de two universes of samples, without magnetic conditioning and with reactivation at 60°C, according to Figure 7, It is observed that the values of the shear stresses of the adhesive formulated without magnetic conditioning and with reactivation by heated air were rather close. Even so the average value of 4.18×10^5 Pa, for reactivated test specimens was 22% higher than in the non-reactivated ones, whose average value was $3{,}42 \times 10^5$ Pa. The results of shear stresses, supported by two universes of test specimens with field and without and reactivation at 60°C, Figure 8 below it is observed that the behavior of the adhesive shear stresses formulated with magnetic conditioning, without reactivation, the average value is 5.94×10^5 Pa, higher compared to the magnetically conditioned and reactivated heated air at 60°C, showing an average value of $2.85\ 10^5$ Pa. With this it was possible to increase the adhesion of the water based polyurethane adhesive for significant values, without thermal reactivation by heated air. The only difference in the production process was the magnetic conditioning. Regarding the magnetic conditioning, the magnetic field changed the size of micelles of the polymeric matrix, in which was found an increase of the viscosity and shear stresses. In the application process, when activated by heated air, the adhesive magnetically conditioned, loses property in the values of shear stresses due to magnetic disorientation provoked by the heating (hot air). Therefore, when the thermal energy is not applied, the shear stresses values exceed

Water-Base				
GROUPS				
	5	6	7	8
Type of exposition I	Without Magnetic conditioning (SCM)	With Magnetic conditioning (CCM)	Without Magnetic conditioning (SCM)	With Magnetic conditioning (CCM)
Type of exposition I I (dose)	15 kGy	15 kGy	25 kGy	25 kGy
Test Specimen		Shear Tension		
1	1.49	6.27	1.96	0.56
2	3.68	7.45	0.67	0.63
3	3.68	6.77	2.26	0.77
4	2.64	9.35	1.37	0.30
5	4.57	7.14	1.32	0.28
6	3.66	5.70	2.79	0.84
Average	3.29	7.11	1.73	0.57

Table 3: Results of Traction Tests -Shear Stresses 10^5 Pa - irradiate samples.

the reactivation by heating processes currently utilized. It is observed in the values shown in Group 1 (Table 2), which represents the product currently used in the market.

Regarding the adhesion capacity, the results of the shear stresses tests for irradiated samples are shown in Table 3: To facilitate the visualization of the behavior of shear stresses supported by the irradiated test specimens, from two universes of samples, without and with the magnetic conditioning (Figures 10 and 11). Figure 10 Tensile test results, for the Group 5 without reactivation, without magnetic conditioning and irradiated to 15 kGy; Group 6 without reactivation, magnetic conditioning and irradiated to 15 kGy. It was observed thus, that the adhesive shear stresses formulated without magnetic conditioning (Group 5), without reactivation and irradiated at 15 kGy, presented average value of 7.11×10^5 Pa. In relation to conditioning magnetically (Group 6), without reactivation, and irradiated at 15 kGy which presented an average value of 3.29 105 Pa. The samples irradiated at 25 kGy, without reactivation magnetic conditioning and without (Group 7), presented an average value of 1.73×10^5 Pa. Regarding the magnetically conditioning (Group 8), without reactivation and irradiated at 25 kGy, which presented an average value of 0.57×10^5 Pa) [22].

Conclusions

Based on this research work, we can verify the behavior of the adhesion strengths submitted by shear stresses in the water-based polyurethane adhesive, under the effect of the magnetic field. Regarding the adhesiveness, it was evident the possibility of increment it with the magnetic conditioning. Comparing the results of the shear stresses tests, of the four groups of initial test specimens, we succeed to approach the result of adhesion capacity, compared with a solvent-based polyurethane adhesive produced currently in the market. The studied formulation proved efficient after the magnetic conditioning and without reactivation which also generated a significant alteration of the viscosity. The variation is increased from 350 mPa.s to 1100 mPa.s, without being necessary to add any other product as a thickener. Therefore, the new concept of the process of water based polyurethane adhesive manufacturing conditioning magnetically, in which the thermal reactivation is not required to promote the union of the substrates, confers adequate properties for this type of adhesive, making it competitive on the market compared with the application process. If the adhesives with the same formulation with magnetic conditioning and without reactivation were compared with an adhesive without magnetic conditioning and reactivation, the increase in the adhesion capacity was of 42.11% with those magnetically conditioning

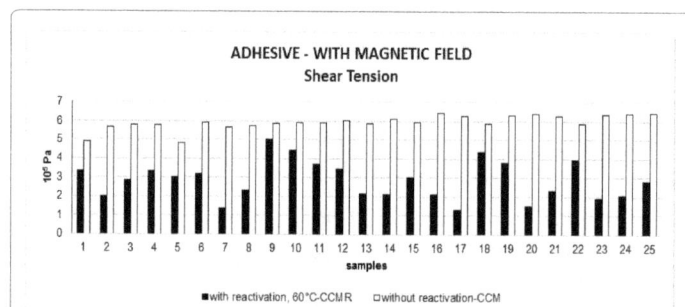

Figure 9: Tensile test results for Group 2, with magnetic conditioning and reactivated at 60°C and Group 4, with magnetic conditioning and without reativation by heated air.

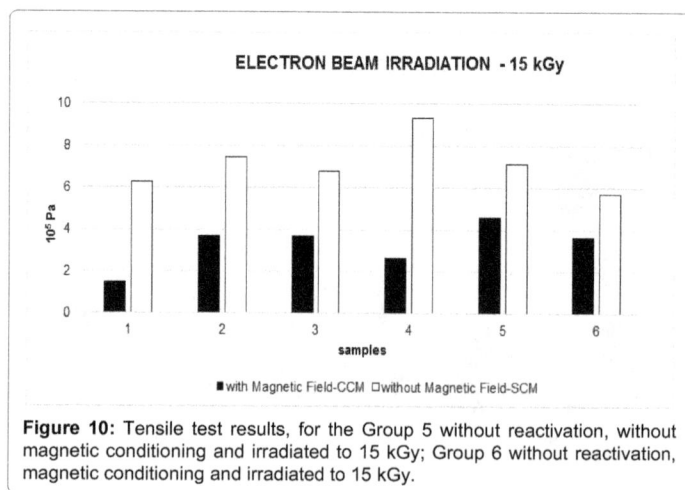

Figure 10: Tensile test results, for the Group 5 without reactivation, without magnetic conditioning and irradiated to 15 kGy; Group 6 without reactivation, magnetic conditioning and irradiated to 15 kGy.

adhesive. The result was far beyond expectations, without any insertion of other raw materials such as, for example, addition of resins to improve adhesive adhesion characteristic. It was observed when there is a magnetic conditioning and the test specimens were submitted to non-ionizing irradiation, regardless the dose, in the tests of the final four groups, the adhesion decreased. If we verify the shear stresses at the adhesive without reactivation, without magnetic conditioning and irradiated at 15 kGy and found an average value of 3.29 105 Pa in the shear stresses tests, i.e., 2.16 times lower compared to the average value found in the test specimens with magnetically conditioned and without reactivation, which was of 7.11 105 Pa. When a non-ionizing irradiation dose of 25 kGy was applied, the shear stresses decrease considerably,

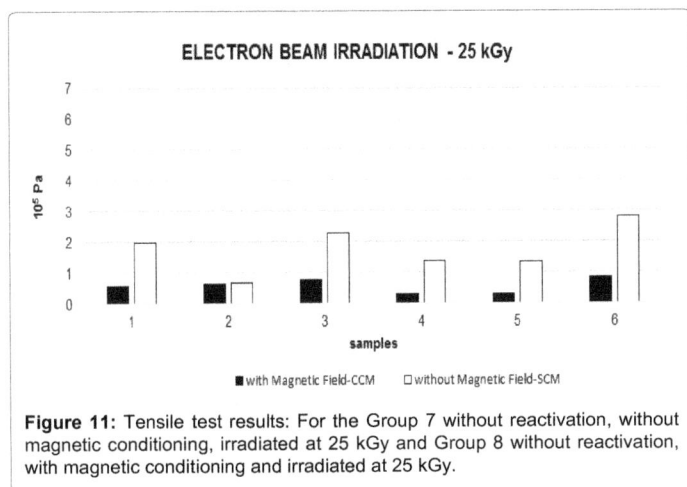

Figure 11: Tensile test results: For the Group 7 without reactivation, without magnetic conditioning, irradiated at 25 kGy and Group 8 without reactivation, with magnetic conditioning and irradiated at 25 kGy.

regardless the process, with or without magnetic conditioning, and have as a consequence, the breaking of the bonds of the polymeric matrix on the test speciments interface. Those adhesives formulated in this work, with magnetic conditioning and without reactivation by heated air, presented the best adhesion capacity in the application process. Because when submitted to heat air, all the acquired magnetic information is lost, thereby decreasing its efficiency in shear stresses and compromising the adhesion of these test specimens. For the production process it is possible to use magnetic conditioning, in which we will have the best result of shear stresses without the need for reactivation of the adhesive by heated air (heat), to carry out the bonding of the substrates in the preparation of the test specimens. This process innovates the system of production of water based polyurethane adhesives, thereby saving thermal energy which reduces the manufacturing costs of operation, making the final product (shoes, belts, bags and similar) more competitive as well as being ecologically friendly. However, the difference in the production process was the magnetic conditioning, more effective without reactivation for heat, in the application of the adhesive. In the future others water based adhesives families can be studied, such as acrylic, polyvinyl chloride, polyvinyl acetate, among others.

Disclosure Statement

No potential conflict of interest was reported by the authors.

Acknowledgments

Our special thanks to the company Brascola, Convestro, Evonik, Rhein Chemie Rheinau. The Professor Dr. Hélio Wiebeck, the Professor Dr. Leonardo Gondim de Andrade Silva, Departamento de Tecnologia das Radiações-IPEN (Instituto de Pesquisas Energéticas Nucleares). The Dra. Vanessa C. Loureiro and the Professor Dra. Esther Galvão.

References

1. Petrie EM (2007) Handbook of Adhesives and Sealants. 2nd edn. McGraw-Hill, New York, USA.

2. Pizzi A, Mittal KL (2003) Handbook of Adhesive Technology. 2nd edn. Marcel Dekker Inc., New York, USA.

3. Skeist I (1989) Handbook of Adhesives. 3rd edn. Van Nostrand Reinhold Co., New York, USA.

4. Liesa F, Bilurbina L, Marcombo SA (1990) Adhesivos Industriales. Barcelona.

5. Belova V (1972) Magnetic treatment of water. Soviet Science Review: Scientific Developments in the USSR. Moscow. n.3.

6. Kronenberg JK (1987) Magnetic Water Treatment Demystified Magnets. Paper Courtesy of Teldon of Canada Ltd., pp: 150-156.

7. Landgraf FJG, Garcia PMP, Poço JG, Giulietti M (2004) Effect Magnetic Field in Aqueous Solutions. In: XVI Brazilian Engineering and Materials Science Congress.

8. Lambert K (1999) Study of factors related to magnetic treatment of calcium carbonate saturated water. Provo (EUA): Brigham Young University, Department of Civil and Environmental Engineering.

9. Costa W (2006) Contribution to the study of the magnetic conditioning of water for industrial purposes. Dissertation, São Paulo BR School University Center of Mauá.

10. Souza EMM (2015) Contact adhesive polychloroprene water-based nano additivated conditioning and magnetically, Thesis. São Paulo (BR) Polytechnic School of the University of São Paulo.

11. Costa W (2012) Behavior of the adhesion forces of silicone adhesive sealant and butyl mastic under the effect of ionizing radiation. Polytechnic School of the University of São Paulo. Department of Metallurgy and Materials.

12. Brazilian Association Industry Chemistry (2011) Statistics segment of glues, adhesives and sealants. 10: 1-19.

13. Calpena EO, Ais FA, Palau AMT (2009) Influence of the chemical structure of urethane-based thickeners on the properties of waterborne polyurethane adhesives. The Journal of Adhesion 85: 665-689.

14. Limiñana MAP, Aís FA, Palau AMT (2007) Influence of the hard-to-soft segment ratio on the adhesion of water-borne polyurethane adhesive. Journal of Adhesion Science and Technology 218: 755-773.

15. Souza MME, Costa W, Silva LG (2016) Behavior of adhesion forces of the aqueous-based polychloroprene adhesive magnetically conditioned. Journal of Adhesion Science and Technology.

16. Costa W, Wiebeck H, Martinelli JR (2012) Improved Adhesion of Butyl Mastic Using Electron Beam Irradiation. Journal of Adhesion Science and Technology 26: 1699-1704.

17. Keberle W, Oertel G, Bayer AG (1970) Anionic polyurethane dispersions and a process for the production thereof. United States Patent US 3,539,483A.

18. Hombach R, Reiff H, Wenzel W, Bayer AG (1984) Aqueous adhesives containing water-dispersible polyisocyanate preparations. United States Patent US 4,433,095A.

19. Maksymkiw M, Haider G, Dochniak MHB (1997) Fuller Licensing & Financing Inc., Water-based adhesive formulation having enhanced characteristics. United States Patent US 5,624,758A.

20. Sagiv E (1997) Olin Corporation. Water-based urethane adhesive. United States patent US 5,688,356A.

21. Bayer (2015) Leverkusen: Coatings, Adhesives and Specialties.

22. Zenner MD, Madbouly SA, Chen JS (2015) Unexpected tackifiers from isosorbide. Chem Sus Chem 83: 448-451.

Investigation of Rate of Photo Degradation of Chlorothalonil, Lambda Cyhalothrin, Pentachlorophenol and Chlropysis on Tomato and Spinach

Mbugua JK*, Mbui D and Kamau GN

University of Nairobi, Nairobi, Kenya

Abstract

Photo-degradation of common pesticides and herbicides on the surface of tomato fruit and spinach leaves were studied. Samples were spiked with 100 mg/ml of lambda cyhalothrin, chlorothalonil, chlorpyriphos and pentachlorophenol standard solutions in acetone. Thy were air dried for 1 minute and exposed to light of various intensities; sunlight, 40 w, 60 w, 75 w and 100 w bulbs after spreading on the surface of spinach and tomatoes for 15, 30, 45 and 60 minutes and then washed in acetone. The pesticide concentration in the samples was analyzed with a UV Visible spectrophotometer. The rate of degradation of chlorothalonil, lambda cyhalothrin, pentachlorophenol and chlorpyriphos was also calculated and rate constants obtained for each residue. The results obtained indicated that the 100 W bulb degradation ranged from 20-95% for all the molecules in both tomato fruit and spinach leave surface. The residues breakdown followed a 1st order kinetics.

Keywords: Residue; Photo-degradation; Chlorothalonil; Dursban; Lambda cyhalothrin; Pentachlorophenol; Spectrophotometry

Introduction

Integrated pest management (IPM) strategies employ the use of different classes of pesticides on agricultural crops. Frequent use of pest control chemicals affects the properties of the soil and ground water by altering the pH thereby inhibiting microbial activities which is important for soil fertility [1,2]. Approximately 0.1% of applied pesticides reach the target pest, which means that the bigger percentage of the pesticide ends up contaminating the environment [3]. Photo-degradation is an abiotic process in the dissipation of residues where molecular excitation by absorption of photons result in organic reaction [4]. The degradation of pesticides is mainly focused on hydrolysis, photo-degradation, and microbial degradation [4-6]. Photo-degradation of pesticides has a significant influence on pesticide residue, efficacy and toxicity and in a big way determines the fate of pesticides on the crop surface and in the atmospheric and water environment [5]. The photochemical degradation properties of pesticides have become an important index in ecological environment safety evaluation of pesticide. Photo-degradation study is required to provide information of photo-degradation for pesticide registration in many countries. Organo-phosphates are normally stable at ambient temperatures and their isomers are formed at elevated temperatures. For example, the P=S(thiono) bond where P is phosphate isomerizes to P=S (thiolo) which are more toxic to mammals [7-9]. Ultra-violet light is a high source of energy, and promotes photo-degradation. Some pesticides are equipped with UV light blockers so as to reduce the rate of photo decomposition of their active ingredient [10]. Since the UV-violet component of sunlight which varies from 200-400 nm degrades the residues, it is vital to study the effect of light on degradation of pesticide molecules to understand their fate in the environment when applied [11,12].

Materials

All chemicals used were of analytical grade quality 99.8% Acetone, 99.8% Acetonitrile and obtained from 64271 Darmstadt, Germany and Aldrich respectively; Distilled water was double distilled into a glass bottle; 99.3% Pentachlorophenol, 99.8% Chlorpyrifos and 99.3% lambda cyhalothrin from labor Dr. Ehrenstorfer; 99.0% Chlorothalonil from Annopol 6 Warsaw Poland, 1% Acetic Acid HPLC grade from Sigma Adrich. UV-Visible spectrophotometer (Shimadzu UV-Visible 1650 PC Shimadzu Scientific Instruments, 7102 River wood Drive Columbia, MD 21046, USA), Analytical balance (Fischer scientific A-160); 1 cm quartz cuvette.

Methodology

Stock solution preparation

1000 ppm stock solution for chlorothalonil, chlorpyrifos and pentachlorophenol were prepared in acetone while distilled water was used for lambda cyhalothrin. 10, 20, 40, 60, 80 and 100 mg/liter solutions were prepared by serial dilution method. The solutions were scanned using UV-Visible spectrophotometer. Plot of absorbance versus concentration for each pesticide standard was made and further used in the degradation study. Five sets of spinach leaves (5 cm wide) and tomatoes were placed in a separate petri dishes. 2 ml each of the acetone-dissolved 1000 mg/l stock solutions was spread on the plant surface and the acetone allowed to evaporate for 1 minute. The set was then exposed to 40 w, 60 w, 75 w and 100 w bulbs which were enclosed in a container to prevent light loss for 4 hours. Another set up was exposed to sunlight for 4 hours. Temperature in each set up was recorded during the exposure. The samples were then allowed to stabilize in the laboratory for 24 hours before washing with 10 ml acetone. The extracts were analyzed for the pesticide levels using UV-Visible spectrometer at 340 nm for chlorothalonil, chlorpyriphos and pentachlorophenol and 254 nm for lambda cyhalothrin. The amount of residue left after exposure was determined. To study the effect of time on the rate of degradation of these pesticide residues, 5 cm by 5 cm of spinach leaves were cut and dipped in 100 ppm standard solutions of chlorothalonil, chlorpyriphos, pentachlorophenol and lambda cyhalothrin prepared in acetone for 2 minutes. The leaves were placed in a Petri dish and acetone allowed to evaporate in air for 1 minute. The setups in the petri dishes were then were exposed to 40 w, 60 w, 75 w and 100 w bulbs for 10, 20, 30 and 60 minutes each. The spiked samples were removed and washed with 5 ml acetone and then residue concentration analyzed with UV-Visible spectrophotometer.

*Corresponding author: Mbugua JK, University of Nairobi, Nairobi, Kenya
E-mail: djames085@gmail.com

Kinetic study

In order to study the rate of degradation of chlorpyriphos, pentachlorophenol, lambda cyhalothrin and chlorothalonil formulations plots of time versus natural logarithm of concentration were made to find the rate of degradation while the half lives period ($t_{0.5}$) were calculated according to equation [13].

$$t_{0.5}=\ln2/k=0.693/k$$

$$k=1/Tx.\ln a/bx$$

where, k=Rate constant, T_x=Time in days, a=Initial residue, b_x=Residue at time (x).

Results and Discussion

The plots of absorbance against concentration were observed to adhere to Beer's law with regression values of 0.969, 0.981, 0.973 and 0.992 for lambda cyhalothrin, pentachlorophenol, chlorothalonil and chlorpyriphos, respectively confirming linearity of the obtained absorbance. Table 1 below show the calibration curves equations with their regression values for the four pesticides.

The rate of degradation

The decrease in pesticides residues concentrations as depicted in Figures 1a-1d confirms the degradation of chlorpyriphos, lambda cyhalothrin, chlorothalonil and pentachlorophenol on exposure to different light intensities on the surface of the spinach leaf. For all the four residues under study, photo degradation is observed to be dependent on time of exposure, light intensity and temperature. The percentage loss is higher in pentachlorophenol, chlorothalonil with chlorpyriphos and lambda cyhalothrin degrading almost at the same rate at 100 w on exposure for 60 minutes. The % loss for 100 w on exposure for 60 minutes is 87, 83, 71 and 71 for Pentachlorophenol, chlorothalonil, lambda cyhalothrin and chlorpyriphos, respectively. The difference in the loss is most probably due to the difference in molecular structure. This corroborates results given were observed that photo degradation of organic molecules is highly influence by the molecular structure. To determine the rate constant of residues, plots of natural logarithm of concentration against time were made. The rate constant is obtained from slope of the plots (Figure 2). The plots confirm that the reactions are first order as expected.

Pesticide	Equation	Regression
Chlorpyriphos	Y=0.001x+0.003	0.992
Pentachlorophenol	Y=0.004+0.009	0.981
Chlorothalonil	Y=0.005x+0.006	0.993
Lambda cyhalothrin	Y=0.010+0.015	0.969

Table 1: Calibration curves equation.

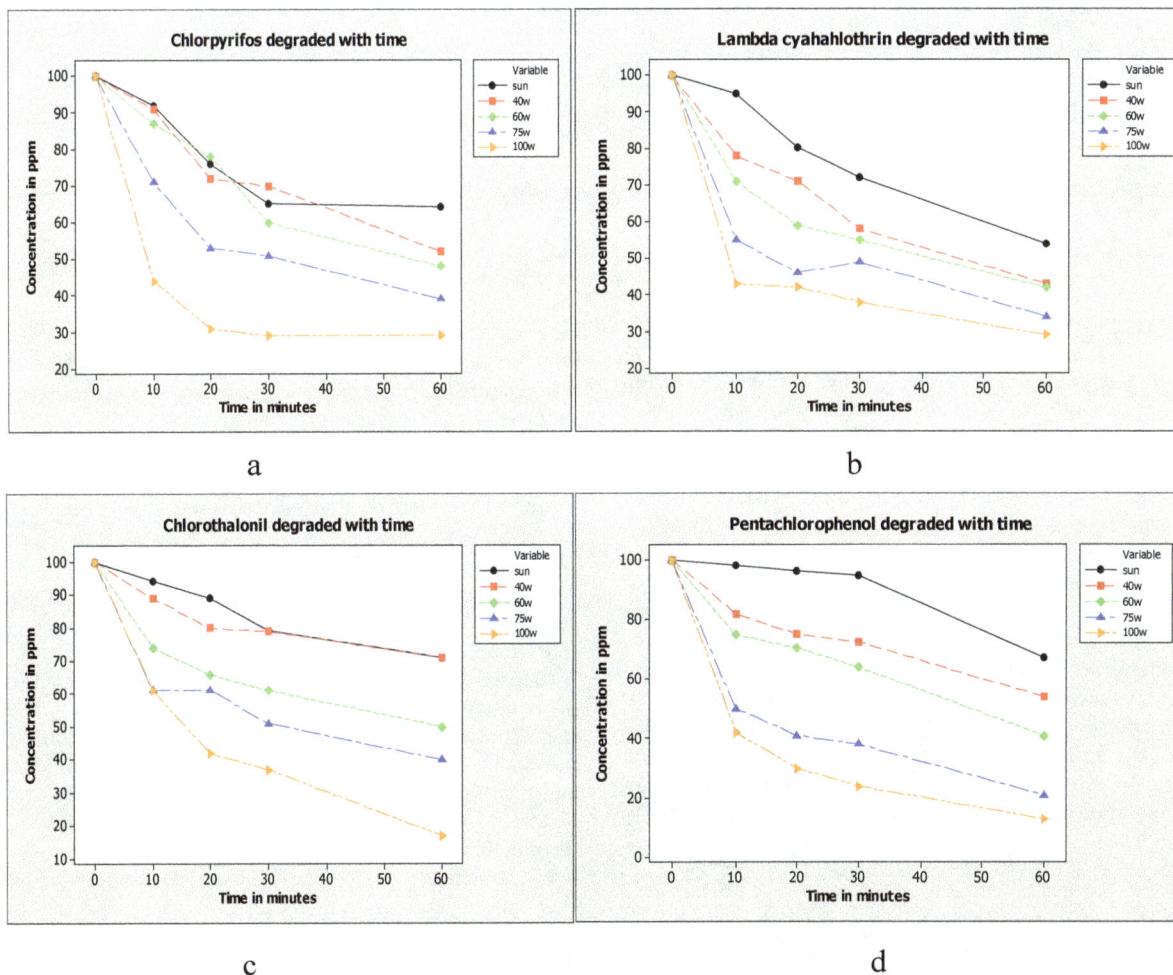

a

b

c

d

Figures 1: Photodegradation trend of (a) chlorpyrifos (b) lambda cyhalothrin (c) chlorothalonil (d) lambda cyhalothrin with time in minutes.

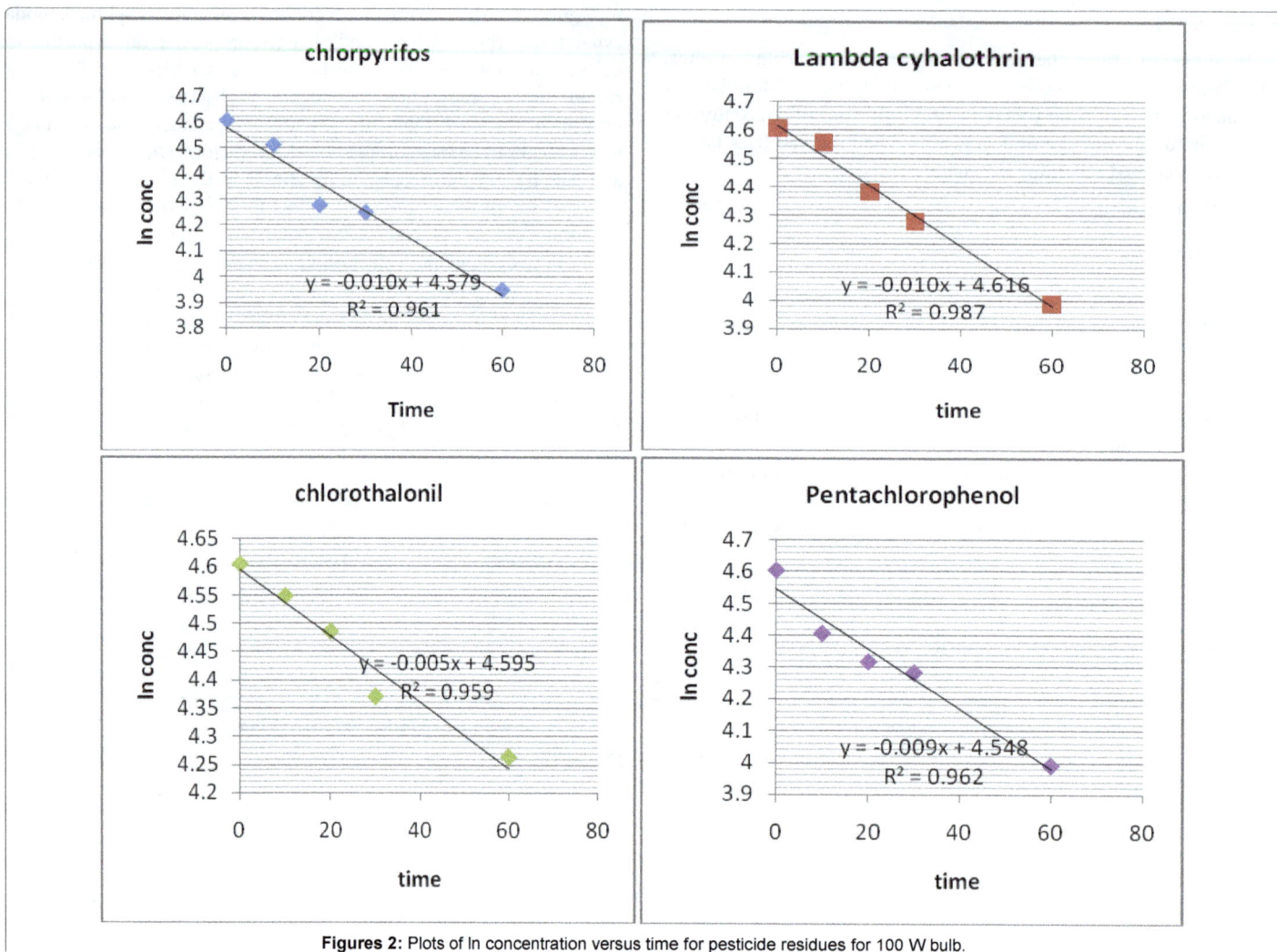

Figures 2: Plots of ln concentration versus time for pesticide residues for 100 W bulb.

Energy calculations

The total amount of energy absorbed by 5 cm by 5 cm leaf surface from the sun on a hot summer can be calculated as follows; Energy=power × time. Electromagnetic wave from the sun is 1.4 kW/m² but only 80% of this reaches the earth surface on a hot summer day, therefore, 80/100 × 1.4=1.12 kW/m²; Area=5 cm × 5 cm=25/10000=0.0025 m² power=0.0025 m² × 1.12 kW/m²=0.0028 kW, Energy=0.0028 kW × 14400 sec=40.32 joules. This means that the amount of energy absorbed by the spinach and tomato when exposed to 40 w, 60 w, 75 w and 100 w bulbs for four hours can be calculated since the surface area and the wattage is known. The amount of energy responsible for breaking down the pesticide molecules are shown in Table 2. The sun has the highest amount of energy reaching the surface of the 0.0025 and 0.002798 m² of the spinach and tomato respectively. From the plots of percentage degradation of the four residues, the sun degraded the least. Several factors account for this degradation trend; the distance from the sun to the molecule surface is 93,000,000 miles meaning the intensity

reaching the crop surface is reduced to a large extent unlike for the bulb which is 30 cm above the surface. Secondly, as noted earlier by Suett temperature is a major factor in the breakdown of pesticides whereby he reported doubling of chemical reactions with 10°C rises in temperature. Photo-degradation from the sun is influence by light intensity while light intensity and temperature influences breakdown of residues. This explains the pattern of photo-degradation (Figure 3).

Lambda cyhalothrin

Exposure of lambda-cyhalothrin spiked spinach and tomatoes samples to sunlight, bulb light of 40 w, 60 w, 75 w and 100 W) for four hours resulted in nearly complete degradation of cyhalothrin with losses greater than 95% of initial amounts applied (Figure 4). The degradation was observed to vary with light intensity. Different wattage bulbs emitted different amount of light intensity. This means that photo-degradation of lambda cyhalothrin on the surface of spinach; tomato and blank petri dish resulted from effects of light intensity. Figure 5 illustrate the amount degraded versus wattage. Lambda cyhalothrin degradation was highest in 100 W bulb as expected from light. This is in agreement with the Stark Einstein law which stipulates that for every photon absorbed a molecule undergoes photochemical reactions. This means that the reaction depends on the number of photons that illuminate a surface (wattage) and the bond being cleaved (molecular structure) in the two plants investigated the half life decreases with the strength of radiation which further agrees with Stark Einstein law; the

Light	Spinach ($t_{1/2}$)	Tomatoes ($t_{1/2}$)
40 w	0.060	0.059
60 w	0.058	0.058
75 w	0.055	0.055
100 w	0.045	0.054

Table 2: Half-lives of lambda cyhalothrin.

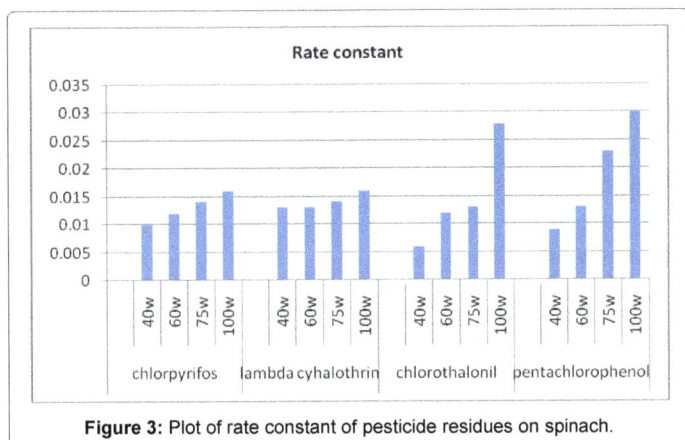

Figure 3: Plot of rate constant of pesticide residues on spinach.

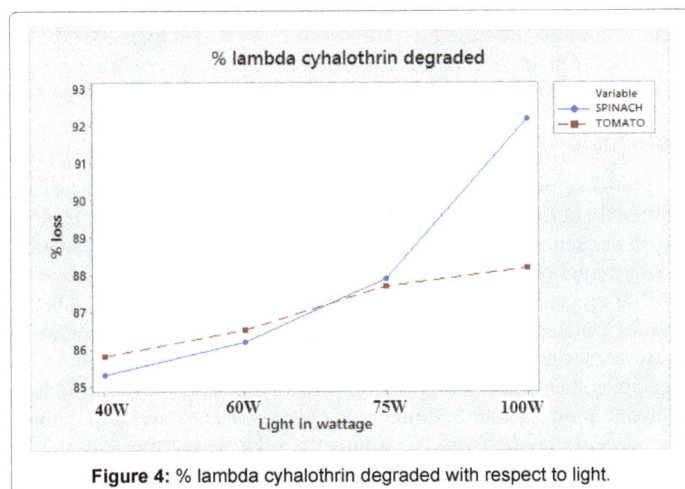

Figure 4: % lambda cyhalothrin degraded with respect to light.

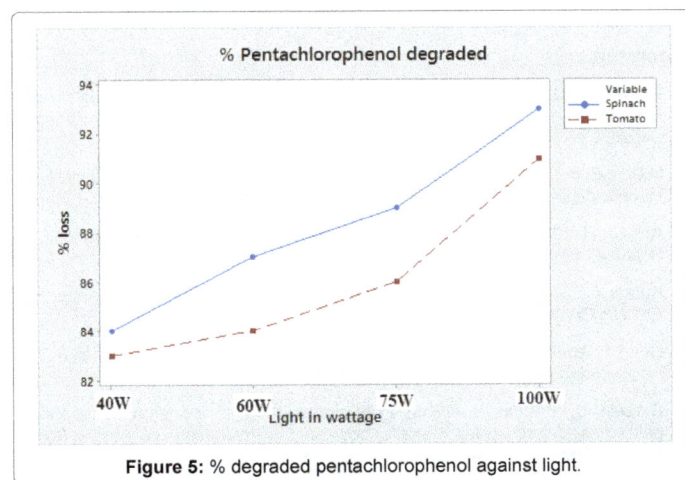

Figure 5: % degraded pentachlorophenol against light.

higher the radiant energy, the higher the number of photons involved, and the higher the amount of molecules undergoing reaction and therefore the shorter the amount of time required for half the entities to photo-degrade. The results obtained for lambda cyhalothrin photo-degradation are consistent with those earlier observed they found that on exposure of Lambda cyhalothrin to UV-light (18 W, 254 nm) for 20 minutes resulted to more than 95% loss of the initial concentrations. The protective waxy coating in tomatoes explain the lower degradation of this molecule on the tomato surface. The degradation followed a first order kinetics with an average half-life of 0.0545 on spinach leaf

surface and 0.0565 on tomato surface. Similar results were observed by Fernandez who obtained a half-life of 4.26 min for lambda cyhalothrin degradation by UV-light. Table 3 below indicate the obtained half-life on exposure of the molecule for 4 hours to different light intensities. Except for the 100 wattage, the half-lives do not seem to depend on the species of the plant under investigation, which may imply that at relatively low wattage, reaction rate only depends on the radiant energy and the temperature [14,15].

Pentachlorophenol

When Pentachlorophenol was exposed to different energy bulbs, the amount of Pentachlorophenol after 4 hours was obtained and the trend of the residues degraded against light (Figure 6) is shown below. As earlier observed for lambda cyhalothrin, the loss of Pentachlorophenol on radiating with light of different intensities was dependent on wattage. The percentage loss was lowest for 40 W bull and is explained by the fact that the number of photons is less compared to 100 W bulb. This means that the number of molecules undergoing photo-degradation is less compared to 60 W, 75 W and 100 W cases. The half-lives for spinach are consistently higher than those of tomatoes. It could be that there are two competing reactions which mean that less Pentachlorophenol is degraded per unit time i.e., the rate of photo-degradation of Pentachlorophenol is lower and thus takes much longer to get to half the initial concentration. The protective coat and lack of chlorophyll in tomato make the half-life less than that of spinach. It is worth noting that half-lives of Pentachlorophenol are much higher than those of lambda cyhalothrin at all instances, which may be due to the structure of Pentachlorophenol (Table 4).

Light	Spinach ($t_{1/2}$)	Tomatoes ($t_{1/2}$)
40 w	0.141	0.115
60 w	0.105	0.096
75 w	0.091	0.089
100 w	0.060	0.055

Table 3: Half-lives of pentachlorophenol on spinach and tomato.

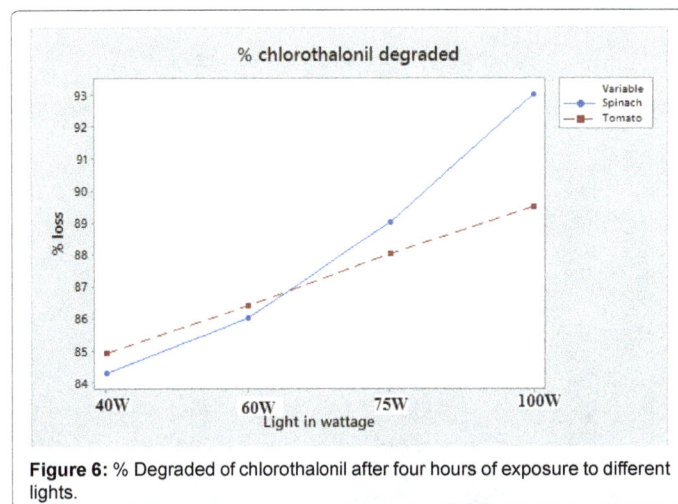

Figure 6: % Degraded of chlorothalonil after four hours of exposure to different lights.

Light	spinach	tomatoes
40 w	0.079	0.080
60 w	0.058	0.058
75 w	0.054	0.054
100 w	0.041	0.041

Table 4: Calculated half-lives of chlorothalonil.

Chlorothalonil

Chlorothalonil is an organochlorine molecule. Just like in case of Pentachlorophenol, the % loss of chlorothalonil on exposure to lights of varying intensities was depended on light intensity as shown in Figure 7. Like lambda cyalothrin the photo degradation of chlorothalonil is found to be independent of species of crop analysed, but only dependent on wattage which could indicate that they have similar photo-degradation patterns. The presence of the cyano group in both species may greatly influence the degradation pattern in the plants in the investigated. Despite the fact that the half-life obtained are comparable to those of lambda cyhalothrin, the amount of chlorothalonil degraded by each of the radiated lights is much lower than that of lambda cyhalothrin (Figure 8). This can be due to the amount of energy required to break of the Cl-C (in benzene) bond (339 KJ/mol); thus eventually this is the amount of energy required to degrade the pesticide even with the same amount of radiation. Once more the half-lives were observed to decreases with wattage due to increased number of photons degrading more molecules (Tables 5 and 6).

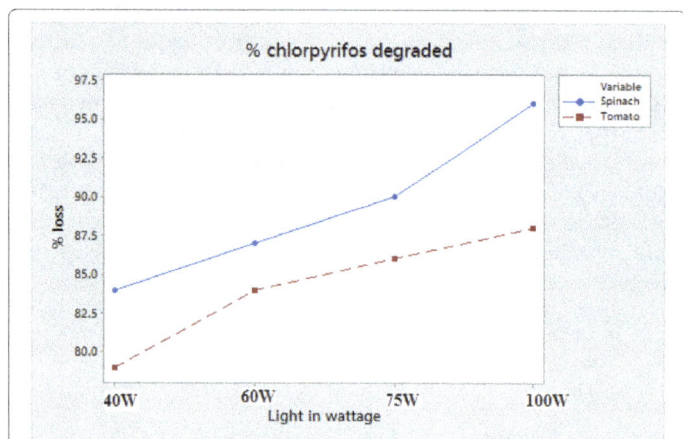

Figure 7: Amount of chlorpyriphos lost after four hours of exposure to different lights.

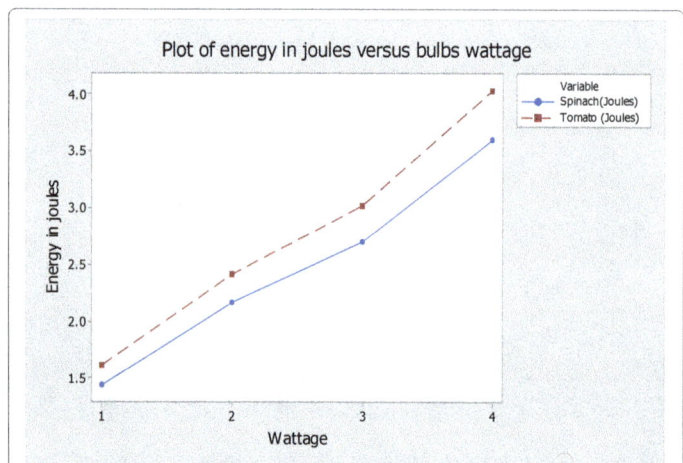

Figure 8: Plot of energy in joules versus (a) light and (b) temperature.

Light	spinach ($t_{1/2}$)	tomatoes ($t_{1/2}$)
40 w	0.065	0.065
60 w	0.064	0.064
75 w	0.062	0.062
100 w	0.042	0.042

Table 5: Calculated half-lives of chlorpyriphos.

Light	Spinach (Joules)	Tomato (Joules)
Sunlight	40.3200	45.0901
40 w	1.4400	1.6105
60 w	2.1600	2.4157
75 w	2.7000	3.0197
100 w	3.6000	4.0263

Table 6: Energy reaching the plant surface.

Chlorpyriphos

Chlorpyriphos is an organophosphate insecticide whose degradation behavior is highly dependent on light intensity and surface of exposure as indicated in Figure 8 below. Like lambda cyhalothrin and chlorothalonil, photodegradation of chlorpyriphos is independent of species of crops analysed but dependent on light radiated. This indicates that the photo degradation mechanism of chlorpyriphos on spinach and tomatoes are similar. Contrary to that, irradiation on blank surface indicates that the half-life of chlorpyriphos is higher than that on the spinach and tomatoes. The half-lives were observed to decrease with wattage due to the photons degrading more molecules.

Conclusion

Photo-degradation of residues on the surface of spinach leaf by different light intensity followed first order kinetic. The degradation rate was in sequence of 100 w>75 w>60>40>sun. The rate of degradation heavily relied on temperature, exposure time and light intensity. The half-life ranged from 0.069 to 0.141 days for spinach and 0.074-0.105 days for tomatoes. The rate of photo-degradation was heavily dependent on the molecular structure of the residue. Pesticide administration on vegetables and other fast growing crops should be done cautiously to avoid food contamination. The by-products of residues photo-degradation may be more hazardous than the parent molecules. This study therefore recommends thorough washing of food stuff prior to use and further research in analysis the structures of photo-degradation by-products.

References

1. Rehman GKMM, Motoyama N (2000) Determination of Chlorpyriphos residue in andosol upland soils using methanol phosphoric acid extraction. J Pesticide Sci 25: 387-391.

2. Malinowski H (2000) Impact of more important factors on the activity of insecticide used in forest protection. I Soil Insecticides 144: 89-100.

3. Ardley JH (1999) Pesticide considerations on environment concern. Agricultural Sciences 12: 21-24.

4. Katagi T (2004) Photodegradation of pesticides on plant and soil surface. Rev Environ Contam Toxicol 182: 1-189.

5. Hong P, Zeng Y (2002) Degradation of pentachlorophenol by ozonation and biodegradability of intermediates. Water Res 36: 4243-4254.

6. Barceló D, Durand B (1993) Photodegradation of the organophosphorus pesticides chlorpyriphos, fenamiphos, and vamidothion in water. Toxicol Environ Chem 38: 183-199.

7. Jong Hyang K, ByoungChul M (1997) Decomposition of organo phosphorous compounds with ultraviolet energy. J of Korean Ind and Eng Chemistry 9: 28-32.

8. Kuperberg JM, Soliman KFA, Stino FJR, Kolta MG (2000) Effects of time of day on chlorpyrifos induced alterations in body temperature. Life Sci 67: 2001-2009.

9. Jack R, Plimmer (2001) Handbook of Pesticide Toxicology. Academic Press, pp: 95-107.

10. Kansas State University Agricultural Experiment Station and Cooperative Extension Service (1990) Factors affecting pesticide behavior and breakdown formulations.

11. Ruzo LO, Casida JE (1982) Pyrethroid photochemistry: intramolecular sensitization and photoreactvity of 3-phenoxybenzyl, 3-phenyl-benzyl and 3-esters. J Agric Food chem 30: 963-966.

12. James JP, Spence JH, Ford JH (1984) Gas chromatographic determination of chlorpyriphos residues in greenhouse vegetables after treatment of potting media with Dursban for imported fire ant. JAOAC 67: 1091-1094.

13. Weerasinghe CA, Matheews JM, Wright RS, Wang RY (1992) Aquatic photo degradation of Albendazole and its major metabolites.2. Reaction quantum yield, photolysis rate, and half-life in the environment. J Agric Food Chem 40: 419-421.

14. Ola MY, Emara S, Aziz AAE (2010) Effect of Heat direct Sunlight and UV-Rays on the Stability of some Chlropyrifos Formulations with Emphasis to their Content of Sulfotep. Nature and Science 8: 229-234.

15. Fernandez AM, Sanchez PL, Lores ML, Lompart M (2007) Alternative sample preparation method for photochemical studies based on solid phase micro-extraction: synthetic pyrethroid photochemistry. J Chromatography Sample Prep 1152: 156-167.

21

Adsorption of Lambda Cyhalothrin on to Athi River Sediments: Apparent Thermodynamic Properties

Sherrif SS*, Madadi V, Mbugua JK and Kamau GN

Department of Chemistry, University of Nairobi, Nairobi, Kenya

Abstract

Lambda-Cyhalothrin is extensively used in agriculture, horticulture, and public health management in Kenya. The pesticide is effective against many vegetable pests, rice and other disease agents. The adsorption of L-Cyhalothrin onto sediments of Athi River was studied using UV-visible and the results were analyzed by fitting data into isotherm models including Freundlich and Dubinin-Radushvich plot to understand the environmental impacts of this pesticide. The models presented different numerical values but the results demonstrated similar characteristics. The study was aimed at determination of adsorption characteristics of L-Cyhalothrin on sediments from Athi River. Different masses of sediments were spiked with different concentrations of L-Cyhalothrin. The mixtures were shaken at varied times to reach equilibrium. The amounts of pesticide adsorbed were determined by analyzing the difference between the UV-visible Spectrophotometer data and the initial concentrations. Line Plots were used to determine other thermodynamic variables. Various isotherm models were used to explain the adsorption characteristics of the Lambda-Cyhalothrin. It stands out that Dubinin-Radushvich (D-R) with average R^2: 0.935, 0.938 and 0.898 fitted best in this experiment while Freudlich R^2: 0.783, 0.899 and 0.812 for all sediment samples, Upstream, Midstream, and Downstream, respectively. The spontaneity of the adsorption process was also realized in ΔG (Gibb's free energy) values as predicted by Freundlich. In both models, Midstream ΔG was negative (-) showing complete spontaneous characteristics. Also in Upstream, ΔG was negative (-) for Freundlich whereas Downstream ΔG was positive (+). Generally, adsorption capacity of Athi river sediments was low due to low mineral contents and total organic Carbon including other physicochemical properties specifically, textures, Nitrogen content including its temperature and moisture contents.

Keywords: Lambda-cyhalothrin; Thermodynamic variables; Dubinin-Radushvich isotherm; Athi River

Introduction

The use of pesticides including insecticides, herbicides, fungicides, rodenticides, and acaricides can be dated before the 16th century when chemicals such as arsenic sulphide, Paris green (Copper acetoarsenite) were used to control Malaria transmissions. Though the use of synthetic organic pesticides began around 1940 Lambda-Cyhalothrin is an artificial insecticide that characterizes the biochemical effects of natural pyrethrin pesticide It is the active ingredient in many insecticides such as Karate, Warrior, and Icon that are commonly applied to control insects and pests in public health emergency, agricultural/horticultural farming and in homes [1]. It is effective against many crop and disease vector insects and ticks. The use of Pesticides including Lambda-Cyhalothrin in livestock and crop farming have led to increased food security and livelihood in many developing countries though uncontrolled excessive use is hazardous to human life as well to the environment. Lambda-Cyhalothrin and other pesticides rapidly disappear upon application in the environment [2] Studies have proven that approximately 2-5% of applied pesticides go to targeted organisms and subsequently the rest end up in the environment [3]. Pesticide substrates in the environment undergo several chemical processes including oxidation, absorption, degradation, halogenations, transfer and other microbial actions [4] Transfer of pesticide residues away from the targeted sites through volatilization, spray drift, agricultural runoff, leaching, or crop removal is a major determining factor in assessing the overall environmental impacts of pesticides [5]. Acute (short-term) human exposure can lead to skin irritation, burn or allergic reaction [6]. It is very toxic particularly to various species of fish and other aquatic organisms.

Materials and Methods

The following instruments, materials and reagents were used: UV-Visible spectrophotometer (UV-1700 Schamadzul), Mini Orbital Shaker, Analytical balance (Fischer scientific A-160), Lambda-Cyhalothrin (karate 25 g/kg pure), Glass bottles, Distilled water, Stop watch, Acetone (90% pure) and River sediment from Athi River.

Procedures

The samples were dried for a week and pulverized to maintaining its originality. The Calcium, Potassium and sodium contents were analyzed using Flame Photometer, whereas other elements including Phosphorus, Magnesium and Manganese were determined calorimetrically, [7] The available total organic carbon content was analyzed by oxidative spectrophotometer, [8] The total N was determined thru distillation followed by Titration, [9]. The PH was also determined simply by a pH meter and other trace metals (Fe, Zn and Cu) using an Atomic Absorption Spectrophotometer (AAS), [10]. The cation ion exchange capacity of the metallic elements at neutral pH was determined by Titration after distillation [11] To conduct the adsorption study, 2, 4, 6, 8, 10, 20, 40, 60, 80, and 100 g/L were prepared and adsorption values determined at the UV-vis between 200 to 900 nm. The existence of the adsorption/desorption equilibrium of 2.0 g, 1.0 g, 0.5 g, 0.2 g and 0.1 g of the soil samples were investigated. The dried sediments were placed in glass bottles and shaken with the pesticide for an hour. The aqueous parts were poured and filtered with a Whitman filter paper to obtain a clear concentrated L-Cyhalothrin solution. The absorbance of Ce and Qe was obtained at the UV-Visible spectrophotometer at 218 nm. ΔG, n and K were found using 0.5 g of each sample mixed with deionized water. The samples were spiked with 50, 40, 30, 20, and 10 ppm of L-Cyhalothrin and shaken for 15 min, 30 min, 45 min and 60 min using

---

*Corresponding author: Sherrif SS, Department of Chemistry, University of Nairobi, Nairobi, Kenya, E-mail: sheriffsalia@gmail.com

an orbital shaker. Similarly, concentration study of L-Cyhalothrin was carried out using UV visible Spectrophotometer.

Results and Discussion

The soils properties have significant impacts on its adsorption capacity. The profile of the soils according to Table 1 on Athi River soil showed that the organic carbon content is low for all three sampling sites, (0.54, 0.91 and 0.75 respectively). This means the adsorption capacity of these samples is generally low thus not proportionately. Another factor that influences the adsorption is the physical dipole-dipole attraction (Van der Waals force) that improves the adsorption capacity for the soil. Below in Table 1 are the sediment profiles of the Athi River.

Concentration study

The prepared standard solution of Lambda-Cyhalothrin behaved well at λ max 218 nm at UV-Visible spectra as shown below in Figure 1. The plot of absorbance against concentration at 218 nm shaped a straight line. This means Beer's Law is obeyed

$$A = \varepsilon\, CL \tag{1}$$

In the above equation, A is the absorbance, L is the path length of the light and C is concentration. The symbol $\acute{\varepsilon}$, is the molar absorption coefficient for the chemical species of interest.

Adsorption isotherm

The process of adsorption is defined as the summative accumulation of a chemical species at the interface of an aqueous solution and a solid phase [12] The chemical species is called adsorbate and the surface at which it accumulates is called adsorbent. The adsorption kinetics of Lambda-Cyhalothrin onto Athi River sediments was carried out by preparing 10 ml of Lambda-Cyhalothrin in various concentrations (10 ppm- 50 ppm) and shaken at varying times (15 minutes, 30 minutes, 45 minutes and 60 minutes) on an orbital shaker. The mixture was filtered and the concentrated residual solution was analyzed using UV-spectrophotometer (SHEMAZU 1700). The amount of the pesticide adsorbed (mg/g) was determined using the Vanderburgh and Van Griekenm's formula [13]

$$Q = v(c_i - c_f)/w \tag{2}$$

Where Q is the amount of solute adsorbed, V is the volume of the absorbate, C_i and C_f are the initial and final concentrations, respectively. W is the weight in gram of the adsorbent and data fitted into freudlich and Dubinin-Radushkevich isotherm models.

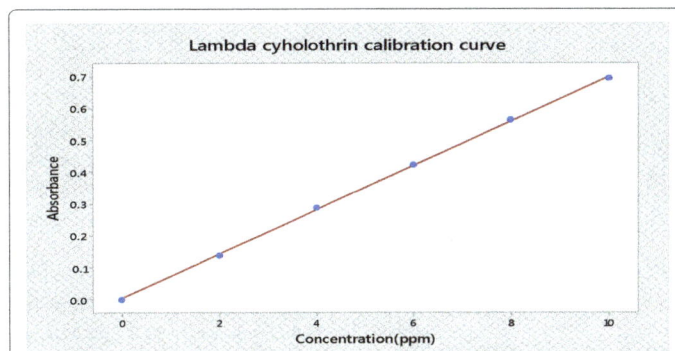

Figure 1: Plot of Concentration from 0-12 ppm of Lambda-Cyhalothrin.

Freundlich isotherm model

The same data obtained from the statistics of the research were exposed to Freundlich isotherm which gave positive (+) ΔG values for sample C (Downstream) and negative (-) ΔG for samples A and B (Up and Mid Streams, respectively). This means that samples A and B underwent spontaneous adsorption using Freudlich isotherm while sample C was nonspontaneous. Thus, the Freundlich Isotherm is characteristic of heterogeneous surface, (Freundlich, 1906). It can be expressed as:

$$q_e = k_f c^{1/n} e \tag{3}$$

From equation 1.2, qe (mg/g) is amount of pesticide adsorbed and Ce (mg/L) is the equilibrium concentration and n is a constant that measures the adsorption of non-linearity between the solute concentration in the solution and the adsorption. n is characteristic of the quasi-Gaussian heterogeneity related to the adsorption surface. Kf (L/g) is the Freundlich isotherm constant. The linearized equation for Freundlich isotherm is expressed as follows:

$$\ln q_e = \ln k^f + 1/\ln c_e \tag{4}$$

From this expression, plots of lnqe versus ln C_e as presented in this linear form of Freundlich Isotherm have slopes at 1/n and intercepts at lnKF. Figures 2-4 below were generated using the data highlighted in Table 2 above. The data fitted strongly in the adsorption of Lambda-Cyhalothrin unto Athi River sediments of the three (3) sampling areas. This can be validated from the regression (r^2) values ranging from 0.742 to 0.992 and the linearity of the plot in Figures 2-4.

Dubinin-Radushkevich (d-r) isotherm model

This model was used to show the apparent adsorption energy heterogeneity at sites of adsorption. This model has fitted well with high solute activities including intermediate concentration range data. This approach was originally applied to differentiate between the physical and chemical adsorptions of metal ions with it main free energy (E) per molecule [14] The equation and its linear forms are expressed below:

$$\ln q_e = \ln q_d - B_D E^2 \tag{5}$$

$$E = RT \ln [1 + 1/c_e] \tag{6}$$

From the above expressions, q_D is adsorption strength of soil/sediment (mg/g), B_D is Dubinin-Radushkevich constant (mol²/KJ²), qe is amount of adsorbate at equilibrium (mg/g), and E is Polanyi potential (Dubinin isotherm constant). T is the absolute temperature (k), R is the universal gas constant (j/molk) and Ce is the adsorbate equilibrium concentration (mg/L). One unique characteristic of this model is that it is temperature dependent. This allows for all suitable data to lie in the same curve called characteristic curve when adsorption data at separate

Field	Upstream (A)	Midstream (B)	Downstream (C)
Sediment depth	2-5 cm	2-5 cm	2-5 cm
Parameters	Value	Value	Value
Total Nitrogen %	0.08	0.11	0.10
Total Org. Carbon %	0.54	0.91	0.75
Phosphorus ppm	8.00	11..00	40.00
Potassium me %	0.82	1.06	0.60
Calcium me%	23.90	30.20	5.10
Magnesium me%	3.28	3.36	1.76
Manganese me%	2.12	2.68	0.61
Copper ppm	1.56	1.55	1.71
Iron ppm	191.00	174.00	39.30
Zinc ppm	5.00	5.94	2.04
Sodium me%	1.00	1.29	0.50
Elect. Cond. Ms/cm	0.93	0.86	N/A

Table 1: Sediment profile of Athi River.

Time (minute)	Sample	LnK^F	KF	R^2		ΔG
15	A	0.8957	1.883	6.5732	75.4	-4,665.27
	B	1.002	1.691	5.4249	82.9	-1,921.94
	C	2.136	-0.322	0.7247	78.1	797.78
30	A	0.9518	1.834	6.2589	86.1	-4,543.88
	B	1.841	0.5828	1.7915	99.2	-1,444.56
	C	1.966	-0.2400	0.7867	90.4	594.62
45	A	0.8852	1.852	6.3726	75	-4,588.48
	B	1.174	1.549	4.7066	84.4	-3,837.67
	C	3.228	-2.2880	0.1015	74.2	5,668.69
60	A	1.053	1.635	5.1295	76.6	-4,050.85
	B	1.249	1.484	4.41055	93.3	-3,676.74
	C	2.930	-2.151	0.1164	82	5,329.27

Table 2: Freundlich adsorption data.

Time (minute)	Sample	Ln qD	BD	qD	R^2	E (KJ/mol)
15	A	3.854	-0.000002	47.1814	91.8	-353,553.39
	B	3.844	-0.000002	46.7120	93.8	-353,553.39
	C	4.421	-0.000006	83.1794	86.4	-117,851.13
30	A	3.795	-0.000001	44.4782	98	-707,106.78
	B	3.839	-0.000002	46.4790	89.6	-353,553.39
	C	4.203	-0.000005	66.8867	96.9	-141,421.36
45	A	3.841	-0.000002	46.5720	91.5	-353,553.39
	B	3.995	-0.000002	54.3258	94.1	-353,553.39
	C	5.139	-0.000010	170.5451	85.7	-70,710.68
60	A	3.952	-0.000002	52.0393	92.6	-353,553.39
	B	3.888	-0.000002	48.8132	97.5	-353,553.39
	C	4.754	-0.000010	116.0475	90.2	-70,710.68

Table 3: Dubinin-Radushkevick isotherm Data.

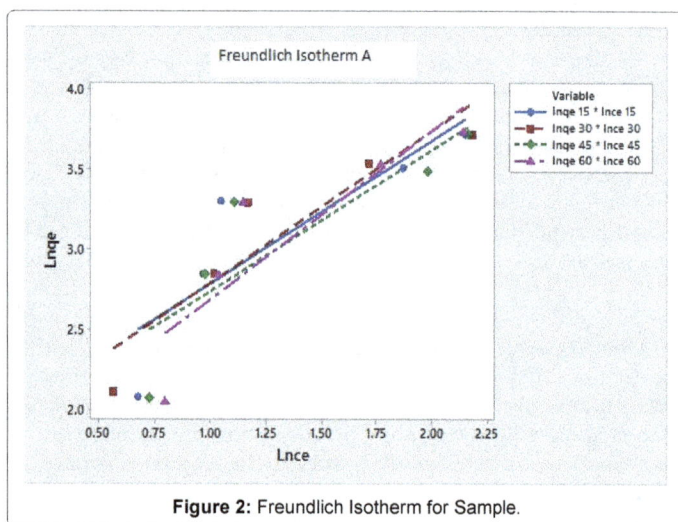

Figure 2: Freundlich Isotherm for Sample.

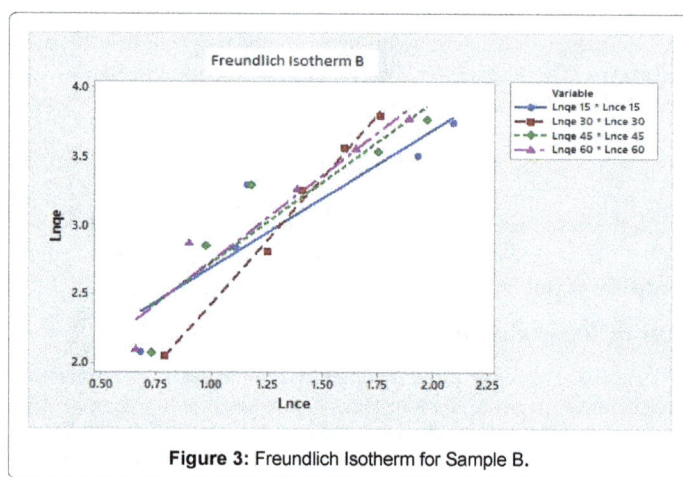

Figure 3: Freundlich Isotherm for Sample B.

temperatures are plotted as function of the amount adsorbed (lnqe) against the energy (E). Table 3 below gives the summary parameters use in D-R data to generate the graphs in Figures 5-7. From Table 3 above, the potential energy (E) can be calculated from the value of B_D using the formula: $lnqe = lnq_D - B_D E^2$. For example, in 15 minutes of shaking time, lnqe 15=3.844 -0.000002 E^2, r^2=0.938. Lnq_D=3.844, B_D=0.000002 $q_D = e^{3.844}$ and E=-353,553.39 kj/mol. Generally, when the adsorption energy value falls below 8 Kj/mol, the adsorption process is said to be characterized by physiosorption, i.e., physical binding of the pesticide to the surface of the soil/sediment. If the value is higher than 8 kj/mol but less than 20 kj/mol, it is said to be characterized by ion exchange and similarly, its values beginning from 20 kj/mol and above are said to be characterized by particle diffusion [15-18]. From Table 3, E (Kj/mol) values are lesser than 8 kj/mol, which deduces that the adsorption process of Lamda-Cyhalothrin was predominantly a physiosorption process. From Figure 5 the data fitted well in Dubinin-Radushkevich isotherm at the various shaking times (15 minutes, 30 minutes, 45 minutes and 60 minutes), for samples A and B whereas sample C has some deviations, though the regression (r^2) values range from 0.857-0.980.

Conclusion

Adsorption of Lambda-Cyhalothrin decreases with increase in mass of loam sediment. The longer the contact time the higher the adsorption. Increase in concentration results in decreased proportion of adsorption. This is because at high initial concentration, the number of moles of Lambda-Cyhalothrin available to the surface area is high,

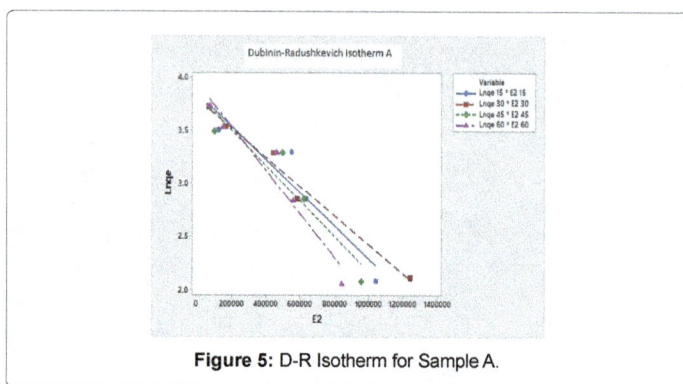

Figure 4: Freundlich Isotherm for Sample C.

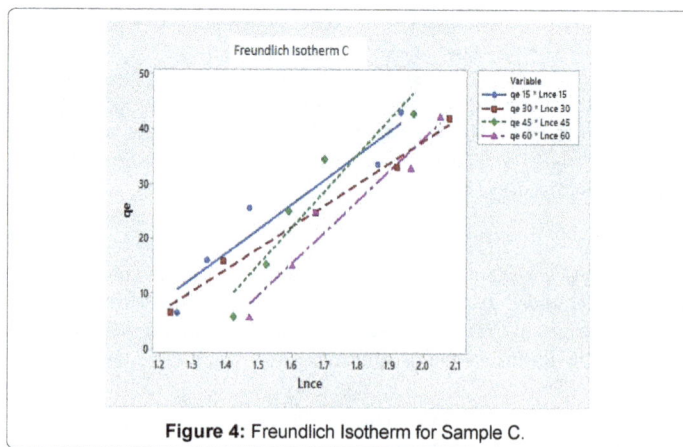

Figure 5: D-R Isotherm for Sample A.

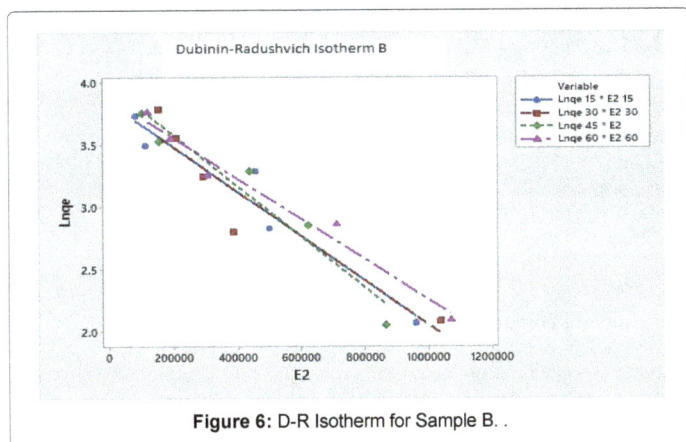

Figure 6: D-R Isotherm for Sample B. .

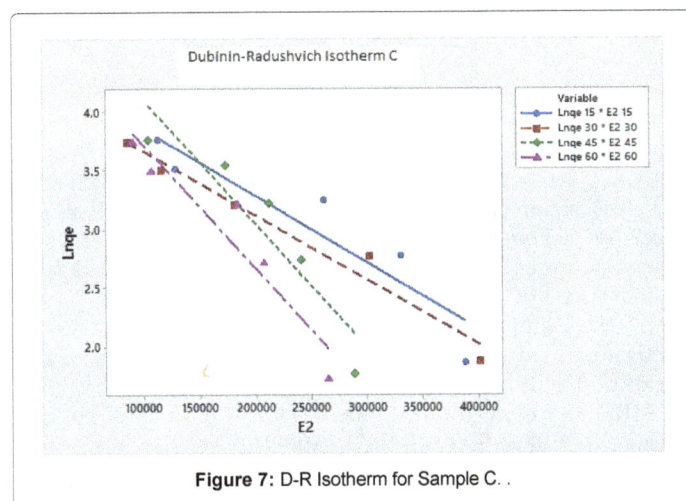

Figure 7: D-R Isotherm for Sample C. .

so functional adsorption becomes dependent on initial concentration. Adsorption of lambda cyhalothrin followed Fredlich-peterson isotherm model with regression values ranging from 0.954 to 0.992. As shaking time increased from 15 minutes to 60 minutes, the value of n increased from 0.1998 to 0.2914. The value of ΔG (Gibb's free energy) was 11.7946 ± 0.3 Kjol/mol which indicated that adsorption of Lambda-Cyhalothrin by Athi River sediments was spontaneous, indicating high affinity to the sediments.

References

1. CDPR (2011) California Air Resource Board, cdpr.ca.gov. Retrieved on November 20.

2. He LM, Troiano J, Wang A, Kean K (2008) Environmental Chemistry Ecotoxicity and Fate of Lambda-Cyhalothrin. Reviews of Environmental Contamination and Toxicology 195: 71-81.

3. FAO/WHO (2012) Evaluation Report based on submission of data from BharaRasayan Limited (TC). In: Food and Agriculture Organization. FAO specifications and evaluations for agricultural pesticides- Lambda-Cyhalothrin, pp: 15-23.

4. Garcia SG, Scheiben D, Binder CR (2011) The weight method: a new screening method for estimating pesticide deposition from knapsack sprayers in developing countries. Chemosphere 82: 1571-1577.

5. Senoro DB, Maravillas SL, Ghafari N, Rivera C (2016) Modeling of the residues Transport of Lambda-Cyhalothrin, Cypermithrin, Malathion and Endosulfan in three different Environmental compartments in the Philippines. Sustainable Environmental Research 26: 168-176.

6. Sharda USA LLC (2013) Sharda Lambda-Cyhalothrin 1EC Insecticide 7217 Lancaster Pike suite A, Hockessin DE 19707.

7. Mehlich A (1953) Determination of P, Ca, Mg, K, Na, and NH_4. North Carolina Soil Test Division, pp: 23-89.

8. Gislason EA, Craig NC (2005) Cementing the foundations of thermodynamics: comparison of system-based and surroundings-based definitions of work and heat. J Chem Thermodynamics 37: 954-966.

9. Jan AP, Marten W, Stephen H (2008) Handbook for Kjeldahl Digestion. pp: 11-42.

10. Yang SY, Chang WL (2005) Use of finite mixture distribution theory to determine the criteria of cadmium concentrations in taiwan farmland soils. Soil Science 170: 55-62.

11. Carroll D (1959) Ion exchange in clays and other minerals. Geological Society of America Bulletin 70: 749-779.

12. Sposito G (1989) The Chemistry of Soils. Oxford University Press, England.

13. Igwe JC, Abia AA (2006) A Bioseparation process for removing heavy metals from waste water using biosorbents. African Journal of Biochemistry 5: 1167-1179.

14. Dubinin MM, Radushkevich LV (1974) Equation of the Characteristic Curve of Activated Charcoal, Proceedings of the Academy of sciences of the USSR. Chemistry Section 55: 331-333.

15. De AK (2010) Environmental Chemistry. 7th edn. New Age International Publishers. New Delhi, India, pp: 198-261.

16. Freundlich HMF (1906) Over the Adsorption in Solution. J Phys Chem 57: 385- 471.

17. Carter MR (2006) Soil Sampling and Methods of Analysis. 2nd edn., pp: 43-49.

18. Tuner RC, Clark JS (1965) Lime potential in acid clay and soil suspensions. Trans Comm II and IV Int. Soc Soil Science, pp: 208-215.

22

Fragmentation Study of Substituted Chalcones: Gas Phase Formation of Benz-1-oxin Cation

Abdul Rauf Raza[1]*, Aeysha Sultan[1], Nisar Ullah[2], Muhammad Ramzan Saeed Ashraf Janjua[1,2] and Khalid Mohammed Khan[3]

[1]*Ibn e Sina Block, Department of Chemistry, University of Sargodha, Sargodha-40100, Pakistan*
[2]*Chemistry Department, King Fahd University of Petroleum and Minerals, Dhahran-31261, Saudi Arabia*
[3]*HEJ Research Institute of Chemistry, International Centre for Chemical & Biological Sciences, University of Karachi, Karachi-75270, Pakistan*

Abstract

The mass spectra of a number of substituted chalcones have been observed to show intense M - X peaks (where X=Cl, Br, OH, OMe), which largely arise through the loss of an ortho-substituent from the ring-A of chalcones. The base peak is attributed to highly resonance stabilized benz-1-oxin cation, which would be formed via modified McLafferty rearrangement in gas phase (70 eV). The exact mass measurement of such fragments and DFT studies supports the formation and stability of benz-1-oxin cation. This protocol may also be conveniently used to distinguish among different positional isomers of substituted chalcones.

Keywords: Chalcone; Mass spectrometry; McLafferty rearrangement; Benz-1-oxin cation; Gas phase intramolecular substitutions

Introduction

Chalcones (1,3-diaryl-2-propen-1-ones) represent a diverse group of natural and synthetic compounds having an array of biological activities. These secondary metabolites serve as key precursors in the synthesis of many biologically important compounds such as benzothiazepine, pyrazolines, 1,4-diketones, and flavones. A longstanding scientific endeavors have demonstrated that chalcones display a broad spectrum of pharmacological activities including antimalarial [1], anticancer [2], antibacterial [3], anti-inflammatory [4], antifungal [5], antipyretic, antimutagenic [6], antioxidant [7], cytotoxic, antitumor [8], etc. Since chemical/biological (re) activities are associated with structural properties of a compound, a detailed structural profile has to be established in order to predict a compound's reactivity/structure before and after metabolism. Among all spectroscopic/spectrometric techniques, mass spectrometry is one of the most frequently employed structural elucidation techniques.

Mass spectrometry (MS) has been widely employed for the structural characterization of chalcones. The simplest chalcone has been reported to display a quite straight forward fragmentation pattern; however, surprisingly, the fragmentation pattern of chalcones changes dramatically with rather small changes in substituents [9]. Numerous efforts have been devoted to structural characterization of chalcones by employing soft-ionization techniques [10-16]. A few of such techniques include: ESI [17], ESI tandem MS [18], CI [19,20], FD [21], FAB [22] and DART [23]. Although a significant number of reports are available on use the soft-ionization for MS of chalcones; the literature concerning use of harsh ionization techniques (like EIMS) for the study of fragmentation pattern of chalcones still remains to be sporadic [24].

Herein, we wish to report a detailed study of mass fragmentation pattern of substituted chalcones, benzylidene indanones and benzylidene tetralone under harsh EIMS conditions. The study also involves calculation of the relative free energy change of the possible fragments / ion structures of proposed benz-1-oxin cation based on computational evidence.

Experimental

The TLC was carried out on pre-coated silica gel (0.25 mm thick layer over Al sheet, Merck) with fluorescent indicator. The spots were visualized under UV lamps (365 nm and 254 nm λ) of 8 W power and/ or KMnO$_4$ dip upon heating. The compounds were purified either on a glass column packed silica gel (0.6 mm-0.2 mm, 60 Å mesh size, Merck) or by crystallization. All solutions were concentrated under reduced pressure (25 mmHg) on a rotary evaporator (Laborota 4001, Heidolph) at 35°C -40°C. Melting points were determined using a MF-8 (Gallenkamp) instrument and are reported uncorrected. The IR-spectra are recorded on Prestige 21 spectrophotometer (Shimadzu) as KBr discs. The ^{1}H (300 MHz, 400 MHz and 500 MHz) and ^{13}C NMR (75 MHz) are recorded on AM-300 MHz, 400 MHz and 500 MHz instrument (Bruker) in CDCl$_3$ using TMS as internal standard.

The chalcones, arylidene tetralones and arylidene indanones were prepared by our reported procedure in excellent yields. All chalcones, arylidene tetralone and arylidene indanones were synthesized by Claisen-Schmidt condensation of the appropriate benzaldehyde and ketone under silica-H$_2$SO$_4$ catalyzed conditions [25].

General procedure for the silica sulfuric acid (SSA) catalyzed synthesis of chalcones

The SSA (0.02 g) was added to a well stirred suspension of ketone (7.53 mmol, 1 eq) and aromatic aldehyde (7.91 mmol, 1.05 eq) and the resulting mixture was heated at 65°C for 1.5 hr. The reaction mixture was cooled to ambient and partitioned between brine (25 mL) and CH$_2$Cl$_2$ (3 × 15 mL). The combined organic extract was washed with brine (3 × 25 mL), dried over anhydrous Na$_2$SO$_4$, filtered and concentrated under reduced pressure to afford the chalcone in excellent yields (78%-93%).

Mass spectrometer experiments

The LR EI-MS studies were carried out on MAT312 machine by a heated inlet system or with a heated-cooled probe with the lowest feasible sample temperature and ionization was carried

*Corresponding author: Abdul Rauf Raza, Ibn e Sina Block, Department of Chemistry, University of Sargodha, Sargodha-40100, Pakistan
E-mail: roofichemist2012@gmail.com

out under electron impact (70 eV) conditions. All the elemental compositions given for the [M]·+ and [M-X]+ were obtained by exact mass measurements carried out by the peak-matching method. The perflourokerosene (PFK) was used as reference for the determination of exact masses. The mass resolution was set at 10,000 (approx) with an external calibration mode.

The HR EIMS experiments were performed on a Thermo Finnigan MAT 95 XP double focusing mass spectrometer (Thermo Fisher Scientific, Germany). The N_2 was used as the sheath gas and auxiliary gas. The EI conditions were as follows: ion source temperature: 250°C; ion source vacuum: 3.4×10^{-6} mbar; analyzer vacuum: 4.1×10^{-7} mbar; DI probe temperature: 40°C up to 360°C; electron energy: 70 eV and accelerating voltage: 5759 V. Data acquisition and analysis were carried out with the Xcalibur software package (ver. 1.4 SR1; Thermo Fisher Scientific). The Gaussian 09 program package was used for all DFT calculations [26].

Results and Discussion

As a step towards green chemistry, the chalcones and benzylene indanone/tetralones were prepared under solvent free conditions by making use of silica-H_2SO_4 [25]. The condensation of aldehydes and ketones at 65°C in the presence of SSA afforded chalcones in high yields (Scheme 1, Table 1).

Reagent and conditions

ArCHO (1.05 eq), SSA, 65°C, 1.5 hr, neat

The formation of chalcones was indicated by a decrease in C=O stretching ύ and a bathochromic shift in the UV spectra of products. The ¹H NMR of the open chain chalcones showed the presence of olefinic protons exhibiting J_{trans}, whereas the benzylidene indanone/tetralone based chalcones displayed a single olefinic proton which appeared as a broad singlet in majority of the cases (Table 1).

The single crystal XRD of crystalline samples confirmed the structures of products beyond doubt (Figure 1) [27]. The Mass fragmentation pattern of simplest chalcone 5 has been studied extensively. It has been reported that in addition to the expected α-cleavage pattern, characteristic for carbonyl compounds, the fragmentation of 5 involve the formation of highly resonance stabilized benz-1-oxin radical 8 via loss of H radical (route a) [15,26] and phenylpropynone 8 through loss of a benzene molecule (route b) (Scheme 2) [15].

On the other hand, the fragmentation of 2-nitrochalcones has been reported to be very simple with suppressed characteristic fragmentation pattern of chalcones. Baldas et al. have explained in detail the effect of presence of NO_2 at C-2 that provides anchemeric assistance for the formation of benzoyl cation [15]. We have recently observed a quite surprising pattern in the fragmentation pathway of chalcones with substituents at C-2 of ring-A. Based on our current findings we wish to disclose that under EI conditions, the fragmentation pattern of chalcones derivatives is not only influenced by the presence of NO_2 group at C-2; any substituent at C-2 affects the fragmentation pattern in a similar manner. Such fragmentation pattern involves the loss of a substituent at C-2 and the resultant cation in turn, appears as a base peak in almost all cases [28].

Entry	X	Y	Z	n	Product	Percentage yield (%)	$\delta_H{}^a$ (J in Hz)		
							H²ᵇ	H³ᵇ	H¹'ᶜ
1a	H	H	Cl	-	2a	87	7.15 (16.2)	7.49 (16.2)	-
1b	Cl	5-NO₂	H	-	2b	95	7.70 (15.9)	8.20 (15.9)	-
1c	OMe	H	Cl	-	2c	83	7.15 (16.0)	7.77 (16.0)	-
1d	Cl	H	Cl	-	2d	88	7.37 (16.4)	7.97 (16.4)	-
1e	OMe	3-OMe	Cl	-	2e	88	7.14 (16.5)	7.78 (16.5)	-
1f	Cl	6-Cl	Cl	-	2f	91	7.31 (16.4)	7.62 (16.4)	-
1g	NO₂	H	Cl	-	2g	89	7.03 (15.9)	7.92 (15.9)	-
1h	Cl	5-NO₂	Cl	-	2h	93	7.66 (15.3)	8.19 (15.3)	-
3a	H	H	H	1	4a	87	-	-	7.93
3b	Br	H	H	1	4b	91	-	-	7.95
3c	Cl	H	H	1	4c	82	-	-	8.03
3d	H	3-Cl	H	1	4d	91	-	-	7.69
3e	H	4-Cl	H	1	4e	90	-	-	7.55
3f	Cl	4-Cl	H	1	4f	79	-	-	7.49
3g	Cl	6-Cl	H	1	4g	86	-	-	7.36
3h	Cl	5-NO₂	H	1	4h	80	-	-	8.01
3i	OMe	H	H	1	4i	87	-	-	8.13
3j	H	3-OMe	H	1	4j	88	-	-	7.63
3k	OMe	3-OMe	H	1	4k	89	-	-	8.03
3l	H	3,4-(OMe)₂	H	1	4l	85	-	-	7.18
3m	OH	3-Me	H	1	4m	81	-	-	7.85
3n	2-OH	H	H	1	4n	78	-	-	7.61
3o	H	4-OH	H	1	4o	82	-	-	7.58-7.63
3p	H	3-NO₂	H	1	4p	85	-	-	8.53
3q	Cl	H	H	2	4q	92	-	-	7.82
3r	Cl	4-Cl	H	2	4r	93	-	-	7.94
3s	Cl	5-NO₂	H	2	4s	91	-	-	8.71-8.73
3t	H	4-OMe	H	2	4t	92	-	-	7.83

ᵃchemical shifts are reported in ppm & coupling constants are reported in Hz.
ᵇdoublet showing J_{trans}.
ᶜbroad singlet was observed in majority of cases.

Table 1: The percenatge yield and δ_H of olefinic protons in synthesized chalcones.

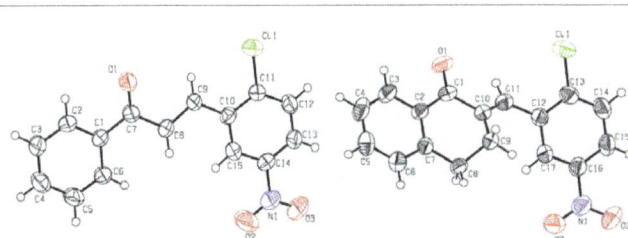

Figure 1: The ORTEP diagram of 2h (left); 4s (right) [27]

In order to ascertain the role of substituent on ring-B on the fragmentation pattern, compound 2a (2`,4`-dichlorinated derivative) was subjected to fragmentation under EI. The resulting spectrum seemed to be very simple with a prominent [M]·+ Along with a base peak, corresponding to 2,4-dichlorobenzoyl cation (m/z 173 amu, 175 amu, 177 amu) appearing in 9:6:1 (100%, 62%, 11% respectively), a characteristic pattern for the dichlorinated compounds. The lack of loss of Cl radical indicated that the substituent on ring B does not influence the fragmentation pattern of chalcones (Figure 2a).

The influence of ring-A substituents on the fragmentation pattern of chalcones derivatives was compared with ring-B substituted chalcones. To this end, chalcone 2b, bearing 2-chloro-5-nitro groups on ring-A

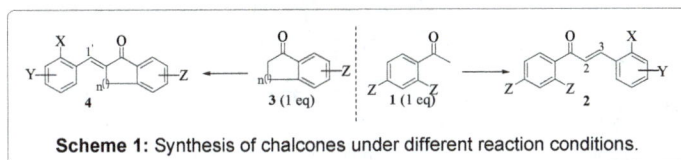

Scheme 1: Synthesis of chalcones under different reaction conditions.

was subjected to fragmentation under EI conditions. The resultant mass spectrum indicated a prominent [M-Cl]$^+$ peak. However, no significant loss corresponding to NO$_2$ was observed (Figure 2b), which indicated that only substituent at C-2 of ring-A influences the fragmentation pattern of such compounds.

The mass spectra of chalcones derivatives containing monosubstitution (at C-2) or disubstitutions (at C-2 and C-3 or C-2 and C-6) on ring-A and 2,4-dichlorosubstitutions on ring-B have revealed that the fragmentation pattern of these derivatives entirely depend on the location of the substituent on ring-A only (Figure 2). On the other hand, 2,4-dichloro substitutions on ring-B don't seem to be playing any significant role in the fragmentation pattern except the formation of fragment ion 6 by α-cleavage. The spectra of such chalcones display the formation of 2,4-dichlorobenzoyl cation (m/z 173 amu, 175 amu, 177 amu) in 9:6:1, appearing either as a base peak or a significant fragment ion. However, without any exception, loss of

substituent at C-2 of ring-A afforded either a significant fragment ion or base peak in almost all cases. In case of chalcones derivatives bearing monosubstitution at C-2 of ring-A, such as compounds 2c (2-OMe) (Figure 3a), the [M-OMe]$^+$ fragment (m/z 275 amu, 277 amu, 279 amu) in 9:6:1 (43%, 28%, 5%) appeared as a prominent peak whereas in case of compound 2d (2-Cl) (Figure 3b), [M-Cl] fragment ion formed the base peak (100%, 70%, 12%) (Table 2).

Similarly, when compounds containing di substitutions on ring-A, for instance 2e (2,3-dimethoxysubstitutions) (Figure 3c) and 2f (2,6-dichloro substitutions) (Figure 3d), were subjected to EI conditions, the same pattern was observed i.e., the cleavage of [M-X] along with α-cleavage which resulted in a prominent 2,4-dichlorobenzoyl cation. However, no [M-NO$_2$] loss was observed when chalcone 2g was subjected to fragmentation under EI conditions. Likewise in case of compound 2h, quite intense peaks due to [M-Cl] were observed at 320, 322, 324 amu in 9:6:1 (97%, 62%, 11%) with the 2,4-dichlorobenzoyl cation forming the base peak (Table 2). In the cases of all disubstituted chalcones no further loss was observed after the first loss of substituent. In contrast to the findings of Baldas and Porter [15], no [M-H] ion was observed in these cases.

The cleavage of a C-X (aryl-X) bond is not a frequently observed mode of fragmentation. Based on our findings, we propose that C-X cleavage occurs via gas phase isomerization of chalcone followed by

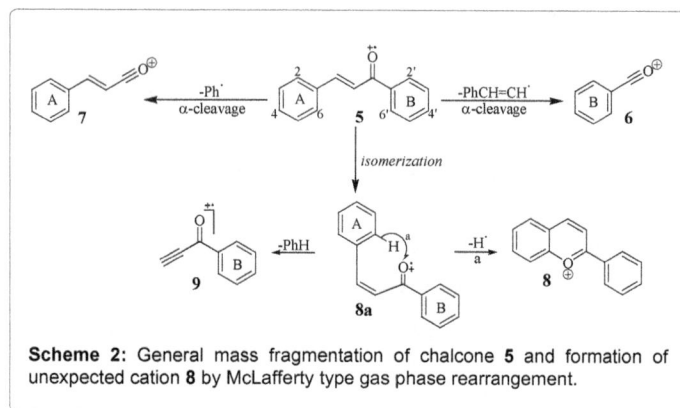

Scheme 2: General mass fragmentation of chalcone 5 and formation of unexpected cation 8 by McLafferty type gas phase rearrangement.

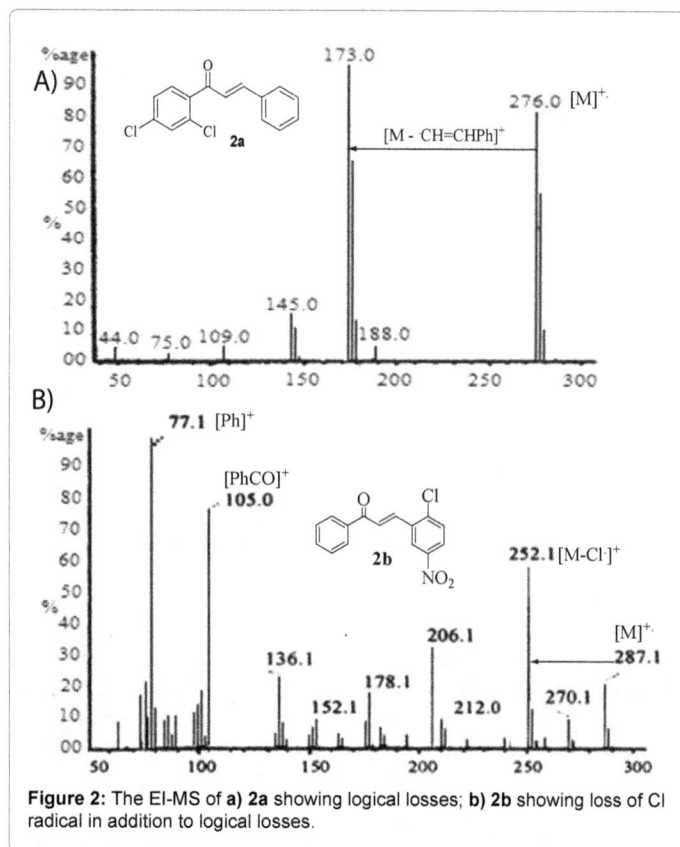

Figure 3: Portions of mass spectrum showing [M-X] in a) 2-methoxy substituted chalcones 2c; b) 2-chloro substituted chalcones 2d; c) 2,3-dimethoxychalcone 2e; d) 2,6-dichlorochalcones 2f.

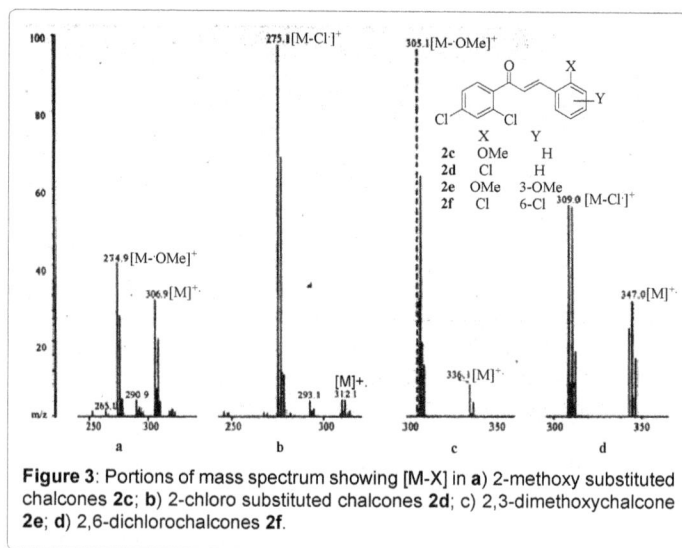

Figure 2: The EI-MS of a) 2a showing logical losses; b) 2b showing loss of Cl radical in addition to logical losses.

	X	Y	Z	[M]$^+$ amu, (%)	[M – X]$^+$ amu, (%)	A amu, (%)
2a	H	H	Cl	276 (82), 278 (55), 280 (9)	-	173 (100), 175 (62), 177 (11)
2b	Cl	5-NO$_2$	H	287 (23), 289 (7)	252 (61)	105 (78)
2c	OMe	H	Cl	306 (32), 308 (21), 310 (4)	275 (43), 277 (28), 279 (5)	173 (100), 175 (69), 177 (11)
2d	Cl	H	Cl	310 (4), 312 (5), 314 (1), 316 (0)	275 (100), 277 (67), 279 (12)	173 (34), 175 (23), 177 (4)
2e	OMe	3-OMe	Cl	336 (7), 338 (4), 340 (0)	305 (100), 307 (64), 309 (14)	173 (45), 175 (34), 177 (8)
2f	Cl	6-Cl	Cl	344 (24), 346 (31), 348 (16), 350 (0), 352 (0)	309 (56), 311 (56), 313 (18), 315 (0)	173 (100), 175 (65), 177 (11)
2g	NO$_2$	H	Cl	321 (0), 323 (0), 325 (0)	275 (15), 277 (10), 279 (2)	173 (100), 175 (68), 177 (12)
2h	Cl	5-NO$_2$	Cl	355 (9), 357 (9), 359 (0), 361 (0)	320 (97), 322 (62), 324 (11)	173 (100), 175 (66), 177 (10)

Table 2: Logical fragments of chalcones formed under EI conditions.

a McLafferty type of rearrangement in which the oxygen of the C=O facilitates the C-X bond cleavage and hence the formation of benz-1-oxin cation (Scheme 3). The peak matching experiment confirmed the [M-X] assignment to be correct.

In order to justify our findings, several highly substituted chalcones were subjected to fragmentation under EI conditions which supported our observation and resulted in the formation of benz-1-oxin as a base peak in almost all cases (Scheme 4, Table 2).

Inspired by the tendency of chalcones to undergo [M-X] cleavage forming benz-1-oxin cation in gas phase EI conditions, we studied the fragmentation behavior of various substituted cyclic chalcones (i.e., arylidene indanones and arylidene tetralones) under EI conditions. Surprisingly, arylidene indanone /tetralones substituted at C-2 exhibited the same pattern of [M-X] cleavage and the resulting cation formed base peak in almost all cases. However, when the substitution at C-2 becomes OH (2-OH), as in **4m**, the fragment ion [M-˙OH] doesn't appear as a prominent peak.

In case of cyclic chalcones substituted at C-3 and/or C-4, [M-X] didn't appear as prominent peak; [M-H˙] forms a reasonably stable fragment instead. The loss of neutral CO was also observed as a minor signal in few cyclic chalcones (Scheme 5, Table 3). In order to confirm that the loss of H radical does not originate from the aliphatic

methylene carbon or ring B, chalcone **4g** (2,6-dichloro substitution on ring A) was subjected to EI conditions. As was expected, no [M-H˙] was observed which confirmed that the generation of [M-H˙] does not takes place from ring B or the methylene carbon but rather occurs from ring A. In addition, it was also established that in case of **4a**, H at C-2 was lost, forming a benz-1-oxin cation; an observation consistent with the findings of Baldas et al. [15].

The peak matching results of some selected acyclic (**2b**, **2e**) and cyclic chalcones (**4b**, **4i**, **4q**, **4r**) were observed to be in close agreement with calculated values, assuring the formation of targeted chalcones. Furthermore, the peak matching results of their corresponding benz-1-oxin [M-X] cation also showed very minute difference (0.4-1.5 millimass unit) in calculated and observed values (Table 4), which confirmed the assignment to be correct.

The DFT calculations were performed by taking into consideration four hypothetical isomers (**9a-d**) of benz-1-oxin cations, which originates from **4d-4g**. The molecular ion of **4d-4g** shows poor abundance (means poor stability) while their corresponding benz-1-oxin cations (isomers **9a**, **9b** and **9d**) shows base peak/high abundance (Table 3), which reflects high stability. The structures of isomers (**9a-d**) were optimized in gas phase using B3LYP functional and 6-311+G(d) basis set. The HOMO (Highest Occupied Molecular Orbital)-LUMO (Lowest Unoccupied Molecular Orbital) analysis was carried out to locate the charge transfer place within the molecule. The visualization of the electron charge was obtained using the Gauss View program. The intramolecular charge transfer has also been confirmed by HOMO-LUMO analysis (Figure 4). The frontier molecular orbitals (HOMO-LUMO) and their properties, such as energy, are very useful for physicist and chemists and are of prime significance in quantum chemistry. The frontier molecular orbitals make use of frontier electron density for predicting the most reactive position in π-electron systems

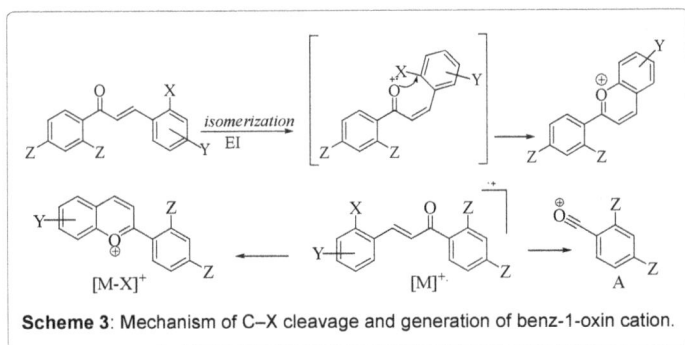

Scheme 3: Mechanism of C–X cleavage and generation of benz-1-oxin cation.

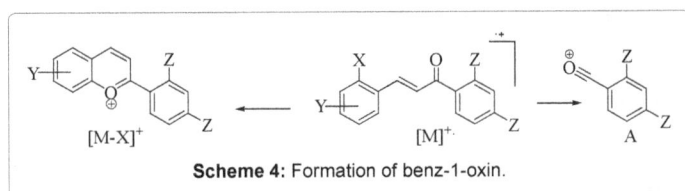

Scheme 4: Formation of benz-1-oxin.

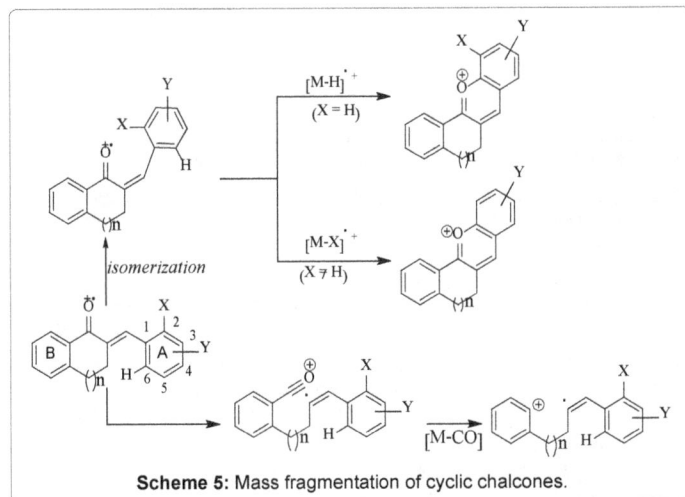

Scheme 5: Mass fragmentation of cyclic chalcones.

	n	X	Y	[M]⁺· amu (% abundance)	[M − X]⁺· amu, (%)	[M − H]⁺ amu (% abundance)	[M − CO]⁺· amu (% abundance)
4a	0	H	H	220 (64)	-	219 (100)	192 (24)
4b	0	Br	H	298 (31), 300 (32)	219 (100)	-	-
4c	0	Cl	H	254 (3), 256 (1)	219 (100)	-	-
4d	0	H	3-Cl	254 (5), 256 (2)	219 (87)	-	-
4e	0	H	4-Cl	254 (33), 256 (13)	219 (25)	-	-
4f	0	Cl	4-Cl	288 (9), 290 (10) 292 (4)	253 (100), 255 (38)		-
4g	0	Cl	6-Cl	288 (4), 290 (3) 292 (0)	253 (100), 255 (45)	-	-
4h	0	Cl	5-NO₂	299 (10), 301 (3)	264 (100)		271 (2), 273 (0)
4i	0	OMe	H	250 (5)	219 (100)	249 (2)	-
4j	0	H	3-OMe	250 (100)	219 (21)	249 (57)	-
4k	0	OMe	3-OMe	280 (17)	249 (100)	-	-
4l	0	H	3,4-(OMe)₂	280 (100)	249 (41)	265 (39)	-
4m	0	OH	3-Me	266 (100)	251 (19)	265 (74)	-
4n	0	2-OH	H	236 (84)	219 (58)	235 (43)	208 (18)
4o	0	H	4-OH	236 (100)	-	235 (71)	208 (30)
4p	0	H	3-NO₂	265 (42)	219 (35)	264 (24)	237 (2)
4q	1	Cl	H	268 (3), 270 (1)	233 (100)	267 (2)	-
4r	1	Cl	4-Cl	302 (4), 303(3), 304 (3)	267 (100), 269 (37)	-	-
4s	1	Cl	5-NO₂	315 (21), 317 (6)	280 (81)	-	-
4t	1	H	4-OMe	264 (86)	233 (37%)	263 (100%)	-

Table 3: Logical fragments of cyclic chalcones formed under EI conditions.

Entry	[M]⁺·			[M – X]⁺·		
	Calculated (formula)ᴬ	Observed	Δ (mmu)	Calculated (formula)ᴬ	Observed	Δ (mmu)
2b	287.0349 ($C_{15}H_{10}O_3NCl$)	287.0343	0.6	252.0661 ($C_{15}H_{10}O_3N$)	252.0655	0.5
2d	336.0320 ($C_{17}H_{14}O_3Cl_2$)	336.0322	0.2	305.0136 ($C_{16}H_{11}O_2Cl_2$)	305.0145	0.9
4b	297.9993 ($C_{16}H_{11}OBr$)	297.9998	0.5	219.0810 ($C_{16}H_{11}O$)	219.0800	1.0
4i	250.0994 ($C_{17}H_{14}O_2$)	250.0981	1.2	219.0810 ($C_{16}H_{11}O$)	219.0799	1.1
4q	268.0655 ($C_{17}H_{13}OCl$)	268.0664	0.9	233.0966 ($C_{17}H_{13}O$)	233.0970	0.4
4r	302.0265 ($C_{17}H_{12}OCl_2$)	302.0256	0.9	267.0577 ($C_{17}H_{12}OCl$)	267.0592	1.5

ᴬ Only the peak matching results of most abundant isotope is reported here.

Table 4: Peak matching results of [M]⁺· and [M – X]⁺· of a few chalcones.

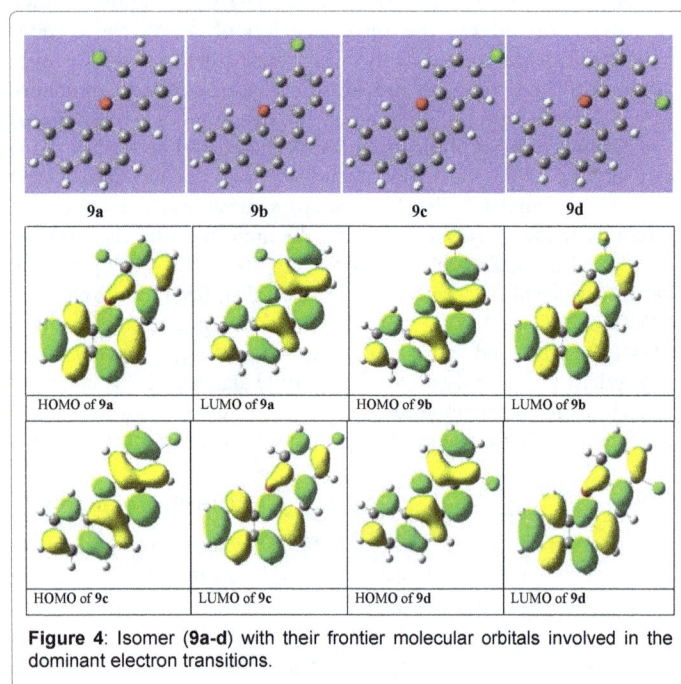

Figure 4: Isomer (**9a-d**) with their frontier molecular orbitals involved in the dominant electron transitions.

System	Point group (P)	Dipole moment (D)	Total energy (a.u.)
1	C_1	2.0863	-1189.2479
2	C_1	2.4483	-1189.2510
3	C_1	2.3739	-1189.2509
4	C_1	1.5970	-1189.2504

Table 5: Point group, dipole moment (Debye) and total bonding energy (a.u.) of **9(a-d)**.

and also explain several types of reactions in conjugated system. The conjugated molecules are characterized by a small HOMO-LUMO separation. Both the HOMO and LUMO are the main orbitals, which take part in chemical stability. The HOMO represents the ability to donate an electron, LUMO as an electron acceptor, represents the ability to gain an electron.

The HOMO and LUMO energy calculated by B3LYP/6-311+G(d) method is shown in Table 5. All isomers (**9a-d**) showed almost equal energy but isomer **9b** may be considered relatively more stable among all (Table 5). It reflects that the presence of a substituent (Y) on variable positions of ring A of **4** does not make that big difference in the energies/stabilization of resulting benz-1-oxin cations. The

major contributor towards the high stability of benz-1-oxin cations is resonance stabilization. It supports our hypothesis as described in Scheme 5.

Conclusion

The gas phase loss/substitution of δ-substituent (substituent at C-2) of ring A of chalcones/arylidene indanones and tetralones has been confirmed to occur via a McLafferty type rearrangement. The hypothesis is supported by HR MS and DFT studies. This report of gas phase formation of benz-1-oxin from δ-substituted chalcones is an important breakthrough that can be employed to identify ortho-substituted chalcones from those with different substitution patterns.

Acknowledgment

The authors are grateful to the Higher Education Commission, Government of Pakistan for generous support of a research project (HEC 20-809), fellowship and International Research Support Initiative Programme (IRSIP) award to Ms Aeysha Sultan (074-141-Ps4-435) and financial support for ESI MS, LR EIMS & NMR-analyses. Nisar Ullah would like to acknowledge the support provided by King Abdul Aziz City for Science and Technology (KACST) through the Science and Technology Unit at King Fahd University of Petroleum & Minerals (KFUPM) for funding this work through project No. 11-BIO2138-04 as part of the National Science, Technology and Innovation Plan.

References

1. Bhattacharya A, Mishra LC, Sharma M, Awasthi SK, Bhasin VK (2009) Antimalarial pharmacodynamics of chalcone derivatives in combination with artemisinin against *Plasmodium falciparum in vitro*. Eur J Med Chem 44: 3388-3393.

2. Ilango K, Valentina P, Saluja G (2010) Synthesis and *in-vitro* anti-cancer activity of some substituted Chalcone derivatives. Res J Pharm Biol Chem Sci 1: 354-359.

3. Hamdi N, Fischmeister C, Puerta MC, Valerga P (2010) A rapid access to new coumarinyl chalcone and substituted chromeno[4,3-c]pyrazol-4(1H)-ones and their antibacterial and DPPH radical scavenging activities. Med Chem Res 20: 522-530.

4. Hsieh HK, Lee TH, Wang JP, Wang JJ, Lin CN (1998) Synthesis and anti-inflammatory effect of chalcones and related compounds. Pharm Res 15: 39-46.

5. Lahtchev KL, Batovska DI, Parushev SP, Ubiyvovk VM, Sibirny AA (2008) Antifungal activity of chalcones: a mechanistic study using various yeast strains. Eur J Med Chem 43: 2220-2228.

6. Torigoe T, Arisawa M, Itoh S, Fujiu M, Maruyama HB (1983) Anti-mutagenic chalcones: antagonizing the mutagenicity of benzo(a)pyrene on *Salmonella typhimurium*. Biochem Biophys Res Commun 112: 833-842.

7. Haraguchi H, Ishikawa H, Mizutani K, Tamura Y, Kinoshita T (1998) Antioxidative and superoxide scavenging activities of retrochalcones in *Glycyrrhiza inflata*. Bioorg Med Chem 6: 339-347.

8. de Vincenzo R, Ferlini C, Distefano M, Gaggini C, Riva A, et al. (2000) *In vitro* evaluation of newly developed chalcone analogues in human cancer cells. Cancer Chemother Pharmacol 46: 305-312.

9. Zhang L, Xu L, Xiao SS, Liao QF, Li Q, et al. (2007) Characterization of flavonoids in the extract of *Sophora flavescens* Ait. by high-performance liquid chromatography coupled with diode-array detector and electrospray ionization mass spectrometry. J Pharm Biomed Anal 44: 1019-1028.

10. Zhang J, Brodbelt JS (2003) Structural characterization and isomer differentiation of chalcones by electrospray ionization tandem mass spectrometry. J Mass Spectrom 38: 555-572.

11. Van De Sande C, Serum JW, Vandewalle M (1972) Studies in organic mass spectrometry-XII: Mass spectra of chalcones and flavanones. The isomerisation of 2'-hydroxy-chalcone and flavanone. Org Mass Spectrom 6: 1333-1346.

12. Menezes JC, Cavaleiro JA, Kamat SP, Barros CM, Domingues MR (2013) Electrospray tandem mass spectrometry analysis of methylenedioxy chalcones, flavanones and flavones. Rapid Commun Mass Spectrom 27: 1303-1310.

13. Tai Y, Pei S, Wan J, Cao X, Pan Y (2006) Fragmentation study of protonated chalcones by atmospheric pressure chemical ionization and tandem mass spectrometry. Rapid Commun Mass Spectrom 20: 994-1000.

14. Baldas J, Porter QN (1979) Mass Spectrometric Studies. XIV. Nitrochalcones. Aust J Chem 32: 2249-2256.

15. Nesmeyanov AN, Zagorevskii DV, Nekrasov YS, Sizoi VF, Postnov VM, et al. (1979) Mass spectrometry of π-complexes of transition metals : XII. Ferrocenyl and cymantrenyl derivatives of chalcones. J Organomet Chem 169: 77-81.

16. Parmar VS, Sharma SK, Vardhan A, Sharma RK, Møller J, et al. (1993) New fragmentation pathways in the electron impact mass spectrometry of derivatized pyrano-1,3-diphenylprop-2-enones. Org Mass Spectrom 28: 23-26.

17. Zhang Y, Zhang P, Cheng Y (2008) Structural characterization of isoprenylated flavonoids from Kushen by electrospray ionization multistage tandem mass spectrometry. J Mass Spectrom 43: 1421-1431.

18. Zhang J, Huo F, Zhou Z, Bai Y, Liu H (2012) The Principles and Applications of An Ambient Ionization Method-Direct Analysis in Real Time (DART). Prog Chem 24: 101-109.

19. Bankova VS, Mollova NN, Popov SS (1986) Chemical ionization mass spectrometry with amines as reactant gases. I-amine chemical ionization mass spectrometry of flavonoid aglycones. Org Mass Spectrom 21: 109-116.

20. Tai Y, Pei S, Wan J, Cao X, Pan Y (2006) Fragmentation study of protonated chalcones by atmospheric pressure chemical ionization and tandem mass spectrometry. Rapid Commun Mass Spectrom 20: 994-1000.

21. Hoffmann B, Hölzl J (1989) Chalcone glucosides from *Bidens pilosa*. Phytochemistry 28: 247-249.

22. Redl K, Davis B, Bauer R (1992) Chalcone glycosides from *Bidens campylotheca*. Phytochemistry 32: 218-220.

23. Prasain JK, Tezuka Y, Jian JL, Tanaka K, Basnet P, et al. (1997) Six Novel Diarylheptanoids Bearing Chalcone or Flavanone Moiety from the Seeds of *Alpinia blepharocalyx*. Tetrahedron 53: 7833-7842.

24. Motiur-Rahman AFM, Attwa MW, Ahmad P, Baseeruddin M, Kadi AA (2013) Fragmentation Behavior Studies of Chalcones Employing Direct Analysis in Real Time (DART). Mass Spectrom Lett 4: 30-33.

25. Grostic MF (1967) Mass spectrometry of organic compounds. By H. Budzikiewicz, C. Djerassi, and D. H. Williams. Holden-Day, Inc., San Francisco, USA.

26. Sultan A, Raza AR, Abbas M, Khan KM, Tahir MN, et al. (2013) Evaluation of silica-H_2SO_4 as an efficient heterogeneous catalyst for the synthesis of chalcones. Molecules 18: 10081-10094.

27. Frisch MJ, Trucks GW, Schlegel HB, Scuseria GE, Robb MA, et al. (2009) Gaussian 09, Revision A.1 2009 (Gaussian Inc.: Wallingford, CT).

28. Ardanaz CE, Traldi P, Vettorin U, Kavkat J, Guidugli F (1991) The ion-trap mass spectrometer in ion structure studies. The case of [M[BOND]H]+ ions from chalcones. Rapid Commun Mass Spectrum 5: 5-10.

23

Development and Characterization of Ternary Hybrid Membranes in Natural Polymers and their Application in the Microbiology Field

Bouhadiba K[1], Djennad MH[1] and Hammadi K[2*]

[1]Laboratory of SEA2M, Department of Chemistry, Faculty of Sciences and Technology, University of Mostaganem, Mostaganem, Algeria
[2]Pedagogical Laboratory of Microbiology, Department of Biology, Faculty of Life and Natural Sciences, University of Mostaganem, Mostaganem, Algeria

Abstract

New ternary membranes have been synthesized and characterized. The membranes have been elaborated with natural polymers such as "corn starch" and "Bentonite" added with activated carbon. The technique of phase reversal induced by a solvent at ambient temperature has been applied for the development of these membranes. A set of characterization techniques which include FTIR, SEM are used for structural, morphological knowledge. The ternary membranes are performed on microbiological separation for the filtration of bacteria (*Escherichia coli*, *Pseudomonas aeruginosa*) and fungus (*Aspergillus niger*, *Penicillium notatium*) and compared to binary membranes which include natural polymers such as "corn starch" and "Bentonite" as compositions. Results showed that the ternary membrane has high performance compared to binary membranes for the filtration *Escherichia coli*, *Pseudomonas aeruginosa* and fungi *Aspergillus niger*.

Keywords: Natural polymers; Synthesis, Characterization; Bacteria; Fungi; Separation

Introduction

Membrane technologies have been widely studied in recent years [1,2]. The key efficient and economical separation of this process is the membrane [3], which is defined as a physical barrier to retain or pass microorganisms. The most important desired characteristics of the membranes are the good permeability, high selectivity and stability [3-5]. Today, the membrane separation must now be considered as powerful advanced technologies and convenient to use [6]. Their installation and implementation requires attention and special expertise [7-9]. The application of membrane filtration technology in the field of microbiology requires the development of new membranes based on natural polymers such as corn starch and bentonite [10]. The use of a biopolymer such as starch can be an interesting solution because such a polymer is relatively inexpensive, abundant and biodegradable. The inversion technique of induced phase by solvent has been applied to the preparation of these membranes [11]. This technique is most suitable for our selected polymers [12,13]. Our work involves the development of two types of membrane, the binary starch and bentonite membrane and ternary membrane by adding commercial activated carbon to the first. The incorporation of the latter material was a need as there was increased retention of certain microorganisms selected during filtration.

Materials and Methods

Natural powder of corn starch is extracted from maize raw material. Two types of commercial bentonite sodium were obtained from BENTAL Company "ENOF" of Mostaganem. A commercial activated carbon (REIDEL) was purchased from Maen (Germany) and the distilled water was used as solvent during membranes fabrications. Bacterial strains (*Escherichia coli*, *Pseudomonas aeruginosa*) were obtained from laboratory hospital of Mostaganem, and fungi strains (*Aspergillus niger*, *Penicillium sp*) were obtained from microbiology laboratory, department of biology of Mostaganem/Algeria.

Membrane preparation

The membrane preparation technique used in our work requires corn starch, bentonite and commercial activated carbon (REIDEL). A series of membranes with various proportions of these three materials was developed. The technique of induced phase inversion by solvent is used to make an asymmetric support with thin layers and high porosity, this method allows the fabrication of flat asymmetric hybrid membranes at different concentrations of polymers and activated carbon. The selected suitable solvent was distilled water for reasons of cost, safety, availability and its chemical affinity with natural polymers. As the corn starch is insoluble in cold water, the corn starch prepared solutions are heated to a temperature of 70°C and stirred magnetically for a period of 90 minutes, after the solubility test of the corn starch, the selected solution in which the concentration is 25 g/L, this solution is called a starch paste. The incorporation of bentonite and activated carbon to the starch solution allows the preparation of ternary and binary mixture whose protocol is as follows. 20 ml of the prepared corn starch solution was added to different amounts of treated/untreated bentonite and various proportions of activated carbon. The mixture is heated under magnetic stirring for two hours to ensure solution homogeneity. 1 ml were collected from the mixture which then spread on a filter paper type Macherey-Nagel on circular diameter of 7 cm using a metal roll. The prepared membranes were dried in an oven and maintained at a moderate temperature, the membranes then wrapped in aluminum foil to sterilize those that have been applied to microbiological filtration. The characterization was performed on selected membranes after filtration microbiological tests.

Microbiological separation

Strains of bacteria such as (*Escherichia coli*, *Pseudomonas aeruginosa*) and fungus (*Aspergillus niger*, *Penicillium* sp) were activated and purified by Gram staining technique before any microbiological test filtration. All conditions were prepared in the laboratory using conventional microbiological techniques: the broth and nutrient agar for bacteria, broth and Sabouraud agar for fungus. After activation of

---

*Corresponding author:** Hammadi Kheira, Department of Chemistry, Faculty of Sciences and Technology, University of Mostaganem, Mostaganem, Algeria
E-mail: kyrabiology@yahoo.fr

the different strains, the bacterial suspensions and those of fungus were reactivated before each test with respecting of the filtration duration (24 to 72 hours) and the incubation temperature (27°C to 30°C). The filtration was carried out in the microbiology laboratory in well sterile conditions to avoid any air contaminations. After the filtration test on wide range of developed membranes, series membranes have been selected in which the bentonite concentration is 100 g/L and that of the activated carbon of ternary is 12.5 g/L. As our goal is to design selective polymeric membranes. The selected series of membranes present the retention property for the selected microorganisms.

Results and Discussion

Membrane characterization

The prepared membranes are characterized using two methods, the top surface morphologies of membranes were observed by scanning electron microscopy (Hitachi TM 1000) which can achieve 30000 nm of resolution. The chemical group an interaction is identified using Fourier Transform Infrared Spectroscopy (FTIR) (SHIMADZU IR Prestige-21), range of wave number is 400-4000 cm^{-1}.

Fourier Transform Infrared Spectroscopy (FTIR)

Fourier transform infrared spectroscopy was utilized to identify vibration modes of different functional groups of prepared membranes M_1, M_2, M_3 and M_4. The results are shown in Figure 1. As can be observed, the four membranes IR Specters are similar in the range of 4000-1000 cm^{-1}, it seems that the peaks intensity decreases during the addition of activated carbon. The band intensity caused by the stretching vibrations Si-O-Si (1035 -1024-1037-1042) and the bands due to deformation vibrations Al-OH (915 cm^{-1} for the membranes M_1 et M_2) were not affected by the addition of activated carbon as shown in Figures 1-4.

Scanning Electron Microscopy (SEM)

As shown on Figures 5 and 6, the photos a, b, c and d corresponding to the membranes M_1, M_2, M_3 and M_4 shows that the surfaces have pores with size and distribution and shape more or less heterogeneous.

Performance evaluation of ternary membranes compared to binary membranes

The microbiological experiments was used to test the membranes M_1, M_2, M_3 and M_4 the ability of these to filtrate the microorganisms;

Escherichia coli, *Pseudomonas aeruginosa* as bacterial test and *Penicillum notatum* and *Aspergillus niger* as a fungal strains. The microorganisms strains was delivered from the laboratory of microbiology; University of Mostaganem, Algeria, after isolation and preparation of the cultures at specified milieu the nutrient broth milieu for bacteria and sabouraud liquid for fungi. The results of the filtration of selected membranes M_1, M_2, M_3 and M_4 are grouped in three (Tables 1-4). The first experiment where the membranes M_1, M_2, M_3, and M_4 have been tested on filtration of *Escherichia-coli* and *Pseudomonas aeruginosa* microorganisms, the binary membrane with treated bentonite M_1 has retained the tow microorganisms, but the results were negatives with not treated membrane M_3 due the effect of treated bentonite on structure of the membrane pore diameter. The retaining of *Escherichia coli* and *Pseudomonas aeruginosa* bacteria has been enhanced after adding the activated carbon on the membrane M_3 for the results of the filtration. *Aspergillus niger* and *Penicillium notatum* in Table 4, the membranes M_3 have showed a negative results contrary with bacteria showed in Table 3. For the membrane M_4, the retention of Aspergillus can be explained by the morphology of this fungi which has a significant mycelium that cannot be passing through diameter holes of the membranes M_3 and M_4. The negative result of membranes M_1, M_2 may be due to the morphology spores released by the fungi.

Conclusion

From this work, a ternary membrane M_2, M_4 were successfully prepared by inversion technique of induced phase by solvent, the fabricated membrane is shown a porous surface with heterogeneous morphology. The investigated ternary membranes M_2, M_4, the membrane M_4 which is the adding of the activated carbon on (Bentonite/corn starch) binary membrane results in enhanced on microbiological separation compared to binary membranes M_1, M_2. The most important characteristics (pore diameter and porous volume) scanning electron microscopy observation showed homogeneous layers and porous without cracking; the same results were showed by Agoudjil et al. in 2014; where they obtained a silica membranes which can be used in Ultrafiltration [14]. We can say this work; is new approach to think about how we can use the "bentonite"; natural polymer "corn starch" and active carbone to prepare polymeric membrane to filtrate the microorganism; furthermore work is on to see permeability of the prepared membrane using other bacteria; fungi and, viruses.

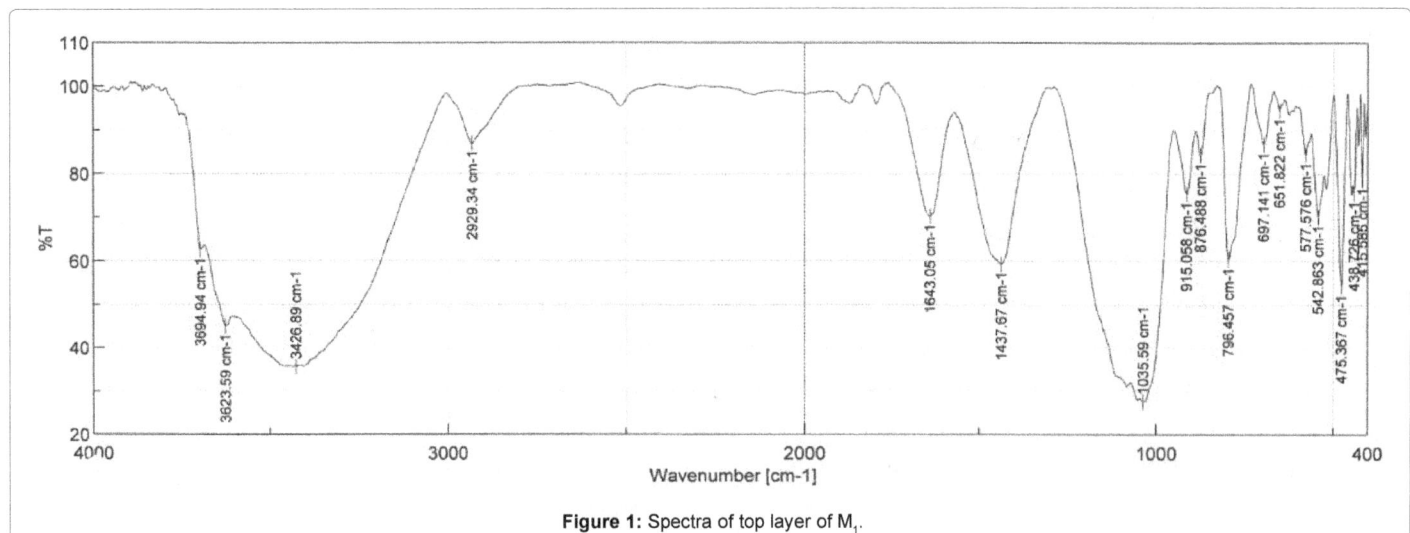

Figure 1: Spectra of top layer of M_1.

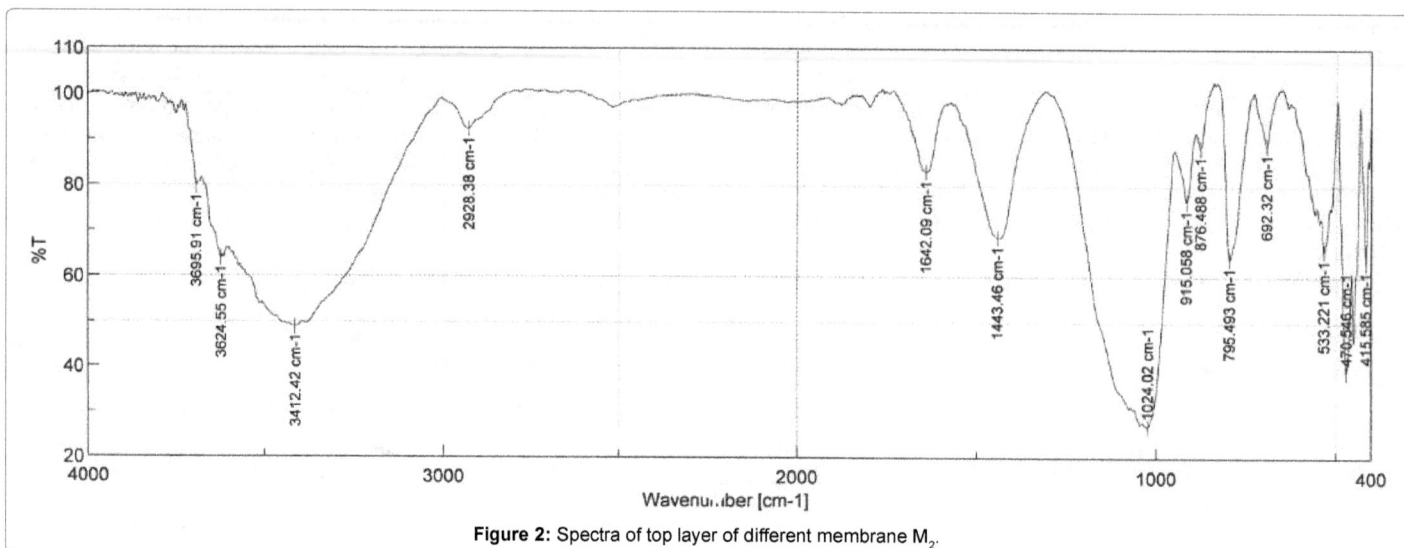

Figure 2: Spectra of top layer of different membrane M₂.

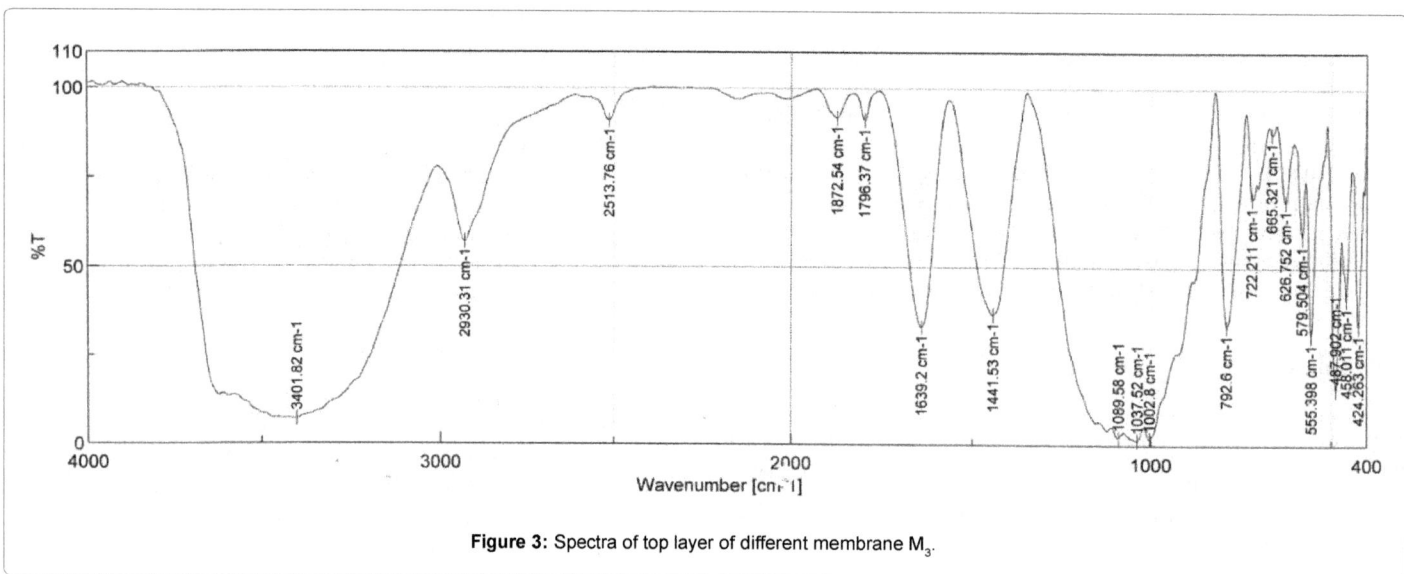

Figure 3: Spectra of top layer of different membrane M₃.

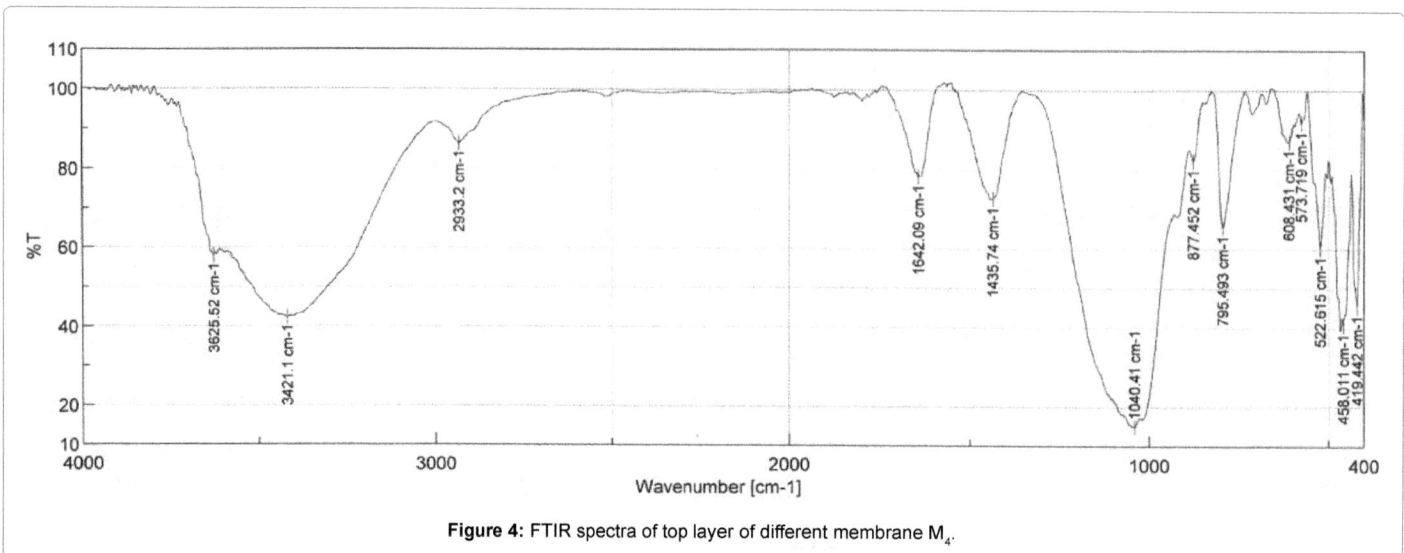

Figure 4: FTIR spectra of top layer of different membrane M₄.

2012/06/25   11:04 L D3.6   x1.2k   50 um

TM1000-0089-06-12   TM1000-0267-04-12

2012/04/30   11:31 L D3.5   x1.8k   50 um

Figure 5: SEM photographs for binary and ternary membranes: (a) M$_1$, (b) M$_2$.

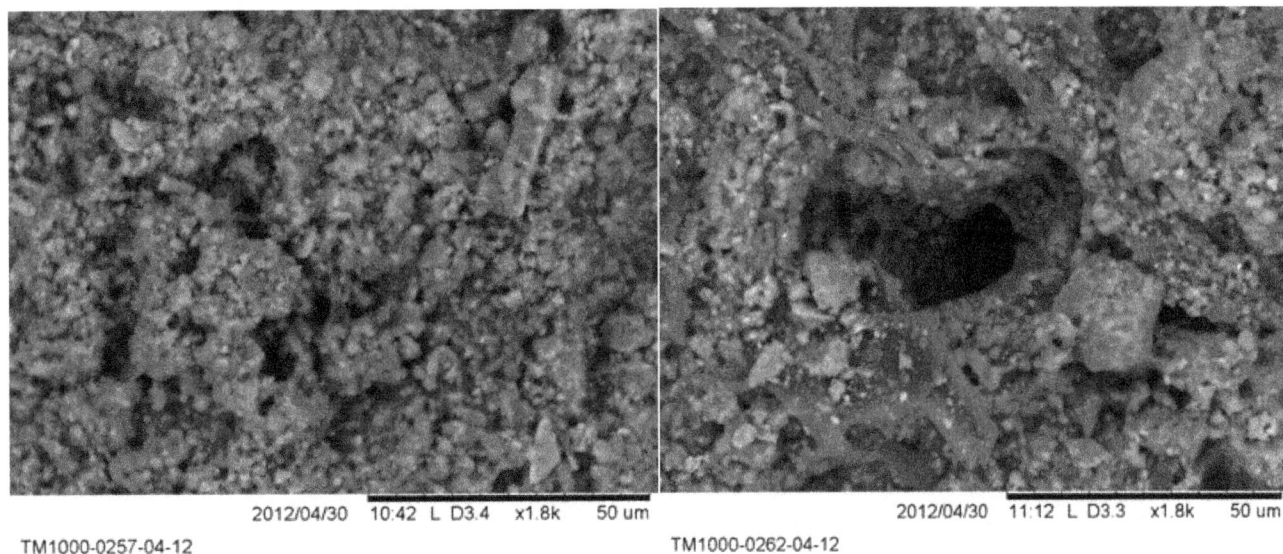

2012/04/30   10:42 L D3.4   x1.8k   50 um

TM1000-0257-04-12   TM1000-0262-04-12

2012/04/30   11:12 L D3.3   x1.8k   50 um

Figure 6: SEM photographs for binary and ternary membranes: (c) M$_3$, (d) M$_4$.

Selected Membrane	Amidon Concentration (g/L)	Bentonite concentration (g/L)	Activited carbon concentration (g/L)
M$_1$	25	100	/
M$_2$	25	100	12.5
M$_3$	25	100	/
M$_4$	25	100	12.5

M$_1$, M$_2$: Synthesized membranes with treated Bentonite; M$_3$, M$_4$: Synthesized membranes with natural Bentonite

Table 1: Different selected membranes.

Bacterial strains membranes	*Escherichia coli*	*Pseudomonas aeroginosa*
M$_1$	-	-
M$_2$	+	-
M$_3$	+	+
M$_4$	-	-

(+) Presence of bacteria in the filtrate; (-) Absence of bacteria in the filtrate.

Table 2: Results of bacteria filtration.

Fungal strains membranes	*Aspergillus niger*	*Penicillium sp*
M$_1$	+	+
M$_2$	+	+
M$_3$	-	-
M$_4$	-	+

(+) Presence of bacteria in the filtrate; (-) Absence of bacteria in the filtrate

Table 3: Results of the *A. niger* and *P. notatum* filtration.

Tested microorganisms	Selective membranes
Escherichia coli	M$_1$, M$_4$
Pseudomonas aeruginosa	M$_1$, M$_2$, M$_4$
Aspergillus niger	M$_3$, M$_4$
Penicillium notatum	M$_3$

Table 4: The retained selective membranes.

References

1. Jean PB (1988) Membrane separation processes. Transport Membrane techniques. Application. Masson, pp: 123-147.

2. CFM 2 Booklet (2002) Micro and ultrafiltration: conduction and pilot tests. Treatment of buckets and effluents.

3. Audinos R, Isoard P (1986) Glossary: Technical terms of membrane processes. Société française de filtration, pp: 142.

4. Medina GY (2015) Study of solubilization phenomena Pre-polymers of epoxy resin in fatty acid esters as Biosolvents. Doctoral Thesis. INP of Toulouse, France.

5. Bessiere Y (2005) Front membrane filtration: highlighting of the critical filter volume for anticipation and control of clogging, Toulouse. Doctoral Thesis.

6. Hassani L, Pontie M, Dach H, Diawara C (2006) Application of membrane technologies in drinking water production. Interests and limits of use. Mali Symposium on Applied Sciences.

7. Micro and Ultrafiltration (2002) Conduct of pilot tests, treatment of water and effluents.

8. Cheruan M (1998) Ultrafiltration and microfiltration handbook. Technomic Publishing Co. Inc., Lancaster, p: 552.

9. Guide to Nano Filtration (2000) CFM Handbook.

10. Maillevialle J, Odendaal PE, Weisner MR (1996) The emergence of Membranes in water and waste water treatment. In: Water Treatment Membrane Process.

11. Drioli E, Criscuoli A, Curcio E (2005) Membrane Contactors: Fundamentals, Applications and Potentialities: Membrane Science and Technology. Membrane Science and Technology, p: 516.

12. Kang YS, Kim SH, Young J, Won J (2004) Influence of the addition of PVP on the morphology of asymmetric polyimide phase inversion membranes: Effect of PVP molecular weight. Journal of Membrane Science 236: 203-207.

13. Roosta A, Mousavi S, Ramazani A, Roshan MA (2006) Correlation of Nitrogen Enriching Polymeric Membranes Performance Developed through Various Methods with their Morphological Structure Revealed via SEM. Iranian Polymer Journal 15: 291-298.

14. Agoudjil N, Lamrani A, Larbot L (2014) Silica porous membranes synthesis and characterization Desalination and Water Treatment. Desalination and Water Treatment 15: 2988-2995.

24

Effects of Fluorine Atoms Amount and Fluorinated Acrylic Chain Length Chemically Attached to Hydroxyl Groups on the Hydrophobic Properties of Cotton Fabrics

Liu X[1], Yang G[1] and Lipik VT[2]*

[1]*Institute for Sport Research, Nanyang Technological University, Singapore*
[2]*School of Material Science and Engineering, Nanyang Technological University, Singapore*

Abstract

A two-step chemical treatment of cotton fabric was performed with the attachment of fluorine moieties on hydroxyl group sites. The hydroxyl groups of cotton were initially acrylated by 2-isocyanaethyl methacrylate. Acrylic monomers, containing four to twelve fluorine atoms, had been used to build polymeric chain directly on the surface of cotton by means of radical polymerization. Fabrics became 10-20% stiffer and microscopy showed a clear change of the cotton surface after the treatment. Coated samples of cotton had shown hydrophobic property with a highest contact angle of 128 degrees. It was found that the increase of molecular weight of fluorinated polyacrylate on the cotton surface lowered contact angle value. The best results in hydrophobicity had been obtained at the molar stoichiometric ratio between the number of hydroxyl groups of cotton and the amount of fluorinated monomer added. This developed method allowed for the direct radical polymerization on cotton fabric, providing good hydrophobic properties with the formation of fluorinated polyacrylate of different molecular weights.

Keywords: Hydrophobicity; Fluorinated coating; Radical polymerization; Fluorinated acrylates; Contact angle

Introduction

Cotton is a popular natural fiber for making comfortable apparel. However, cotton absorbs water due to its abundance of hydroxyl (–OH) groups, thus reducing its mechanical and thermoinsulation properties. Various methods have been used to obtain water repellency on cotton surfaces, such as coating with water-resistant materials like nanoparticles [1] paraffin wax [2] silicon resin [3] or fluorocarbon [4]. Coating via the chemical transformation of –OH group is the widely known method for enhancing water resistance of cotton fibers and fabrics. Chemical transformation of –OH groups has some disadvantages such as long reaction time, involvement of non-ecologically friendly chemicals or solvents, necessity of special reactors as compared to methods using simple wetting in emulsions or coating with polymers. However, chemical transformation of –OH groups of cotton provides a better control of the process and higher washing durability as compared to physical coating due to formations of chemical bonds between cotton surfaces and water-repellent agents. These water-repellent agents could be in the forms of single molecules, oligomers or polymers. Among known chemical methods, fluorine is the best element to be used to lower the surface free energy and make fabric hydrophobic [5,6]. Fluorine has a small radius and a high electronegativity, thus the covalent bond between fluorine and carbon is extremely stable. When fluorine is replaced by other elements such as H and C, in the order $-CF_3 < CF_2H < CF_2 < CH_3 < CH_2$, the surface free energy is increased. Therefore, at the present, the impregnation of textiles with fluorine containing polymer dispersions or solutions is the most applied technique [7]. The increase in the number of fluorine-carbon bonds in the functional group attached to cotton will result in lower surface energy and higher hydrophobicity [8]. However, it has limitations such as poor adhesion between the substrate and the fluorine-based hydrophobic coating due to the weak van der Waals force [9]. Hence, chemical attachment of fluorinated water repellent agent to a cotton surface is a prominent direction that allows for the creation of a stable hydrophobic surface. And fluorinated acrylates are the most common and convenient substances for hydrophobic coating [10]. Currently, chemical attachment of fluorine-containing agents to the cotton surface is being developed by three different approaches. The first direction encompasses synthesis of various small molecules and polymers, which can react with –OH groups on the cotton surface. For example, Schondelmaier [11] had synthesized fluoroalkylsilane molecules with hydrophobic and hydrophilic parts, where the hydrophilic parts of the molecules react with –OH groups. An interesting acrylate-based fluorinated agent with six fluorine atoms monomer had also been synthesized by Yang [12]. The chemistry was realized in three steps and the final product has shown good water-repellent properties when applied on cotton fabrics. A clear benefit for industrial application is that this treatment of cotton can be realized within a short time. A good water-repellent property was obtained at the usage of diblock copolymer in which one part provides water-repellent property whereas another part bearing alkoxysilane or epoxyde group can be chemically attached to the OH group on the cotton surface [13]. The second approach consists of the development of one-step methods of the fluorinated coating of cotton fabric. For example, a one-step process of attachment of a fluorinated monomer to the surface of cotton had been shown by Cai [14] who used irradiation-induced graft polymerization for direct deposition of nonafluorohexyl-1-acrylate on the cotton surface. In another example, direct fluorination of cotton had been done using elemental fluorine [15]. Another direction consists in the modification of –OH groups on a cotton surface into other functional groups, which are to be used for attachment of different water repellent agent, can be done. For example, Yu [16] successfully oxidized the primary –OH groups of cotton by laccase/TEMPO treatment, obtaining aldehyde instead of –OH group, which were then used for the reaction with octadecylamine.

*Corresponding author: Vitali L, School of Material Science and Engineering, Nanyang Technological University, Singapore, E-mail: vitali@ntu.edu.sg

In the first deep work, devoted to fluorination of cotton [17], authors studied hydrophobic property of cotton coated by fluorocontaining chemicals with chain length up to C19, which were fixed on the cotton surface by different resins. In our work, following the above-mentioned third approach, we, targeting to make wider its possibility, modified the cotton surface, transforming –OH groups into acrylic functional groups. This allows for the further polymerization of fluorinated acrylic monomers and the attachment of macromolecules directly onto the surface of a cotton fabric with variation of chain length of fluorinated polymer.

Materials and Methods

Chemicals

The chemicals, 2-isocyanatoethyl methacrylate, 2,2,3,3-Tetrafluorofluoropropyl methacrylate (TFPA), 2,2,3,4,4,4-Hexafluorofluorobutyl acrylate (HFBA), 2,2,3,4,4,5,5-Octafluoropentyl acrylate (OFPA) and 2,2,3,3,4,4,5,5,5,6,6,7,7-Dodecafluoroheptyl acrylate (DFHA), dibutyltin dilaurate, azobisisobutyronitrile (AIBN), dichloromethane and dimethylformamide were purchased from Sigma-Aldrich. All chemicals were used as received.

Modification of cotton fabric

Plain woven cotton fabric samples, with an area of about 25 cm² each, were washed three times with distilled water and dried at 120°C for 2 hours before use. Four different fluorinated acrylic monomers of different chain lengths were used to investigate the effect that chain lengths of fluorine-containing acrylic attached to a cotton surface has on hydrophobicity. The monomers were attached via a two-step chemical reaction. The first step is acrylation, where –OH groups on cotton are transformed into an acrylic group. The second step is radical polymerization, where the fluorinated monomers are polymerized and attached, forming macromolecules on the cotton surface.

Acrylation of cotton: The modification of –OH groups of the cotton into acrylic had been done with reference to the published procedure (Scheme 1). For the synthesis of the intermediate product, cotton fabric strips measuring 2 cm by 5 cm were immersed in dichloromethane solvent in a two-neck flask, followed by addition of 2-isocyanatoethyl methacrylate (acrylating agent) and dibutyltin dilaurate catalyst [18]. The molar ratio of –OH groups of cotton to acrylating agent was 1:10 and the amount of catalyst added was 1% of the molar amount of the acrylating agent. The two-neck flask was kept under nitrogen gas for 5 days under room temperature after which the cotton fabric was rinsed with ethanol and water multiple times. The cotton samples were then dried in the oven at 70°C for 24 hours.

Radical polymerization and the attachment of fluorinated monomers onto cotton surface: Fluorinated monomers with different amounts of fluorine (2,2,3,3-Tetrafluorofluoropropyl methacrylate, 2,2,3,4,4,4-Hexafluorofluorobutyl acrylate, 2,2,3,4,4,5,5-Octafluoropentyl acrylate and 2,2,3,3,4,4,5,5,5,6,6,7,7-Dodecafluoroheptyl acrylate) were attached to an acrylic group on the fabric via radical polymerization (Scheme 2). Polymerization had been done in dimethylformamide for 24 hours at 70°C. Three different ratios (1:1, 1:5, 1:10) of acrylic groups on the cotton surface to fluorinated acrylate monomers had been used in the experiment, targeting polymers with different molecular weights attached on the cotton surface. The quantity of fluorinated monomers was calculated under the assumption that all the –OH groups of cotton had been acrylated during the treatment with 2-isocyanatoethyl methacrylate. The amount of catalyst – AIBN – remained constant for all experiments at a ratio of 1:50 to the molar amount of monomer taken for the polymerization. After polymerization, the

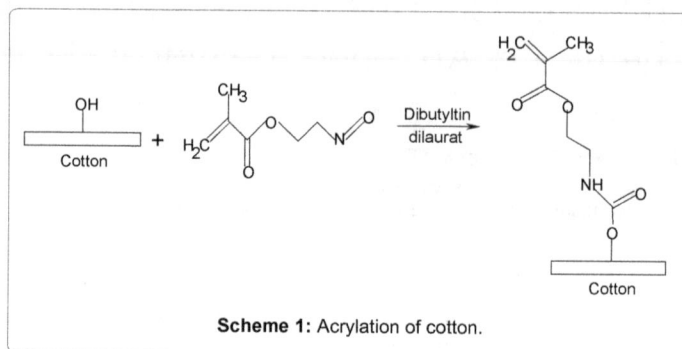

Scheme 1: Acrylation of cotton.

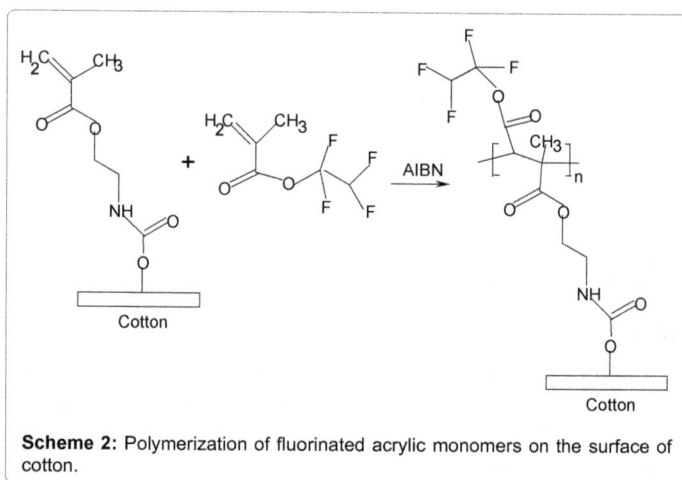

Scheme 2: Polymerization of fluorinated acrylic monomers on the surface of cotton.

samples were washed with acetone and with ethanol subsequently. The cotton samples were then dried in an oven at 70°C.

Characterization

Attenuated Total Reflectance Fourier Transform Infrared (ATR-FTIR): In order to determine the change in the functional groups after the chemical treatment of cotton fabric, Attenuated Total Reflectance Fourier Transform Infrared (ATR-FTIR) (Perkin Elmer Frontier) was used.

Contact angle: Contact angle (CA) measurements were carried out using Contact Angle Dataphysics OCA 20 (Dataphysics, Germany) instrument and 6 uL deionized water droplets at room temperature. Five measurements at different positions on the cotton fabric were used for the determination of the average contact angle values. All the values reported were static contact angles, and measurements were taken 30 seconds after the water droplets were applied onto the cotton fabric.

Scanning Electron Microscopy (SEM)/Energy Dispersive X-ray Spectroscopy (EDX): Scanning Electron Microscopy (FESEM, JSM-6340F) was used to observe the morphology of cotton samples before and after treatment while the energy-dispersive X-ray spectroscopy EDX was carried out by JEOL-6360 under the scanning voltage of 20 kV. The amount of fluorine contented was calculated through area mapping.

Mechanical analysis: The tensile test had been performed at room temperature with Mechanical Tester MTS C43. Each sample was prepared with a size of 10 mm by 40 mm, with a thickness of around 0.25 mm. A load cell of 100 N was used. Five measurements for each type of fabric were performed.

Effects of Fluorine Atoms Amount and Fluorinated Acrylic Chain Length Chemically Attached to Hydroxyl...

143

Results

Characterization of obtained cotton samples

Successful deposition of hydrophobic coatings on the cotton sample was evaluated by comparing the FTIR spectra of coated cotton samples with the untreated sample. FTIR spectra of untreated and treated cotton fabrics are presented in Figure 1. The peak near 1730 cm^{-1} was attributed to the stretching vibration of the carbonyl group of acrylates monomers attached to the fabric. Because of a small amount of fluorinated polymer on the cotton surface and overlapping of peaks typical for –CF$_3$ group, there are no further clearly distinguishable differences between treated and untreated cotton. All presented SEM photos were taken using cotton samples obtained using the ratio of 1:1 of–OH groups to fluorinated monomer because this ratio has shown higher values of a contact angle. The SEM image of the untreated cotton sample (Figure 2) had shown a smooth surface with characteristic parallel ridges. The primary wall structure was also obvious from the SEM image. The SEM image of the treated cotton samples did not exhibit the fine fibril structure that was present in the starting material. Thin laminates and patches with irregular edges were observed. These may have resulted from the rearrangement of surface cellulose due to the formation of a thin fluorinated layer in the primary wall. To characterize the chemical change of cotton after the treatment, quantitative EDX analysis of elements were made using the same samples chosen for SEM. The fluorine content found on the surface of the differently treated cotton is presented in Table 1. As seen in Table 1, the amount of fluorine on the surface of cotton had increased from 0.46% for sample coated with TFPA up to 0.90% for the sample coated with DFHA. Theoretically, if all –OH groups were modified and covered by fluorinated monomers, the amount of fluorine in the treated cotton should be 6.2% wt. for TFPA and up to 12.9% wt. for DFHA. Therefore, the degree of fluorination of acrylated –OH group can be approximately estimated to be 7%.

Water repellent property

The average contact angles of the differently treated cotton fabrics and different ratios of the number of double bonds on the cotton surface to the amount of fluorinated monomer are presented in Table 2. It should be noted that the untreated contact angle of cotton fabrics could not be measured because the water droplet was quickly absorbed into the fabric within 4 seconds. This is due to the high hydrophilic properties of cotton. The intermediate product, where acrylic groups are attached on the surface of cotton, has a contact angle

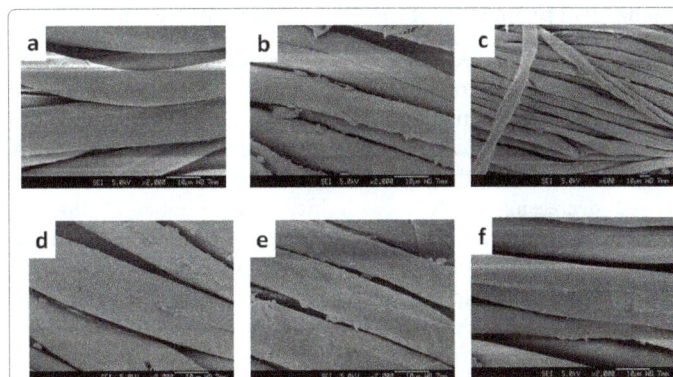

Figure 2: SEM pictures of a) initial cotton fabric and fabrics treated with b) 2,2,3,3-tetrafluoropropyl methacrylate c) high resolution of b), d) 2,2,3,4,4,4-hexafluorobutyl acrylate e) 2,2,3,3,4,4,5,5-octafluoropentyl acrylate f) 2,2,3,3,5,5,5,6,6,7,7-docecafluoroheptyl acrylate.

Type of fluorinated monomer	Amount of fluorine % mass
TFPA	0.46
HFBA	0.78
OFPA	0.79
DFHA	0.90

Table 1: Amount of fluorine in the treated cotton samples.

Type of fluorinated monomer	Average contact angle (°) Ratio of Double bond on surface: Fluorinated monomer		
	1:1	1:5	1:10
TFPA	107.7	103.6	89.5
HFBA	113.2	105.5	101.9
OFPA	123.6	122.6	107.8
DFHA	128.2	122.8	115.4

Table 2: Contact angle values for differently treated cotton fabrics.

of 98.3°, which confirmed that there had been a modification of the –OH groups. However, it was still lower than that of fluorine-treated cotton fabric, due to the absence of hydrophobic tail group (C-F). It can be seen that the cotton treated with longer chains of fluorinated monomers were more hydrophobic as compared to that with shorter chains. This might be due to nonpolar carbon-fluorine (C-F) chains decreasing the overall polarity of the molecule. In addition, longer carbon chains increase the hydrophobicity of the fabric surface, resulting in greater contact angle. As a general trend, the surface energy of cotton fabric decreases as the carbon chain lengths and amount of fluorine atoms increase. In addition, the introduction of long fluorine chains creates rougher surfaces on the cotton fibers, and this plays a positive role in hydrophobicity by decreasing the surface free energy as well. The increase in the molar ratio of the fluorinated monomer has an interesting influence on hydrophobicity. The higher amount of fluorinated monomers we add, the lower hydrophobicity of fabric we obtain. A large molecular chain on the cotton surface could create active sites between the fluorinated monomer chain and the substrate. In particular, a 1:10 molar ratio could create a multiple-linking system, as was mentioned by Colleoni [19], in which the influence of the fluorine chains steric hindrance is balanced by an increase in the number of links between the coating and the cotton fabric sample substrate. Long macromolecule with fluorine atoms on the surface of cotton is also more inclined to make some compact structures such as crystals, coils or globules where fluorine atoms could be hidden and become unavailable to impact hydrophobicity.

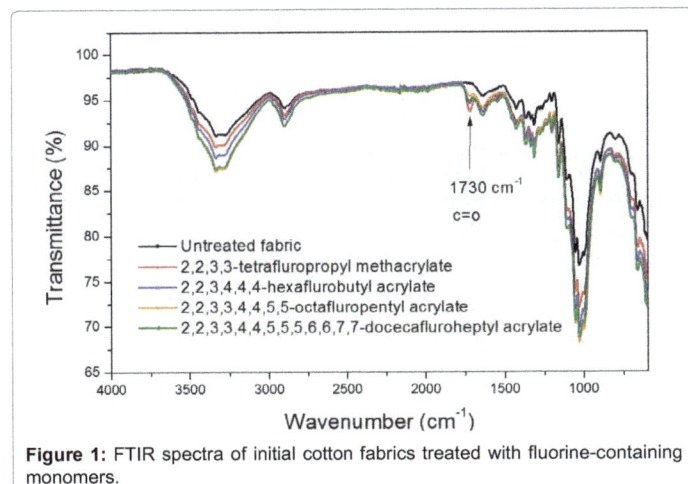

Figure 1: FTIR spectra of initial cotton fabrics treated with fluorine-containing monomers.

Mechanical properties of modified fluorinated cotton

It had been observed that there were slight differences between the mechanical properties of untreated and treated cotton. Young moduli of all samples were fluctuating within the same range. Untreated cotton had the lowest modulus (201.0 N/mm²) while the sample with TFPA treatment had shown the highest (328.4 N/mm). The treated samples showed decreasing modulus with increasing number of fluorine up to 272.0 N/mm² for DFHA treated sample. Measurements of the mechanical properties of the untreated and treated cotton fabrics are given in Figure 3. From the graphs, it could be seen that samples of fabric became stiffer after treatment, with the maximum stress value being 60% higher for fabrics treated with TFPA as compared to the untreated fabric. It had also been observed that the treated cotton fabrics had a loss in strain and elongation at break values almost twice as compared to untreated fabric. It could be related to the adhesion between cotton fibers. In addition, the formation of polymeric bridges between cotton fibers due to polymerization could also be possible. This phenomenon had also been mentioned by Maity [15]. An interesting observation was the value of stress connected with the amount of fluorine atoms in monomers used for treatment. Monomers with higher fluorine content used for treatment leads to lower stiffness and stress value of the fabric. It could likely be related to the crystallinity of polymer formed on the cotton surface. Lower fluorine content and shorter length of monomer tends to result in higher crystallinity at the same degree of polymerization.

Conclusion

A thin coating with fluorinated polymer chemically attached to the cotton fabric was shown. The amount of fluorine in the final fabric is lower than 1%wt, which is lower when compared to other similar treatments, and yet the fabric is showing good water-repellent property. The cotton that was treated with 2,2,3,3,4,4,5,5,5,6,6,7,7-docecafluoroheptyl acrylate obtained the highest contact angle of 128°. Although modification of cotton can allow for better hydrophobicity, mechanical properties need not be sacrificed. It had been found that by making the length of the attached polymer longer, not only will the mechanical properties not be affected significantly but at the same time it gives the best result in terms of hydrophilicity. It can also be said that there will be a smaller environmental impact as less fluorine will go into the environment during the utilization steps of fluorinated cotton fabric

as well as the usage. The developed method opens a wide area allowing for coating with layers, using a combination of acrylic monomers for coating, making different lengths of hydrophobic polymers on the cotton, and creating other properties, for example antimicrobial, of fabric.

References

1. Xue CH, Jia ST, Zhang J, Tian LQ, Chen HZ, et al. (2008) Preparation of super hydrophobic surface on cotton textile. Sci Tech of Adv Mat 9: 1-7.

2. Shosha MH, Hilw ZH, Aly AA, Amr A, Al I, et al. (2008) Paraffin wax emulsion as water repellent for cotton/polyester blended fabric. J Ind Text 37: 315-325.

3. Przybylak M, Maciejewski H, Dutkiewicz A, Dabec I, Novicki M, et al. (2016) Fabrication of superhydrophobic cotton fabric by a simple chemical modification. Cellulose 23: 2185-2197.

4. Li Y, Zheng X, Xia Z, Lu M (2016) Synthesis of fluorinated block copolymer and superhydrophobic cotton fabric preparation. Prog Org Coat 97: 122-132.

5. Holmquist H, Schellenberger SI, Vander V, Peters GM, Leonards PEG, et al. (2016) Properties performance and associated hazards of state of the art durable water repellent (DWR) chemistry for textile finishing. Environment International 91: 251-264.

6. McKeen LW (2006) Fluorinated coating and finishes handbook. Wiliam Andrew Publication.

7. Mahltig B, Bottcher H (2002) Modified Silica Sol Coatings for Water-Repellent Textiles. J Solgel Sci Technol 27: 43-52.

8. Basu BJ, Bharathidasan T, Anandan C (2013) Superhydrophobic oleophobic PDMS-silica nanocomposite coating. Ice Science 1: 40-51.

9. Chen W, Fadeev AY, Hsieh MC, Oner D (1999) Ultra hydrophobic and Ultra Lyophobic Surfaces: Some Comments and Examples. Langmuir 15: 3395-3399.

10. Yao W, Li Y, Huang X (2014) Fluorinated polymethacrylate: Synthesis and properties. Polymer 55: 6197-6211.

11. Schondelmaier D, Cramm S, Klingeler R, Morenzin J, Zilkens C, et al. (2002) Orientation and Self-Assembly of Hydrophobic Fluoroalkylsilanes. Langmuir 18: 6242-6245.

12. Yang Y, Shen J, Zhang L, Li X (2015) Preparation of a novel water and oil-repellent fabric finishing agent containing a short perfluoroalkyl chain and its application in textile. Mater Res Innov 19: 401-404.

13. Shi Z, Wyman I, Liu G, Hu H, Zou H, et al. (2013) Preparation of water-repellent cotton fabric from fluorinate di block copolymers and evaluation of their durability. Polymer 54: 6406-6414.

14. Cai R, Deng B, Jiang H, Yu Y, Yu M, et al. (2012) Radiation induced graft polymerization of a fluorinated acrylate onto fabric. Radiat Phys Chem 81: 1354-1356.

15. Maity J, Kothary P, Rear EA, Jacob C (2010) Preparation and comparison of hydrophobic cotton fabric obtained by direct fluorination and micellar polymerization of fluoro monomers. Ind Eng Chem Res 49: 6075-6079.

16. Yu Y, Wang Q, Yuan J, Fan X, Wang P, et al. (2016) Hydrophobic modification of cotton fabric with octadecylamine via laccase/TEMPO mediated grafting. Carbo hyd. Polym 13: 549-555.

17. Grajeck EJ, Petersen WH (1962) Oil and water repellent fluorochemical finishes for cotton. Textile Research Journal 32: 320-331.

18. Kumar UN, Kratz K, Wagermaier W, Behl M, Lendlein A, et al. (2010) Non-contact actuation of triple-shape effect in multiphase polymer network nanocomposites in alternating magnetic field. J Mater Chem 20: 3404-3415.

19. Colleoni C, Guido E, Migani V, Rosace G (2015) Hydrophobic behaviour of non-fluorinated sol–gel based cotton and polyester fabric coatings. J Ind Text 44: 815-834.

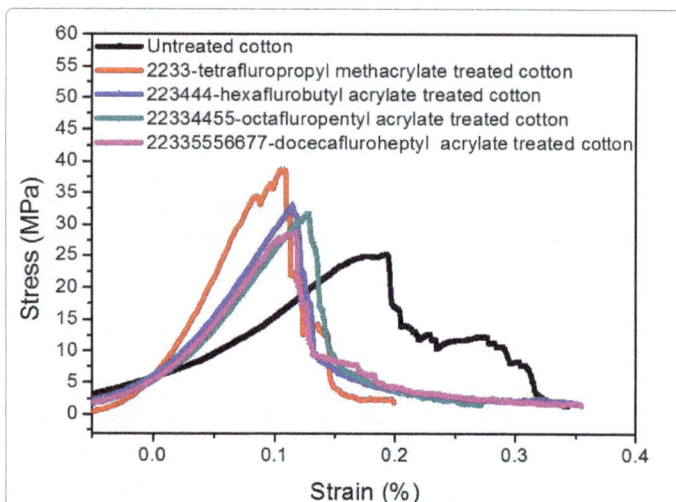

Figure 3: Stress-strain plot of untreated cotton fabric and fabric after treatment with different fluorine-containing monomers.

25

Determination of Tellurium in Tellurium-Containing Organic Compounds by Microwave Plasma-Atomic Emission Spectrometry

Lastovka AV[1,2], Fadeeva VP[1,2], Bazhenov MA[1,2] and Tikhova VD[1*]

[1]NN Vorozhtsov Novosibirsk, Institute of Organic Chemistry, Siberian Branch of Russian Academy of Sciences, Russia
[2]Novosibirsk National Research State University, Russia

Abstract

A method for determining tellurium in tellurium-containing organic compounds (TOC) by a microwave plasma-atomic emission spectrometer Agilent 4100 has been suggested. One of the sample decomposition methods - the oxygen flask combustion or the acid decomposition in a heating block-can be applied. Telluradiazole derivatives and some other TOC with 20 to 58% tellurium content have been analyzed. Elements such as nitrogen, sulfur, potassium, selenium present in the TOC do not prevent from determining tellurium. The relative error of the analysis is 1-5%.

Graphical Abstract (GA)

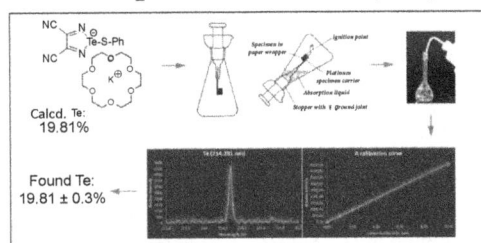

Telluradiazole derivatives and some other tellurium-containing organic compounds with 20 to 58% tellurium content have been analysed by MP-AES spectrometer Agilent 4100

Keywords: Tellurium containing organic compounds; Sample decomposition; Microwave plasma atomic; Emission spectrometry; Agilent 4100

Introduction

Chemistry of tellurium organic compounds (TOC) has intensively been developing in recent decades. The increased interest in this field is due to widely use of these substances as catalysts in organic reactions (e.g., rubber vulcanization), inhibitors of metal corrosion, insecticides, fungicides, components of special photographic materials, pharmaceuticals etc [1-5]. In view of great practical importance of these compounds for science and technology, a large number of analytical methods for the determination of tellurium in various objects were developed: gravimetric methods based on the precipitation of elemental tellurium by treating with inorganic and organic reducing agents, [6-8]; titrimetric methods based on redox reactions using iodide-anion, salts of iron(II), chromium(II) or titanium(III) as reducing agents, as well as potassium permanganate or dichromate as oxidizing agents for Te (IV), [9-11]; electrochemical methods, which allow one to determine the tellurium and selenium at simultaneous presence with a sensitivity being 10^{-5} M, [12-14]; spectrophotometric methods based on the formation of complexes with sulfur-containing organic reagents (diethyldithiocarbamate, thiourea, and their derivatives, bismuthiol II, and so on [15,16] or on the formation of ion pairs of tellurium acido complexes with organic bases-pyrazolone derivatives and rhodamine dyes [17-19]. Three spectrophotometric methods were used to determine small tellurium content in the environment (water, plant material, soil), and thin telluride films [20]. These methods are based on the formation of three differently colored Te-complexes. Magenta color was obtained by oxidation of 4-bromophenyl-hydrazine by tellurium in a basic medium followed by condensation of the product with N-(1-naphthyl) etylenediamine dihydrochloride. Red color was obtained by oxidation of 3-methyl-2-benzothiazolinone hydrazone hydrochloride followed by condensation with the chromotropic acid. Orange color

was obtained by oxidation of 2,3-dimethoxystrichnidin-10-one in an acid medium. Beer's law was obeyed in the range of 1.0–25 mg mL^{-1} for magenta product, 0.7–20 mg mL^{-1} for red, and 0.3–15 mg mL^{-1} for orange. The authors noted that these methods were simpler and more sensitive than the other known ones, and standard deviations were 0.2–0.3%. Several dyes were proposed for the extraction-photometric determination of tellurium: victoria blue 4R, brilliant green, rhodamine 4C. The latter dye was used, for example, in the analyses of tellurium salt solutions and semiconductor films [21]. The extraction-photometric redox process was described for the selective determination of selenium(IV) and tellurium(IV) with the detection limit of $5 \cdot 10^{-5}$ mg L^{-1}, based on different oxidability of analytes by antimony(V) ion associates in acidic or alkaline media [22]. The method does not require either preconcentrating and separating analytes, or masking the associated components in analyte mixtures.The possibility of using benzal green as a reagent for the rapid determination of tellurium(IV) is shown by Amelin et al [23]. The dye was immobilized on viscose fabrics and it formed ion associate with molibdotellurous heteropolyacid with different color intensity. The range of detectable concentrations is 0.01-0.1 mg L^{-1} of Te(IV). Currently, spectrometric methods are most commonly used for the tellurium determination. Atomic absorption spectrometry was used in the analysis of geological samples (ores,

---

*Corresponding author: Tikhova VD, NN Vorozhtsov Novosibirsk, Institute of Organic Chemistry, Siberian Branch of Russian Academy of Science, Russia
E-mail: tikhova@nioch.nsc.ru

minerals, rocks) with tellurium content 0.0-100 mg g^{-1}, [24], coal, [25], in lead and lead alloys [26], in the environment (water, plants, biological materials) [27-29]. Such techniques as atomic emission spectrometry (AES) with inductively coupled (ICP) or microwave (MP) plasma, in conjunction with mass spectrometry, [30-33], atomic fluorescence spectrometry, [34-36], the optical emission spectrometry [37] are used for the tellurium and tellurium trace analysis in various objects. In most of these works, tellurium is converted to the hydride before the spectrometric measurements to increase the sensitivity of the method. The atomic emission method with continuous hydride generation was used for the simultaneous determination of Sb, As, Bi, Ge, Se, and Te in a set of sediment samples [38]. The detection limit for Te was 6.5µg L^{-1}. The tellurium determination was made in the gold concentrate by ICP-AES [39]. It was shown that tellurium may be separated from the majority of interfering components (such as Ag, Pb, Zn, Si, Ca, Mg, Al, Fe) in the sample by coprecipitation with arsenic to eliminate the matrix effect. The detection limit of Te was 4.23 µg g^{-1}. Zhang et al. [40] investigated the possibility of group preconcentration of As, Cd, Se, and Te for the subsequent simultaneous determination of these elements by ICP-AES in the technogenic raw materials. Sorption of arsenic, selenium, tellurium, and cadmium oxyanions on iron, lanthanum, and magnesium hydroxides was used. The relative standard deviation of the method was 9–2.5% under the content of determined elements from 10^{-4} to 10^{-1}% mass. It was found that the presence of iron (> 500 µg mL^{-1}) can lead to 10% decrease of the Te (214.2 nm) signal. When new organotellurium compounds are synthesized they should be completely described. It requires quantitative determination of so-called basic biogenic elements (carbon, hydrogen, nitrogen) as well as tellurium. Unlike small tellurium content in the above-mentioned inorganic, biological and ecological objects of analysis, tellurium content in synthetic TOC may be up to tens of percent's. The first necessary step in TOC analysis is a complete decomposition of organic matter. In the few published studies on the analysis of TOC, they used wet decomposition in mineral acids, [41-43], burning in a stream of oxygen at 900950°C or the oxygen flask combustion. Afterwards, tellurium was determined by the methods of gravimetry potentiometry, [42], spectrophotometry [43]. The main disadvantages of all these methods are their long duration and tediousness. MP-AES method for the determination of tellurium in TOC has not previously been used, although it is characterized by high accuracy and selectivity, rapidity, efficiency, wide range of detectable concentrations, and it does not involve converting tellurium in a single analytical form. The aim of this work is finding the conditions for sample preparation (decomposition) and developing a procedure for quantitative determination of tellurium in synthetic multi-element tellurium-containing organic compounds by microwave plasma-atomic emission spectrometry using the Agilent 4100 spectrometer.

Experimental

Instruments. Analysis was performed with Agilent 4100 MP-AES (Australia) microwave plasma – atomic emission spectrometer. The sample supply system was a self-adjusted one-piece quartz torch. Plasma was generated by igniting with auxiliary argon and maintained with nitrogen. The configuration of the plasma was vertically-oriented with axial observation and computer-control of the plasma viewing position. A Curny-Terner monochromator, self-adjusted, with a 600 mm focal length was used. Holographic gitter with 2400 lines mm^{-1} was used. The spectral range was 178–780 nm and spectral resolution was 0.050 nm. A High-speed CCD detector, having 532 × 128 pixels, with a quantum efficiency of >90% was used; it was cooled to 0°C with a Peltier element. A double-pass glass cyclonic spray chamber and a One-Neb micronebulizer were used. Hot Block™ Digestion System (25-Well

100 ml) Agilent 4107 (Australia) was applied for acid decomposition of organic samples. The samples and reagents were weighed with Mettler Toledo AT20 (Switzerland) microanalytical balance and Sartorius CP224S (Germany) analytical balance. Reagents. Solutions of H_2O_2, KNO_3, HCl, HNO_3 (by Reakhim) were prepared from reagent-grade and high-purity chemicals. Oxygen gas from a cylinder with a purity of 99.7% was used for burning. The specimen holder was made of platinum wire (∅=0.6–1.0 mm) with a purity of 99.99%. Blue ribbon ash-free filters were cut into 3 × 3 cm squares with 1-cm branches, impregnated with a saturated KNO_3 solution and dried; an H-type polyethylene film; glycerol, polyurethane foam, twice-distilled water. Reference substance – Certified reference standard sample of tellurium(IV) ions solution-MCO 0529: 2003 (GSO 6082-91) (Russia).

Results and Discussion

Analytical wavelength selection. Comparative measurements of the Te standard solutions were made at 3 wavelengths: 214.281; 225.902; 238.578 nm. The calibration curves were constructed followed by measurement of the Te solution (5 ppm) (**Table 1**). It is clear from **Table 1** that only two emission lines are appropriate for the measurements. Atomic line 214.281 nm was selected due to its maximal intensity, but the measurement at this wavelength should be carried out after the full nitrogen purge of monochromator. Although this emission line is influenced by vanadium, tantalum, and cadmium, however, such elements are not encountered in the studied synthetic organic compounds. The study of sample decomposition. A tellurium-containing organic compound, 3,4-dicyano-1,2,5-telluradiazole (C_4N_4Te), was used to study the methods of sample preparation. Composition and structure of this substance is confirmed by elemental analysis, NMR, IR and MS spectrometry. The purity of the compound is 99.9%. When using the oxygen flask combustion, we tested 3 different solutions for absorbing the combustion products: 1) 10 mL of 0.1 M HCl; 2) 3 mL of HNO_3 (conc.)+5 mL H_2O+2 mL H_2O_2; 3) 10 mL of HNO_3 (conc.). The best results were obtained with absorbing mixtures 1) and 3). When using the nitric acid (8 mL HNO_3 (conc.)+2 mL of 30% H_2O_2) decomposition in the heating block, we experimentally selected the temperature of the reaction (from 60 to 130 °C) and the duration (45 to 60 min) of the process. Finally, satisfactory results were obtained after decomposition of the sample for one hour at 130°C. The following results were obtained in the discussed conditions (P=0.95; n=6): with the oxygen flask combustion –55.5 ± 1.7%; with the acid decomposition in the heating block –55.5 ± 0.3%. The calculated tellurium content in 3,4-dicyano-1,2,5-tellurodiazole is 55.1%. Therefore, both methods can be used to decompose the TOC when analyzing tellurium.

Procedures for the decomposition of tellurium-containing organic compounds

The oxygen flask combustion: A portion of 0.1M HCl (10 mL) was placed into a 500 mL Erlenmeyer's flask which was filled with oxygen from a cylinder. A weighed portion of a tellurium-containing compound (~2.0–3.0 mg) wrapped into a small square of a potassium nitrate-impregnated ash-free filter with a piece of polyethylene film was fixed on a Pt wire; the branch of the filter was ignited, placed into the flask, and burned. The combustion having been completed, the flask was thoroughly shaken for 5 min and placed into a fridge for 40 min to complete the absorption of combustion products. The acid decomposition in a heating block: A weighed portion of a tellurium-containing compound (~2.0–3.0 mg) was placed in polypropylene 100 mL vessel. HNO_3 (8 mL) and 30% H_2O_2 (2 mL) were added to the probe. The vessel was covered with a lid (not tightly) and placed in a heating block HotBlock™ Digestion System, Agilent 4107. It took 60 min at 130°C to complete decomposition of the substance.

λ, nm	214.281	225.902	238.578
I, rel.	4885.9	1147.2	2127.8
The equation of the calibration curve	565.81 c_i-3.46	(80.98 c_i)/(1-0.04 c_i)	201.37 c_i-4.81
Approximation	linear	rational	linear
r (correlation coefficient)	1.00000	1.00000	0.99999
Found Te, ppm	5.04 ± 0.02	4.33 ± 0.11	5.08 ± 0.04

Table 1: The results of determining 5 ppm Te at three emission lines (P=0.95; n=7).

Substance	Empirical formula	Calculated, % Te	Found, % Te	
			The oxygen flask combustion	The acid decomposition in heating block
1-Thiophenyl-3,4-dicyano-1,2,5-tellurodiazole 18-crown-6 potassium salt complex	$C_{22}H_{29}KN_4O_6STe$	19.81	19.81 ± 0.3	19.92 ± 0.3
4,4'-(Tellurobis(propane-3,1-diyl)) bis(2,6-di-tert-buthylphenol)	$C_{34}H_{54}O_2Te$	20.50	20.72 ± 0.6	*
Tetrabuthylammonium 1-bromo-3,4-dicyano-1,2,5-telluradiazolide	$C_{20}H_{36}N_5BrTe$	23.0	23.3 ± 1.0	23,6 ± 1.7
bis(Diethylamino-thioxo-methanesulfenyl)tellane	$C_{10}H_{20}N_2S_4Te$	30.08	30.22 ± 0.13	29.9 ± 0.3
Potassium 1-isoselenocyanato-3,4-dicyano-1,2,5-telluradiazolide	C_5N_5KSeTe	34.0	*	33.5 ± 1.2
3,4-dicyano-1,2,5-tellurodiazole	C_4N_4Te	55.1	55.5 ± 1.7	55.5 ± 0.3
2,2'-Ditellanediyldianiline	$C_{12}H_{12}N_2Te_2$	58.07	57.99 ± 0.18	57.89 ± 0.05

Table 2: Results of tellurium determination in TOC using two methods of sample preparation (P=0.95, n=5).

After the decomposition, the content of the flask or vessel were quantitatively transferred to a 100 mL volumetric flask rinsing its walls, stopper, wire or vessel with twice-distilled water and diluted to the mark. The tellurium content in the resulting solution was determined by MP AES using a preliminarily constructed calibration curve. Construction of calibration curves. The calibration curve was made using the solutions obtained with dilution of Certified reference standard sample of tellurium(IV) ions solution MCO 0529:2003 (GSO 6082-91) (Russia). We used 4 concentrations of Te(IV)–1, 5, 10, 15 ppm. The intensity of Te emission line at 214.241 nm was recorded on the spectrometer. The data were processed using the Agilent MP Expert software (version 1.4.0.4317). The conditions of measurements on the spectrometer were as follows: pressure in the spray chamber was 100 kPa, uptake delay–15 s, read time–3 s, stabilization time–15 s, pump speed–15 ppm, number of replicates–3, automatic background correction. A linear intensity dependence of the recorded signal ($r \geq 0.99998$) was observed for reference substance solutions in this concentration range. Various tellurium-containing organic compounds have been analyzed according to the developed technique (**Table 2**).

Conclusion

Thus, the method for the tellurium determination has been proposed. It combines the oxygen flask combustion or acid decomposition followed by measuring Te content by the MP-AES method. Such elements as nitrogen, sulfur, potassium, selenium, included in the TOC do not influence the quantitative determination of tellurium. Statistical processing was performed on the results obtained by the decomposition of 5–6 weighed portions of each substance. The relative error is in the range of 1–5%.

Acknowledgements

We are grateful to Prof. AV Zibarev and employees of the Laboratory of Heterocyclic Compounds (NIOCH SB RAS) for the samples provided for our study.

References

1. Comasseto J, Barrientos AR (2000) Add a Little Tellurium to Your Synthetic Plans. Aldrichimica Acta 33: 66-102

2. Petragrani N, Stefani H (2005) Advances in Organic Tellurium Chemistry. Tetrahedron 61: 1613-1679.

3. Petragrani N, Stefani H (2007) Tellurium in Organic Synthesis. London, UK: Academic Press.

4. Princival J, Dos SA, Comasseto J (2010) Reactive Organometallics from Organo tellurides: Application in Organic Synthesis. J Braz Chem Soc 21: 2042-2054.

5. Wei C, Lu W, Huaping Xu (2015) Selenium/Tellurium Containing Polymer Materials in Nanobiotechnology. Nano Today 10: 717-736.

6. Wang J, Cheng K (1970) Precipitation of Tellurium with Bismuthiol II. Microchem J 15: 607-621.

7. Asuero A (1979) Analytical Uses of Phenylhydrazines and Phenylhydrazones. I Microchem J 24: 217-233.

8. Singh N, Rastogi K, Agrawal R. (1985) Separation and Determination of Selenium(IV) and Tellurium(IV) by using Morpholine-4-carbodithioate. J Indian Chem Soc 62: 394-396.

9. Johnson R, Frederickson D (1952) Titration of Quadrivalent Tellurium with Thiosulfate. Anal Chem 24: 866-867.

10. Geiersberger K, Durst A (1952) A contribution to tellurium selenium analysis. Z Analyt Chem 135: 11-14.

11. Lenher V, Wakefield H (1923) The Volumetric Determination of Tellurium by Dichromate Method. J Am Chem Soc 45: 1423-1425.

12. Ivanova Z, Ignatenko E, Tarasova V (1973) Potentiometric and Amperometric Titration of Selenium and Tellurium using Ascorbic Acid. J Anal Chem 28: 1980-1984.

13. Locatelli C (2005)Overlapping Voltamperic Peaks - an Analytical Procedure for Simultaneous Determination of Trace Metals. Application to Food and Environmental Matrices. Anal Bioanal Chem 381: 1073-1081.

14. Terpinski E (1988) Spectrophotometric Determination of Trace Amounts of Tellurium(IV) with Tetramethylthiourea. Analyst 113: 1473-1475.

15. Yoshida H, Taga M (1966) Spectrophotometric Determination of Small Amounts of Tellurium with Bismuthiol II. Talanta 13: 185-191.

16. Nazarenko I, Ermakov A (1971) Analiticheskaya Chimiya Selena i Tellura. (Analytical Chemistry of Selenium and Tellurium). Moscow, USSR.

17. Balogh I, Andruch V(1999) Comparative Spectrophotometric Study of the Complexation and Extraction of Tellurium with Various Halide Ions and N,N'-di(acetoxyethyl)indocarbocyanine. Analyt Chim Acta 386: 161-167.

18. Suvardhan K, Krishna P, Puttaiah E, Chiranjeevi P (2007) Spectrometric Determination of Tellurium(IV) in Environmental and Telluride Film Samples. J Analyt Chem 62: 1032-1039.

19. Kish P, Balog I, Andrukh V, Goloib M (1990) Complex Formation and Solvent Extraction of Chlorotellurite with Cationic Violet. Zh Anal Khim 45: 915-919.

20. Sergeev G, Shlyapunova E. (2006) Highly Sensitive Determination of Selenium(IV) and Tellurium(IV) using Extraction-Photometric Redox Method. Analitika i Kontrol 10: 195–199.

21. Amelin V, Koroleva O (2009) Fabrics and Papers Modified with Analytical Reagents for the Test Determination of Selenium(IV) and Tellurium(IV). J Anal Chem 64: 1275-1278.

22. Torgov V, Vall G, Demidova M, Yatsenko V(1995) Extraction-Atomic Absorption Method for the Determination of Arsenic, Antimony, Selenium and Tellurium in Geological Samples. Chemical Geology 124: 101-107.

23. Oda S, Arikawa Y (2005) Determination of Tellurium in Coal Samples by Means of Graphite Furnace Atomic Absorption Spectrometry after Coprecipitation with Iron(III) Hydroxyde. Bunseki Kagaku 54: 1033-1037.

24. Mesko M, Pozebon D, Flores E, Dressler V (2004) Determination of Tellurium in Lead and Lead Alloy using Flow Injection–Hydride Generation Atomic Absorption Spectrometry. Analyt Chim Acta 517: 195-200.

25. Najafi N, Tavakoli H, Alizadeh R, Seidi S (2010) Speciation and Determination of Ultra Trace Amounts of Inorganic Tellurium in Environmental Water Samples by Dispersive Liquid-Liquid Microextraction and Electrothermal Atomic Absorption Spectrometry. Analyt Chim Acta 670: 18-23.

26. Kaplan M, Cerutti S, Salonin J, Gasguez J, Martinez L, et al. (2005) Preconcentration and Determination of Tellurium in Garlic Samples by Hydride Generation Atomic Absorption Spectrometry. J of AOAC International 88: 1242-1246.

27. Grotti M, Abelmoschi M, Soggia F, Frache R (2003) Determination of Ultratrace Elements in Natural Waters by Solid-Phase Extraction and atomic spectrometry methods. Anal Bioanal Chem 375: 242-247.

28. Yu C, Cai Q, Guo Z, Yang Z, Khoo S, et al. (2003) Speciation analysis of tellurium by solid-phase extraction in the presence of ammonium pyrrolidine dithiocarbamate and inductively coupled plasma mass spectrometry. Analyt Bioanal Chem 376: 236-242.

29. Quadros D, Borges D (2014) Direct analysis of alcoholic beverages for the determination of cobalt, nickel and tellurium by inductively coupled plasma mass spectrometry following photochemical vapor generation. Microchem J 116: 244-248.

30. Yang G, Zheng J, Tagami K, Uchida S (2013) Rapid and sensitive determination of tellurium in soil and plant samples by sector-field inductively coupled plasma mass spectrometry. Talanta 116: 181-187.

31. Yu S, Zhang H, Jin Q (1999) Study on determination of Tellurium by MIPAES. Gaodeng Xuexiao Huaxue Xuebao 11: 84-86.

32. Cava MP, Cervera M, Pastor A, Guardia M (2003) Hydride generation atomic fluorescence spectrometric determination of ultratraces of selenium and tellurium in cow milk. Analyt Chim Acta 481: 291-300.

33. Wang F, Zhang G (2011) Simultaneous Quantitative Analysis of Arsenic, Bismuth, Selenium, and Tellurium in Soil Samples using Multi-channel Hydride-Generation Atomic Fluorescence Spectrometry. Appl Spectr 65: 315-319.

34. Chen Y, Alzahrani A, Deng T, Belzile N (2016) Valence Properties of Tellurium in Different Chemical Systems and Its Determination in Refractory Environmental Samples using Hydride Generation Atomic Fluorescence Spectroscopy. Analyt Chim Acta 905: 42-50.

35. Kaplan M, Cerutti S, Moyano S, Olsina R(2004) On-line Preconcentration System by Coprecipitation with Lanthanum Hydroxide using Packed-bed Filter for Determination of Tellurium in Water by ICP-OES with USN. Instrumentation Sciences&Technology 32: 423-431.

36. Morrow A, Wiltshire G, Hursthouse A (1997) An Improved Method for the Simultaneous of Sb, As, Bi, Ge, Se, and Te by Hydride Generation ICP-AES: Application to Environmental Samples. Atomic Spectroscopy 18: 23-28.

37. Zhang H, Ni W, Xiao F, Mao X (2016) Determination of Selenium and Tellurium in Gold Concentrate by Arsenic Coprecipitation-Inductively Coupled Plasma Atomic Emission Spectrometry. Yejin Fenxi/Metallurgical Analysis 36: 32-36.

38. Doronina M, Shiryaeva O, Filatova D, Baranovskaya V (2013) Determination of Arsenic, Cadmium, Selenium, and Tellurium in the Technogenic Raw Materials after the Sorption Concentration on Hydroxydes using Atomic Emission Spectrometry with Inductively Coupled Plasma. Zavodskaya Laboratoriya Diagnostika Materialov 79: 3-7.

39. Kruse F, Sanftner R, Suttle J (1953) Volumetric Determination of Tellurium in Organic Compounds. Anal Chem 25: 500-502.

40. Masson M (1976) Microdetermination of Selenium, Tellurium, and Arsenic in Organic Compounds. Mikrochim Acta 65: 399-411.

41. Thavornyutikarn P (1973) Tellurium Analysis in Organotellurium Compounds by Atomic Absorption Spectroscopy. J Organomet Chem 51: 237-239.

42. Anisimova G, Klimova V (1980) Tellurium Determination in Organic Compounds. J Anal Chem 35: 607-609.

43. Clark E, Turaihi M (1976) A Rapid Procedure for the Determination of the Tellurium Content of Organotellurium Compounds. J Organomet Chem 118: 55-58.

26

Synthesis, Characterization and Antioxidant Activity of Carvacrol Containing Novel Thiadiazole and Oxadiazole Moieties

Suresh DB, Jamatsing DR, Pravin SK and Ratnamala SB*

School of Chemical Sciences, North Maharashtra University, Jalgaon, Maharashtra, India

Abstract

Carvacrol is a well-known antioxidant found in the extract of various angiospermic plants. The purpose of present research is to synthesize new carvacrol derivatives associated with heterocycles namely 1,3,4-thiadiazole and 1,3,4-oxadiazole to explore their extraordinary potential in medicine and agriculture. Structures of newly synthesized compounds were confirmed by spectroscopic techniques such as FT-IR, ^1H and ^{13}C NMR and LC-MS. Finally, synthesized derivatives were evaluated for their *in-vitro* antioxidant activity by using radical scavenger DPPH assay. All the compounds exhibited remarkable antioxidant activity, out of which compound showed better or similar antioxidant activity compared to standard compound ascorbic acid.

Keywords: Carvacrol; 1,3,4-thiadiazole; 1,3,4-oxadiazole; Antioxidant; DPPH; Ascorbic acid

Introduction

Several Reactive Oxygen Species (ROS) are important cellular components, enzymatically generated in aerobic living organisms, which show significant role in various physiological and pathological processes [1,2]. In contrast, the accumulation of excessive ROS, mostly due to external influences such as radiation, cigarette smoke, ultraviolet light, drugs, pathogens, etc can cause damage upon cellular macromolecules like, DNA, proteins and lipids, therefore contributing to the development of various diseases such as, atherosclerosis, myocardial infarction, ischemia, epilepsy, diabetes mellitus, anemia and carcinogenesis [3-5]. To overcome this, investigates aimed at the synthesis of new antioxidants with better properties from a pharmacological point of view have been performed [6] have synthesized of 1,3,4-oxadiazoles containing 4-(methyl sulfonyl) benzyl moiety and evaluated their *in-vitro* antioxidant activity [7]. Similarly, have reported that 1,3,4-oxadiazole tagged thieno [2,3-d] pyrimidine derivatives having significant radical scavenger activity [8].

The prevalent existence of the heterocycles in bioactive natural products, drugs, and agrochemicals has made them as important synthetic targets [9]. Five-membered heterocyclic compounds; oxadiazoles and thiadiazoles have attracted significant interest in medicinal chemistry, pesticide chemistry, polymer sciences, material science and they are the building blocks of new molecular systems for biologically active molecules [10]. The Nitrogen-oxygen heterocycles are also of synthetic interest as they constitute an important class of natural and non-natural products and many of them exhibit useful biological activities [11,12]. 1,3,4-oxadiazoles are biologically versatile compounds displaying a variety of biological effects which include antifungal [13], bactericidal [14], analgesic and anti-inflammatory [15]. 1,3,4-thiadiazoles also possess various biological properties such as antitumor, anticonvulsant, antihypertensive, anesthetic, antibacterial and cardiotonic activities [16-18]. Carvacrol (5-isopropyl-3-methylphenol) is a major constituent of oregano oil [19]. Likewise, the phenolic monoterpenoids found in essential oils of many plants also possess different biological activities [20,21]. Therefore, we choose carvacrol as starting material for our research. Keeping in view the importance of oxadiazole and Thiadiazole in biology, and in continuation of our ongoing research on biologically active molecules, we have prepared oxadiazole and thiadiazole derivatives of naturally occurring phenolic antioxidant [22,23]. In this regard, it is important to search for and synthesize new classes of compounds that have antioxidant properties.

Experimental

Melting points of all the synthesized compounds were determined by open capillary method. The reaction was monitored by ascending thin layer chromatography (TLC) and was performed on 200 µm thick aluminum sheets having silica gel 60 F_{254} as adsorbent. The solvent system used for developing the TLC plate was hexane and ethyl acetate (3:1). Spots were visualized under UV-light. ^1H nuclear magnetic resonance (^1H NMR) and ^{13}C NMR spectra were scanned at 400 MHz and 100 MHz respectively on Varian Mercury YH-300 FT NMR in DMSO-d_6. Chemical shift values (δ) are given in parts per million (ppm).

The antioxidant activity of the synthesized compounds was performed by 2,2-diphenyl-1-picrylhydrazyl (DPPH) free radical scavenging assay [24]. All the well characterized derivatives of carvacrol were dissolved to prepare a stock solution of 1 mg/mL using DMSO. Fifty microliter solutions of the compounds were added to 1 mL of a 0.1 mM solution of DPPH in methanol. After 2 h, absorbance values were measured at 517 nm. Ascorbic acid was used as standard.

Synthesis

The synthesis of the compounds was carried out according to the procedure outlined in Scheme 1. Firstly, carvacrol (Figure 1) was converted to its 4-nitroso derivatives (Supplementary Figures 1 and 2) by treating ethanolic solution of phenol with concentrated hydrochloric acid and sodium nitrite at 0°C [25]. The nitroso compound was reduced in an ammonical solution by passing H_2S gas [26]. The 4-aminophenol (Supplementary Figure 3) prepared according to Scheme 1 was further reacted with acetic anhydride to form acetamide (Supplementary Figure 4) derivative [27]. Acetamide was treated with ethyl-bromoacetate to obtain ester type compound (Supplementary Figure 5) [28] and further converted to their hydrazide compound (Supplementary Figure 6) by reported methods [29].

---

*Corresponding author: Ratnamala S Bendre, School of Chemical Sciences, North Maharashtra University, Jalgaon, Maharashtra, India
E-mail: bendrers@gmail.com

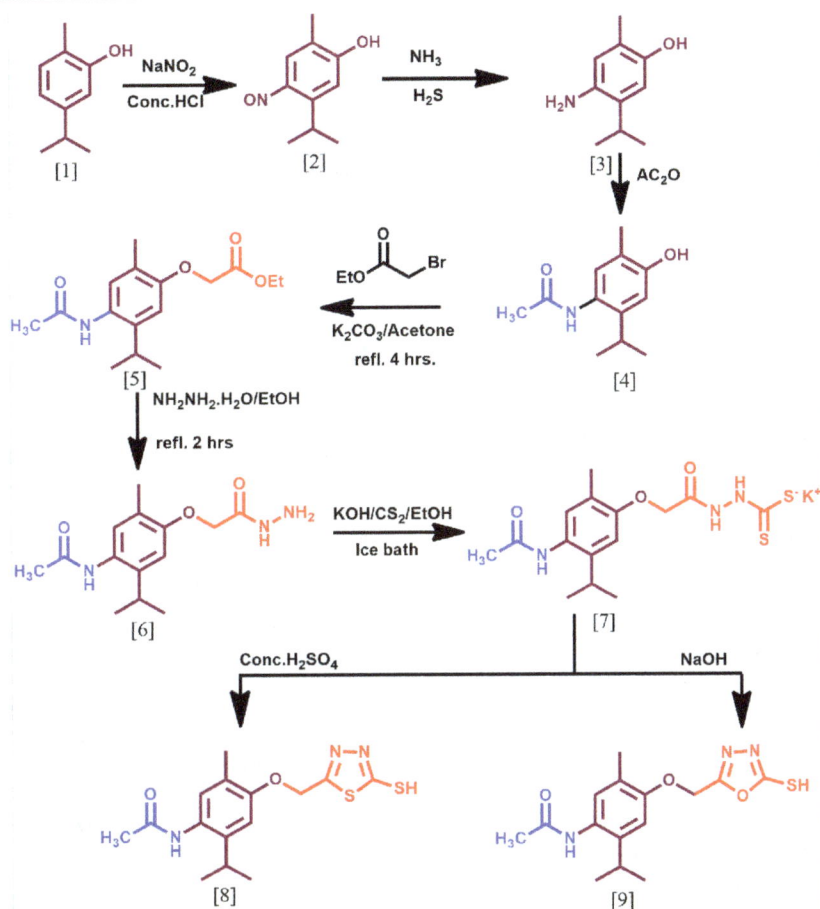

Scheme 1: Reaction program.

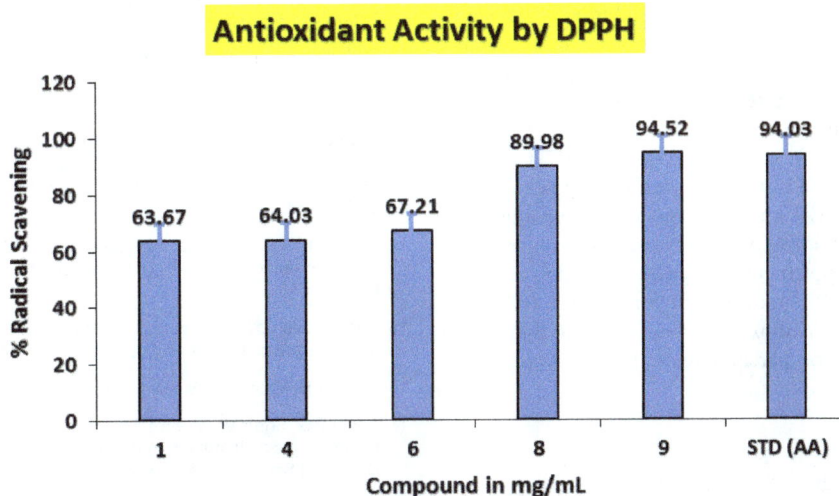

Figure 1: *In vitro* antioxidant activity of synthesized compounds by using DPPH radical scavenging assay (where Std (AA): ascorbic acid).

General procedure for synthesis of potassium-2-(2-(4-acetamido-5-isopropyl-2-methylphenoxy) acetyl) hydrazine carbodithiaoate

The recrystallized product of hydrazide (0.0035 moles) in the absolute ethanol (50 ml) was transferred to a conical flask and this reaction mixture was kept in ice bath on magnetic stirrer, care was taken to keep the temperature below 0°C for 1 hr. To this reaction mixture KOH (0.0071 moles) was added and it was further stirred vigorously for 10 min. Then CS$_2$ (0.0071 moles) was added drop by drop after 5-10 min interval. After 1 hr. excess amount of CS$_2$ was added and then precipitate formed is nothing but the hydrazine salt.

The whitish colored salt is filtered off and dried under UV light [30] (Supplementary Figure 7).

General procedure for synthesis of N-(2-isopropyl-4-((5-mercapto-1,3,4-thiadiazol-2-yl)-5-methoxy)-5-methylphenyl) acetamide

The dry and pure product of hydrazide salt is taken in the round bottom flask and the cold conc. H_2SO_4 at 5°C is poured in the round bottom flask. This reaction mixture was stirred on magnetic stirrer at room temperature for about 5-6 hrs. Thereafter the reaction mixture was poured in a beaker containing crushed ice and stirred vigorously for 5-10 min. The pink colored product formed is filtered off and washed with cold water and dried in UV light [30] (Supplementary Figure 8).

General procedure for synthesis of N-(2-isopropyl-4-((5-mercapto-1,3,4-oxadiazol-2-yl)-5-methoxy)5-methylphenyl) acetamide

The dry and pure product of hydrazide salt is placed in round bottom flask and NaOH solution was added into it. This reaction mixture was stirred on magnetic stirrer for about 5-6 hrs. After 5-6 hrs. the stirring was stopped and this reaction mixture was poured in beaker containing crush ice pieces and stirred for 5-10 min. to solidify the product. This product is filtered out and washed with cold water and dried in UV light [30] (Supplementary Figure 9).

Results and Discussion

In-vitro antioxidant activity

Antioxidant activity of organic molecules is related to their electron or hydrogen atom donating ability to DPPH radical, so that they become stable diamagnetic scaffolds. The interaction of synthesized compounds with stable DPPH free radical indicates their free radical scavenging ability. The reduction ability of DPPH radicals was determined by decline in their absorbance at 517 nm enthused by antioxidants [31]. Majority of the tested compounds in these series showed good interaction with the DPPH radical at 1 mg/mL concentration. The scavenging effects of all the synthesized compounds on DPPH radical are presented as % inhibition in (Supplementary Figure 1). DPPH radical scavenging activity of the synthesized compounds exhibited outstanding results as compared to the standard Ascorbic acid. Maximum DPPH radical scavenging activity was observed in compounds (Supplementary Figures 8 and 9) (89.98 and 94.52%) which is higher or comparable with standard antioxidant ascorbic acid (94.03%) at the same concentration. It appears that compounds (Supplementary Figures 1, 4 and 6) are also significant scavengers of the DPPH radical with % inhibition (63.67%, 64.03%, 67.21% respectively). In Antioxidant results showed that good efficacy and derivatization of the parent compound has resulted in good antioxidant efficacy.

Conclusion

In conclusion, we have achieved a convenient protocol for the synthesis 1,3,4-thiadiazole and oxadiazole incorporated carvacrol moiety in good yield and evaluated their in vitro antioxidant activity by using DPPH radical scavenger assay. Our antioxidant screening results indicate that exciting DPPH radical scavenging activity was observed in compounds (Supplementary Figures 8 and 9) in comparison with standard ascorbic acid. The lead compounds emerging with the most potent antioxidant activity in this study (Supplementary Figures 8 and 9) will be further structurally modified towards the discovery of a compound with optimal antioxidant activity. These results may also provide some significance guidance for the development of new class antioxidant.

Acknowledgements

The Authors extend their appreciation to UGC SAP (DSA-I) New Delhi for funding the research work.

References

1. Dautréaux B, Toledano MB (2007) ROS as signaling molecules mechanisms that generate specificity in ROS homeostasis. Nat Rev Mol Cell Biol 8: 813-824.

2. Ischiropoulos H (2002) Introduction to serial reviews reactive nitrogen species, tyrosine nitration and cell signaling 1, 2. Free Radical Biology and Medicine 33: 727.

3. Butterfield DA, Drake J, Pocernich C, Castegna A (2001) Evidence of oxidative damage in Alzheimer's disease brain: central role for amyloid β-peptide. Trends in molecular medicine 7: 548-554.

4. Uttara B, Singh AV, Zamboni P, Mahajan RT (2009) Oxidative stress and neurodegenerative diseases: a review of upstream and downstream antioxidant therapeutic options. Current neuropharmacology 7: 65-74.

5. Adly AA (2010) Oxidative stress and disease: an updated review. Res J Immunol 3: 129-145.

6. Miliovsky M, Svinyarov I, Prokopova E, Batovska D (2015) Synthesis and antioxidant activity of polyhydroxylated trans-restricted 2-arylcinnamic acids. Molecules 20: 2555-2575.

7. Vittal S, Poojary B, Bansal P, Nandagokula C (2011) Synthesis, characterization, and antioxidant activity of some 1, 3, 4-oxadiazoles carrying 4-(methylsulfonyl) benzyl moiety. Der Pharma Chemica 3: 138-146.

8. Kotaiah Y, Harikrishna N, Nagaraju, K, Rao CV (2012) Synthesis and antioxidant activity of 1, 3, 4-oxadiazole tagged thieno [2, 3-d] pyrimidine derivatives. European Journal of Medicinal Chemistry 58: 340-345.

9. He DH, Zhu YC, Yang ZR, Hu AX (2009) Synthesis and Characterization of Novel Stilbene Derivatives with 1, 3, 4-Oxadiazole Unit. Journal of the Chinese Chemical Society 56: 268-270.

10. Tactics for Management (1986) National Academic Press, Washington DC, p: 157.

11. Adib M, Jahromi AH, Tavoosi N, Mahdavi M, Bijanzadeh HR, et al. (2006) Microwave-assisted efficient, one-pot, three-component synthesis of 3, 5-disubstituted 1, 2, 4-oxadiazoles under solvent-free conditions. Tetrahedron letters 47: 2965-2967.

12. Swinbourne JF, Hunt HJ, Klinkert G (1987) An Efficient One Pot Synthesis of 4H-Pyrrolo [3, 2, ij] quinolines. Advances in Heterocyclic Chemistry 23: 103-170.

13. Pete UD, Zade CM, Bhosale JD, Tupe SG, Chaudhary PM, et al. (2012) Hybrid molecules of carvacrol and benzoyl urea/thiourea with potential applications in agriculture and medicine. Bioorganic & medicinal chemistry letters 22: 5550-5554.

14. Kumbhar PP, Dewang PM (2001) Eco-friendly Pest Management Using Monoterpenoids. I. Antifungal Efficacy of Thymol Derivatives. J Sci Ind Res 60: 645-648.

15. Ultee A, Slump RA, Steging G, Smid EJ (2000) Antimicrobial activity of carvacrol toward Bacillus cereus on rice. Journal of Food Protection 63: 620-624.

16. Baser KH, Vashi BS, Mehta DS (1995) Indian J Chem Sect B 34 9: 802-808.

17. Vashi BS, Shah VH (1996) Synthesis and biological screening of substituted thymolyl thiazolidinones and thymolyl azetidinones. Journal of the Indian Chemical Society 73: 491-492.

18. Vashi BS, Mehta DS, Shah VH (1996) synthesis of 2, 5-disubstituted-1, 3, 4-oxadiazole, 1, 5-disubstituted-2-mercapto-1, 3, 4-triazole and 2, 5-disubstituted-1, 3, 4-thiadiazole derivatives as po tential antimicrobial agents. Indian journal of chemistry. Sect. B: Organic chemistry, including medical chemistry 35: 111-115.

19. Hajimehdipoor, H, Shekarchi M, Khanavi M, Adib N, Amri M, et al. (2010) A validated high performance liquid chromatography method for the analysis of thymol and carvacrol in Thymus vulgaris L. volatile oil. Pharmacognosy magazine 6: 154-158.

20. Bagul SD, Rajput JD, Tadavi SK, Bendre RS (2016) Design synthesis and

biological activities of novel 5-isopropyl-2-methylphenolhydrazide-based sulfonamide derivatives. Research on Chemical Intermediates, pp: 1-12.

21. Rajput JD, Bagul SD, Tadavi SK, Karandikar PS, Bendre RS (2016) Design, Synthesis, and Biological Evaluation of Novel Class Diindolyl Methanes (DIMs) Derived from Naturally Occurring Phenolic Monoterpenoids. Medicinal chemistry.

22. Quiroga PR, Asensio CM, Nepote V (2015) Antioxidant effects of the monoterpenes carvacrol, thymol and sabinene hydrate on chemical and sensory stability of roasted sunflower seeds. J Sci Food Agric 95: 471-479.

23. Yanishlieva NV, Marinova EM, Gordon MH, Raneva VG (1999) Antioxidant activity and mechanism of action of thymol and carvacrol in two lipid systems. Food Chem 64: 59-66.

24. Burda S, Oleszek W (2001) Antioxidant and antiradical activities of flavonoids. J Agric Food Chem 49: 2774-2779.

25. Vashi BS, Mehta DS, Shah VH (1995) Synthesis and Biological Activity of 4-Thiazolidinones, 2-Azetidinones 4-Imidazolinone Derivatives Having Thymol Moiety. Chem Inform 26.

26. Nargund LVG, Reddy GRN, Hari Prasad V (1996) Synthesis and antibacterial activity of a series 1-aryl-2-mercapto-5-4-acetamidophenoxy) methyl-1, 3, 4-triazoles, thiadiazoles and 2-4-(acetamido-phenox t) carbonyl-3, 4, 5-trisubstituted-pyrazoles. Indian journal of chemistry. Sect. B: Organic chemistry, including medical chemistry 35: 499-502.

27. Ellis F, Osborne C (2002) Paracetamol a curriculum resource. Royal Society of Chemistry.

28. Wagle, Adhikari AV, Kumari NS (2008) Synthesis of some new 2-(3-methyl-7-substituted-2-oxoquinoxalinyl)-5-(aryl)-1, 3, 4-oxadiazoles as potential non-steroidal anti-inflammatory and analgesic agents. Indian journal of chemistry. Section B, Organic including medicinal 47: 439.

29. Maslat AO, Abussaud M, Tashtoush H, Talib M (2002) Synthesis, antibacterial, antifungal, and genotoxic activity of bis-1, 3, 4-oxadiazole derivatives. Polish J Pharmacol 54: 55-60.

30. Aggarwal N, Kumar R, Dureja P, Khurana JM (2012) Synthesis of Novel Nalidixic Acid-Based 1, 3, 4-Thiadiazole and 1, 3, 4-Oxadiazole Derivatives as Potent Antibacterial Agents. Chemical biology & drug design 79: 384-397.

31. Soare JR, Dinis TC, Cunha AP, Almeida L (1997) Antioxidant activities of some extracts of *Thymus zygis*. Free radical research 26: 469-478.

27

New Method for Spectrophotometric Determination of Lisinopril in Pure Form and in Pharmaceutical Formulations

Mohammad Shraitah and Malek MS Okdeh*

Department of Chemistry, Techreen University, Lattakia, Syria

Abstract

An accurate, simple, fast and cheap spectrophotometric method has been developed for the determination of lisinopril in pharmaceutical pure and dosage forms. The method is based on the reaction of Alizarin with primary amine present in the lisinopril in the presence of 80% ethyl alcohol. This reaction produces a complex Red colored product which absorbs maximally at 434 nm. Beer's law was obeyed in the range of 4.415-300.23 µg/mL with molar absorptivity of 1.619×10^3 L mole^{-1}cm^{-1} Sandell's sensitivity 0.272 µg.cm^{-2}. The effects of variables such as temperature, heating time, concentration of color producing reagent, and stability of color were investigated to optimize the procedure. The results are validated statistically. The proposed method was applied to commercially available tablets, and the results were Pharmaceutical formulations..

Keywords: Lisinopril; Alizarin; Spectrophotometry

Introduction

Lisinopril (S)-1-[N-[1-(ethoxycarbonyl)-3 phenylpropy1]-L-alanyl]-L-proline are Angiotensin-Converting Enzyme (ACE) has been widely used for the treatment of hypertension and heart failure. The analytical profiles of the drugs have been reviewed [1,2]. Enalapril maleate has been assayed by spectrophotometric [3-7], potentiometric [8,9], HPLC [10-14] and ^{1}H-NMR [15] methods. In tablets, lisinopril dihydrate has been determined by GC [16,17], spectrophotometric [18-21], colorimetric and fluorimetric [17] procedures. Capillary electrophoresis has been used to separate closely related ACE inhibitors and to quantities them in their pharmaceutical preparations [22,23] and stripping voltammetric method [24]. Quite a few researchers have dealt with the development of methods that quantify lisinopril in biological media. Methods that include Polarographic, spectrophotometric [25,26] even today because of its inherent simplicity, sensitivity, visible spectrophotometry is the technique of choice selectivity, accuracy, precision and cost-effectiveness. LNP in pharmaceuticals has been assayed based on reaction with N-bromosuccinimide and the charge transfer complexation reaction [27].

Sodium hypochlorite-phenyl hydrazine [7], 1-fluoro-2,4-dinitrobenzene [28] and ascorbic acid [29]. Most of these methods employ organic solvents as reaction medium, require longer heating times, use expensive reagents, and/or are less sensitive (Table 1). Of the various regents used in the assay of LNP in pharmaceuticals, ninhydrin has been employed by quite a few researchers. For example, Rehman et al. [29] used ninhydrin in DMF medium for kinetic spectrophotometric determination of LNP by initial rate and fixed-time procedures. Both methods showed linear response over 50 µg/mL LNP. The reagent in the same organic solvent medium (DMF) but involving heating was used by Raza et al. [30] to quantify LNP in 10-150 µg/mL range. Rajashekaran and Udayavani [31] assayed LNP in the 10-40 µg/mL range by measuring the coloured product formed between ninhydrin and LNP in acetone medium at elevated temperature. The common feature of all the three methods using ninhydrin [29-31] is the use of organic solvent as the reaction medium which quite often is undesirable.

Experimental

Reagents and apparatus

-Lisinopril (100.03% pure reference substance, produced by Lupin, India)

-Stock solution (1 mg/mL): 100 mg lisinopril was dissolved in 20% ml water and 80% ethyl alcohol in a 100 mL volumetric flask.

-Stock solution (1 mg/mL): 100 mg Alizarin was dissolved in 20% ml water and 80% ethyl alcohol in a 100 mL volumetric flask.

Buffer solution

Different buffer Solution used 0.2M Acetate buffer, 0.2M Ammonium buffer, 0.2M borate buffer and 0.2M (pH=2.0-12.0) universal Britton buffer solution.

-FeCl$_3$ Solution

Ferric chloride solution 1% dissolved in alkaline weak medium from -Ammonium hydroxide (1.0×10^{-4}M).

-Analytical balance

-UV-Vis Spectrophotometer Model SP3000 OpTMA from Korea

Principle of the method

We studied the best volume and concentration of the Lisinopril, Alizarin, universal Britton buffer at pH=8.0, Ferric chloride solutions on the formation red complex, and added 0.4 ml Ferric chloride solution 1% dissolved in weak NH$_4$OH determined at λ_{max}=434 nm.

Lisinopril-Alizarin method

To different aliquots of Alizarin solution corresponding to 0.5-7.0 ml^{-1} was transferred into a series of 10 ml volumetric flasks. 0.5-6.0 ml of Lisinopril solution and Universal buffer Britton solution pH=8.0 were added to each flask diluted to volume with 1:2 H$_2$O:C$_2$H$_5$OH. The solution was heated in a water bath at 40 ± 1°C (5 min), respectively. The mixtures were cooled and the volume was completed to 10 mL with mixture solvent measured after 10 min of mixing against reagent blank [32,33].

Analysis of pharmaceutical formulations

20 tablets were accurately weighted finely powdered and dissolved

---

***Corresponding author:** Malek MS Okdeh, Department of Chemistry, Techreen University, Lattakia, Syria, E-mail: dr.malekokdeh@yahoo.com

into sufficient volume of mixture solvent. The mixture was stirred well and filtered through Whatman filter paper No. 42 and the filtrate was diluted with mixture solvent added universal Britton buffer pH=8.0 and 0.4 ml $FeCl_3$ 1% in alkali weak medium from NH_4OH (1×10^{-4} M) in 10 ml volumetric flask. The mixtures were cooled and the volume was completed to 10 mL with mixture solvent and absorbance was measured after 7 min of mixing against reagent blank [34].

Results and Discussion

Preliminary investigations have been shown that Lisinopril react with Alizarin in buffer Britton solution 0.1M at pH=8.0 in presence catalytic reagent as ferric chloride 0.40 ml with Concentration (1×10^{-3}M) to give red coloured complex which absorbs at λ_{max}=434 nm as shown in Figure 1 [30,35].

The optimum reaction conditions for quantitative determination of the ion pair complexes were established via number of preliminary experiments. Several parameters such as amount of buffer added, reagent concentration, temperature, heating time, sequence of addition and color stability. It was observed that complete color development was attained at 40 ± 1°C (7 min) (Figure 2). The effect of Alizarin concentration on the color development was investigated 3 ml of Alizarin reagent produced maximum color intensity (Figure 3) [36].

Stoichiometric relationship

A series of solutions were prepared by mixing equimolecular proportions while keeping the total molar concentration constant in all cases and reagent concentration within range 100-800 µM or complex Lisinopril-Alizarin (LNP-ALZ) solutions changed the volume of Lisinopril (VLNP) and Volume of Alizarin was kept constant (VALZ) within range from (1.0-8.0 ml) and the total volume was kept constant in all these series are equal to (LNP+ALZ=9.0 ml) The absorbance values were then plotted against the mole fraction (VLNP)/(VLNP+VAZ) or VAZ/(VLNP+VAZ). The stoichiometry of the reaction between Lisinopril and Alizarin at selected conditions (Figure 4) was observed. The stoichiometry of the reaction between drugs and at the selected conditions was established by the molar ratio method. In this method 0.4 mL of 1% $FeCl_3$ in alkali weak from NH_4OH medium is and 0.05 ml buffer Universal Britton pH=8.0 kept constant and variable concentrations of drugs (5.0×10^{-4}M) were added [34,35]. The absorbance was measured at λ_{max} against blank solution prepared in the same manner. The absorbance values were then plotted against the molar ratio [Alizarine]/[Lisinopril] (Figure 5).

Developed color was stable up to 72 hours which was considered sufficient time for an analysis (Figure 6). Beer's law was obeyed in the range of 4.415-300.23 µg/ml. More than 99% recover of Lisinopril was obtained in the presence of possible excipients and ingredient in lisinopril formulations (Tables 1 and 2) [35].

Optical characteristics and statistical data for the regression equation of the proposed method are given in Table 1. Commercial formulation was successfully analyzed for the lisinopril by the proposed method and the results are compared with reference method (20) (Table 3) did not exceed the theoretical values, which indicates the absence of any difference between the methods compared. The proposed method gives good results for lisinopril in pure and pharmaceutical formulations [35,36].

Conclusion

The proposed method for the estimation of Lisinopril using

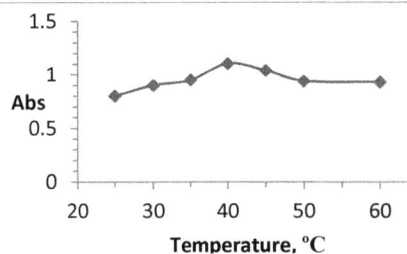

Figure 2: Effect of temperature on the color development..

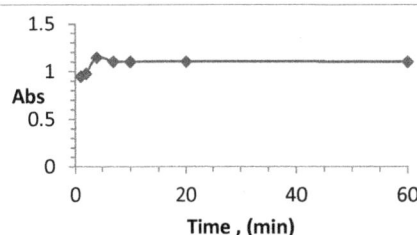

Figure 3: Effect of heating time on color intensity

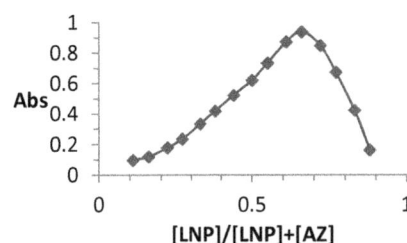

Figure 4: The mole fraction of (VLNP)/(VLNP+VAZ).

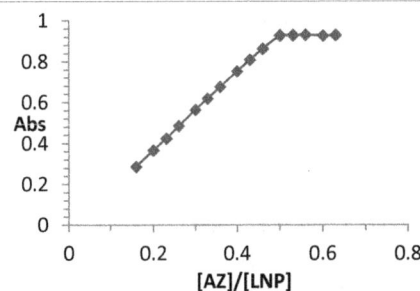

Figure 5: The molar ratio of [Alizarin]/[Lisinopril].

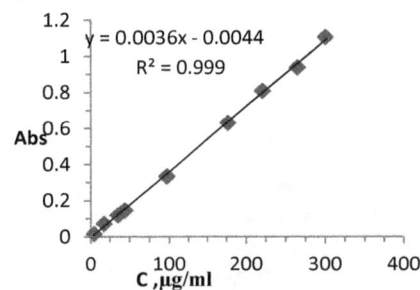

Figure 6: The molar ratio of [Alizarin]/[Lisinopril].

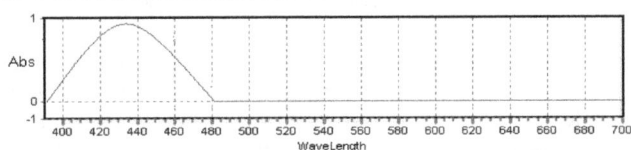

Figure 1: Absorption spectrum of lisinopril-alizarin formation.

Drug samples (µg/ml) Amount taken	Found (µg/ml)	Stander devation SD	R.S.D %	Detection limit (µg/ml)	Analytical Error SD/ (n)$^{1/?}$	Relative Recovery (%) R
4.415	4.503	0.155	3.44	4.503 ± 0.191	0.069	101.99
17.666	17.723	0.150	0.84	17.723 ± 0.185	0.067	100.32
35.321	35.414	0.164	0.46	35.414 ± 0.202	0.073	100.26
44.152	44.240	0.174	0.39	44.240 ± 0.213	0.077	100.19
97.134	97.276	0.217	0.22	97.276 ± 0.269	0.097	100.14
176.608	176.801	0.272	0.15	176.801 ± 0.335	0.121	100.10
220.760	220.602	0.281	0.12	220.602 ± 0.347	0. 125	99.92
264.912	264.802	0.275	0.10	264.802 ± 0.338	0.122	99.95
300.233	300.116	0.277	0.09	300.116 ± 0.341	0.123	99.96

Five independent analyses

Table 1: Test of precision and accuracy of the proposed method.

Parameter	Value
λmax	432 nm
Beer's law limit (µg/mL)	4.415-300.233
Molar abs orptivity (L mole^{-1} cm^{-1})	1.619 × 10^3
Sandell's s ens itivity (µg/mL per 0.001 A)	0.273
Regres s ion equation (Y*)	
Slope (m)	0.003
Intercept (c)	0.004
Correlation coefficient	0.999
Relative Standard Deviation**	3.44
Limit of Detection (µg/mL)***	2.08
Limit of quantitation (µg/ml)	6.94

*Y=mx+C; Where x is the concentration of analyte (µg/mL) and Y is absorbance unit; **: Calculated from six determinations; ***: Calculated as per ICH guidelines

Table 2: Optical characteristics and statistical data for the regression equation of the proposed method.

S. No.	Reagents	λ$_{max}$	Linear Dynamic µg mL^{-1}	Reaction time	Molar absorptivity (ε)Lmol^{-1}cm^{-1}	LOD	LQP	References
1	Alizarine	432	4.415-300.23	7 min at 40°C	1.619 × 10^3	-	-	This Work
2	Dichlone	580	40-120	10 min at rt	2.6 × 10^3	-	-	[20]
3	Acetylacetone + Formaldehyde	356	6.0-42.0	10 min at 100°C	9.62× 10^3	-	-	[20]
4	2,4- dinitrofluorobenzene	400	8.0-120.0	30 min at 80°C	-	1.16	3.87	[28]
5	Phenylhydraizine	362	40-200	20 min at 85°C	-	-	-	[8]
6	7-chloro-4-nitrobenzo-2-oxa-1, 3-diazole	470	20.0-560	30 min at 70°C	-	0.27	0.891	[36]
7	Ninhydrin .	410	10-40	10 min at 100°C	1.845 × 10^3	-	-	[31]
8	As corbic acid method	530	5-50	15 min at 100°C	4.548 × 10^3	0.349	1.152	[29]
9	Ninhydin kinetic method							
a)	Initial rate method	595	10-50	Immed iately after mixing the reagent at rt	-	0.118	0.389	[29]
b)	Rate cons tant method	595	10-40	-do-	-	2.839	9.369	[29]
c)	Fixed time method	595	5-50	10 min at rt	4.70 × 10^3	1.03	3.399	[29]

rt: Room temperature

Table 3: Comparision of the proposed methods with existing spectrophotometric methods for the assay of lisinopril in pharmaceutical formulations.

Alizarin is advantages over many of the reported methods. The methods are rapid, simple and have good sensitivity and accuracy. Proposed method makes use of simple reagent, which an ordinary analytical laboratory can afford. The high recovery percentage and low relative Standard deviations reflect the high accuracy and precision of the proposed method. The method are easy, applicable to a wide range of concentration, besides being less time consuming and depend on simple reagent which are available, thus offering economic and acceptable method for the routine determination of Lisinopril in its formulations.

New Method for Spectrophotometric Determination of Lisinopril in Pure Form and in Pharmaceutical...

157

References

1. Lancaster SG, Todd PA (1988) Lisinopril. A preliminary review of its pharmacodynamic and pharmacokinetic properties, and therapeutic use in hypertension and congestive heart failure. Drugs 35: 646-669.

2. The United States Pharmacopoeia (2000) 24th Revision. Asian Edition. United States Pharmacopoeial Convention, Inc., Twinbrook Parkway, Rockville, MD, USA.

3. Kato T (1985) Flow-injection spectrophotometric determination of enalapril in pharmaceuticals with bromothymol blue. Anal Chim Acta 175: 339-344.

4. Blaih SM, Abdine HH, El-Yazbi FA, Shaalan RA (2000) Spectrophotometric determination of enalapril maleate and ramipril in dosage forms. Spectrosc Lett 33: 91-102.

5. Dhake AS, Kasture VS, Sayed MR (2002) Spectrophotometric method for simultaneous estimation of amlodipine besylate and enalpril maleate intablets. Indian Drugs 39: 14-17.

6. Ayad MM, Shalaby AA, Abdellatef HE, Hosny MM (2002) Spectrophotometric and AAS determination of ramipril and enalapril through ternary complex formation. J Pharm Biomed Anal 28: 311-321.

7. Razen J, Senica D (2001) Concentration of lisinopril purified by liquid chromatography-A comparison between reverse osmosis and evaporation. Acta Chim Slov 48: 597-612.

8. El-Gindy A, Ashour A, Abdel-Fattah L, Shabana MM (2001) Spectrophotometric and HPTLC-densitometric determination of lisinopril and hydrochlorothiazide in binary mixtures. J Pharm Biomed Anal 25: 923-931.

9. Leis HJ, Fauler G, Raspotnig G, Windischhofer W (1999) An improved method for the measurement of the angiotensin-converting enzyme inhibitor lisinopril in human plasma by stable isotope dilution gas chromatography/negative ion chemical ionization mass spectrometry. Rapid Commun Mass Spectrom 13: 650-653.

10. Leis HJ, Fauler G, Raspotnig G, Windischhofer W (1998) Quantitative determination of the angiotensin-converting enzyme inhibitor lisinopril in human plasma by stable isotope dilution gas chromatography/negative ion chemical ionization mass spectrometry. Rapid Commun Mass Spectrom 12: 1591-1594.

11. Quin X, Nquen DT, Dominic P (1993) Separation of lisinopril and its RSS diastereoisomer by miceller electrokinetic chromatography. J Liquid Chromatogr 16: 3713-3734.

12. Hillaret S, VandenBopscha W (2000) Optimization of capillary electrophoretic separation of several inhibitors of the angiotensin-converting enzyme. J Chromatogr A 895: 33-42.

13. Gotti R, Andrisano V, Cavrini V, Bertucci C, Furlanetto S (2000) Analysis of ACE-inhibitors by CE using alkylsulfonic additives. J Pharm Biomed Anal 22: 423-431.

14. Rajasekaran A, Murugesan S (2001) Polarographic studies of Lisinopril. Asian J Chem 13: 1245-1246.

15. Ouyang J, Baeyens WR, Delanghe J, Van der Weken G, Calokerinos AC (1998) Cerium (IV)-based chemiluminescence analysis of hydrochlorothiazide. Talanta 46: 961-968.

16. Worland PJ, Jarrott B (1986) Radioimmunoassay for the quantitation of lisinopril and enalaprilat. J Pharm Sci 75: 512-516.

17. Yuan AS, Gilbert JD (1996) Time-resolved fluoroimmunoassay for the determination of lisinopril and enalapril at in human serum. J Pharm Biomed Anal 14: 773-781.

18. Atmaca S, Tatar S, Iskender G (1994) Spectrophotometric determination of lisinopril in tablets. Acta Pharm Turc 36: 13-16.

19. Iskender G, Yarenei B (1995) A spectrophotometric method for the determination of lisinopril in tablets. Acta Pharm Turc 37: 5-8.

20. El-Yazbi FA, Abdine HH, Shaalan RA (1999) Spectrophotometric and spectrofluorometric methods for the assay of lisinopril in single and multicomponent pharmaceutical dosage forms. J Pharm Biomed Anal 19: 819-827.

21. El-Emam AA, Hansen SH, Moustafa MA, El-Ashry SM, El-Sherbiny DT (2004) Determination of lisinopril in dosage forms and spiked human plasma through derivatization with 7-chloro-4-nitrobenzo-2-oxa-1,3-diazole (NBD-Cl) followed by spectrophotometry or HPLC with fluorimetric detection. J Pharm Biomed Anal 34: 35-44.

22. Aruna DP, Mallikarjuna RGPV, Krishna PKMM, Sastry CSP (2003) Four simple spectrophotometric determination of lisinopril in pure state and in tablets. Indian J Pharm Sci 65: 296-299.

23. Ozer D, Senel H (1999) Determination of lisinopril from pharmaceutical preparations by derivative UV spectrophotometry. J Pharm Biomed Anal 21: 691-695.

24. Prasad CVN, Saha RN, Parimoo P (1999) Simultaneous Determination of Amlodipine-Enalapril Maleate and Amlodipine-Lisinopril in Combined Tablet Preparations by Derivative Spectrophotometry. Pharm Pharmacol Commun 5: 383-388.

25. Erk N (1998) Combined study of the ratio spectra derivative spectrophotometry, derivative spectrophotometry and Vierordt's method applied to the analysis of lisinopril and hydrochlorothiazide in tablets. Spectrosc Lett 31: 633-645.

26. Melby LR, Harder RJ, Hertler WR, Mahler W, Benson RE, et al. (1962) Substituted Quinodimethans. II. Anion-radical Derivatives and Complexes of 7,7,8,8-Tetracyanoquinodimethan. J Am Chem Soc 84: 3374-3387.

27. Rahman N, Anwar N, Kashif M (2005) Application of pi-acceptors to the spectrophotometric determination of lisinopril in commercial dosage forms. Farmaco 60: 605-611.

28. Paraskevas G, Atta-Politou J, Koupparis M (2002) Spectrophotometric determination of lisinopril in tablets using 1-fluoro-2,4-dinitrobenzene reagent. J Pharm Biomed Anal 29: 865-872.

29. Rehman N, Singh M, Hoda MN (2005) Optimized and validated spectrophotometric methods for the determination of lisinopril in pharmaceutical formulations using ninhydrin and ascorbic acid. J Braz Chem Soc 16: 1001-1009.

30. Raza A, Ansari TM, Rehman AU (2005) Spectrophotometric determination of lisinopril in pure and pharmaceutical formulations. J Chin Chem Soc 52: 1055-1059.

31. Rajasekaran A, Udayavani S (2001) Spectrophotometric determination of lisinopril in pharmaceutical formulations. J Indian Chem Soc 78: 485-486.

32. European Pharmacopoeia (2007) European Directorate for Quality medicine and health care. Monograph number 1120. p: 2277.

33. Basavaiah K, Tharpa K, Hiriyanna SG, Vinay KB (2009) Spectrophotometric determination of lisinopril in pharmaceuticals using ninhydrin-a modified approach. J Food & Drug Anal 17: 93-99.

34. Rahman N, Haque SM (2008) Optimized and validated spectrophotometric methods for the determination of enalapril maleate in commercial dosage forms. Anal Chem Insights 3: 31-43.

35. Rahman N, Siddiqui MR, Azmi SNH (2007) Spectrophotometric determination of lisinopril in commercial dosage forms using N-bromosuccinimide and chloranil. Chem Anal (Warsaw) 52: 465-480.

36. Shama SA, Amin AS, Omara H (2011) Spectrophotometric Microdetermination of Some Antihypertensive Drugs in Pure Form and in Pharmaceutical Formulations. J Chil Chem Soc 56: 566-570.

28

Interaction of Fluorescent 2-(1-Methoxynaphthalen-4-Yl)-1-(4-Methoxyphenyl)-4, 5-Diphenyl-1H-Imidazole with Pristine Zno, Cu-Doped Zno and Ag-Doped Zno Nanoparticles

P Ponnambalam[1], S Kumar[1,2*] and P Ramanathan[3]

[1]Department of Chemistry, Bharathiar University, Coimbatore-641046, Tamilnadu, India
[2]Department of Chemistry, Thiruvalluvar College of Engineering and Technology, Tamilnadu, India
[3]Department of Chemistry, Thanthai Hans Roever College (Autonomous), Perambalur, Tamilnadu, India

Abstract

A sensitive 2-(1-methoxynaphthalen-4-yl)-1-(4-methoxyphenyl)-4, 5-diphenyl-1H-imidazole (MNMPI) fluorescent sensor for nanoparticulates like ZnO, Cu-doped ZnO and Ag-doped ZnO has been designed and synthesized. Facile preparation of ZnO, Cu-doped ZnO and Ag-doped ZnO nanoparticles by sol-gel method using PVP K-30 as templating agents is reported and characterised by powder X-ray diffraction (XRD), scanning electron microscopy (SEM), UV-Visible spectroscopy and photoluminescence spectroscopy (PL). The synthesized sensor release is enhanced by nanocrystalline pristine ZnO but is suppressed by Cu-doped ZnO and Ag-doped ZnO nanoparticles. The suppression of fluorescence is additional by copper than by silver doping. The LUMO and HOMO energy gap of MNMPI associated with Cu-doped ZnO are lowers compared to those of pristine ZnO and thus red shift compared to that with pristine ZnO. The average crystallite sizes of ZnO, Cu-doped ZnO and Ag-doped ZnO have been deduced as 32 nm, 36 nm and 26 nm and calculated surface area for ZnO, Cu-doped ZnO and Ag-doped ZnO are 30.04 m^2/g, 40.66 m^2/g and 29.37 m^2/g respectively. The observed enhanced absorbance with the distributed semiconductor nanoparticle is due to adsorption of MNMPI on semiconductor surface. This is because of the efficient transfer of electron from the excited state of the MNMPI to the conduction band of the semiconductor nanoparticle.

Keywords: MNMPI; ZnO; Cu-doped ZnO; Ag-doped ZnO; XRD; SEM; UV; PL

Introduction

The report as Xia is for Polymer-stabilized nano ZnO with blue emission [1] and the cell imaging is obtained by tunable photoluminescence with and the ZnO@polymer core-shell nanoparticles [2,3]. Using single crystals or polycrystalline of Co^{2+}: ZnO prepared by pellet sintering many scholars identified the observable photo response of Co-doped ZnO. They establish that the Ni-doped ZnO unfilled spheres exhibited only feeble ferromagnetism at 300 K whereas Co-doped ZnO hollow exhibited ferromagnetism at room temperature. Not due to any cobalt oxide phase formation or any metallic Co isolation the observed nature of ferromagnetism was intrinsic. Heterocyclic imidazole moieties have also attracted significant attention because of their unique optical properties [4] and for their use in preparing functionalized materials [5]. Nanoparticles can be used as drug carriers because they have enormous surface area and due to their submicron size they can efficiently be taken up by the cells [6]. ZnO is an attractive semiconductor material with wide direct band gap (3.37 eV), large exciton binding energy (60 meV) and a hexagonal structure and have significant applications in optoelectronics, sensors and actuators [7-9]. Hence, ZnO is one of the most attractive platforms for binding enzyme and shows potential material for a wide range of biosensor applications. Au, Ag, or Pt noble metal layered ZnO is significant for photoelectron transfer (PET) in the bulk and interface of ZnO semiconductors [10]. Under clarification of UV light, the exciton absorption bands of ZnO are strongly bleached due to the accumulation of conduction band electrons [11]. Thus, the effectiveness of both the photocatalysis and photoelectric energy conversion can be significantly enhanced by depositing noble metals on the surface of ZnO [12]. The properties and applications of noble metal ZnO nanostructured materials are also determined by its morphology, structure and the organization of nanostructured ZnO architectures [13-16]. ZnO based various ceramics were synthesised by liquid phase sintering of ZnO powder of different sizes and morphologies. The HOMO and LUMO

potentials for the considered sensor must match with the conduction and valence band edges of the semiconductor nanocrystals [17].

Experimental

Materials and methods

Benzil, 4-methoxyaniline, 4-methoxynaphthaldehyde, ammonium acetate and borontrifluoride ethylethartate were purchased from Sigma Aldrich. Zinc acetate (Sd fine), polyvinylpyrrolidone (PVP K-30, Himedia), ammonia (Qualigens). The solvents used for spectral measurements were of spectroscopic grade and purchased by Hi-media. Distilled ethanol and deionized distilled water were employed for the experiments.

Synthesis 2-(1-methoxynaphthalen-4-yl)-1-(4-methoxyphenyl)-4, 5-diphenyl-1H-imidazole

The product 2-(1-methoxynaphthalen-4-yl)-1-(4-methoxy phenyl)-4, 5-diphenyl-1H-imidazole was prepared by refluxing benzil (1 mmol), 4-methoxyaniline (1 mmol), 4-methoxynaphthaldehyde (1 mmol) and ammonium acetate (1 mmol) in ethanol (20 mL) for 2 h, borontrifluoride ethylethartate (1 mol%) acting as a catalyst (Scheme-1). The progress of the reaction was followed by TLC. After completion of the reaction, the mixture was cooled, dissolved in acetone and filtered. The product was

*Corresponding author: Kumar S, Professor and Head, Department of Chemistry, Thiruvalluvar College of Engineering and Technology, Tamilnadu, India
E-mail: ramanathanp2010@gmail.com

purified by column chromatography with benzene: ethyl acetate (9:1) as the eluent.

M.P. 296°C. Anal. calcd. for $C_{33}H_{26}N_2O_2$: C, 82.13; H, 5.43; N, 5.81. Found: C, 82.11; H, 5.41; N, 5.80. ^{1}H NMR (400 MHz, CDCl$_3$): δ 3.55 (s, 3H), 3.88 (s, 3H), 6.44 (d, J=8.4 Hz, 2H), 6.62 (d, J=8.0 Hz, 1H), 6.73 (d, J=8.4 Hz, 2H), 7.12-7.18 (m, 9H), 7.29-7.44 (m, 4H), 7.82 (dd, J=6.4 Hz, 1H), 8.13 (d, 1H). ^{13}C NMR (400 MHz, CDCl$_3$): δ 54.18, 54.57, 102.98, 112.76, 117.88, 121.03, 124.33, 124.34, 124.46, 125.91, 126.12, 126.55, 127.11, 127.17, 127.32, 127.46, 127.74, 127.78, 129.01, 129.28, 129.97, 132.14, 132.36, 135.40, 145.36, 155.39, 157.70. MS: m/z. 482.57 [M +].

Synthesis of nanocrystalline oxides by Sol-gel method

Zinc nitrate (0.1 g) solution [with or without Cu(NO$_3$)$_2$/AgNO$_3$] in 10 ml 0.01 M PVP K-30, newly prepared solution of 1:1 aq. NH$_3$ was added slowly to reach a pH of 7, under continuous stirring. The stirring was sustained for another 30 min to get a gel. The formed glassy like white gel was allowed to age overnight. It was filtered and washed with water and ethanol several times, dried at 100°C for 12 h and calcinated at 500°C for 3 h to pale grey solid.

Spectral measurements

The ^{1}H NMR and proton decoupled ^{13}C NMR spectra were recorded to using a Bruker 400 MHz NMR spectrometer operating at 400 MHz and 100 MHz, respectively. The UV-vis absorption and emission spectra were recorded with PerkinElmer Lambda 35 spectrophotometer and PerkinElmer LS55 spectrofluorimeter, respectively. The powder X-ray diffractogram (XRD) was recorded with a PAN analytical X'Pert PRO diffractometer using Cu Kα rays at 1.5406 Å with a tube current of 30 mA at 40 kV. A JEOL JSM 10LV scanning electron microscope (SEM) equipped with a highly sensitive backscattered detector and low vacuum secondary detector was used to get the SEM image of the sample. The UV-vis absorption spectra were recorded with PerkinElmer Lambda 35 spectrophotometer.

Results and Discussion

XRD analysis of ZnO, Cu-doped ZnO and Ag-doped ZnO nanoparticles

X-ray diffraction patterns (XRD) of pristine ZnO, Cu-doped ZnO and Ag-doped ZnO nanoparticles obtained by sol-gel method (Figure 1). All the diffraction patterns match with the JCPDS pattern of Zincite

(89-7102). The crystal structures of pristine ZnO and doped ZnO are primitive hexagonal with crystal constants a and b as 3.249 A° and c as 5.025 A°. In the case of doping with copper, as the radii of Zn^{2+} and Cu^{2+} are comparable, Cu^{2+} can replace Zn^{2+} in the lattice without change in the lattice parameters. The XRD of Ag-doped ZnO reveals the presence of metallic silver in face centered crystal lattice whereas the Cu-doped ZnO fails to provide any peak other than those of ZnO. The observed peak at 38.2° is characteristic of the 111-peak of face centered cubic phase of metallic silver. The Ag$^+$ ion is better (radius 1.22 A°) than that of Zn^{2+} (0.72 A°) and hence cannot be included in to the ZnO lattice.

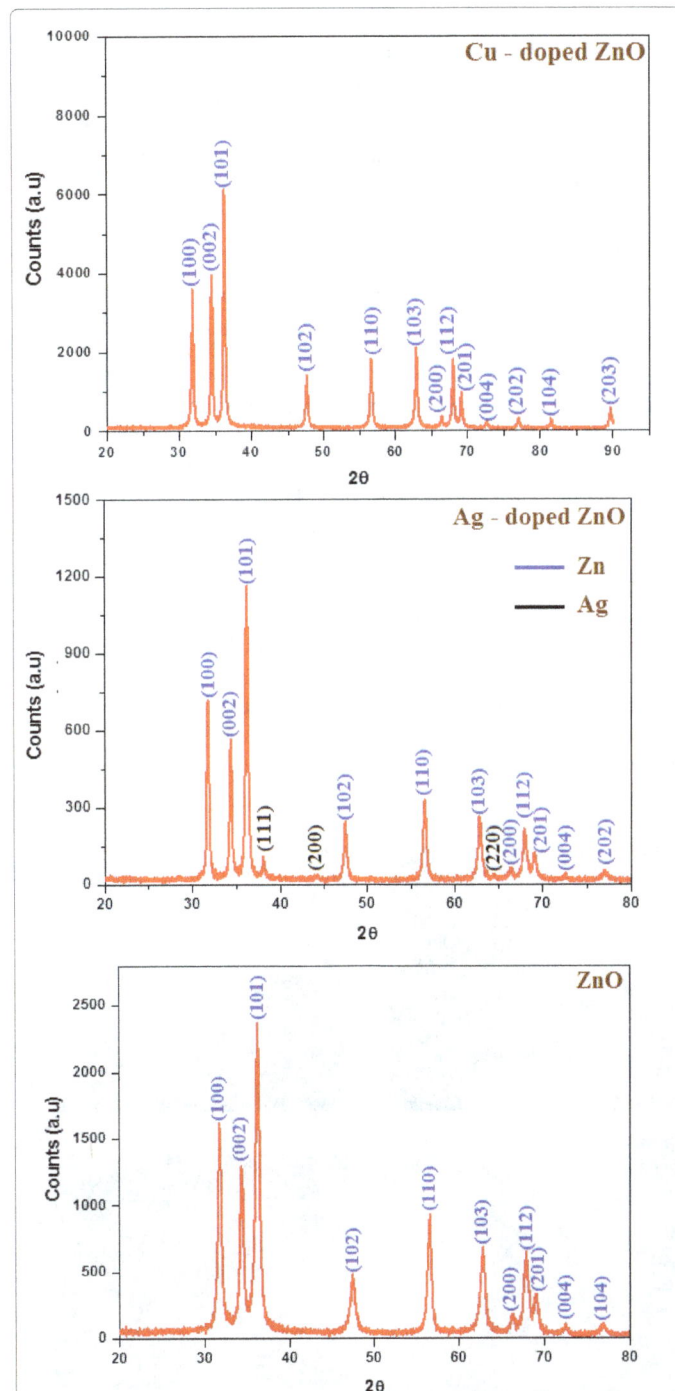

Figure 1: X-ray diffraction patterns (XRD) of pristine ZnO, Cu-doped ZnO and Ag-doped ZnO.

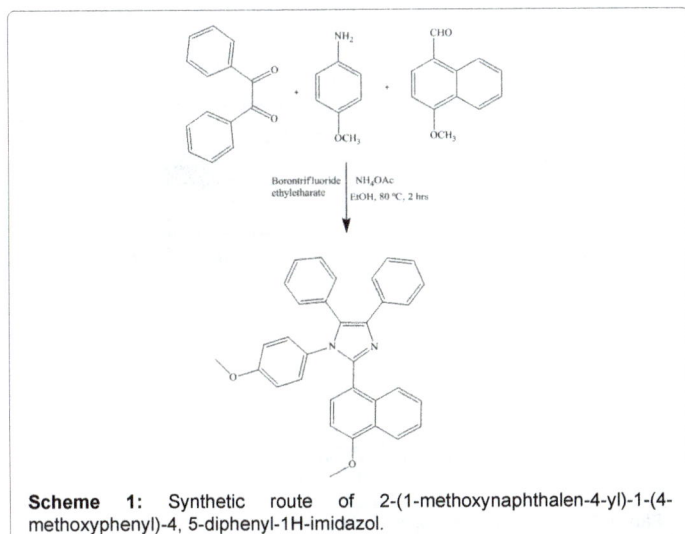

Scheme 1: Synthetic route of 2-(1-methoxynaphthalen-4-yl)-1-(4-methoxyphenyl)-4, 5-diphenyl-1H-imidazol.

Hence silver preferentially desire to segregate around the ZnO grain boundaries. The average crystallite sizes (L) of the sol-gel synthesized ZnO, Cu-doped ZnO and Ag-doped ZnO have been deduced as 32 nm, 36 nm and 26 nm, respectively. They have been obtained from the full width at half maximum (FWHM) of the most intense peaks of the individual crystals using the Scherrer equation, $L=0.9\ \lambda/\beta cos\theta$, where λ is the wavelength of the X-rays used, θ is the diffraction angle and β is the full width at half maximum of the peak. The calculated surface area for ZnO, Cu-doped ZnO and Ag-doped ZnO are 30.04 m²/g, 40.66 m²/g and 29.37 m²/g, respectively.

SEM and EDS analysis of ZnO, Cu-doped ZnO, Ag-doped ZnO nanoparticles and imidazole-ZnO complex

The SEM images of pristine ZnO, Cu-doped ZnO and Ag-doped ZnO nanoparticles are displayed in Figure 2. The particles are flower like use of PVP as templating agent provides finite morphology. The EDS spectra are shown in Figure 3 confirm the existence of zinc, oxygen, copper and silver signals which implies the purity of the synthesized ZnO and Cu-doped ZnO nanoparticles.

Absorption and emission behaviours of imidazole with ZnO, Cu-doped ZnO, Ag-doped ZnO nanoparticles

The photoluminescence spectra of pristine ZnO, Cu-doped ZnO and Ag-doped ZnO nanoparticles have been recorded at room temperature. They are shown in Figure 4. The pristine ZnO, Cu-doped ZnO and Ag-doped ZnO nanoparticles exhibit near band gap emission (NBE) and deep level emission (DLE). The DLE arises due to different intrinsic and extrinsic structural defects in all the nanoparticles [18]. The NBE originates from the recombination of free photogenerated electrons and holes. These UV emissions concur with the absorption edges deduced from the Kubelka-Munk plots. Pristine ZnO display the emission around 418 nm was observed [19,20]. The emission originates from the electron transition from the superficial donor level of oxygen vacancies to the valence band (VB) and electron transition from the superficial donor level of zinc interstitials to the VB [21]. This emission energy corresponds to the electron change from deep-level donor of the

ionized oxygen vacancies to the VB. The absorption spectra of MNMPI in the existence of pristine ZnO, Cu-doped ZnO and Ag-doped ZnO nanoparticles distributed at different loading and also in their absence are displayed in Figure 5. The nanoparticles enhance the absorbance of MNMPI remarkably. The observed enhanced absorbance with the distributed semiconductor nanoparticle is due to adsorption of MNMPI on semiconductor surface. This is because of the efficient transfer of electron from the excited state of the MNMPI to the conduction band of the semiconductor nanoparticle. The emission spectra of MNMPI in the existence of pristine ZnO, Cu-doped ZnO and Ag-doped ZnO nanoparticles distributed at different loading and also in their absence are displayed in Figure 4. The pristine ZnO nanoparticle enhances the emission of MNMPI. The enhanced emission with the dispersed semiconductor nanoparticles is due to the adsorption of MNMPI on semiconductor surface. Fluorescence enhancement is due to the formation of complex [MNMPI-nanoparticulate ZnO]. Doping of ZnO by Ag and Cu shows that the dopants inhibit the fluorescence enhancement by ZnO. Figure 6 shows the linear variation of log [F_0-F/F_0] vs. [nanoparticles] and the calculated binding constant (K) is given in Table 1. The order of binding constant (K) is Cu-doped ZnO>ZnO>Ag-doped ZnO. The binding constant (K) of the imidazole with ZnO and Cu-doped ZnO are in the order of 10^7 whereas that with

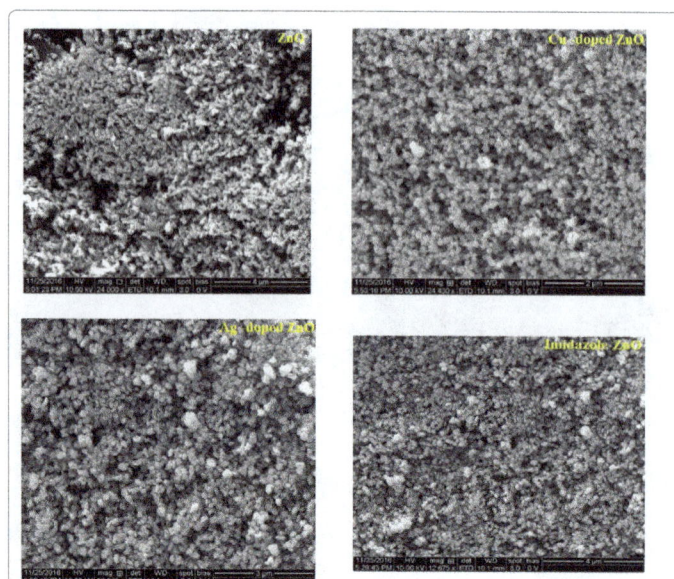

Figure 2: SEM images of pristine ZnO, Cu-doped ZnO, Ag-doped ZnO and MNMPI-ZnO.

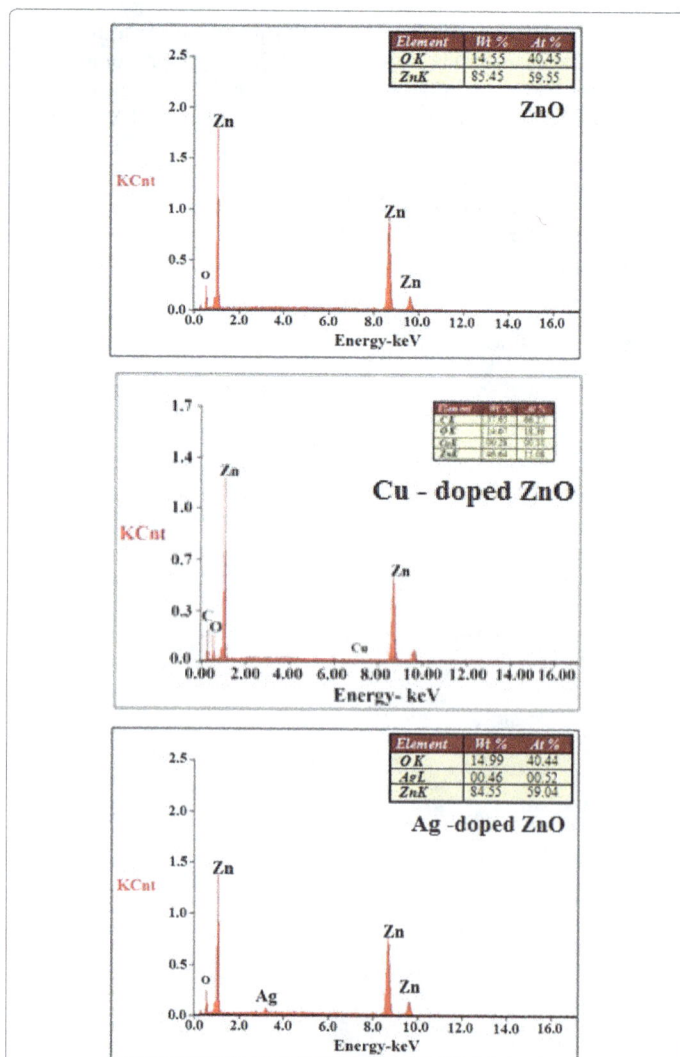

Figure 3: EDX spectra of pristine ZnO, Cu-doped ZnO and Ag-doped ZnO.

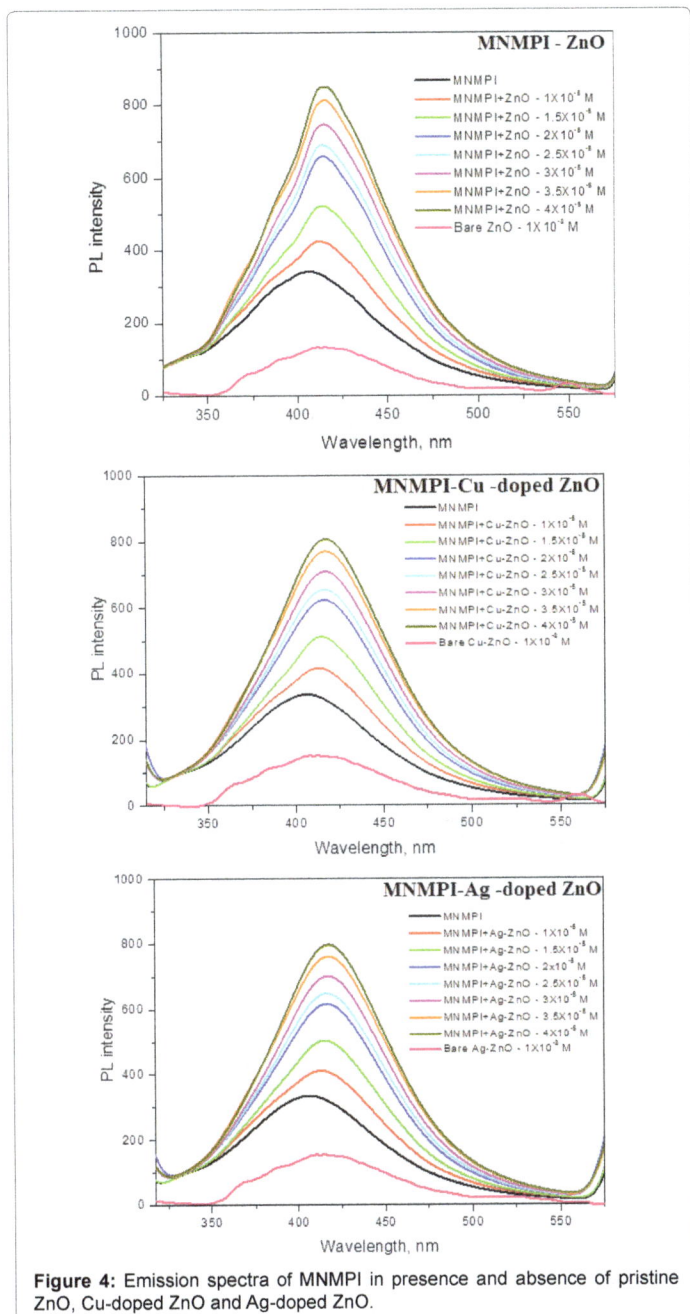

Figure 4: Emission spectra of MNMPI in presence and absence of pristine ZnO, Cu-doped ZnO and Ag-doped ZnO.

levels have been calculated using the equations [21], HOMO=-e $(E_{ox}+4.71)$ (eV); LUMO=-e $(E_{red}+4.71)$ (eV). On the origin of the relative HOMO and LUMO energy levels of an inaccessible MNMPI molecule along with the conduction band and valence band edges of

Complex	τ	k_r	k_{nr}	K	n
MNMPI	3.72	0.80	1.9	-	-
MNMPI... ZnO	3.33	0.90	2.1	2.98×10^8	0.97
MNMPI... Cu-ZnO	3.21	0.69	2.6	9.89×10^9	0.98
MNMPI... Ag-ZnO	3.42	0.79	2.2	9.02×10^7	0.91

Table 1: Photoluminescence (τ, ns), radiative (k_r, 10^8 s^{-1}), non-radiative (k_{nr}, 10^8 s^{-1}), binding constant (K), binding sites (n).

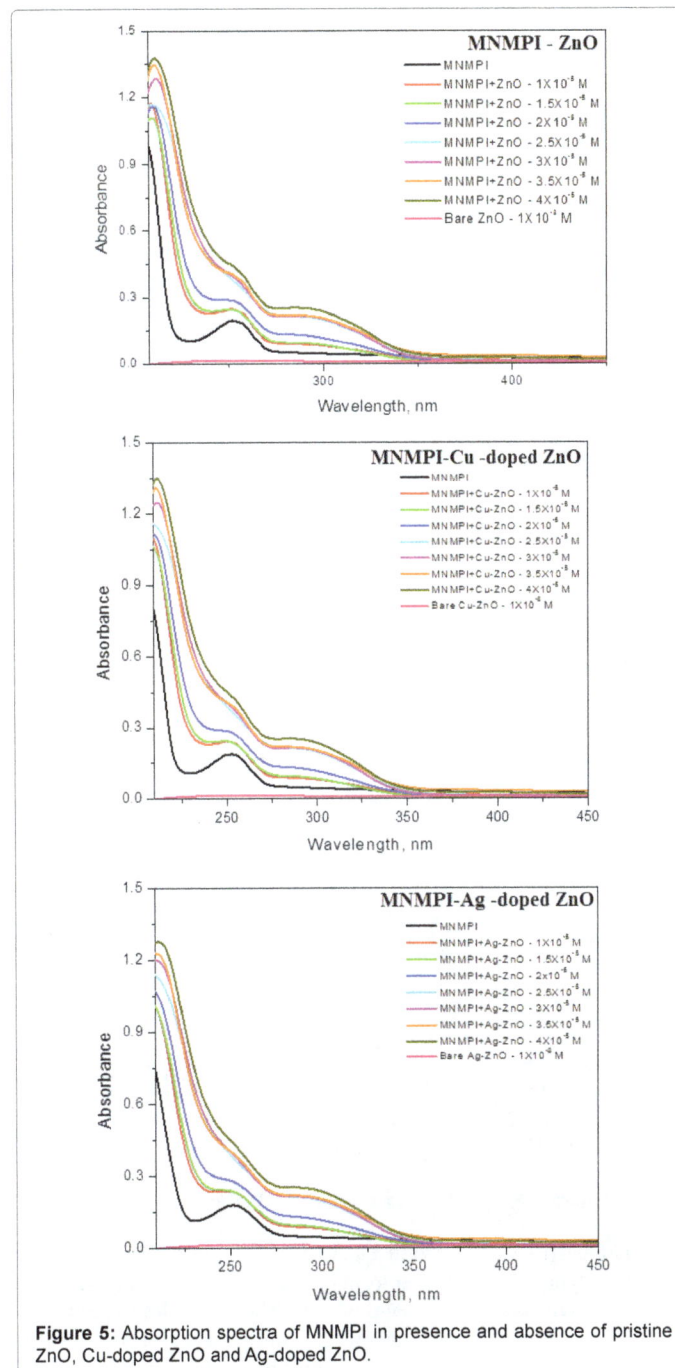

Figure 5: Absorption spectra of MNMPI in presence and absence of pristine ZnO, Cu-doped ZnO and Ag-doped ZnO.

Ag-doped ZnO is far less. This is because in Ag-doped ZnO, Ag^0 is deposited on the surface of the crystal thereby inhibiting the binding of the imidazole with ZnO. In Cu-doped ZnO, Cu^{2+} is likely to be present in the cationic sites or the interstitial positions thereby not influence the binding of imidazole with ZnO. Both the dopants suppress the enhancement of fluorescence and the inhibition is more by copper than by silver doping. The possible reason is Cu^{2+} in Cu-doped ZnO may bind with the imidazole and this binding could be much stronger than that by Zn^{2+}. The binding of Ag with imidazole is not as strong as that of Cu^{2+} or Zn^{2+} and thus less binding constant.

HOMO-LUMO energy levels of imidazole with ZnO, Cu-doped ZnO, and Ag-doped ZnO nanoparticles

From the onset oxidation potential (E_{ox}) and the onset reduction potential (E_{red}) of MNMPI derivative, HOMO and LUMO energy

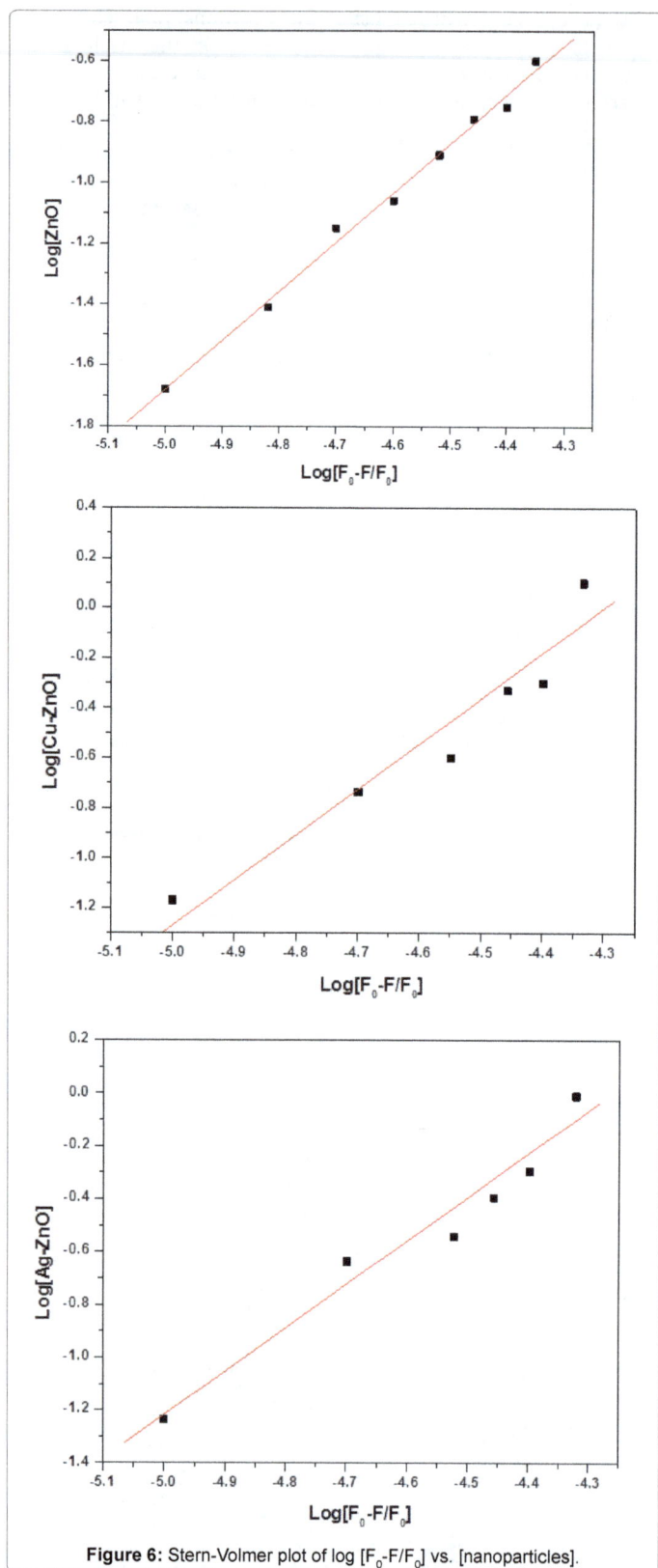

Figure 6: Stern-Volmer plot of log [F_0-F/F_0] vs. [nanoparticles].

nanocrystals is also possible because the electron in the LUMO of the excited imidazole is of higher energy compared to that in the CB of ZnO nanocrystals [22]. This should lead to quenching of fluorescence of MNMPI. However, in differing to the expectations, enhancement of fluorescence is observed in the presence of ZnO nanocrystals. This may be because of lowering of the HOMO and LUMO energy levels of MNMPI due to adsorption on ZnO nanoparticles. The polar ZnO covering enhances the delocalisation of the π electrons and lowers the HOMO and LUMO energy levels due to adsorption. The chemical

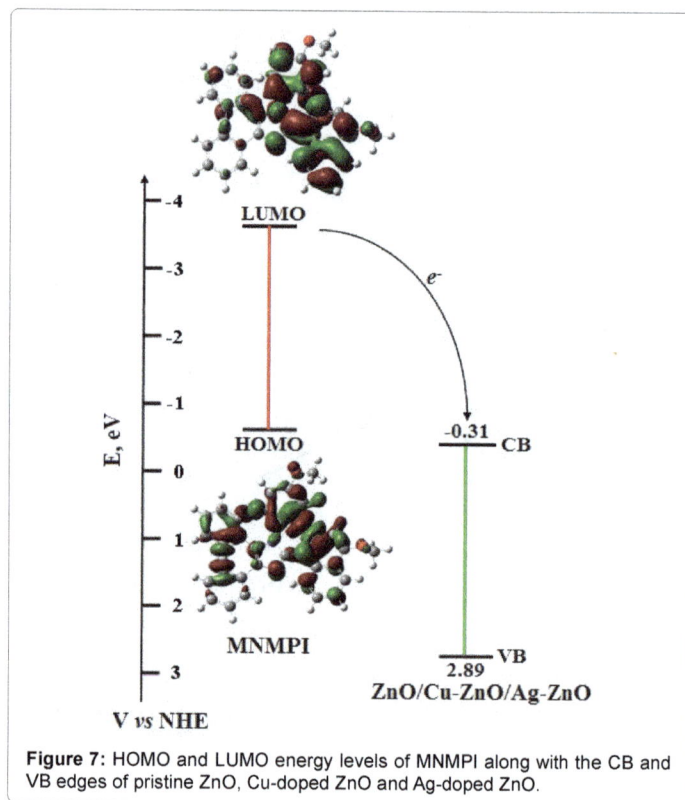

Figure 7: HOMO and LUMO energy levels of MNMPI along with the CB and VB edges of pristine ZnO, Cu-doped ZnO and Ag-doped ZnO.

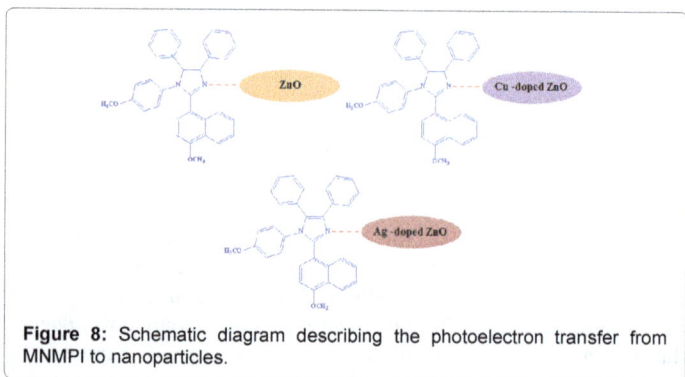

Figure 8: Schematic diagram describing the photoelectron transfer from MNMPI to nanoparticles.

affinity between the nitrogen atom of the imidazole and zinc ion on the surface of the nano-oxide may be a reason for strong adsorption of the imidazole on nanoparticle causes the enhancement.

Binding interaction of imidazole with ZnO, Cu-doped ZnO, Ag-doped ZnO nanoparticles

The binding strength of MNMPI through its azomethine nitrogen with Cu^{2+} in the doped ZnO is likely to be stronger than that with Zn^{2+} in pristine ZnO (Figure 8). The LUMO and HOMO energy gap of

ZnO nanoparticles as shown in Figure 7, the electron injection would be thermodynamically allowed from the excited singlet of the MNMPI derivative to the conduction band of ZnO. The energy levels presented in Figure 7 reveals the enhancement of fluorescence of MNMPI by ZnO nanocrystals. Electron transfer from the excited imidazole to the

MNMPI associated with Cu-doped ZnO are lowers compared to those of pristine ZnO. This inference stems from the observed red shift of the fluorescence of MNMPI on binding with Cu^{2+}-doped ZnO compared to that with pristine ZnO. The emission of MNMPI is blue shifted on association with Ag-doped ZnO. The doped silver is likely to be present as metallic silver nano deposit on the surface of ZnO nanoparticles. The metallic silver on the surface of ZnO nanoparticles is likely to interrelate with MNMPI through its azomethine nitrogen. The interaction is likely to be weak as Ag^0 is involved instead of its ionic form. The deposited silver on the surface of ZnO nanocrystals act as a shield making the MNMPI-Zn^{2+} of ZnO interaction less possible. The energy gap between HOMO and LUMO of the MNMPI-Ag-doped ZnO becomes larger compared to that with complex MNMPI-ZnO.

Conclusion

In conclusion, a sensitive MNMPI fluorescent sensor for nanoparticulate ZnO has been designed and synthesized. Facile preparation of ZnO, Cu-doped ZnO and Ag-doped ZnO nanoparticles by sol-gel method using PVP K-30 as templating agents is reported and characterised by X-ray diffraction, energy dispersive X-ray, UV-visible diffuse reflectance and photoluminescence spectra. MNMPI is adsorbed on the surface of semiconductor nanoparticle during azomethine nitrogen. The polar ZnO surface enhances the delocalisation of the π electrons and lowers the HOMO and LUMO energy levels due to adsorption. The LUMO and HOMO energy gap of MNMPI associated with Cu-doped ZnO are lowers compared to those of pristine ZnO and thus red shift compared to that with pristine ZnO. The energy gap between HOMO and LUMO of the complex MNMPI-Ag-doped ZnO becomes larger compared to that with complex MNMPI-ZnO and the emission compared to those of pristine ZnO. The conduction band energy position determines the electron transfer from excited state MNMPI to the ZnO, Cu-doped ZnO and Ag-doped ZnO nanoparticles.

Acknowledgments

The authors are very thankful to guide Dr. S. Kumar, Professor and Head, Department of Chemistry, Thiruvalluvar College of Engineering and Technology, Vandavasi for moral support in my studies. Instrumentation facilities are will be provided by Department of Chemistry, Annamalai University, Annamalainagar, Chidambaram.

References

1. Xiong HM, Wang ZD, Xia YY (2006) Polymerization initiated by inherent free radicals on nanoparticle surfaces: A Simple method of obtaining ultrastable (ZnO) polymer core-shell nanoparticles with strong blue fluorescence. Adv Mater 18: 748-751.

2. Xiong HM, Xu Y, Ren QG, Xia YY (2008) Stable aqueous ZnO@polymer core-shell nanoparticles with tunable photoluminescence and their application in cell imaging. J Am Chem Soc 130: 7522-7523.

3. Subramanian V, Wolf EE, Kamat PV (2003) Green emission to probe photoinduced charging events in ZnO-Au nanoparticles. Charge distribution and Fermi-level equilibration. J Phys Chem B 107: 7479-7485.

4. Hung WS, Lin JT, Chien CH, Tao T, Sun SS, et al. (2004) Highly phosphorescent bis-cyclometalated iridium complexes containing benzoimidazole-based ligands. Chem Mater 16: 2480-2488.

5. Nakashima K (2003) Lophine derivatives as versatile analytical tools. Biomed Chromatogr 17: 83-95.

6. Wiseman A (1985) Handbook of Enzyme Biotechnology, Horwoodi, Chichester.

7. Masuda Y, Kato K (2008) High c-axis oriented stand-alone ZnO self-assembled film. Crystal Growth and Design 8: 275-279.

8. Zhang XM, Lu MY, Zhang Y, Chen LJ, Wang ZL, et al. (2009) Fabrication of a high-brightness blue-light-emitting diode using a ZnO-nanowire array grown on p-GaN thin film. Adv Mater 21: 2767-2770.

9. Topoglidis E, Cass AEG, Regan BO, Durrant JR (2001) Immobilisation and bioelectrochemistry of proteins on nanoporous TiO_2 and ZnO films. J Electroanal Chem 517: 20-27.

10. Chen S, Ingram RS, Hostetler MJ, Pietron JJ (1998) Gold nanoelectrodes of varied size: Transition to molecule-like charging. Science 280: 2098-2101.

11. Ozgur U, Alivov I, Liu C, Teke A (2005) A comprehensive review of ZnO materials and devices. J Appl Phys 98: pp:041301.

12. Huang MH, Mao S, Feick H, Yan HQ (2001) Room-temperature ultraviolet nanowire nanolasers. Science 292: 1897-1899.

13. Wang YW, Zhang LD, Wang GZ (2002) Catalytic growth of semiconducting zinc oxide nanowires and their photoluminescence properties. J Cryst Growth 234: 171-175.

14. Feng X, Feng L, Jin M, Zhai J (2004) Reversible super-hydrophobicity to super-hydrophilicity transition of aligned ZnO nanorod films. J Am Chem Soc 126: 62-63.

15. Law M, Greene LE, Jhonson JC (2005) Nanowire dye-sensitized solar cells. Nat Mater 4: 455-459.

16. Karunakaran C, Jayabharathi J, Kalaiarasi V, Jayamoorthy K (2014) Characterization and electronic spectral studies of 2-(naphthalen-1-yl)-4,5-diphenyl-1H-imidazole bound Fe_2O_3 nanoparticles. Spectrochimica Acta Part A 120: 84-87.

17. Wang X, Zhang Q, Wan Q, Dai G (2011) Controllable ZnO architectures by ethanolamine-assisted hydrothermal reaction for enhanced photocatalytic activity. J Phys Chem C 115: 2769-2775.

18. Becker J, Raghupathi KR, Pierre J (2011) Tuning of the crystallite and particle sizes of ZnO nanocrystalline materials in solvothermal synthesis and their photocatalytic activity for dye degradation. J Phys Chem C 115: 13844-13850.

19. Jing L, Qu Y, Wang B, Li S (2006) Review of photoluminescence performance of nano-sized semiconductor materials and its relationships with photocatalytic activity. J Sun Sol Energy Mater Sol Cells 90: 1773-1787.

20. Hou J, Huo L, He C, Yang C (2006) Synthesis and absorption spectra of Poly (3-(phenylenevinyl) thiophene)s with conjugated side chains. Macromolecules 39: 594-603.

21. He WY, Li Y, Xue C, Hu ZD (2005) Effect of Chinese medicine alpinetin on the structure of human serum albumin. Bioorg Med Chem 13: 1837-1845.

22. Kavarnos GJ, Turro N (1986) Photosensitization by reversible electron transfer: Theories, experimental evidence, and examples. Chem Rev 86: 401-449.

29

Sulfamic Acid: An Efficient and Recyclable Solid Acid Catalyst for the Synthesis of Quinoline-4-Carboxylic Acid Derivatives in Water

Yahya S, Beheshtiha SH, Majid M* and Dehghani M

Department of Chemistry, School of Science, Alzahra University, Vanak, Tehran, Iran

Abstract

A simple, cost-effective, environmentally, and convenient procedure for the synthesis of quinoline-4-carboxylic acid derivatives is described through a one-pot MCR condensation of pyruvic acid, various aniline derivatives, and differently substituted aryl aldehydes in the presence of sulfamic acid (NH_2SO_3H) (SA) as an efficient and recyclable catalytic system under mild reaction conditions in water.

Keywords: Quinoline-4-caboxylic acid derivatives; One-pot MCR; Sulfamic acid; Recyclable catalyst; Green chemistry

Introduction

Quinoline [1] 1-aza-napthalene and benzo[b]pyridine are nitrogen containing heterocyclic aromatic compounds. Quinolines are very important compounds due to their wide spectrum of biological activities behaving as anti-malarial, anti-bacterial, anti-fungal, anti-asthmatic, anti-hypertensive, anti-inflammatory, anti-platelet activity [2-9]. In addition quinolines have been employed in the study of bioorganic and bio organo-metallic processes [10,11]. Considering the significant applications in the fields of medicinal, bioorganic, industrial, and synthetic organic chemistry, there has been tremendous interest in developing efficient methods for the synthesis of quinolines. Some derivatives of such as quinoline-4-carboxylic acid elicited profound changes in the morphology of typical tips of *Botrytis cinerea* [12] Quinoline-4-carboxylic acids are one of the most important series of quinoline derivatives because they exhibit a wide variety of medicinal effects and are applied as active components in industrial antioxidants [13,14]. Meanwhile, quinoline-4-carboxylic acids are the key precursors for the synthesis of other useful quinoline derivatives [15]. Despite remarkable efforts in the last decade [16-19] the development of effective methods for the synthesis of quinoxaline ring is still an important challenge and much in demands. In recent years, many methods for the synthesis of these quinoline acids have successively been reported. The conventional synthesis involves the Pfitzinger reaction [20,21]. Doebner reaction [22,23], Friedlander and Combes methods [24-31]. However, these synthetic methods require expensive and hazardous solvents and chemicals, as well as harsh reaction condition. Therefore, the development of simple, convenient, and environmentally benign approaches for the synthesis of quinolines is still desirable. In recent years, the use of solid acids as heterogeneous catalysts has received tremendous interest in various areas of organic synthesis [32-34]. Heterogeneous solid acids are advantageous over conventional homogeneous acid catalysts as they can be easily recovered from the reaction mixture by simple filtration and can be reused after activation or without activation and more importantly without appreciable lose in activities. SA has emerged as a substitute for conventional Bronsted- and Lewis acid catalysts. This catalyst is an amino acid containing sulfur element with mild acidity. It is fast coming up as a stable, white odorless crystalline, commercially available, also not volatile and corrosive, thus can be considered as green catalyst in organic synthesis [35-37]. Interestingly, sulfamic acid exists not only in its amino sulfonic acid form, but also as $H_3N^+SO^{-3}$ zwitterionic units [38] immiscible with commonly employed non-polar organic solvents [35]. Also, it is a white crystalline solid [39]. In recent years, SA has been extensively used as an efficient heterogeneous catalyst for acid catalyzed reactions, such as, acetalization, [40] esterification, [41-43] functional group protections and deprotections [44] the Michael addition [45]

amino Diels–Alder reactions [46], Biginelli condensations [47] and Beckmann rearrangement [48].

As the aforesaid conditions are not compatible with heat- or acid-sensitive substrates, there is a need to develop an effective synthesis of quinoxalines employing more ecofriendly conditions and catalysts [49-60]. With considering green chemistry principles, especially using water greenest and most abundant solvent [61-70]. We have recently reviewed the application of Doebner reaction in synthesis of heterocycles including quinolone-4-carboxylic acids [71]. We also frequently used Sulfamic acid as an efficient catalyst in different organic transformations [72-80] and published a review highlighting the application of Sulfamic acid in organic synthesis [81,82]. Armed with these experiences, in response to the need for the facile, efficient and green synthesis of quinolone-4-carboxylic acid derivatives, herein we wish to report the high yielding, one pot MCRs synthesis of this heterocyclic *via* condensation of pyruvic acid, various amines and aryl benzaldehyde in the presence of SA as a relatively inexpensive and available catalyst in water under mild reaction conditions. These eco-friendly protocols offer several advantages such as green, convenient and cost-effective procedures with high yield, shorter reaction time, simpler work-up, recovery, and reusability of solid acid heterogeneous catalyst (SA) in subsequent reactions.

Results and Discussion

To optimize the reaction conditions, a mixture of pyruvic acid (1.2 mmol), aniline (1.1 mmol), and benzaldehyde (1.0 mmol) was selected as a model reaction (Scheme 1) and was examined under different conditions. Initially, we examined various catalysts such as NH_2SO_3H, $H_6P_2Mo_{18}O_{62}$, HNO_3, DABCO, Nano-Fe_3O_4 and also tested the un-catalyzed reaction (Table 1). Because of type of reaction that needs acidic or basic catalyst, or in another word needs a stimulus to begin a reaction and also base on previous work [82] we choice these catalysts to examine to in this reaction (Scheme 1). As shown in Table 1, among the catalysts tested, NH_2SO_3H(SA) gave the highest yield for the corresponding quinoline derivative (Table 1) [83-85].

---

*Corresponding author: Majid M, Department of Chemistry, School of Science, Alzahra University, Vanak, Tehran, Iran, E-mail: mmh1331@yahoo.com

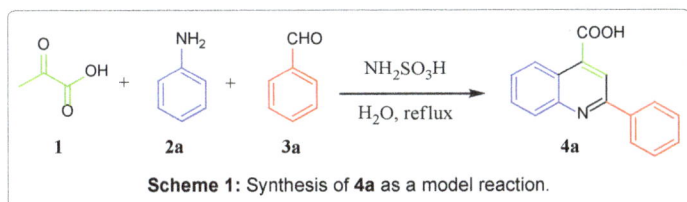

Scheme 1: Synthesis of **4a** as a model reaction.

Entry	Catalyst (mol %)	Time (hr)	Yield[b] (%)
1	NH_2SO_3H	3	90
2	$H_6P_2Mo_{18}O_{62}$	3	85
3	HNO_3	3	81
4	DABCO	4.5	59
5	Nano-Fe_3O_4	6	68
6	-	18	Trace

Table 1: Synthesis of 4a in presence of different catalysts in reflux condition. Synthesis of 4a in presence of different catalysts in reflux condition. a 1.2 mmol (1), 1.1 mmol (2a) and 1.0 mmol (3a) in presence of 3 mol% of catalyst and 5 mL of H_2O. [b]Refers to the isolated yield.

The other catalysts examined such as HNO_3 and $H_6P_2MO_{18}O_{62}$ were also effective but generally not as much (Table 1). To find a suitable solvent, the model reaction was carried out in different solvents (Table 2). The best result was achieved by conducting the reaction in the presence of 3 mol% of SA under reflux conditions in water (Table 2) [82,86]. It was also found out that the best result is achieved in the presence of 3 mol% of catalyst(SA) under reflux condition (Table 3). To establish, the substrate scope of this methodology, we reacted divergent benzaldehydes, and amines with pyruvic acid under the optimized reaction conditions to obtain the corresponding qiounoline-4-carboxylic acids in excellent yields (Scheme 2; 4a-i). All compounds were known and their structures were confirmed by comparison of physical and spectral data with those of already reported (Table 4). The results were listed in Table 4 [82]. A proposed mechanism for the SA-catalyzed formation of quinoline-4-carboxylic acid derivatives is proposed and shown in Scheme 3. The mechanism of this reaction using other catalysts was suggested and explained in recent years [90-93]. The proposed Sulfamic acid-catalyzed reaction involves the 1,2- addition of the aniline to aromatic aldehyde to form adducts **5**, which followed by elimination of H_2O to achieve intermediate **6**. The latter is subjected to Mannich reaction with pyruvic acid which followed by intermolecular cyclization, to afford the intermediate **8**, which in turn dehydrated and then dehydrated and oxidized to give the corresponding quinoline **4** [82] (Scheme 3). From the green chemistry points of view, experiments concerning the recycling and reuse of the catalyst (SA) was carried out. The insolubility of the catalyst SA in water made the catalyzed reaction, heterogeneous, thus the catalyst is separated by a simple filtration. Upon the completion of the reaction, monitored and observed with TLC, the catalyst SA was separated by filtration and reused. As shown in Figure 1, the catalyst could efficiently have recovered even after 4 cycles without suffering any significant drop in its catalytic activity or the yield of reaction (Figure 1) [82,83]. In summary, an extremely facile, high-yielding method for the preparation of quinoline-4-carboxylic acid derivatives under mild reaction conditions, *via* a one-pot MCR, in the presence of an efficient, highly effective, economically, and reusable solid acid heterogeneous catalyst in was developed. This catalyst could easily be separated from the reaction mixture with high yields and purity and can be directly reused after simple extraction. In addition, using water as green solvent and operational simplicity are the merits of this method. The procedure has many advantages such as short reaction time, the requirement of small amount of catalyst, mild acidity conditions, giving virtually quantitative yields and being performed under environmentally friendly conditions.

Experimental Section

Materials and equipment's

Chemicals and solvents were purchased from Merck-Aldrich and used directly with high-grade quality, without any purification. TLC

Entry	Solvent	Time (hr)	Temperature (°C)	Yield[b] (%)
1	H_2O	3	Reflux	90
2	CH_3CH_2OH	6.5	Reflux	80
3	CH_3CN	18	Reflux	Trace
4	-	26	100	69

Table 2: Synthesis of 4a in presence of different solvents. Synthesis of 4a in presence of different solvents. a 1.2 mmol (1), 1.1 mmol (2a) and 1.0 mmol (3a) in presence of 3 mol% of catalyst (Sulfamic acid) and 5 mL of solvent. [b]Refers to the isolated yield.

Entry (3212)	Catalyst mol %	Time (hr)	Temperature (°C)	Yield[b] (%)
1	2	4.5	Reflux	78
2	3	3	Reflux	90
3	4	3	Reflux	90
4	5	3	Reflux	90
5	2	5	100	62
6	3	5.5	100	50
7	2	8.5	120	43
8	4	9	130	75
9	3	4	70-80	71
10	3	5.5	90-100	77
11[c]	3	8	110-120	80

Table 3: Synthesis of 4a in presence of different amount of catalyst and temperature. Synthesis of 4a in presence of different amount of catalyst (Sulfamic acid) and temperature. a 1.2 mmol (1), 1.1 mmol (2a) and 1.0 mmol (3a). [b]Refers to the isolated yield. [c]In solvent free condition.

Entry	R^1	R^2	Product	Time (hr)	Yield[b] (%)	Mp (°C)/Lit. Mp [Ref]
1	H	H	4a	3	90	211-212/212-214 [82]
2	4-Me	H	4b	4	88	213-214/218-219 [83]
3	4-NO_2	H	4c	11	43	233-235/238-239 [84]
4	3-Cl	H	4d	4	80	256-259/256-260 [85]
5	H	4-OMe	4e	7	83	213-215/216-217 [86]
6	H	4-OH	4f	3	89	268-270/271-273 [87]
7	H	4-NO_2	4g	3	90	319-322/324 [88]
8	4-Me	4-OMe	4h	3.5	79	231-233/235-237 [89]
9	4-Me	4-NO_2	4i	5.5	91	263/265-266 [82]

Table 4: Synthesis of quinoline-4-carboxylic acid derivatives (4a-i) under optimized conditions Synthesis of quinoline-4-carboxylic acid derivatives (4a-i) under optimized conditions. a 1.2 mmol (1), 1.1 mmol (2) and 1.0 mmol (3) in presence of 3 mol% of catalyst (Sulfamic acid) and 5 mL H_2O at reflux condition. [b]Refers to the isolated yield.

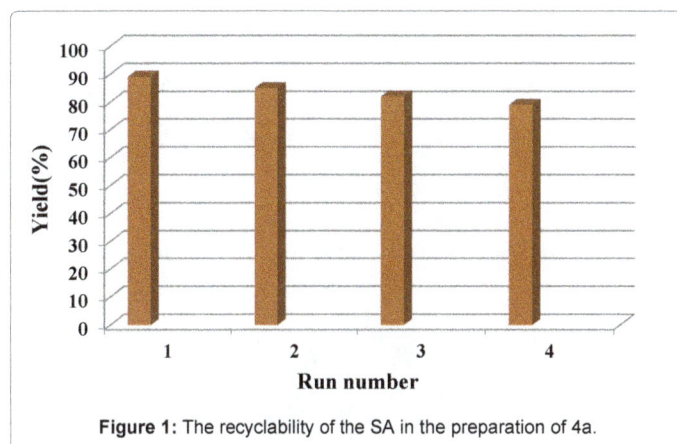

Figure 1: The recyclability of the SA in the preparation of 4a.

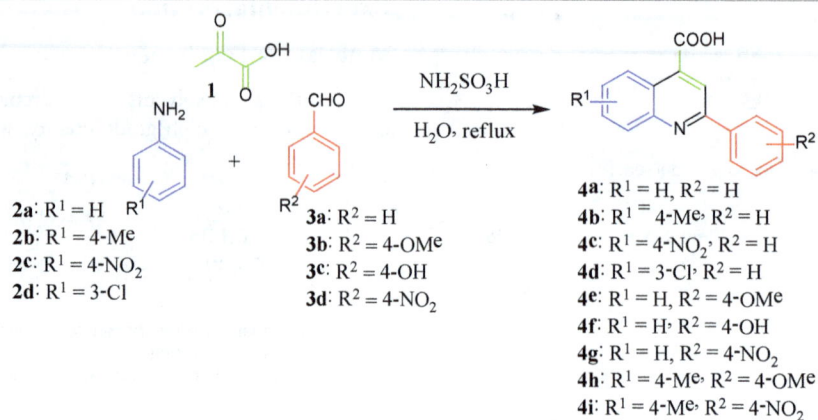

Scheme 2: Synthesis of quinoline-4-carboxylic acid derivatives.

4a: R^1 = H, R^2 = H
4b: R^1 = 4-Me, R^2 = H
4c: R^1 = 4-NO$_2$, R^2 = H
4d: R^1 = 3-Cl, R^2 = H
4e: R^1 = H, R^2 = 4-OMe
4f: R^1 = H, R^2 = 4-OH
4g: R^1 = H, R^2 = 4-NO$_2$
4h: R^1 = 4-Me, R^2 = 4-OMe
4i: R^1 = 4-Me, R^2 = 4-NO$_2$

2a: R^1 = H
2b: R^1 = 4-Me
2c: R^1 = 4-NO$_2$
2d: R^1 = 3-Cl

3a: R^2 = H
3b: R^2 = 4-OMe
3c: R^2 = 4-OH
3d: R^2 = 4-NO$_2$

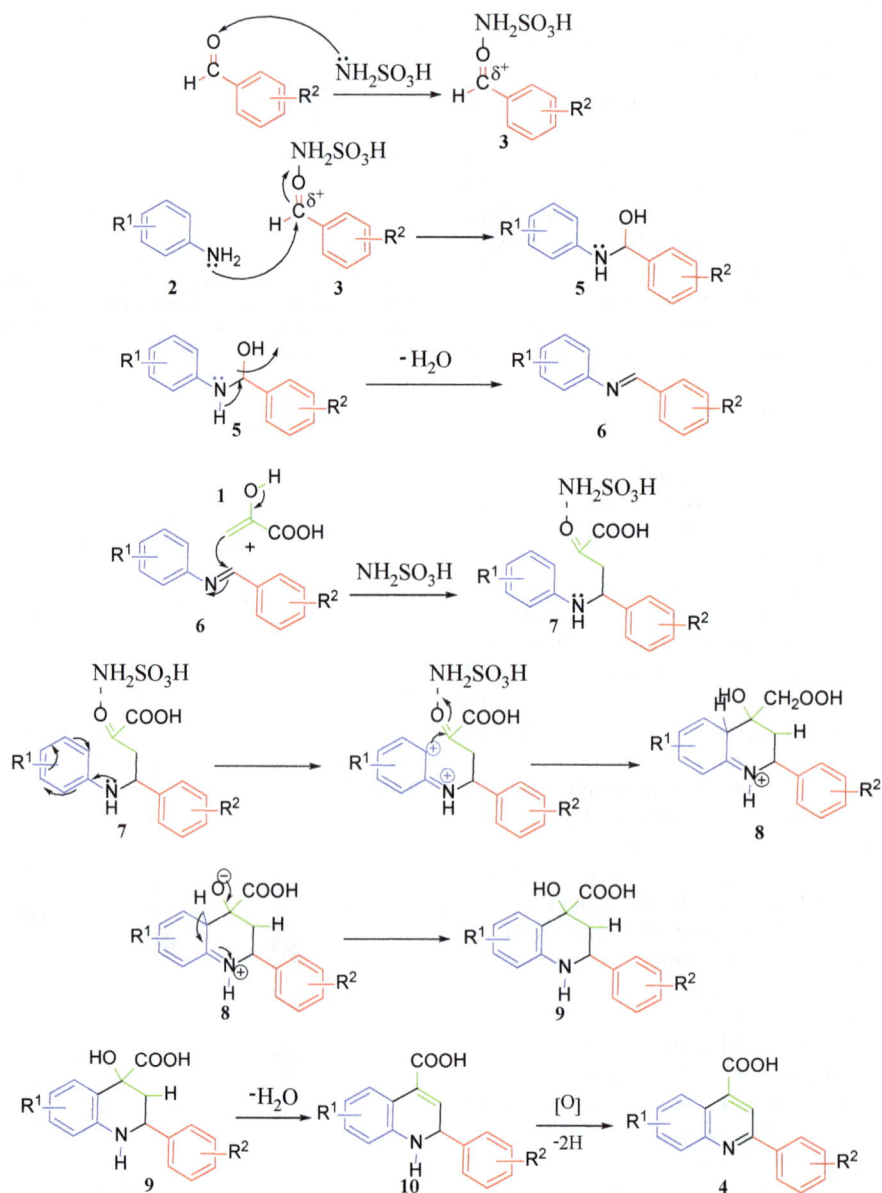

Scheme 3: Proposed mechanism of the synthesis of **4**.

analyses were done using percolated TLC silica gel 60 F$_{254}$ (Merck) plates. The ^{1}H NMR spectra were recorded by Bruker Ultrashield 500 MHz respectively advance instrument, with CDCl$_3$ used as solvent. Proton resonances are designated as singlet (s), doublet (d), triplet (t) and multiplet (m). FT-IR spectra were recorded using KBr disks on FTIR Bruker Tensor 27 instrument in the 500-4000 cm^{-1} region. The vibration transition frequencies are reported in wave numbers (cm^{-1}). Band intensities are assigned as weak (w), medium (m), and strong (s). Melting points were measured using a capillary tube method with a Barnstead Electrothermal 9200 apparatus. All products were known and identified by comparison of their physical and spectroscopic data with those of authentic compounds and found being identical. All yields refer to isolated products.

The synthesis of quinoline-4-carboxylic acid derivatives (4a-i)

General procedure: A mixture of pyruvic acid (1.2 mmol), amine (1.1 mmol), and benzaldehyde (1.0 mmol) in 5 ml of water was well stirred with NH$_2$SO$_3$H (3 mol%). This mixture was magnetically stirred under reflux for appropriate time according to Table 4. The progress of reaction was monitored by TLC (7:3, n-hexane/ethyl acetate). Upon completion of the reaction, the mixture was cooled to room temperature. After filtration, the catalyst which reminded in the aqueous phase can be recovered simply by removing the water through heating and the aqueous phase, and stored for use in subsequent run under the same conditions without any treatment.

The remaining solution is concentrated by increasing slightly CH$_2$Cl$_2$, and was magnetically stirred for 5 min. The precipitate was filtered to give quinoline-4-carboxylic acid as a crude. This crude was recrystallized with acetic acid to get pure product (Table 4; 4a-i) All the triazoles are known compounds, and their physical data were found to be identical with those of authentic samples [82-89].

Compound 4a: m.p.: 211-212°C [82]. IR (KBr): n=3429, 3335, 3042, 1679, 1590, 1487,1378 cm^{-1}; ^{1}H NMR (500 MHz, CDCl$_3$): d=8.76 (d, 1H, J=8.5 Hz), 8.59 (s, 1H), 8.22 (d, 1H, J=8.5 Hz), 8.36 (d, 2H, J=7.4 Hz), 7.30 (t, 1H, J=7.5 Hz), 7.53 (t, 1H, J=7.5 Hz), 7.95 (m, 2H), 6.50 (m, 1H).

Compound 4b: m.p.: 213–214°C [87]. ^{1}H NMR (500 MHz, CDCl$_3$): d=7.50 (s, 1H), 8.55 (s, 1H), 8.20 (m, 2H), 8.10 (d, 1H, J=8.4 Hz), 7.36 (d, 1H, J=8.4 Hz), 7.70 (t, 2H, J=7.4 Hz), 7.22 (m, 1H), 2.62 (s, 3H).

Compound 4c: m.p.: 233–235°C [88]. ^{1}H NMR (500 MHz, CDCl$_3$): d=8.86 (d, 1H, J=8.9 Hz), 7.45 (m, 2H), 7.00–7.40 (m, 5H), 8.15 (m, 1H).

Compound 4d: m.p.: 256–259°C [89]. ^{1}H NMR (500 MHz, CDCl$_3$): d=8.52 (s, 1H), 8.20–8.39 (m, 2H), 7.49 (m, 5H), 7.89 (d, 1H, J=9.1 Hz).

Compound 4e: m.p.: 213–215°C [90]. ^{1}H NMR (500 MHz, CDCl$_3$): d=8.20 (s, 1H), 7.23 (d, 1H, J=8.9 Hz), 7.10 (d, 2H, J=8.0 Hz), 7.42 (t, 1H, J=7.0 Hz), 7.30 (d, 2H, J=8.9 Hz), 8.47 (d, 1H, J=8.0 Hz). 6.76 (t, 1H, J=7.1 Hz), 3.72 (s, 3H).

Compound 4f: m.p.: 268–270°C [91]. ^{1}H NMR (500 MHz, CDCl$_3$): d=8.23 (d, 1H, J=8.7 Hz), 7.51 (s, 1H), 8.22 (m, 3H), 7.50 (t, 1H, J=7.3 Hz), 8.38 (d, 2H, J=8.7 Hz), 6.62 (t, 1H, J=7.3 Hz).

Compound 4g: m.p.: 319-322°C [92].^{1}H NMR (500 MHz, CDCl$_3$): d=8.25 (d, 1H, J=8.0 Hz), 7.35 (s, 1H), 8.83 (m, 2H), 7.13 (d, 1H, J=8.1 Hz),7.78 (t, 1H, J=7.6 Hz), 7.56 (t, 1H, J=7.5 Hz), 7.00 (m, 2H).

Compound 4h: m.p.: 231-233°C [93]. ^{1}H NMR (500 MHz, CDCl$_3$): d=8.22 (s,1H), 8.41 (s, 1H), 8.29 (d, 2H, J=8.3 Hz), 7.16 (d, 1H, J=8.2 Hz), 7.85(d, 1H, J=8.3 Hz), 7.25 (d, 2H, J=8.4 Hz), 2.90 (s, 3H), 2.62 (s, 3H).

Compound 4i: m.p.: 263°C [82]. ^{1}H NMR (500 MHz, CDCl$_3$): d=8.95 (s,1H),7.52 (s, 1H), 8.47 (m, 4H), 8.49 (d, 1H, J=8.7 Hz), 7.23 (d, 1H, J=9.4 Hz),2.85 (s, 3H).

Acknowledgements

The authors are thankful to Alzahra University Research Council for partial financial assistance.

References

1. Abadi AH, Hegazy GH, Zaher AAE (2005) Synthesis of novel 4-substituted-7-trifluoromethylquinoline derivatives with nitric oxide releasing properties and their evaluation as analgesic and anti-inflammatory agents. Bio org Med Chem 13: 5759-5765.

2. Larsen RD, Corley EG, King AO, Carrol JD, Davis P, et al. (1996) Practical Route to a New Class of LTD4 Receptor Antagonists. Org Chem 61: 3398-3405.

3. Chen YL, Fang KC, Sheu JY, Hsu SL, Tzeng CC, et al. (2001) Synthesis and antibacterial evaluation of certain quinolone derivatives. J Med Chem 44: 2374-2377

4. Roma G, Braccio MD, Grossi G, Mattioli F, Ghia M, et al. (2000) Synthesis and antioxidant properties of novel quinoline–chalcogenium compounds. Eur J Med Chem 35: 1021

5. Kalluraya B, Sreenivasa S (1998) Synthesis and pharmacological properties of some quinoline derivatives. Farmaco II 53: 399-404.

6. Doube D, Blouin M, Brideau C, Chan C, Desmarais S (1998) Bioorg Med Chem Lett 8: 1255.

7. Ko TC, Hour MJ, Lien JC, Teng CM, Lee KH (2001) Synthesis of 4-alkoxy-2-phenylquinoline derivatives as potent antiplatelet agents. Bioorg Med Chem Lett 11: 279-282.

8. Ferrarini PL, Mori C, Badwneh M, Manera C, Martinelli A, et al. (1997) Heterocycl Chem 34: 1501.

9. Maguire MP, Sheets KR, McVety K, Spada AP, Zilberstein A, et al. (1994) A new series of PDGF receptor tyrosine kinase inhibitors: 3-substituted quinoline derivatives. J Med Chem 37: 2129-2137.

10. Saito I, Sando S, Nakatani K (2001) Improved selectivity for the binding of naphthyridine dimer to guanine-guanine mismatch. Bioorg Med Chem 9: 2381-2385.

11. He C, Lippard SJ, Am J (2001) Chem Soc 40: 1414.

12. Strigacova J, Hudecova D, Varecka L, Lasikova A (2000) Some biological properties of new quinoline-4-carboxylic acid and quinoline-4-carboxamide derivatives. Folia Microbiol (Praha) 45: 305-309.

13. Michael JP (2005) Quinoline, quinazoline and acridone alkaloids. Nat Prod Rep 22: 627-646.

14. Michael JP (2004) Quinoline, quinazoline and acridone alkaloids. Nat Prod Rep 21: 650-668.

15. Yu XY, Hill JM, Yu G, Yang Y (2001) A series of quinoline analogues as potent inhibitors of C. albicans prolyl tRNA synthetase. Biol Med Chem Lett 11: 541-544.

16. Gobec S, Urleb U (2004) Sci. Synth. 16: 845.

17. Kim SY, Park KH, Chung YK (2005) Manganese(IV) dioxide-catalyzed synthesis of quinoxalines under microwave irradiation. Chem Commun 10: 1321-1323.

18. Raw SA, Wilfred CD, Taylor RJK (2003) Preparation of quinoxalines, dihydropyrazines, pyrazines and piperazines using tandem oxidation processes. Chem Commun 18: 2286-2287.

19. Antoniotti S, Dunach E (2002) Direct and catalytic synthesis of quinoxaline derivatives from epoxides and ene-1, 2-diamines. Tetrahedron Lett 43: 3971-3973.

20. Hoi NPB, Xuong ND, Lavit D (1953) Fluorine-containing analogs of 4-hydroxypropiophenone. J Org Chem 18: 910-915.

21. Atwell GJ, Baguley BC, Denny WA (1989) Potential antitumor agents. 57. 2-Phenylquinoline-8-carboxamides as "minimal" DNA-intercalating antitumor agents with in vivo solid tumor activity. J Med Chem 32: 396-401.

22. Dobner VO, Gieseke M, Liebigs J (1887) Ann Chem 242: 290.

23. Allen CFH, Spangler FW, Webster ER (1951) 2-aminopyridine and the doebner reaction. J Org Chem 16: 17-20.

24. Jones G, Katritzky AR, Rees CW, Scriven EFV (1996) Pergamon, New York, USA 5: 167.

25. Cho CS, Oh BH, Kim TJ, Shim SC (2000) Synthesis of quinolines via ruthenium-catalyzed amine exchange reaction between anilines and trialkyl amines. Chem Commun, pp: 1885-1886.

26. Jiang B, Si YC (2002) Zn(II)-Mediated Alkynylation–Cyclization of o-Trifluoroacetyl Anilines: One-Pot Synthesis of 4-Trifluoromethyl-Substituted Quinoline Derivatives. J Org Chem 67: 9449-9451.

27. Skraup H (1880) Chem Ber 13: 2086.

28. Friedlander P (1882) Chem Ber 15: 2572.

29. Mansake RHF, Kulka M (1953) Skraup Synthesis of Quinolines. Org React 7: 59.

30. Linderman RJ, Kirollos SK (1990) Tetrahedron Lett 31: 2689-2798.

31. Theoclitou ME, Robinson LA (2002) Novel facile synthesis of 2, 2, 4 substituted 1, 2-dihydroquinolines via a modified Skraup reaction. Tetrahedron Lett 43: 3907-3910.

32. Clark JH (2002) Solid acids for green chemistry. Acc Chem Res 35: 791-797.

33. Helwani Z, Othman MR, Aziz N, Kim J, Fernando WJN (2009) Solid heterogeneous catalysts for transesterification of triglycerides with methanol: a review. Appl Catal A: General 363: 1-10.

34. Climent MJ, Corma A, Iborra S (2010) Heterogeneous catalysts for the one-pot synthesis of chemicals and fine chemicals. Chem Rev 111: 1072-1133.

35. Wang B (2005) Sulfamic acid: a very useful catalyst. Synlett 2005: 1342-1343.

36. Bo W, Ming YL, Shuan SJ (2003) Ionic liquid-regulated sulfamic acid: chemoselective catalyst for the transesterification of β-ketoesters. Tetrahedron Lett 44: 5037-5039.

37. Wang B, Gu Y, Luo C, Yang T, Yang L, et al. (2004) Sulfamic acid as a cost-effective and recyclable catalyst for liquid Beckmann rearrangement, a green process to produce amides from ketoximes without waste. Tetrahedron Lett 45: 3369-3372.

38. Harbison GS, Kye YS, Penner GH, Grandin M, Monette M (2002) ^{14}N Quadrupolar, ^{14}N and ^{15}N Chemical Shift, and ^{14}N-^{1}H Dipolar Tensors of Sulfamic Acid. J Phys Chem B 106: 10285-10291.

39. Nonose N, Kubota M (1998) Determination of metal impurities in sulfamic acid by isotope dilution electrothermal vaporization inductively coupled plasma mass spectrometry. J Anal At Spectrom 13: 151-156.

40. Jin TS, Sun G, Li YW, Li TS (2002) An efficient and convenient procedure for the preparation of 1, 1-diacetates from aldehydes catalyzed by H_2NSO_3H. Green Chem 4: 255-256.

41. Yiming L, Huaxe ML (1988) 39: 407.

42. Jin L, Ru-Qi Z (2000) Hecheng Huaxue 8: 364.

43. Wang B, He J, Sun RC (2010) Carbamate synthesis from amines and dialkyl carbonate over inexpensive and clean acidic catalyst-Sulfamic acid. Chinese Chem Lett 21: 794-797.

44. An LT, Zou JP, Zhang LL, Zhang Y (2007) Sulfamic acid-catalyzed Michael addition of indoles and pyrrole to electron-deficient nitroolefins under solvent-free condition. Tetrahedron Lett 48: 4297-4300.

45. Nagarajan R, Magesh CJ, Perumal PT (2004) Inter-and intramolecular imino diels-alder reactions catalyzed by sulfamic acid: a mild and efficient catalyst for a one-pot synthesis of tetrahydroquinolines. Synthesis 2004: 69-74.

46. Li JT, Han JF, Yang JH, Li TS (2003) An efficient synthesis of 3, 4-dihydropyrimidin-2-ones catalyzed by NH_2SO_3H under ultrasound irradiation. Ultrason Sonochem 10: 119-122.

47. Heravi MM, Alishiri T (2014) Dimethyl acetylenedicarboxylate. Adv Heterocycl Chem 113: 1.

48. Heravi MM, Talaei B (2014) Ketenes as privileged synthons in the synthesis of heterocyclic compounds, part 1: three-and four-membered heterocycles. Adv Heterocycl Chem 113: 143-244.

49. Heravi MM, Khaghaninejad S, Mostofi M (2015) Pechmann reaction in the synthesis of coumarin derivatives. Adv Heterocycl Chem 112: 1.

50. Heravi MM, Khaghaninejad S, Nazari N (2015) Bischler–Napieralski reaction in the synthesis of isoquinolines. Adv Heterocycl Chem 112: 183-226.

51. Heravi MM, Talaei B (2015) Chapter Three-Ketenes as Privileged Synthons in the Syntheses of Heterocyclic Compounds Part 2: Five-Membered Heterocycles. Adv Heterocycl Chem 114: 147-225.

52. Heravi MM, Vavsari VF (2015) Chapter Two-Recent Advances in Application of Amino Acids: Key Building Blocks in Design and Syntheses of Heterocyclic Compounds. Adv Heterocycl Chem 114: 77-145.

53. Khaghaninejad S, Heravi MM (2014) Paal-Knorr Reaction in the Synthesis of Heterocyclic Compounds. Chem Inform, p: 45.

54. Heravi MM, Zadsirjan V (2015) Chapter Five. Recent Advances in the Synthesis of Benzo [b] furans. Adv Heterocycl Chem 117: 261-376.

55. Heravi MM, Talaei B (2016) Chapter Five. Ketenes as Privileged Synthons in the Synthesis of Heterocyclic Compounds. Part 3: Six-Membered Heterocycles. Adv Heterocycl Chem 118: 195-291.

56. Heravi MM, Asadi S, Lashkariani BM (2013) Recent progress in asymmetric Biginelli reaction. Molecular diversity 17: 389-407.

57. Heravi MM, Hashemi E, Beheshtiha YS, Kamjou K, Toolabi M, et al. (2014) Solvent-free multicomponent reactions using the novel N-sulfonic acid modified poly (styrene-maleic anhydride) as a solid acid catalyst. J Mol Catal A Chem 392: 173-180.

58. Heravi MM, Mousavizadeh F, Ghobadi N, Tajbakhsh M (2014) A green and convenient protocol for the synthesis of novel pyrazolopyranopyrimidines via a one-pot, four-component reaction in water. Tetrahedron Lett 55: 1226-1228.

59. Nemati F, Heravi MM, Elhampour A (2015) Magnetic nano-Fe_3O_4@ TiO_2/Cu_2O core–shell composite: an efficient novel catalyst for the regioselective synthesis of 1, 2, 3-triazoles using a click reaction. RSC Adv 5: 45775-45784.

60. Heravi MM, Sadjadi S, Oskooie HA, Shoar RH, Bamoharram FF (2008) Heteropolyacids as heterogeneous and recyclable catalysts for the synthesis of benzimidazoles. Catal Commun 9: 504-507.

61. Heravi MM, Zadsirjan V, Bakhtiari K, Oskooie HA, Bamoharram FF (2007) Green and reusable heteropolyacid catalyzed oxidation of benzylic, allylic and aliphatic alcohols to carbonyl compounds. Catal Commun 8: 315-318.

62. Heravi MM, Sadjadi S, Oskooie HA, Shoar RH, Bamoharram FF (2008) The synthesis of coumarin-3-carboxylic acids and 3-acetyl-coumarin derivatives using heteropolyacids as heterogeneous and recyclable catalysts. Catal Commun 9: 470-474.

63. Heravi MM, Hashemi E, Beheshtiha YS, Ahmadi S, Hosseinnejad T (2014) PdCl 2 on modified poly (styrene-co-maleic anhydride): A highly active and recyclable catalyst for the Suzuki–Miyaura and Sonogashira reactions. J Mol Catal A: Chem 394: 74-82.

64. Heravi MM, Tajbakhsh M, Ahmadi AN, Mohajerani B (2006) Zeolites. Efficient and eco-friendly catalysts for the synthesis of benzimidazoles. Monatsh Chem 137: 175-179.

65. Heravi MM, Vazin Fard M, Faghihi Z (2013) Heteropoly acids-catalyzed organic reactions in water: doubly green reactions. Green Chem Lett Rev 6: 282-300.

66. Mirsafaei R, Heravi MM, Ahmadi S, Moslemin MH, Hosseinnejad T (2015) In situ prepared copper nanoparticles on modified KIT-5 as an efficient recyclable catalyst and its applications in click reactions in water. J Mol Catal A: Chem 402: 100-108.

67. Heravi MM, Khorasani M, Derikvand F, Oskooie HA, Bamoharram FF (2007) Highly efficient synthesis of coumarin derivatives in the presence of H14 [NaP$_5$W$_{30}$O$_{110}$] as a green and reusable catalyst. Catal Commun 8: 1886-1890.

68. Heravi MM, Bakhtiari K, Zadsirjan V, Bamoharram FF, Heravi OM (2007) Aqua mediated synthesis of substituted 2-amino-4H-chromenes catalyzed by green and reusable Preyssler heteropolyacid. Bioorg Med Chem Lett 17: 4262-4265.

69. Heravi MM, Asadi SH, Azarakhshi F (2014) Curr Org Chem 11: 701.

70. Heravi MM, Alinejhad H, Derikvand F, Oskooie HA, Baghernejad B, et al. (2012) NH$_2$SO$_3$H and H$_6$P$_2$W$_{18}$O$_{62}$·18H$_2$O-Catalyzed, Three-Component, One-Pot Synthesis of Benzo [c] acridine Derivatives. Syn Commun 42: 2033-2039.

71. Heravi MM, Saeedi M, Beheshtiha YS, Oskooie HA (2011) One-pot chemoselective synthesis of novel fused pyrimidine derivatives. Chem Heterocycl Compd (NY) 47: 737-744.

72. Saeedi M, Beheshtiha YS, Heravi MM (2011) Synthesis of Novel Tetrahydro [4,5] imidazo [2,1-b]-chromeno [4,3,2-de] quinazoline and Benzothiazol-2-ylaminoxanthenone Derivatives. Heterocycles 83: 1831-1841.

73. Heravi MM, Saeedi M, Beheshtiha YS, Oskooie HA (2011) One-pot synthesis of benzochromeno-pyrazole derivatives. Molecular diversity 15: 239-243.

74. Heravi MM, Alinejhad H, Bakhtiari K, Oskooie HA (2010) Sulfamic acid catalyzed solvent-free synthesis of 10-aryl-7, 7-dimethyl-6, 7, 8, 10-tetrahydro-9H-[1, 3]-dioxolo [4, 5-b] xanthen-9-ones and 12-aryl-9, 9-dimethyl-8, 9, 10, 12-tetrahydro-11H-benzo [a] xanthen-11-ones. Molecular diversity 14: 621-626.

75. Heravi MM, Bakhtiari K, Alinejhad H, Saeedi M, Malakooti R (2010) MCM-41 Catalyzed Efficient Regioselective Synthesis of β-Aminoalcohol under Solvent-free Conditions. Chin J Chem 28: 269-272.

76. Heravi MM, Derikvand F, Ranjbar L (2010) Sulfamic acid–catalyzed, three-component, one-pot synthesis of [1,2,4] triazolo/benzimidazolo quinazolinone derivatives. Synth Commun 40: 677-685.

77. Heravi MM, Ranjbar L, Derikvand F, Alimadadi B (2008) Three-component one-pot synthesis of 4, 6-diarylpyrimidin-2 (1H)-ones under solvent-free conditions in the presence of sulfamic acid as a green and reusable catalyst. Molecular diversity 12: 191-196.

78. Heravi MM, Ranjbar L, Derikvand F, Bamoharram FF (2007) Sulfamic acid as a cost-effective catalyst instead of metal-containing acids for the one-pot synthesis of β-acetamido ketones. J Mol Catal A Chem 276: 226-229.

79. Heravi MM, Baghernejad B, Oskooie HA (2009) Application of sulfamic acid in organic synthesis-A short review. Curr Org Chem 13: 1002-1014.

80. Wang LM, Hu L, Chen HJ, Sui YY, Shen W (2009) One-pot synthesis of quinoline-4-carboxylic acid derivatives in water: Ytterbium perfluorooctanoate catalyzed Doebner reaction. J Fluor Chem 130: 406-409.

81. Wang L, Han J, Sheng J, Tian H, Fan Z (2005) Rare earth perfluorooctanoate [RE (PFO)3] catalyzed one-pot Mannich reaction: three component synthesis of β-amino carbonyl compounds. Catal Commun 6: 201-204.

82. Iijima H, Kato T, Söderman O (2000) Variation in degree of counterion binding to cesium perfluorooctanoate micelles with surfactant concentration studied by [133]Cs and ^{19}F NMR. Langmuir 16: 318-323.

83. Furó I, Sitnikov R (1999) Order parameter profile of perfluorinated chains in a micelle. Langmuir 15: 2669-2673.

84. Dvinskikh SV, Furo I (2000) Order parameter profile of perfluorinated chains in a lamellar phase. Langmuir 16: 2962-2967.

85. Boa AN, Canavan SP, Hirst PR, Ramsey C, Stead AM, et al. (2005) Synthesis of brequinar analogue inhibitors of malaria parasite dihydroorotate dehydrogenase. Bioorg Med Chem 13: 1945-1967.

86. McCloskey CM (1952) 6-Nitrocinchophen and Related Substances. J Am Chem Soc 74: 5922-5924.

87. Rapport MM, Senear AE, Mead JF, Koepfli JB (1946) The Synthesis of Potential Antimalarials. 2-Phenyl-α-(2-piperidyl)-4-quinolinemethanols. J Am Chem Soc 68: 2697-2703.

88. Lindwall HG, Bandes J, Weinberg I (1931) Preparation of certain brominated cinchophens. J Am Chem Soc 53: 317-319.

89. Belen'kaya RS, Boreko EI, Zemtsova MN, Kalinina MI, Timofeeva MM, et al. (1981) Synthesis and antiviral activity of 2-[aryl (hetaryl)] quinoline-4-carboxylic acids. Pharm Chem J 15: 171-176.

90. Schneider W, Pothmann A (1941) Über die Lösungsfarben von Phenolbetainen der Chinolinreihe. Berichte der deutschen chemischen Gesellschaft (A and B Series) 74: 471-493.

91. Boykin Jr DW, Patel AR, Lutz RE (1968) Antimalarials. IV. New synthesis of. alpha.-(2-pyridyl)-and. alpha.-(2-piperidyl)-2-aryl-4-quinolinemethanols. J Med Chem 11: 273-277.

92. Wu YC, Liu L, Li HJ, Wang D, Chen YJ (2006) Skraup-Doebner-von Miller quinoline synthesis revisited: Reversal of the regiochemistry for γ-aryl-β, γ-unsaturated α-ketoesters. J Org Chem 71: 6592-6595.

93. Denmark SE, Venkatraman S (2006) On the mechanism of the Skraup-Doebner-Von Miller quinoline synthesis. J Org Chem 71: 1668-1676.

30

Synthesis and Analytical Studies of 3-(4-Acetyl-3-Hydroxyphenyl) Diazenyl)-4-Amino-N-(5-Methylisoxazol-3-Yl)Benzene Sulfonamide with Some Metals

Mohauman Mohammad Al-Rufaie*

Department of Chemistry, College of Science, Kufa University, Iraq

Abstract

Synthesis new organic azo dye as reagent 3-((4-acetyl-3-hydroxyphenyl) diazenyl)-4-amino-N-(5-methylisoxazol-3-yl)benzene sulfonamide (SDA) and Analytical Study of Co(II), Ni(II) and Cu(II), metals complexes. The reagent and its complexes were characterized by elemental analysis, UV-Vis, and molar conductivity measurements. The data show that the complexes have the composition of [MR]X2 type. The conductivity data for all complexes are consistent with those expected for an electrolyte. Octahedral environment is suggested for metal complexes.

Keywords: New azo compound; Analytical studies; Metals complexes; Sulfonamide (SDA)

Introduction

Azo compounds are compounds bearing the functional group (R—N≡N—R'), in which R and R' can be either alkyl or aryl [1]. Aryl azo compounds are more stable than alkyl azo compounds (R and R' aliphatic) [2]. One example is diethyldiazene (Et—N=N—Et). At elevated temperatures, the carbon-nitrogen (C-N) bonds in certain alkyl azo compounds cleave with loss of nitrogen gas to generate radicals like azo bisisobutyl nitrile (AIBN) (Scheme 1) [3].

Aryl azo compounds are stable and have a broad range of colors [4], including yellow, orange, red, brown, and blue. The colors differences are caused by different substituents on the aromatic rings which lead to differences in the extent of conjugation of the π-system in the azo compound. In general, the less extensive the conjugated π-system of a molecule, the shorter the wavelength of visible light it will absorb as shown below:

Colorless → yellow → orange → red → green → blue

(Shortest wavelength)   (Longest wavelength)

Azo compounds constitute one of the largest classes of industrially synthesized organic compounds, for their widespread applications in many areas of dye-stuff industry, pharmacy and dissymmetry due to the presence of azo (-N=N-) linkage 1-3 [1,2]. It can simply be defined as any class of artificial dyes that contains the azo group (-N=N-). Describing a dye molecule as nucleophiles will be known as auxochromes, while the aromatic groups are called chromospheres. The dye molecule is often described as a chromogen [3]. Azo compounds are highly colored and have been used as dyes and pigments for a long time. A large number of (N,N')-donor reagent azo compounds have been prepared in the last years [4]. These are the largest group of organic dyes [5]. A number of these azo dyes have been used as chelating reagents in addition of the uses as reagents in analytical chemistry [6]. The present study reports the preparation, spectral characterization and analytical study of new azo reagent (SDA) and metal complexes. Synthesis of most azo dyes involves diazotization of primary aromatic amines, followed by coupling with one or more nucleophiles [7] (Scheme 2).

Azo compounds are important structures in the medicinal and pharmaceutical fields. Furthermore, azo dye compounds also have a lot of applications in industry and photodynamic therapy as well as photosensitive species in photographic or electro photographic systems and they are dominant organic photoconductive materials [8].

Experimental

Apparatus and materials

All reagents and solvents were obtained from Fluke, The Merck and BDH. The melting points were determined on an Electro thermal, melting point 9300. Elemental analyses were carried out by means of Micro analytical unit of 1108 C.H.N.S Elemental analyzer while the UV-Vis. Spectra recorded in ethanol on Shimaduz model 1650PC. Molar conductance measurements were determined in DMF by using an Alpha Digital conductivity meter model 800. pH measurements were carried out using pH-meter Hanna. The metal content of the complex was measured by using atomic absorption technique by Perkin-Elmer model 2280.

Preparation of the reagent (SDA)

The reagent (SDA) was prepared according to the following general procedure [9,10] 4-amino-2-Hydroxy acetophenone (0.01 mol) (1.511 g) was dissolved in (3 ml) concentrated hydrochloric acid and (15 ml) distilled water. The mixture was cooled at (0-5°C) in ice-water bath. Then a solution of sodium nitrite (0.01 mol) dissolved in (5 ml) of distilled water was cooled at (0-5°C). This solution was added a drop wise to the mixture with stirring at the same temperature. The resulting

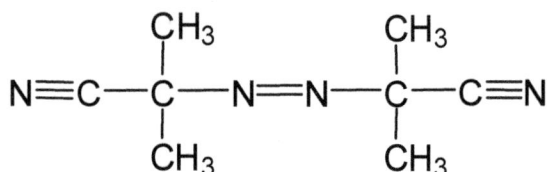

(AIBN)

Scheme 1: Azo bisisobutyl nitrile.

*Corresponding author: Mohauman Mohammad Al-Rufaie, Department of Chemistry, College of Science, Kufa University, Iraq
E-mail: mohaumanmajeed@yahoo.com

diazonium chloride solution was mixed with Sulfamethoxazole (2.5 g, 0.01 mol) dissolved in (200 ml) alkaline ethanol cooled below 0°C. After leaving in the refrigerator for 24 hr, the mixture was acidified with dilute hydrochloric acid until pH=5. The precipitate was filtered off, and re-crystallized twice from hot ethanol, and dried in a vacuum desiccator and shown in the following Scheme 3 [11].

Preparation of metal complexes

The metal complexes were prepared by the mixing of 50 ml ethanolic solution of ($CoCl_2.6H_2O$, $NiCl_2.6H_2O$ and $CuCl_2.2H_2O$) with the 50 ml of ethanolic solution of reagent in (1:1) (metal:reagent) ratio. The resulting mixture was refluxed for 2 h. Colored product appeared on standing and cooling the above solution. The precipitated complexes were filtered, washed and re-crystallized with ethanol several times and dried over anhydrous $CaCl_2$ in desiccators [12,13].

Results and Discussion

The analytical data for the reagent and complexes together with some physical properties are summarized in Table 1 [14]. The analytical data of the complexes correspond well with the general formula $[MR]$ X_2 where M= Co(II), Ni(II), and Cu(II), R=(SDA).

Absorption spectra

The absorption spectra in aqueous ethanolic solution 50% (V/V) were studied for the prepared complexes showed a bath chromic shift

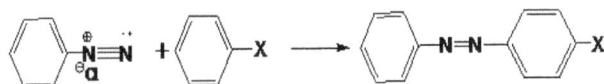

X= electron - donating group   [25]

example : -OH , -NH₂ , -NHR , -NR₂

Scheme 2: Amino and hydroxyl-groups that are commonly used coupling components.

ranging about (84-207) nm. The absorption spectra of reagent (SDA) and Co(II), Ni(II) and Cu(II) chelat complexes is shown in Figures 1-4 [15].

Effect of pH

The effect of acidity of the absorbance values of the complexes was studied in the 50%(v/v) ethanolic by changing the pH value of the solution and the results is shown in Figures 5-7, where demonstrated that the best absorbance of Co(II), Ni(II) and Cu(II) (SDA) system is in the range (6.5-8). The reagent formed stable complexes with metal ions at same pH.

Effect of time

Also the reaction is complete in 5 min at room temperature and remains stable for about 180 min. This show the reagent (SDA) strong coordination with metal ions in this time. The results are shown in Figure 8.

Metal: Reagent ratio

The (metal: reagent) ratios of complexes were determined by molar ratio method at fixed concentration and pH at wavelengths of maximum absorption. The results are given in Table 2, the reagent was found to form (2: 1) chelates with all metal ions.

Calculation of the metal complexes stability constant

Stability constants are obtained spectrophotometrically by measuring the absorbance of solutions of ligand and metal mixture at fixed wavelength λ_{max} and pH values. The degree of formation of the complexes is obtained according to the relationship [16], $\beta=(1-\alpha)/(4\alpha3c2)$, and $\alpha=(Am-As)/Am$, where As and Am are the absorbance's of the partially and fully formed complex respectively at optimum concentration. The calculated β and Log β values for the prepared complexes are recorded in Table 2.

Conclusions

In this present study we report the preparation characterization

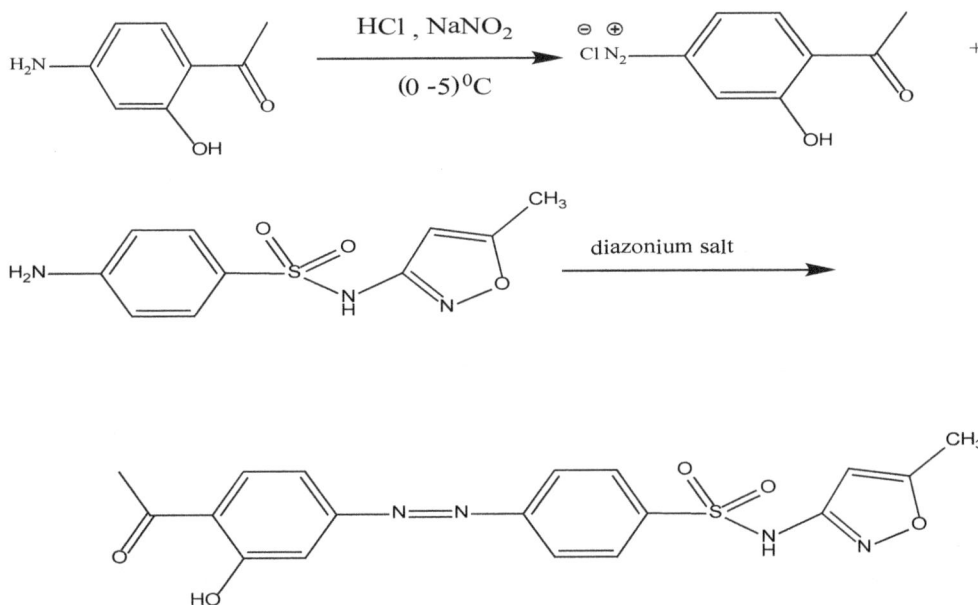

Scheme 3: Scheme of the azo-coupling reaction.

S No	Compound color	MP (°C)	Yield%	Molecular formula	Found (Calc.) %				
					C	H	N	S	M
1	R=(SDA) Brown	250-255	91	$C_{18}H_{16}N_4O_5S$	52.287 (52.42)	3.838 (3.91)	20.025 (20.38)	7.57 (7.77)	-
2	Co-SDA Brown-Red	272-264	82	$C_{18}H_{16}N_4O_5S\,Cl_2Co$	45.101 (45.39)	3.091 (3.17)	17.469 (17.64)	6.523 (6.73)	5.97 (6.19)
3	Ni-SDA Green	206-218	89	$C_{18}H_{16}N_4O_5S\,Cl_2Ni$	45.211 (45.40)	3.03 (3.17)	17.512 (17.65)	6.52 (6.73)	6.021 (6.16)
4	Cu-SDA Blue	223-211	75	$C_{18}H_{16}N_4O_5S\,Cl_2Cu$	44.873 (45.17)	2.98 (3.16)	17.441 (17.56)	6.49 (6.70)	6.28 (6.64)

Table 1: Analytical data and physical properties of the reagent (SDA) and complexes.

S No	Metal ions and color	pH	wave length (λ_{max}) nm	molar conc. × 10^{-5} M	β $L^2.mol^{-2}$	log β	Molar conduc. $S.mol^{-1}.cm^2$
1	Co-SDA Brown-Red	8	625	1	$6.83 × 10^{10}$	10.83	66.23
2	Ni-SDA Green	7.5	510	3.4	$6.39 × 10^{10}$	10.80	63.91
3	Cu-SDA Blue	6.5	608	5	$9.83 × 10^{11}$	11.99	69.21

Table 2: Metal: reagent stability constant value (β), molar conductivity, optimal concentration and wave length.

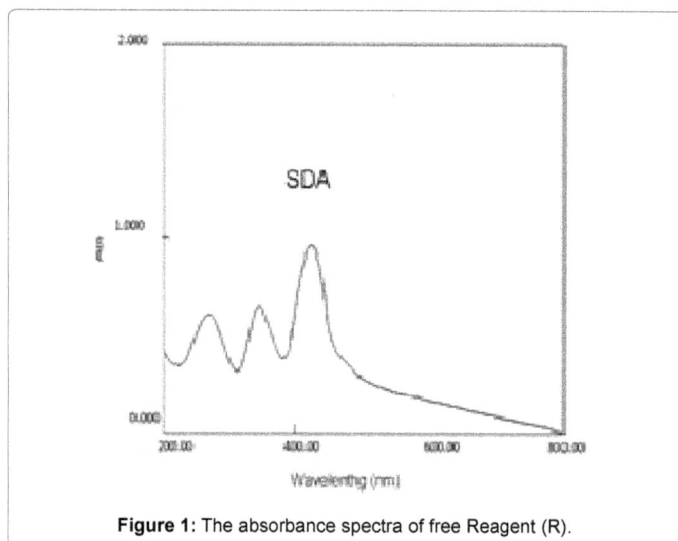

Figure 1: The absorbance spectra of free Reagent (R).

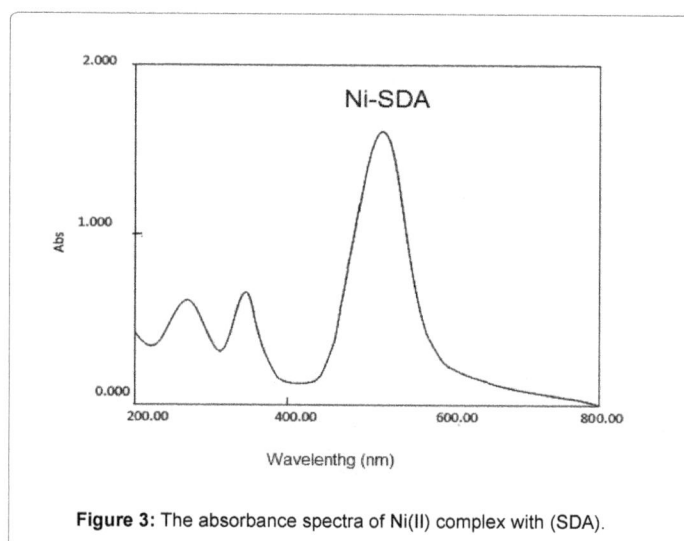

Figure 3: The absorbance spectra of Ni(II) complex with (SDA).

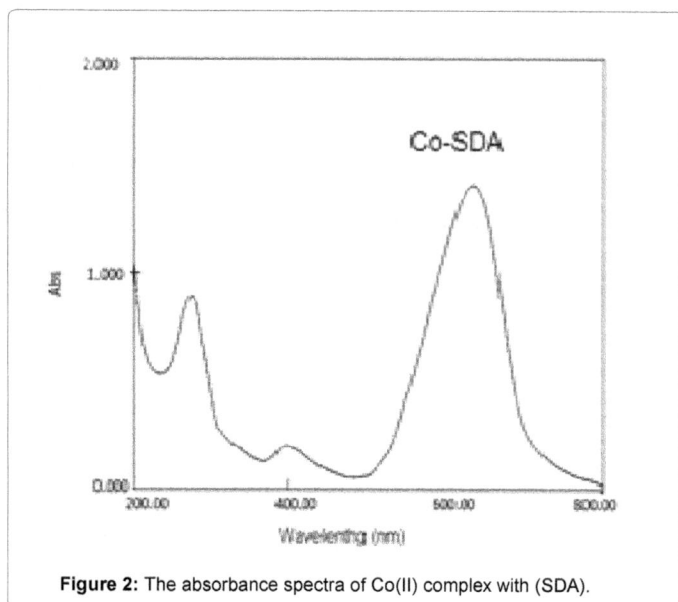

Figure 2: The absorbance spectra of Co(II) complex with (SDA).

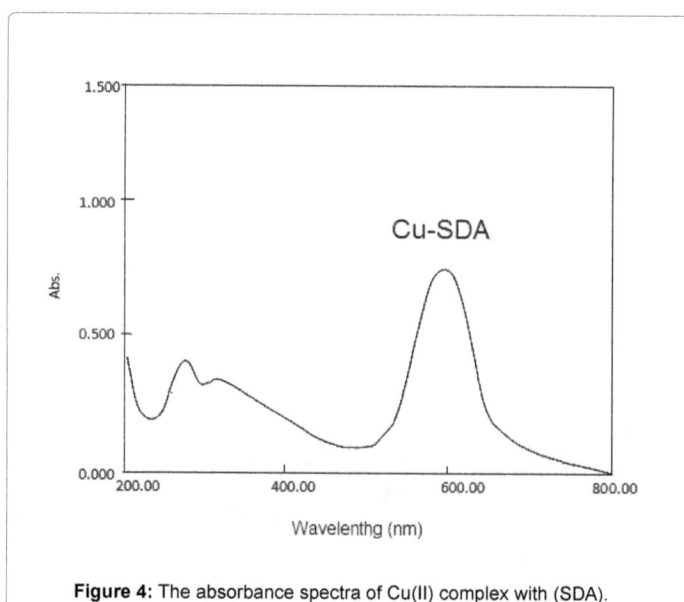

Figure 4: The absorbance spectra of Cu(II) complex with (SDA).

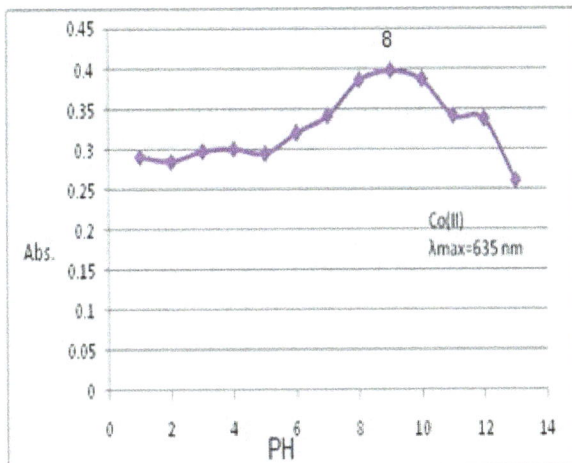

Figure 5: Effect of acidity on Co(II) complex absorbance.

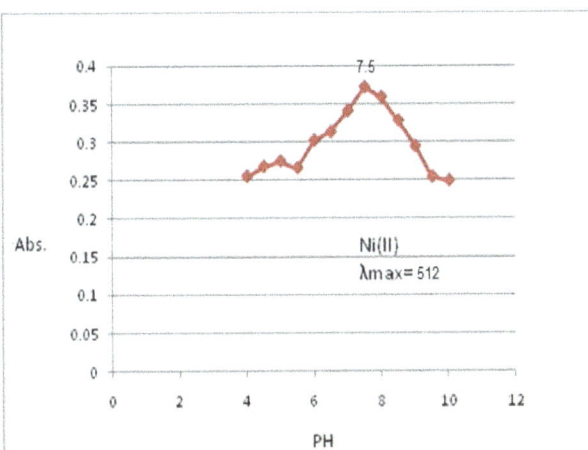

Figure 6: Effect of acidity on Ni(II) complex absorbance.

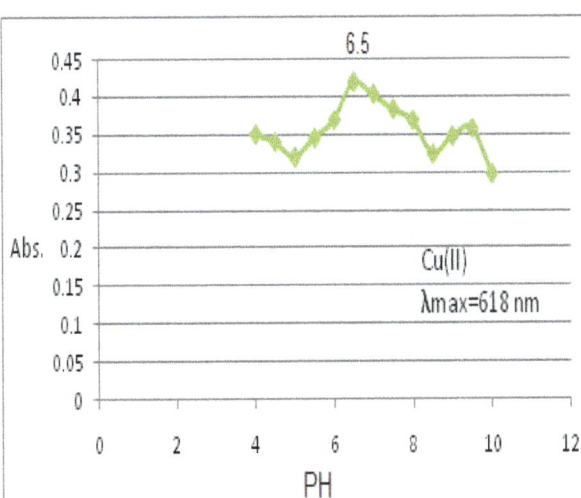

Figure 7: Effect of acidity on Cu(II) complex absorbance.

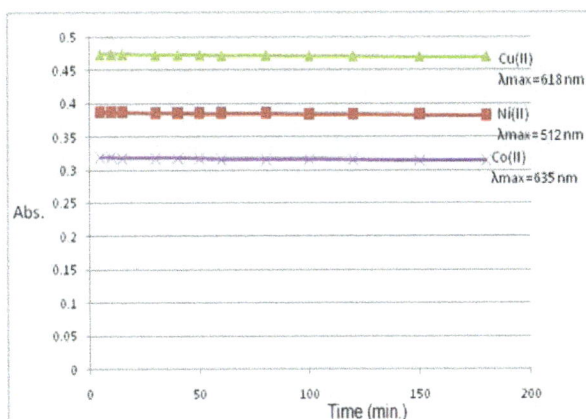

Figure 8: Effect of time on the absorbance of Co(II), Ni(II) and Cu(II) metal complexes at room temperature.

and spectroscopy study of new azo reagent derived from imidazole and its complex with Cu(II) metal ion. The isolated complex was characterized by available techniques. The aryl azo reagent (SDA) behaves as a bidentate chelating agent and coordinating through the N_2 atom of imidazole and another nitrogen atom of azo group which is the farthest of imidazole ring to form five-member metalo ring and oxygen atom for cobalt and nickel complexes. The coordination of the metal ion Cu(II) with reagent(SDA) are to give hexa coordinated show octahedral stereochemistry.

References

1. Rathod KM, Thaker NS (2013) Synthesis and Antimicrobial Activity of Azo Compounds Containing m-Cresol Moiety. Chem Sci Trans 2: 25-28.

2. armion DM (1999) Hand book of colorant. In: Handbook of US colorants. Foods, Drugs, Cosmetics and Medical devices. 3rd edn. Wiley, New York, USA. pp: 23-26.

3. Al-Rubaie LAR, Mhessn RJ (2012) Synthesis and Characterization of Azo Dye Para Red and New Derivatives. E J Chem 9: 465-470.

4. Mathur T, Ray U, Baruri B, Sinha C (2005) Tetrahedral manganese(II) complexes of 1-alkyl-2-(arylazo)imidazoles. X-ray crystal structure of [Mn(HaaiMe)₄](ClO₄)₂·DMF (HaaiMe = 1-methyl-2-(phenylazo)imidazole). J Coord Chem 58: 399-407.

5. Mehdi RT, Ali AM (2005) Preparation and Characterization of New Azo imidazole ligand and Some Transition Metal Complexes. National Journal of Chemistry 20: 540-546.

6. Ali AM, Mehdi RT (2005) Synthesis and Spectroscopic Studies of Some Transition Metal Complexes with 2-[(3-Iodophenyl) azo]-4, 5-diphenyle imidazole. Al-Mustansirya J Sci 16: 1-11.

7. Dinda J, Ray U, Mostafa G, Lu TH, Usman A, et al. (2003) Copper(I)–azoimidazoles: a comparative account on the structure and electronic properties of copper(I) complexes of 1-methyl-2-(phenylazo)imidazole and 1-alkyl-2-(naphthyl-(α/β)-azo)imidazoles. Polyhedron 22: 247-255.

8. Hihara T, Okada Y, Morita Z (2007) A semiempirical molecular orbital study of the photo-reactivity of monoazo reactive dyes derived from γ- and J-acids. Dyes and Pigments 73: 141-161.

9. Skelton R, Dubois F, Zenobi R (2007) A MALDI sample preparation method suitable for insoluble polymers. Anal Chem 72: 1707-1710.

10. Steter JR, Pontólio JO, Campos MLAM, Romero JR (2008) Modified Electrodes Prepared with Polyphenolic Film Containing Ruthenium Complex and Metal Ligand Anchored by Azo Covalent Bond. J Braz Chem Soc 19: 660-666.

11. Ofomaja AE, Ho YS (2007) Equilibrium sorption of anionic dye from aqueous solution by palm kernel fibre as sorbent. Dyes and Pigments 74: 60-66.

12. Salwinska E (2009) Nitroimidazoles. Part X. Synthesis of 1-aryl-4-nitroimidazoles from 1, 4-dinitroimidazoles and primary aromatic amines. Pol J Chem 64: 813-817.

13. Das D, Sinha C (1998) Palladium azoimidazoles. Solvatochromic studies of catecholato complexes. J Trans Met Chem 23: 517-522.

14. Reddy BK, Kumar JR, Reddy KJ, Sarma LS, Reddy AV (2003) A rapid and sensitive extractive spectrophotometric determination of copper(II) in pharmaceutical and environmental samples using benzildithiosemicarbazone. Anal Sci 19: 423-428.

15. Shibata S, Furukawa M, Nakashima R (1976) Syntheses of azo dyes containing 4,5-diphenylimidazole and their evaluation as analytical reagents. Anal Chim Acta 81: 131-141.

16. Gung BW, Taylor RT (2004) Parallel Combinatorial Synthesis of Azo Dyes: A Combinatorial Experiment Suitable for Undergraduate Laboratories. J Chem Educ 81: 1630.

31

Comparative Study for the Analysis of Cefixime Trihydrate and its Degraded Products by Two RP-HPLC Methods, One its Official and Other Developed Validated Method

Elsadig HK* and Abdalfatah MB

Department of Pharmaceutical Chemistry, College of Pharmacy, Prince Sattam Bin Abdulaziz University, Riyadh, Al-kharj, Saudi Arabia

Abstract

The aim of the present work was aimed to carry out comparative study between method (1) and method (2) for separation of cefixime trihydrate and its degraded products by using two different mobile phases, keeping the other parameters such as stationary phase, column condition, wavelength, and device. Mobile phase for method (1) consist of a solution of 0.03 M Tetra butyl ammonium hydroxide (pH 6.5) and acetonitrile with a ratio of 3:1 respectively while Mobile phase for method (2) consist of a mixture of 0.1 M sodium dihydrogen phosphate monohydrate solution (pH 2.5) and methanol with a ratio 3:1 respectively. To study the degraded products sample was subjected to Sun light, UV light, and thermal effects. From data obtained proved the method (2) gave less retention time for the separation of drug with a larger number of decomposed products being detected compared by method (1).

Keywords: Cefixime trihydrate; Comparative study; RP-HPLC method; Cefixime trihydrate; Stability indicating method

Introduction

Cefixime is used to treat a wide variety of bacterial infections. This medication is known as a cephalosporin antibiotic. It works by stopping the growth of bacteria. Cefixime is a broad spectrum cephalosporin antibiotic and is commonly used to treat bacterial infections of the ear, urinary tract, and upper respiratory tract. The bactericidal action of Cefixime is because of the inhibition of cell wall synthesis. It binds to one of the penicillin binding proteins (PBPs) which inhibits the final trans peptidation step of the peptidoglycan synthesis in the bacterial cell wall, thus inhibiting biosynthesis and arresting cell wall assembly resulting in bacterial cell death [1-3]. Cefixime is an orally active 3rd generation cephalosporin which exerts its bactericidal action against both gram positive and gram negative organism by in bacterial cell wall synthesis. Chemically, Cefixime trihydrate name is (6R,7R)-7-[[2-(2-amino-1,3-thiazol-4-yl)-2-(carboxymethoxyimino) acetyl] amino]-3-ethenyl-8-oxo-5-thia-1-azabicyclo [4.2.0]oct-2-ene-2-carboxylic acid trihydrate (Figure 1) and it's molecular formula is $C_{16}H_{15}N_5O_7S_2.3H_2O$ and molecular weight is 507.50 g/mol. It is a white powder that is freely soluble in water (1 g/5 ml) and stable in air, heat and acid solutions, while it is unstable in alkaline medium and light.

Materials and Methods

Materials (Chemicals and reagents)

Cefixime trihydrate was donated from DSM Company, all chemicals and regents used were of a HPLC grade. Tetra butyl ammonium hydroxide 40% aqueous solution, Sodium dihydrogen phosphate monohydrate were obtained from AppliChem, Germany. Methanol and acetonitrile were obtained from fisher scientific UK Limited, UK. Water (HPLC gradient grade) supplied from Panreac, E.U. Orthophasophoric acid 85% was obtained from BDH, England.

Instrument and equipment

a. HPLC instruments a water Breeze 2 system, consisting of binary pump series 1525, UV/VIS detector 2489, and auto sampler series 2707.

b. Sensitive balance, A and D Company limited, Japan.

c. 827 pH lab. metrohm ion analysis, Herisau/ Switzerland.

Experimental

Preparation of mobile phase for method (1) of 0.03 M tetra butyl ammonium hydroxide solution (pH 6.5)

Solution was prepared by weighing 8.2 g Tetra Butyl Ammonium hydroxide (or 20 ml of Tetra Butyl Ammonium hydroxide 40% aqueous solution) and dissolving into 800 ml of distilled water and adjusted to pH 6.5 with 10% ortho phospharic acids and diluted up to 1000 ml with distilled water, and mix with acetonitrile with a ratio of 4:1 respectively and degassed [4].

Preparation of mobile phase for method (2) of 0.1 M sodium dihydrogen phosphate monohydrate solution (pH 2.5)

Solution was prepared by weighing 13.67 g of sodium dihydrogen phosphate monohydrate and dissolving into 900 ml of distilled water and adjusted to pH 2.5 with diluted orthophospharic acid and diluted up to 1000 ml with distilled water, and mix with methanol with a ratio of 3:1 respectively and degassed [5].

Chromatographic conditions used for the analysis of cefixime trihydrate and its degraded

Mobile phase: 1 and 2, Flow Rate, 1.0 ml/min, Injection volume=50 µl, Column=Waters Spherisorb® 5.0 µm ODS2 250 mm × 4.6 mm ID, Temperature=Room temperature (Ambient), Detection wave length at 254 nm.

Preparation of standard stock solution

Stock standard solution having concentration 100 µg/ml was prepared by dissolving pure drug of cefixime trihydrate in water, injected into the chromatographic column (Figures 2-4; Table 1).

*Corresponding author: Elsadig HK, Department of pharmaceutical chemistry, College of Pharmacy, Prince Sattam Bin Abdulaziz University, Riyadh, Al-kharj, Saudi Arabia, E-mail: elsadigk@yahoo.com

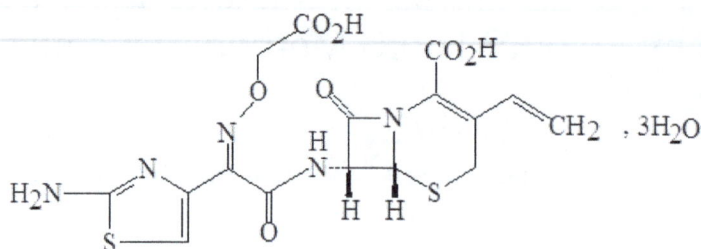

Figure 1: Structural formula of Cefixime Trihydrate.

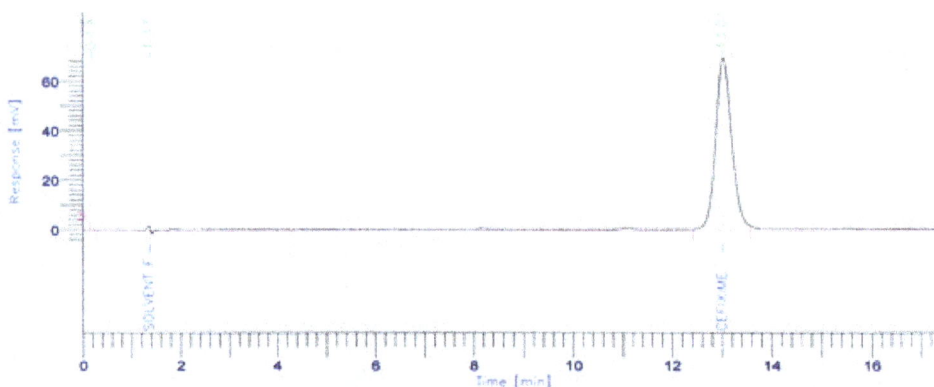

Figure 2: HPLC chromatogram test for the analysis of cefixime trihydrate reference standard by method 1.

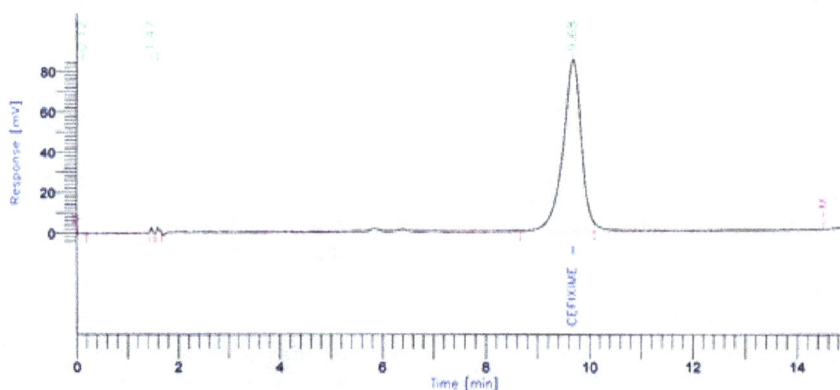

Figure 3: HPLC chromatogram test for the analysis of cefixime trihydrate reference standard by method.

Preparation of degradation product of cefixime

Preparation of cefixime trihydrate solid sun decomposition product: About 5.0 grams of cefixime trihydrate solid were placed between two glass plates (20 ×20 cm), sealed with gum tape and directly exposed to sunlight for six months (March to August). Samples were taken every month and tested for degradation by HPLC (Table 2 and Figure 5).

Preparation of cefixime trihydrate solution UV decomposition product:100 μg / ml of cefixime trihydrate solution in water were prepared and transferred to a stoppered tube. The solutions were placed under UV radiation at λ 254 nm. Samples were taken at 30, 60, 90, 120, and 150 minutes and tested for degradation by HPLC (Table 3 and Figures 6 and 7).

Preparation of cefixime trihydrate solid thermal decomposition at 100°C: Few grams of CEF-3H$_2$O solid were placed in a petridish and put it in oven at 100°C. Samples were taken every hour and tested for degradation by HPLC (Table 4 and Figures 8 and 9).

Preparation of cefixime trihydrate solution thermal decomposition at 100°C for 45 minutes: Solution of cefixime trihydrate (10 mg / 100 ml water) was prepared. The flask was placed into a water-path thermostatic at 100°C for 45-minute (Table 5; Figure 10).

Results and Discussion

Under optimization condition for the RP-HPLC method 1 and method 2 with keeping others fixed and changeable mobile phase, the separation chromatogram obtained of cefixime trihydrate

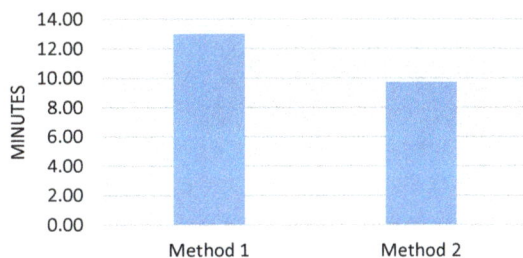

Figure 4: Time vs. method 1 and 2 for the separation of cefixime trihydrate reference standard.

Parameters	Retention Time RT
Method 1	13:03
Method 2	10:15

Table 1: Analysis of cefixime trihydrate reference standard by the method 1 and method 2.

Interval time/month	Method 1% Remaining content	Method 2% Remaining content
1 Month	79.72	78.56
2 Months	64.71	64.51
3 Months	47.45	48.05
4 Months	22.99	22.57
5 Months	11.62	11.65
6 Months	8.11	8.15

Table 2: Analysis of decomposed cefixime trihydrate solid form by sunlight using method 1 and 2.

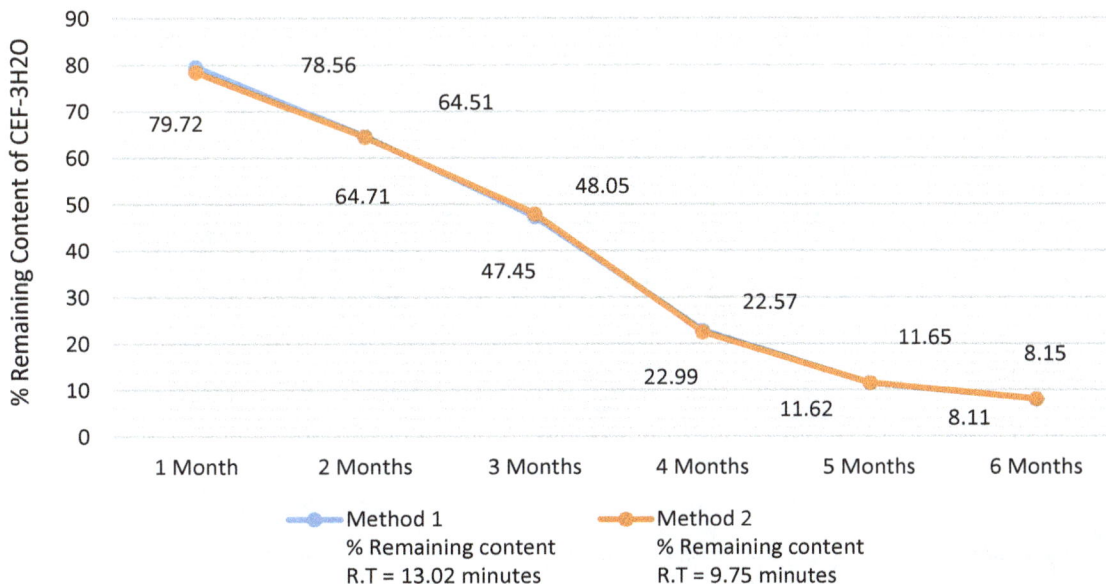

Figure 5: Remaining content of cefixime trihydrate subjected to sunlight analysis by method 1 and 2 vs. Intervals time.

reference standard was appear in Figures 2 and 3, it was found that cefixime trihydrate separation at 9.75 minute by using method 2 while separation at 13.03 minute by using method 1 [6-9]. Discrimination developed validated method 2 versus method 1 illustrated that from the assay test, the retention time of CEF-3H$_2$O less than that time obtained by method 1 were 9.75, 13.03 minutes respectively (Table 1 and Figure 4). There's very little noticeable change when testing for degradation of cefixime trihydrate solid under the influence of sunlight by using the analysis methods 1 and 2 (Table 2 and Figure 5). From the results obtained for the tested degraded products under stress condition of UV for cefixime trihydrate solution, two methods were given two degradation products but better resolution by the method 2 (Table 3 and Figures 6 and 7). The analysis of cefixime trihyrate solid thermal decomposed at 100°C by using method 1 and 2, revealed that method

Results	Method (1)		Method (2)	
	Content %	RT	Content %	RT
Decomposed (1)	17.46%	4.62	18.83%	2.97
Decomposed (2)	23.34%	11.05	22.88%	15.04
Remaining CEF-3H$_2$O	26.89%	13.03	27.19%	9.75

Table 3: Analysis of decomposed cefixime trihydrate solution by UV-light using method 1 and 2.

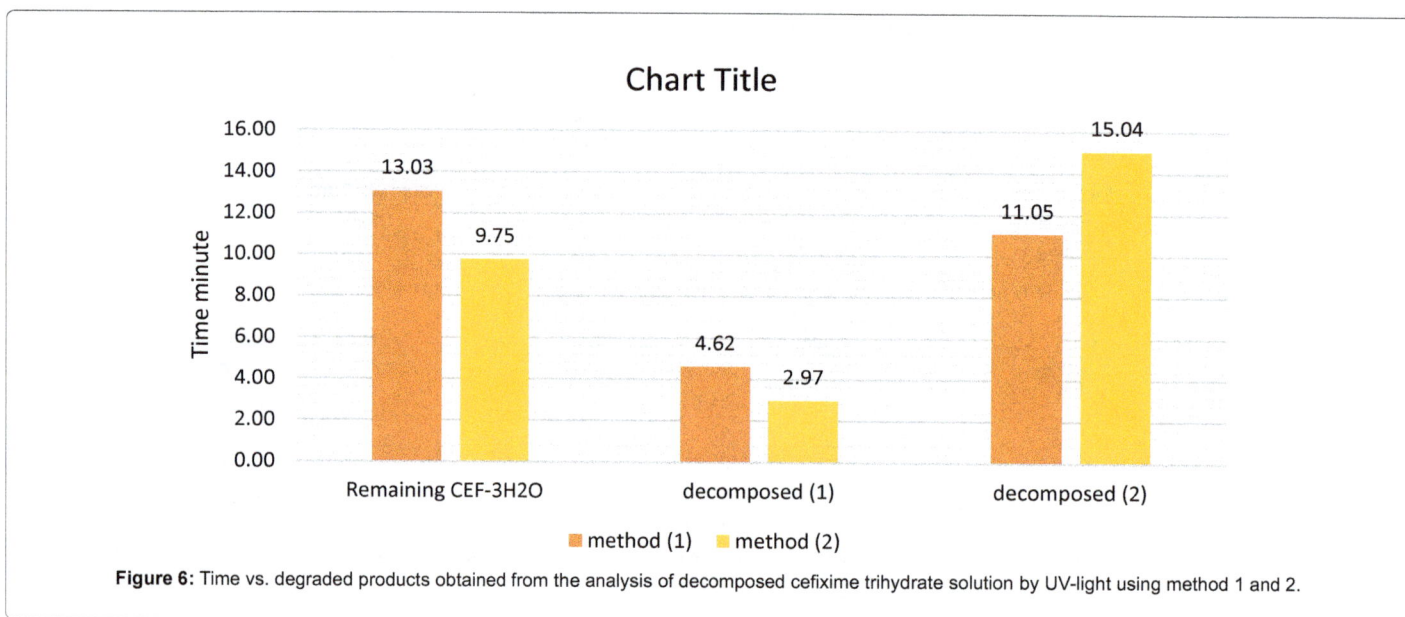

Figure 6: Time vs. degraded products obtained from the analysis of decomposed cefixime trihydrate solution by UV-light using method 1 and 2.

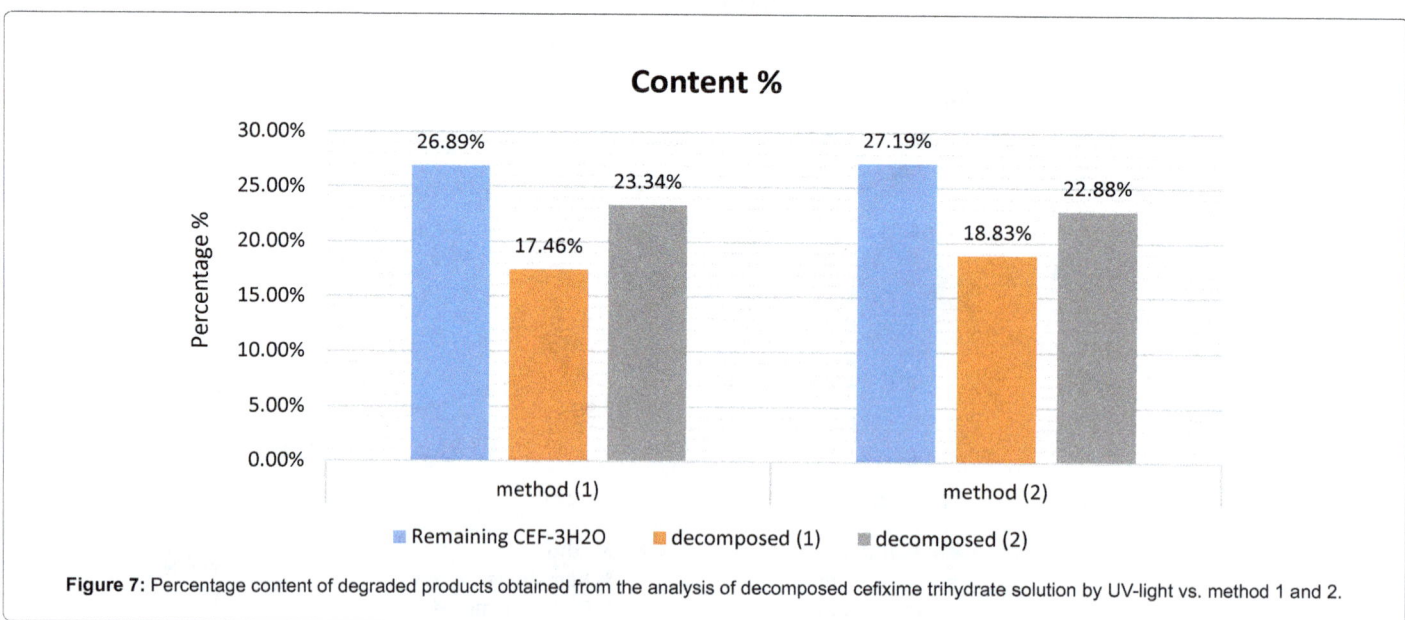

Figure 7: Percentage content of degraded products obtained from the analysis of decomposed cefixime trihydrate solution by UV-light vs. method 1 and 2.

Results	Method (1)		Method (2)	
	Content %	RT	Content %	RT
Decomposed (1)	8.50%	4.51 min	7.89%	3.91 min
Decomposed (2)	12.71%	6.38 min	14.86%	4.216 min
Decomposed (3)	-	-	8.20%	5.899 min
Remaining CEF-3H$_2$O	71.06%	13.014 min	66.34%	9.750 min

Table 4: Analysis of decomposed cefixime trihydrate solid by thermal effect at 100°C using method 1 and 2.

R.T %

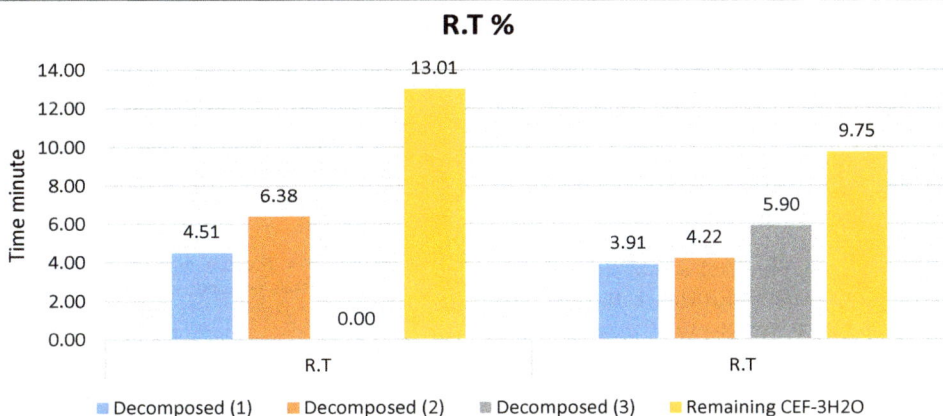

Figure 8: Time vs. degraded products obtained from the analysis of decomposed cefixime trihydrate solid by thermal effect at 100°C using method 1 and 2.

Content %

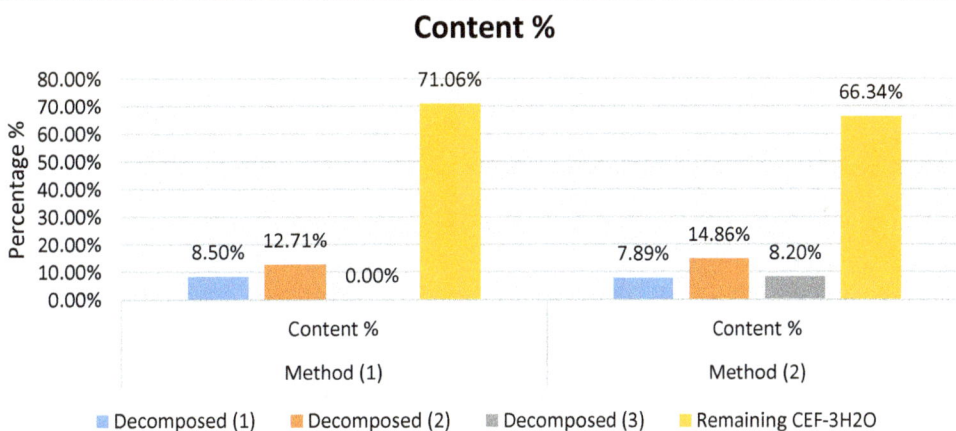

Figure 9: Percentage content of degraded products obtained from the analysis of decomposed cefixime trihydrate solid by thermal effect at 100°C vs. method 1 and 2.

Results	Method (1)		Method (2)	
	Content %	RT	Content %	RT
Decomposed (1)	8.90%	1.78 min	7.86%	14.07 min
Decomposed (2)	37.61%	11.05 min	36.97%	14.99 min
Remaining CEF-3H$_2$O	51.43%	13.02 min	51.69%	9.70 min

Table 5: Analysis of decomposed cefixime trihydrate solution by thermal effect at 100°C using method 1 and 2.

R.T

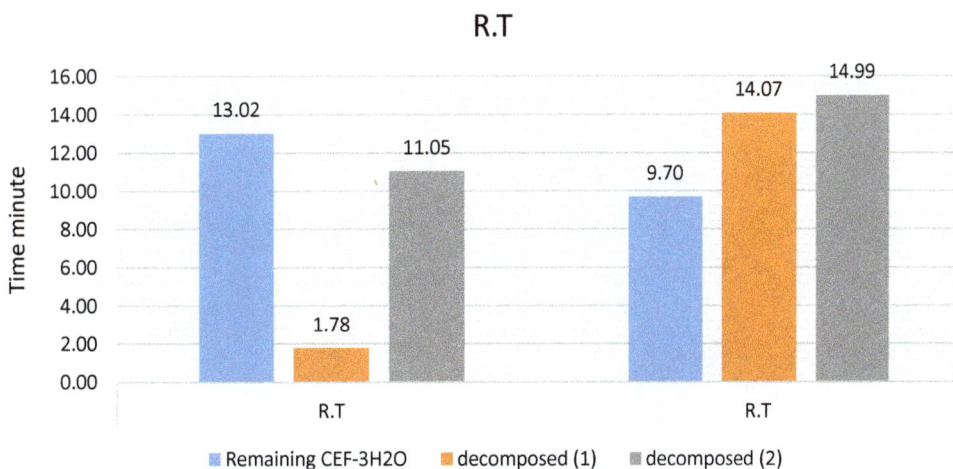

Figure 10: Time vs. degraded products obtained from the analysis of decomposed cefixime trihydrate solution by thermal effect at 100°C using method 1 and 2.

2 detected three decomposed products while method 1 was detect only two decomposed (Table 4 and Figures 8 and 9). There is no noticeable quantitatively and qualitatively change for the analysis of cefixime trihydrate solution thermal decomposed by using method 1 and 2.

Conclusion

This work described the evaluation of analytical method 1 and method 2. The method 2 described enables the quantification and qualification of cefixime trihydrate and its degraded products compared to the method 1. The data obtained demonstrate good precision proves the reliability of the method 2. Hence, the method 2 can be used routinely for qualitative and quantitative estimation of cefixime trihydrate and it can also be use as stability indicating method.

References

1. British Pharmacopoeia Commission (2017) The British Pharmacopoeia. Volume I, the Stationary Office Ltd, London, England, UK, p: 451.

2. Adam EHK, Ahmed EM, Barakat IE (2012) Development and Validation of a High-Performance Liquid Chromatography Method for Determination of Cefixime Trihydrate. International Journal of Pharmaceutical Sciences and Research 3: 469-473.

3. British Pharmacopoeia (2010) British Pharmacopoeia. Commission 4: 410.

4. The United States Pharmacopeia 34, National Formulary 29 (2011) United States Pharmacopeia Convention on 2011. 2: 2220.

5. Martindale M (2009) Royal Pharmaceutical Society of Great Britain. RPS Publishing, UK.

6. Meyer VR (1994) Practical high performance liquid chromatography. J Wiley and Sons, Chichester, London, UK 4: 10-25.

7. Amin MI, Bryan JT (1973) Kinetic factors affecting stability in aqueous formulation. Pharm Sci 62: 1768-1771.

8. Food and drug Administration (2000) Analytical Procedures and Methods Validation: Chemistry, Manufacturing, and Controls Documentation; Availability, Federal Register (Notices) 65: 52776-52777.

9. International Conference on Harmonization (2005) Tripartite Guideline Validation of analytical procedure text and methodology. Q2R1.

32

Synthesis A Reagent [3-Hydroxy 4- (1-Azo-2,7-Dihydroxy) Naphthalene Sulfonic Acid] and Used it for Determination of Flourometholone in Bulk and Pharmaceutical Formulations by Spectrophotometric Method

Amir Alhaj Sakur[1], Malek Okdeh[2]* and Banana Al Fares[1,2]

[1]Department of Analytical and Food Chemistry, Faculty of Pharmacy, University of Aleppo, Syria
[2]Department of Analytical Chemistry, Faculty of Science, Teshreen University, Lattakia, Syria

Abstract

A Reagent [3-hydroxyl 4- (1-azo-2,7-dihydroxyl) naphthalene Sulfonic acid] (AAN)] has been synthesized for the determination of flourometholone (FLU) in pure form and in ophthalmic suspensions (drops) by a simple, sensitive and extraction-free spectrophotometric method. The method is based on the formation of yellow colored complex between FLU and ANN maximum at 416 nm. The stoichiometry of the complex in either form was found to be (1:1). Reaction conditions were optimized to obtain the maximum color intensity. Beer's law was obeyed in the concentration ranges of 0.5-17.0 µg/mL. The limit of quantification (LOQ) was 0.14 µg/mL and molar absorptivity (ε) values was 38555 L/mol·L^{-1}cm^{-1}. The proposed method has been applied successfully to the analysis of FLU in pure form and in its dosage forms and no interference was observed from common excipients present in pharmaceutical formulations. Statistical comparison of the results with the reference method showed excellent agreement and indicated no significant difference in accuracy and precision.

Keywords: Spectrophotometry; Synthesis; Flourometholone

Introduction

Flourometholone (FLU) Systematic name is 9-Fluoro-11β, 17-dihydroxy-6α-methylpregna-1,4-diene-3,20-dione (Figure 1). Flourometholone is an ophthalmic suspension 0.1% which is a topical anti-inflammatory agent for ophthalmic use. Flourometholone is indicated for the treatment of corticosteroid-responsive inflammation of the palpebral and bulbar conjunctiva, cornea and anterior segment of the globe [1].

The assay of FLU in pure and dosage forms requires more investigation. The different analytical methods that have been reported for its determination include HPLC [2], with UV by using 1,4-Dihydrazinophthalazine as reagent [3], derivative spectrophotometry and HPLC [4], UV spectrophotometric by using methanol and sulfonic acid Buffer pH=3 [5], HPTLC [6].

The aim of this work is to synthesis organic reagents and used it in spectroscopic analytical study for the determination of Flourometholone (FLU) through complexion with new complex dye [3- hydroxyl 4-(1- azo -2, 7- hydroxyl) naphthalene Sulfonic acid] (AAN) (Figure 2) in dichloromethane medium has been applied, either in laboratory samples or in dosage forms, without any interference from the excipients that are normally present in formulations.

Experimental

Apparatus

Infrared Spectrometer (FTIR) from company Bruker (Germany) model ALPHA, (LC-MS) from company (Shimadzu) UFLC Shimadzu model LC MS-2010 EV. Melting point KRUSS (Germany) model CE-KSP1, spectrophotometric measurements were made in Jasco company (Japan) model V650, UV-Visible spectrophotometer with 1.00 cm quartz cells. Ultrasonic processor model power sonic 405 was used to sonicate the sample solutions.

The diluter pipette model DIP-1 (Ependorf), having 100 µL sample syringe and five continuously adjustable pipettes covering a volume range from 20 to 5000 µL (model Piptman P, Gilson), were used for preparation of the experimental solutions.

Reagents and solutions

Pharmaceutical form of Flourometholone (FLU 99.88%) was received from Univision Pharmaceutical Co. Ltd. (China). A stock solutions of FLU (2.0×10^{-4} M) were prepared by dissolving the appropriate weight of FLU in 70 mL dichloromethane and the volume were diluted to the mark 100 mL in calibrated flask with dichloromethane and take from the last solution 1 mL to the calibrated flask 10 mL too with the same solvent. Working standard solutions were prepared from suitable dilution of the standard stock solution.

Working standard solution was prepared daily by added different volumes of stock solutions to 2 mL of reagent BCG (1×10^{-4}M) diluting to 10 mL with dichloromethane.

The concentration of FLU (0.5, 1.0, 2.0, 3.0, 5.0, 7.0, 9.0, 12.0, 15.0, 17.0 µg.mL^{-1}) were used for the analysis of FLU by the spectrophotometric method. The method was based on formatting complex between Synthesis AAN dyes and FLU in dichloromethane medium. The colored product was quantified spectrophotometrically using absorption bands at 416 nm for complex of (FLU-AAN) and at 416 nm for (FLU-AAN).

A dye [3- hydroxyl 4-(1- azo -2,7- di hydroxyl) naphthalene Sulfonic acid] (AAN) (1×10^{-4}M) prepared by shaking 10.42 mg of AAN dye in 100 mL dichloromethane to dissolve and made up to mark with dichloromethane in a 250 mL calibrated flask.

Flourometholone ophthalmic suspensions brand name is Fumeron, Fumeron Fort (Rama Pharma Company for pharmaceutical industry,

*Corresponding author: Malek Okdeh, Department of Analytical Chemistry, Faculty of Science, Teshreen University, Lattakia, Syria
E-mail: dr.malekokdeh@yahoo.com

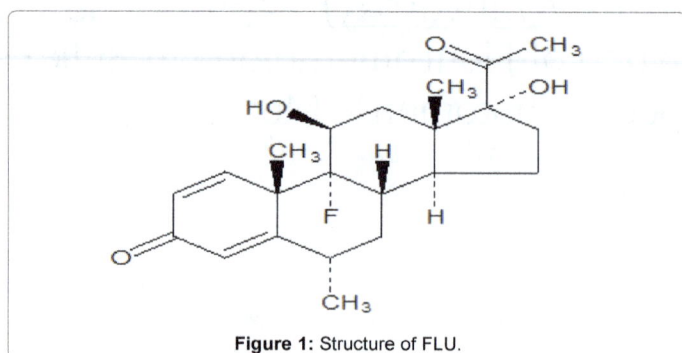

Figure 1: Structure of FLU.

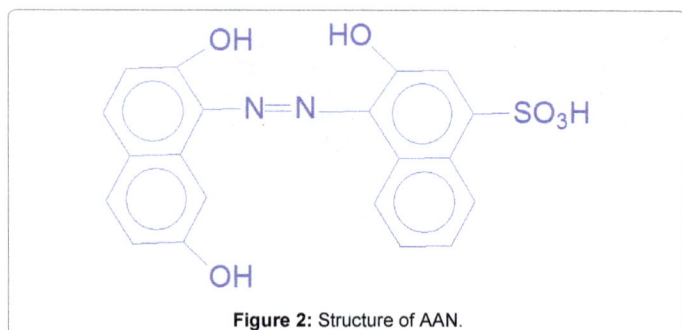

Figure 2: Structure of AAN.

Syria) containing 100 mg and 250 mg in 1 mL, Fludrop, Fludrop Fort (Obary Company, Syria) containing 100 mg and 250 mg in 1 mL, Methoflor (Dyamond pharmaceutical industry, Syria) containing 100 μg in 1 mL. And also there is FLORA-T (Medico pharmaceutical industry, Syria) containing 100 μg in 1 mL from local medical stores. All reagents and solvents were of analytical grade.

Synthesis reagent AAN

The synthesis Reagent is AAN according the following diazotization reaction (Figure 3) [7].

Spectrophotometric procedure

Increasing volumes of FLU working standard solution were transferred into series of 10 mL volumetric flasks that contain 2 mL of AAN reagent (1×10^{-4}M). Solutions were mixed gently and allowed to stand at room temperature. Volumes were made up to mark with dichloromethane and mixed before the spectra were recorded at 416 nm against reagent blank that had been treated similarly.

Determination of FLU/Dye stoichiometric relationship

The composition ratio of drug FLU to dye (AAN) of the colored complex was determined using the molar ratio and continuous variation methods.

Procedure for pharmaceutical samples

An accurately volume amount of the sterile ophthalmic suspensions (drops) equivalent to 75.29 μg of FLU was transferred into 10 mL volumetric flask and added 2 mL of AAN, and diluted with dichloromethane up to the mark. After then the spectra was recorded at 416 nm against reagent blank.

Results and Discussion

Dye structure identification (AAN)

Physical properties: Weight: 410 g/mol. **Appearance:** the dye (AAN) appears red color powder. The solutions in water are stable. **Solubility:** The dye (AAN) free Soluble in water, alcohol, dimethylformamide; insoluble in acetone. **Melting point:** 111-116°C.

UV-Visible spectrophotometry: The Figure 4 shows the spectra of solution (0.01%) of dye AAN in ethanol at λ_{max} = 498 nm.

IR Spectrophotometry: The Figure 5 shows the FTIR spectra of potassium bromide disk the Distinctive peak

ν =1513 cm^{-1} (N=N)

ν =3411 cm^{-1} (O=H)

ν =1620 cm^{-1} (C=C)

LC-MS: The Figure 6 shows the separation chromatogram of AAN by using mobile phase (Water: Methanol) (30:70); after separation MS achieved after applying negative and positive volt (Figure 7). The result shows that the weight of reagent is 410 g/mol.

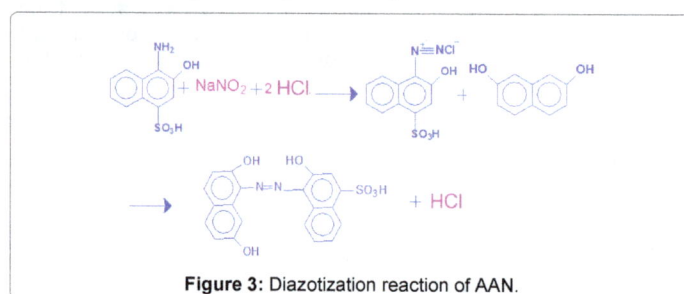

Figure 3: Diazotization reaction of AAN.

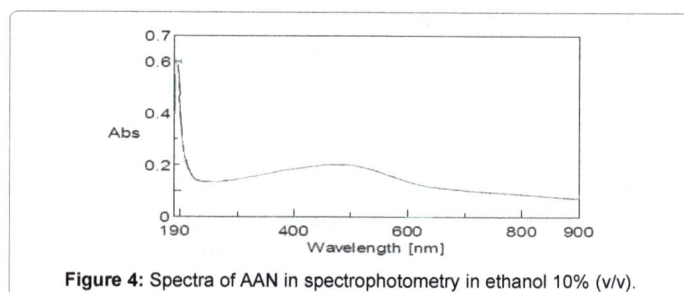

Figure 4: Spectra of AAN in spectrophotometry in ethanol 10% (v/v).

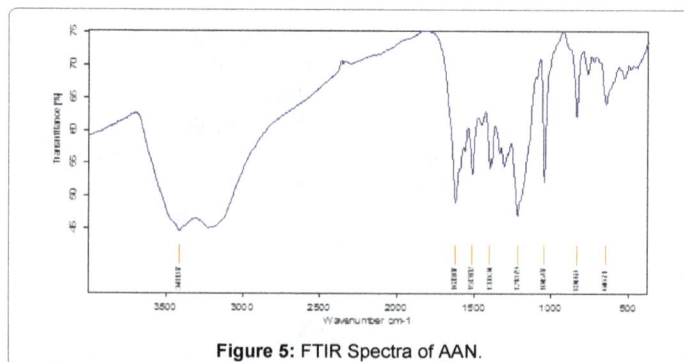

Figure 5: FTIR Spectra of AAN.

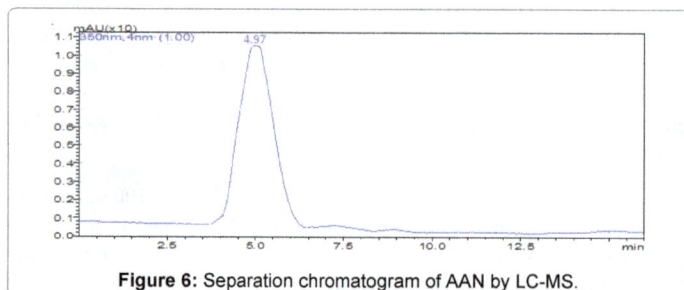

Figure 6: Separation chromatogram of AAN by LC-MS.

Optimization of reaction conditions for (FLU-AAN)

Solvent effect: In order to select a suitable solvent for preparation of the reagent solutions used in the study, the reagents were prepared separately in different solvents such as, chloroform methanol, dichloromethane and dichloromethane, and the reaction of FLU with AAN was followed. Dichloromethane was suited for the complete forming with AAN. Similarly, the effect of the diluting solvent was studied for the method and the results showed that none of the solvents except dichloromethane formed sensitive and stable colored in method. Therefore, dichloromethane was used for dilution throughout the investigation. Dichloromethane was preferred as the most suitable solvent because in this medium, the reagent blank gave negligible blank absorbance and the formed ion-pair complex was found to exhibit higher sensitivity and stability. In other solvents, the reagent blank yielded high absorbance values.

Effect of reaction time and stability: The optimum reaction time for the development of color at ambient temperature ($25 \pm 2°C$) was studied and it was found that the complex forming after added the reagent and no time necessary for the complete formation of ion-pair complexes in a method giving yellow colored solutions have maximum absorbance at λ_{max}. The formed color was stable for more than 24 h in method.

Effect of dye concentration: The influence of the concentration of AAN on the intensity of the color developed at the selected wavelength and constant drug concentration was studied. As shown in Figure 8, the constant absorbance readings were obtained between (0.25-5 mL) of ($1 \times 10^{-4}M$) of AAN, 2 mL of each AAN was used for methods A and B, respectively.

Stoichiometric ratio: Molar ratio method [8]: The stoichiometry of (FLU: Dye) complex by molar ratio method according to following equation: $A_{max} = f([FLU]/[Dye]$, confirms that the ratio of complex FLU:AAN is equal to 1:1 (Figure 9).

Job's method [9]: In order to establish the stoichiometry of FLU and dye (AAN) complex by Job's method of continuous variations was applied. The plot reached a maximum value at a mole fraction of 0.5 which indicated the formation of 1:1 (FLU: Dye) complex (Figure 10) between FLU and AAN.

Validation of the proposed method

Under the optimum experimental conditions, standard calibration curve was constructed at ten concentration levels (n=5) (Figure 11). The correlation coefficient was 0.9999 for method A and 0.9998 indicating very good linearity, over the concentration range of 0.5-17.0 µg/mL. The intercept, slope, limit of detection (LOD), and limit of quantitation (LOQ) are summarized in Table 1. LOD and LOQ values were calculated as $3.3S_b/m$ and $10S_b/m$, respectively where molar absorptivity of regression (Table 1).

The results obtained are summarized in Table 2. The low values of Relative Standard Deviation (RSD) indicate good precision and reproducibility of the method. The average percent recoveries obtained were 98.70-101.53% for AAN, indicating good accuracy of the method.

The repeatability of proposed methods were estimated by measuring five replicate samples of each concentration of flourometholone prepared in one laboratory on the same day. The precision expressed as the Relative Standard Deviation (RSD%) ranged from 0.45% to 3.66% for the smallest concentration, indicating good precision (Table 2).

Application to ophthalmic suspension (eye drops)

The proposed method was applied to the determination of FLU in eye drops. The results in Table 3 showed that the methods are

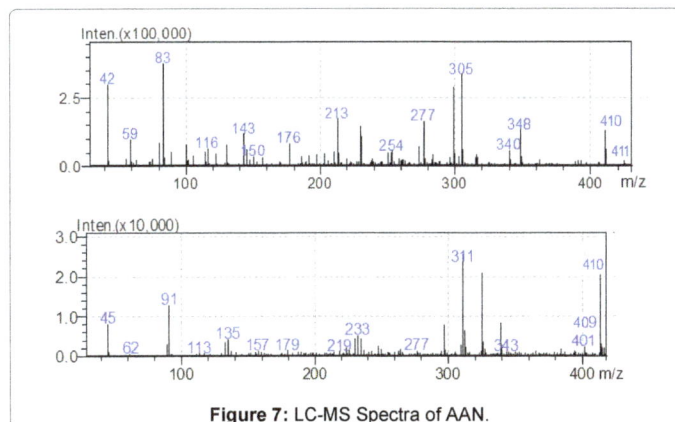

Figure 7: LC-MS Spectra of AAN.

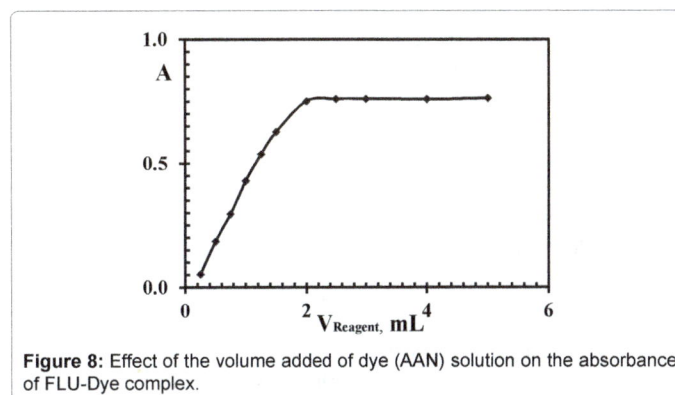

Figure 8: Effect of the volume added of dye (AAN) solution on the absorbance of FLU-Dye complex.

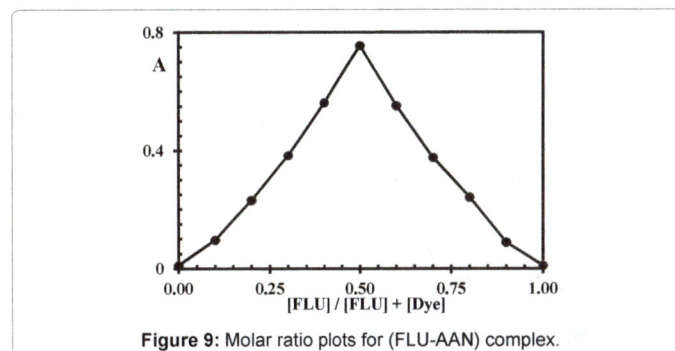

Figure 9: Molar ratio plots for (FLU-AAN) complex.

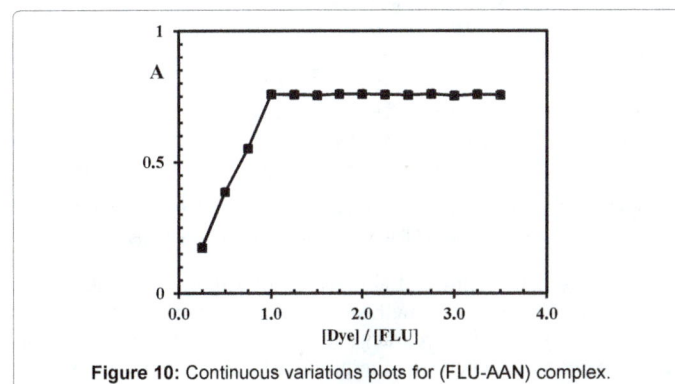

Figure 10: Continuous variations plots for (FLU-AAN) complex.

successful for the determination of ZMT and that the excipients in the dosage forms do not interfere. A statistical comparison of the results for determination of ZMT from the same batch of material by the proposed and reference method is shown in Table 4. The results agreed

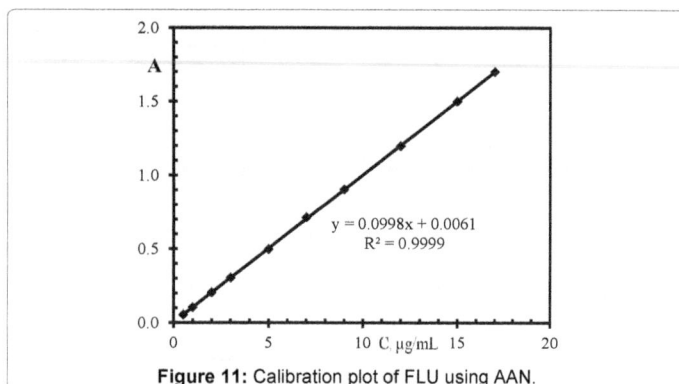

Figure 11: Calibration plot of FLU using AAN.

Parameter	Result
λ_{max} (nm)	416
Linear range (µg/mL)	0.5-17.0
Slope	0.0998
Molar absorptivity (Ɛ), L/ moL.cm	3.8555×10^4
Intercept	0.0061
Correlation coefficient	0.9999
Limit of detection (µg/mL)	0.14
Limit of quantification (µg/mL)	0.45

Table 1: Statistics and analytical parameters of FLU determination by AAN.

Taken FLU (µg/mL)	[a]Found FLU (µg/mL)	SD (µg/mL)	RSD%	Confidence limit	Recovery %
0.5	0.50	0.018	3.60	0.50 ± 0.022	99.52
1.0	1.00	0.020	2.04	1.00 ± 0.025	99.54
2.0	1.99	0.024	1.20	1.99 ± 0.030	99.75
3.0	3.00	0.030	1.00	3.00 ± 0.037	99.93
5.0	4.94	0.055	1.11	4.94 ± 0.068	98.70
7.0	7.11	0.030	0.42	7.11 ± 0.037	101.53
9.0	9.01	0.036	0.40	9.01 ± 0.045	100.12
12.0	11.96	0.066	0.55	11.96 ± 0.082	99.67
15.0	14.98	0.139	0.92	14.98 ± 0.172	99.90
17.0	17.01	0.076	0.45	17.01 ± 0.095	100.03

[a]Average of five determination ± Confidence limit.

Table 2: Precision for determination of FLU in pure form using proposed method.

Product	Taken FLU µg / 1 mL	Found* FLU µg /1 mL	SD (µg/ml)	RSD%	Recovery R%
Fumeron	10.00	99.28	0.04	0.56	99.28
Fumeron Fort	250.00	251.57	0.05	0.65	100.63
Fludrop	100.00	100.16	0.06	0.80	100.16
Fludrop Fort	250.00	250.75	0.05	0.65	100.30
Methoflor	100.00	101.58	0.07	0.95	101.58
Flora-T	100.00	102.17	0.05	0.62	102.17

aAverage and standard deviation of five determinations for the proposed method.

Table 3: Results of the estimation of FLU in eye drops

well with the label claim and also are in agreement with the results obtained by the reference method. Statistical analysis of the results using Student's t-test for accuracy and F-test for precision revealed no significant difference between the proposed and reference method at the 95% confidence level with respect to accuracy and precision.

Accuracy is judged by comparing the results obtained from the presently proposed method that has been applied on commercial ophthalmic drops, with those obtained from a reference method such as HPLC. The resulted values were statistically compared with each

other Table 4 using t- and F-tests. With respect to t- and F-tests, no significant differences were found between the calculated values of both the proposed and the reported methods at 95% confidence level.

Conclusion

The proposed method for the estimation of FLU using synthesis reagent AAN are advantages over many of the reported methods. The methods are rapid, simple and have good sensitivity and accuracy. Proposed method makes use of simple reagent, which an ordinary analytical laboratory can afford. The high recovery percentage and low relative standard deviation reflect the high accuracy and precision of the proposed method. The method are easy, applicable to a wide range of concentration, besides being less time consuming and depend on simple reagent which are available, thus offering economic and acceptable methods for the routine determination of FLU in its formulations.

Product	*Recovery % ± SD	
	Suggested method Using AAN	Pharmacopoeial method
Fumeron 100 µg / 1 mL	99.28 ± 0.04 t=1.29 F=1.43	100.86 ± 0.98 t=1.24
Fumeron Fort 250 µg / 1 mL	100.63 ± 0.05 t=2.48 F=1.18	101.11 ± 0.45 t=1.87
Fludrop 100 µg / 1 mL	100.16 ± 0.06 t=1.69 F=1.12	100.34 ± 0.76 t=1.97
Fludrop Fort 250 µg / 1 mL	100.30 ± 0.05 t=2.21 F=1.61	99.96 ± 0.88 t=1.95
Methoflor 100 µg / 1 mL	101.58 ± 0.07 t=1.72 F=1.38	101.21 ± 0.54 t=1.67
Flora-T 100 µg / 1 mL	102.17 ± 0.05 t=1.46 F=1.52	101.93 ± 0.77 t=1.83

*Average of five determinations for four degree and confidence limit 95%, t=2.776, F=6.26

Table 4: Comparative between tow suggested methods and pharmacopoeial method.

References

1. Physician's Desk Reference (PDR) (2009) Micromedex Thomson Health Care, 63rd USA.

2. Jonvel P, Andermann G (1983) Determination of fluorometholone purity by very high-performance liquid chromatography. Analyst 108: 411-414.

3. Altuntas TG, Korkmaz F, Nebioglu D (2000) Determination of tetrahydrozoline hydrochloride and fluorometholone in pharmaceutical formulations by HPLC and derivative UV spectrophotometry. Pharmazie 55: 49-52.

4. Vladimirov S, Cudina O, Agbab D, Zivanov-Stakic D (1996) Spectrophotometric Determination of Fluorometholone in Pharmaceuticals Using 1,4-Dihydrazinophthalazine. Anal Lett 29: 921-927.

5. Narendra A, Deepika D, Annapurna MM (2014) New Spectrophotometric Methods for the Quantitative Analysis of Fluorometholone in Ophthalmic Suspensions. Chem Sci Trans 3: 445-449.

6. Saleh SS, Lotfy HM, Hassan NY, Elgizawy SM (2013) A comparative study of validated spectrophotometric and TLC- spectrodensitometric methods for the determination of sodium cromoglicate and fluorometholone in ophthalmic solution. Saudi Pharm J 21: 411-421.

7. Tewari KS, Mehrotra SN, Vishnoi NK (1955) Organic Chemistry. Vikas Publishing House Pvt. Ltd.

8. Vosburgh WC, Cooper GR (1941) The Identification of Complex ions in Solution by Spectrophotometric measurements. J Am Chem Soc 63: 437-442.

9. Miller JC, Miller JN (1993) Statistics in Analytical Chemistry. 3rd edn. Ellis Horwood, Chichester, p: 119.

33

Improvements in the High-Performance Liquid Chromatography and Extraction Conditions for the Analysis of Oxidized Fatty Acids Using a Mixed-Mode Spin Column

Takao Sanaki[1], Takuji Fujihara[1], Ryo Iwamoto[2], Takeshi Yoshioka[1*], Kenichi Higashino[1], Toru Nakano[1] and Yoshito Numata[1]

[1]*Shionogi Innovation Center for Drug Discovery, Shionogi & Co., Ltd., Sapporo 001-0021, Japan*
[2]*Business-Academia-Collaborative Laboratory, Graduate School of Pharmaceutical Science, The University of Tokyo, Tokyo 113-0013, Japan*

Abstract

Lipidomics by liquid chromatography/mass spectrometry has been used for a better understanding of the roles of oxidized fatty acids in the development of various diseases. However, further work is required to improve the sample preparation process and the peak tailing of cysteinyl-leukotrienes. In this study, we evaluated various mobile phases and extraction conditions. The addition of phosphoric acid to the mobile phase improved the peak tailing of cysteinyl-leukotrienes. The extraction conditions were also optimized by spin-column possessing an anion-exchange and reversed-phase properties. The extraction efficiency of the modified extraction system was examined using 62 lipids, and 13 deuterated lipids were investigated to evaluate matrix effects and recovery from mouse lung homogenate samples. Extraction efficiencies of ≥70% were obtained for almost all of the lipids. Good results with standard deviations of <15% were obtained for the matrix effects and recovery. Finally, the efficiency of our extraction method was compared with those of several conventional methods, and those of leukotriene C4 was improved significantly using our method. Moreover, the proportion of variance between our method and the conventional methods was >0.99 for all the lipids tested. This newly developed method therefore represents a powerful tool to analyze lipids.

Keywords: Lipid; LC–MS/MS; Spin column; Extraction; Matrix effect; Reversed-phase solid-phase extraction

Introduction

The beneficial effects of polyunsaturated fatty acids for the prevention of cardiovascular disease were first recognized in the late 1960s following a series of epidemiological studies. At that time, thin-layer chromatography was widely used for the measurement of lipids in biological samples [1,2]. Although this method is suitable for distinguishing between the different classes of lipid, it is not suitable for gathering data pertaining to the individual lipid species. Since the 1960s, specific radioimmunoassay [3,4] and gas chromatography/mass spectrometry [5-7] techniques have been developed for the detection of trace quantities of the metabolites of polyunsaturated fatty acids. Furthermore, data collected using these methods have provided researchers with a deeper understanding of the various biological responses elicited by the prostaglandins, thromboxanes and Leukotriene's (LTs) derived from numerous fatty acids, including Arachidonic Acid (AA), Eicosapentaenoic Acid (EPA) and Docosahexaenoic Acid (DHA). For example, the results of several studies have suggested that the resolvins produced by the reactions of the omega-3 fatty acids EPA and DHA can reduce cellular inflammation by inhibiting the production and transportation of inflammatory cells and chemicals at lesion sites [8-13]. Compared with eicosanoids, resolvins are stored in relatively small amounts in the human body, making them difficult to detect with conventional analytical techniques. Prior to the development of liquid chromatography-tandem mass spectrometry (LC-MS/MS) methods for the study of lipids, it was therefore difficult to achieve the detection of resolvins in human samples. The results of our previous studies involving the LC-MS/MS analysis of resolvins and eicosanoids [10,14-17] revealed that the analysis of eicosanoids such as cysteinyl-leukotrienes (Cys-LTs) was severely complicated by carryover and peak tailing issues. These issues were attributed to the adsorption of the eicosanoids on to the metallic surfaces of the analytical system [18], and we believe this could be resolved by optimizing the LC conditions.

Reversed-Phase Solid-Phase Extraction (RP SPE) methods are widely used for the extraction, purification and enrichment of the oxidized fatty acids present in biological samples [18-26], and RP SPE cartridges were used in our previous studies for this purpose [10,14-17]. However, in terms of the analytical process, the combination of different separation modes, such as two-dimensional High-Performance Liquid Chromatography (HPLC), has been shown to be more effective for the purification of biological samples than the combination of the same separation modes. Therefore, we were concerned about the possibility of ion suppression effects resulting from the different types of lipid as we used the same separation modes for the lipid extraction and HPLC processes. Although anion-exchange SPE has been used for the preparation of a wide variety of analytical samples, it is not suitable for the extraction of polar lipids such as prostaglandins and thromboxanes [26]. With this in mind, reversed-phase/anion-exchange SPE (mixed-mode SPE) was evaluated in the current study for the extraction of polar lipids from biological samples. Although mixed-mode SPE has been reported as a novel extraction approach for fatty acid metabolites [27,28], this technique has only been applied to a limited number of lipids and a detailed method has not yet been established.

In this study, we evaluated the use of a mixed-mode SPE system for the extraction of lipids and compared the performance of this method with that of several RP SPE systems using mouse lung homogenate samples.

*****Corresponding author:** Takeshi Yoshioka, Shionogi Innovation Center for Drug Discovery, Shionogi & Co., Ltd., Sapporo 001-0021, Japan
E-mail: takeshi.yoshioka@shionogi.co.jp

Materials and Methods

Ethical approval of the study protocol

This study was conducted in accordance with the guidelines set by Shionogi Innovation Center (Sapporo, Japan).

Chemicals

HPLC-grade methanol (MeOH) and isopropanol (IPA), ammonium acetate, butylated hydroxytoluene, hexane, methyl formate, sodium chloride, dipotassium hydrogen phosphate and potassium dihydrogen phosphate were purchased from Wako Pure Chemical Industries, Ltd. (Osaka, Japan). Hydrochloric acid (HCl) was purchased from Nacalai Tesque, Inc. (Kyoto, Japan). HPLC-grade acetonitrile and phosphoric acid were obtained from Kanto Chemical Co., Inc. (Tokyo, Japan). Liquid chromatography-mass spectrometry grade Formic Acid (FA) was purchased from Sigma-Aldrich (St. Louis, MO, USA). Isoflurane was obtained from DS Pharma Animal Health Co., Ltd. (Osaka, Japan). Lipid standards were purchased from Cayman Chemical Company (Ann Arbor, MI, USA) (Table 1). Ultrapure water was obtained using a Milli-Q system (Millipore, Billerica, MA, USA).

Animals

Male C57BL/6 mice were purchased from Clea Japan Inc. (Tokyo, Japan). All mice were maintained in microisolator cages, where they were exposed to a 12-h light-dark cycle and provided with ad libitum access to standard food and water.

Preparation of standard solutions

Primary stock solutions of 62 analytes and 13 Internal Standards (IS) were prepared in MeOH. Five mixed working solutions (100 pg/μL; AA, EPA and DHA were at 3000 pg/μL) were prepared from the primary stock solutions according to the 5 analytical methods in Table 1. An IS working solution (300 pg/μL; AA-d8 and DHA-d5 at 1000 pg/μL) was prepared from the primary stock solutions by dilution with 10 mmol/L butylated hydroxytoluene in MeOH. Calibration standards were prepared freshly when required by diluting the five mixed working solutions with MeOH to 0.1, 0.3, 1, 3, 10, 30 and 100 pg/μL (final concentrations of AA, EPA and DHA were all 3, 10, 30, 100, 300, 1000 and 3000 pg/μL in these solutions). These calibration standards contained the IS at 30 pg/μL (final concentrations of AA-d8 and DHA-d5 were both 100 pg/μL).

Extraction efficiency of oxidized fatty acids

The lipid extraction process was performed on a MonoSpin™ C18-AX system (GL Sciences Inc., Tokyo, Japan) according to the manufacturer's instructions. Before extraction, the column was pre-activated with 300 μL of methanol and 300 μL of 20 mmol/L potassium phosphate solution (pH 7.0), followed by centrifugation at 9,000 × g for 1 min at 4°C. The calibration standards of each concentration in MeOH (100 μL) and 300 μL of 20 mmol/L potassium phosphate solution (pH 7.0) were mixed and placed directly on the pre-activated spin column. The column was centrifuged at 9,000 × g for 1 min at 4°C and then washed with 300 μL of a 5% NaCl solution by centrifugation. Finally, the column was placed into a new silicon-coated tube and the analytes adsorbed onto the column were eluted twice with 300 μL of 5% NaCl/MeOH (1/9, v/v). The eluant was dried under a gentle stream of nitrogen at 40°C to give a residue, which was reconstituted in 100 μL of MeOH. A 5 μL sample of this solution was then collected and injected into the LC-MS/MS system. The extraction efficiency was assessed in triplicate using the calibration standards at 3, 10 and 30 pg/μL (AA, EPA and DHA concentrations were all 30, 100 and 300 pg/μL in these

solutions). The extraction efficiency was calculated using the following equation, with the peak areas of the IS with and without extraction denoted as A and B, respectively:

$$\text{Extraction efficiency (\%)} = \frac{A}{B} \times 100$$

Matrix effect (ME) and recovery (RE) from lung homogenates

The ME and RE properties were assessed by analyzing the IS solutions in triplicate according to a previously published method [29]. The three different types of IS solution used in this study were prepared as follows.

The Type 1 IS solution was prepared by diluting the neat IS working solution ten times with MeOH. After mixing, the solution was transferred into a vial and a 5 μL sample was injected directly into the LC-MS/MS system.

The Type 2 IS solution was prepared using an IS working solution that had been spiked after the extraction process. After being treated with isoflurane, 36-week-old mice were sacrificed and samples of their lung tissue were extirpated. Lung homogenates (25% mass fraction) were prepared in ultrapure water using Micro Smash™ MS-100R (Tomy Seiko Co., Ltd., Tokyo, Japan) according to the manufacturer's instructions and pooled. Lung homogenate aliquots of 25, 50, 100 and 200 μL (containing 6.25, 12.5, 25 and 50 mg of lung tissue, respectively) were diluted with MeOH (MeOH volume = nine times the volume of the homogenate). The resulting mixtures were then vigorously vortexed for 10 min before being centrifuged at 9,000 × g for 5 min at 4°C. The upper layer was transferred to a silicon-coated tube, where it was dried under a gentle stream of nitrogen at 40°C to give a residue, which was reconstituted in 100 μL of MeOH. This MeOH solution was then extracted according to the protocol in the "Extraction efficiency of oxidized fatty acids" section. After drying the eluent, the resulting residue was reconstituted in 10 μL of IS working solution and 90 μL of MeOH. A 5 μL sample of the reconstituted solution was then injected into the LC-MS/MS system for analysis.

The Type 3 IS solution was prepared using an IS working solution that had been spiked prior to the extraction process. A 10 μL sample of the IS working solution was added to an aliquot of the lung homogenate (25, 50, 100 or 200 μL) along with MeOH (MeOH volume = nine times the volume of the lung homogenate), and the resulting mixture was vortexed prior to being subjected to the lipid extraction process described in the Type 2 protocol. After drying the eluent, the resulting residue was reconstituted in 100 μL of MeOH and 5 μL of the reconstituted solution was injected into the LC-MS/MS system for analysis.

The ME and RE values were calculated using the following equations, with the peak areas of the IS in the Type 1, 2 and 3 denoted as C, D and E, respectively:

$$\text{ME (\%)} = \frac{D}{C} \times 100,$$

$$\text{RE (\%)} = \frac{E}{D} \times 100$$

Lipid extraction using RP SPE

The lipid extraction process was performed using MonoSpin™ C18 columns (GL Sciences Inc.) and Sep-Pak tC18 cartridges (500 mg/6 cc; Waters Corp., Milford, MA, USA). For the MonoSpin™ C18 column, the column was equilibrated with 300 μL of MeOH and 300 μL of H_2O prior to the extraction process. A 10 μL sample of the IS working solution was then added to 50 μL of the lung homogenate together with 450 μL of MeOH, and the resulting mixture was vigorously vortexed

No.	Lipid	Precursor Ion	Product ion	Declustering Potential	Collision energy	Retention Time (min)	Method	Internal standard
01	AA	303	259	-80	-15	7.44	1	AA-d8
02	5-HETE	319	115	-100	-25	5.60	1	15-HETE-d8
03	8-HETE	319	155	-100	-25	5.30	1	15-HETE-d8
04	9-HETE	319	151	-100	-25	5.35	1	15-HETE-d8
05	11-HETE	319	167	-100	-25	5.12	1	15-HETE-d8
06	12-HETE	319	179	-100	-25	5.19	1	15-HETE-d8
07	15-HETE	319	219	-100	-15	4.91	1	15-HETE-d8
08	16-HETE	319	189	-80	-25	4.59	1	15-HETE-d8
09	17-HETE	319	247	-100	-25	4.53	1	15-HETE-d8
10	18-HETE	319	261	-80	-25	4.47	1	15-HETE-d8
11	5;6-EET	319	191	-100	-15	5.99	1	5;6-EET-d11
12	8;9-EET	319	155	-100	-15	5.84	1	15-HETE-d8
13	11;12-EET	319	167	-100	-25	5.72	1	15-HETE-d8
14	14;15-EET	319	219	-100	-15	5.48	1	15-HETE-d8
15	LTB$_4$	335	195	-100	-25	3.38	1	14;15-DHET-d11
16	LXA$_4$	351	115	-80	-25	2.22	1	LXA$_4$-d5
17	PGD$_2$	351	189	-80	-25	3.19	2	PGD$_2$-d4
18	PGE$_2$	351	189	-80	-25	3.10	2	PGE$_2$-d4
19	PGF$_{2\alpha}$	353	193	-80	-35	3.16	2	PGE$_2$-d4
20	6-keto-PGF$_{1\alpha}$	369	163	-80	-35	2.09	2	TXB$_2$-d4
21	TXB$_2$	369	195	-80	-25	2.86	2	TXB$_2$-d4
22	dhk-PGD$_2$	351	175	-80	-35	3.64	2	PGD$_2$-d4
23	dhk-PGE$_2$	351	235	-80	-35	3.34	2	PGE$_2$-d4
24	15-deoxy-PGJ$_2$	315	271	-80	-25	5.86	2	15-deoxy-PGJ$_2$-d4
25	EPA	301	257	-80	-15	6.74	3	AA-d8
26	5-HEPE	317	115	-80	-25	4.85	3	15-HETE-d8
27	8-HEPE	317	127	-80	-25	4.56	3	15-HETE-d8
28	9-HEPE	317	167	-80	-25	4.61	3	15-HETE-d8
29	11-HEPE	317	167	-80	-25	4.42	3	15-HETE-d8
30	12-HEPE	317	179	-80	-25	4.50	3	15-HETE-d8
31	15-HEPE	317	219	-80	-25	4.33	3	15-HETE-d8
32	18-HEPE	317	215	-80	-25	4.13	3	15-HETE-d8
33	8;9-EpETE	317	69	-80	-25	5.11	3	15-HETE-d8
34	11;12-EpETE	317	167	-80	-15	5.01	3	15-HETE-d8
35	14;15-EpETE	317	207	-80	-15	4.94	3	15-HETE-d8
36	17;18-EpETE	317	215	-80	-15	4.72	3	15-HETE-d8
37	DHA	327	283	-80	-15	7.21	4	DHA-d5
38	4-HDoHE	343	101	-80	-25	5.71	4	15-HETE-d8
39	7-HDoHE	343	141	-80	-25	5.32	4	15-HETE-d8
40	8-HDoHE	343	109	-80	-25	5.34	4	15-HETE-d8
41	10-HDoHE	343	153	-80	-25	5.11	4	15-HETE-d8
42	11-HDoHE	343	149	-80	-25	5.17	4	15-HETE-d8
43	13-HDoHE	343	193	-80	-15	4.99	4	15-HETE-d8
44	14-HDoHE	343	205	-80	-15	5.04	4	15-HETE-d8
45	16-HDoHE	343	233	-80	-25	4.87	4	15-HETE-d8
46	17-HDoHE	343	201	-80	-25	4.90	4	15-HETE-d8
47	20-HDoHE	343	285	-80	-15	4.73	4	15-HETE-d8
48	16;17-EpDPE	343	193	-80	-15	5.57	4	15-HETE-d8
49	19;20-EpDPE	343	241	-80	-15	5.31	4	15-HETE-d8
50	17R-RvD1	375	141	-80	-25	2.09	4	14;15-DHET-d11
51	PD1	359	153	-80	-25	3.00	4	14;15-DHET-d11
52	7S-MaR	359	113	-80	-25	2.91	4	14;15-DHET-d11
53	9-HODE	295	171	-100	-25	4.87	5	9-HODE-d4
54	13-HODE	295	195	-100	-25	4.75	5	9-HODE-d4
55	9;10-EpOME	295	171	-100	-25	5.56	5	9-HODE-d4
56	12;13-EpOME	295	195	-100	-25	5.44	5	9-HODE-d4
57	9-HOTrE	293	171	-100	-25	4.16	5	9-HODE-d4
58	13-HOTrE	293	195	-100	-25	4.16	5	9-HODE-d4
59	13-HOTrEr	293	193	-80	-25	4.29	5	9-HODE-d4

60	5-HETrE	321	115	-100	-25	6.34	5	15-HETE-d8
61	8-HETrE	321	157	-100	-25	5.52	5	15-HETE-d8
62	15-HETrE	321	221	-80	-25	5.27	5	15-HETE-d8
63	AA-d8	311	267	-120	-25	7.40	-	-
64	15-HETE-d8	327	226	-80	-15	4.86	-	-
65	5;6-EET-d11	331	202	-100	-15	5.95	-	-
66	LXA$_4$-d5	356	115	-80	-25	2.19	-	-
67	PGD$_2$-d4	355	193	-80	-25	3.18	-	-
68	PGE$_2$-d4	355	193	-80	-25	3.08	-	-
69	TXB$_2$-d4	373	173	-80	-35	2.86	-	-
70	15-deoxy-PGJ$_2$-d4	319	275	-120	-25	5.85	-	-
71	DHA-d5	332	288	-80	-15	7.19	-	-
72	9-HODE-d4	299	172	-100	-25	4.84	-	-
73	14;15-DHET-d11	348	207	-80	-25	3.76	-	-
74	LTC$_4$-d5	630	272	-80	-35	4.69	-	-
75	LTD$_4$-d5	500	177	-120	-25	4.86	-	-

HETE: Hydroxyeicosatetraenoic Acid; EET: Epoxyeicosatrienoic Acid; LT: Leukotriene; LXA4: Lipoxin A4; PG: Prostaglandin; TX: Thromboxane; dhk-PG: 13;14-dihydro-15-keto Prostaglandin; HEPE: Hydroxyeicosapentaenoic Acid; EpETE: Epoxyeicosatetraenoic Acid; HDoHE: Hydroxydocosahexaenoic Acid; EpDPE: Epoxydocosapentaenoic Acid; RvD1: Resolven D1; PD1: Protectin D1; MaR: Maresin; HODE: Hydroxyoctadecadienoic Acid; EpOME: EpoxyOctadecenoic Acid; HOTrE: HydroxyOctadecatrienoic Acid; HETrE: Hydroxyeicosatrienoic Acid; DHET: Dihydroxyeicosatrienoic Acid

Table 1: Optimized selected reaction monitoring pairs and parameters of the oxidized fatty acids.

for 10 min before being centrifuged at 9,000 × g for 5 min at 4°C. The upper layer was transferred to a silicon-coated tube, where it was dried under a gentle stream of nitrogen at 40°C to give a residue, which was reconstituted in 100 μL of MeOH. This MeOH solution was then mixed with 900 μL of aqueous HCl (pH 3.5) and placed directly onto the conditioned column. The column was centrifuged at 9,000 × g for 1 min at 4°C and then washed with 300 μL of H$_2$O and 300 μL of hexane by centrifugation. Finally, the column was placed into a new silicon-coated tube, and the analytes adsorbed onto the column were eluted twice with 300 μL of methyl formate. The eluent was then dried under a gentle stream of nitrogen at 40°C to give a residue, which was reconstituted in 100 μL of MeOH. A 5 μL sample of the reconstituted solution was then introduced into the LC-MS/MS system for analysis.

For the Sep-Pak tC18 cartridge, the cartridge was equilibrated with 12 mL of MeOH and 12 mL of H$_2$O prior to the extraction process. A 10 μL sample of the IS working solution was then added to 50 μL of the lung homogenate together with 450 μL of MeOH. After vortex mixing, 500 μL of the supernatant was collected and diluted with 4.5 mL of aqueous hydrochloric acid (pH 3.5). This acidified solution was then rapidly loaded onto the conditioned cartridge, which was washed with 12 mL of H$_2$O. A 6 mL portion of hexane was then added and the analytes were eluted with 9 mL of methyl formate. The resulting eluent was dried under a gentle stream of nitrogen at 40°C to give a residue, which was reconstituted in 100 μL of MeOH. A 5 μL sample of the reconstituted solution was then introduced to the LC-MS/MS system for analysis.

Comparison of mixed-mode SPE with RP SPE using mouse lung homogenate samples

The ratio of the process efficiency between the mixed-mode SPE and RP SPE systems was calculated using mouse lung homogenate samples. Mouse lung homogenate was extracted in 50 μL aliquots according to the Type 3 protocol (Matrix effect and recovery from lung homogenates) or "Lipid extraction using RP SPE" protocol. The extracted samples were analyzed by LC-MS/MS and the ratio of the process efficiency between the mixed-mode SPE and RP SPE systems was calculated using the following equation, with the peak areas of the IS in the mixed-mode SPE and RP SPE systems denoted as F and G, respectively:

$$\text{The ratio of process efficiency (\%)} = \frac{G}{F} \times 100$$

Furthermore, the correlation coefficient of the measured results for the endogenous oxidized fatty acids between the mixed-mode SPE and RP SPE systems was evaluated using 10 and 50 μL aliquots of the mouse lung homogenate.

Lipid analysis by LC-MS/MS

The compounds were separated on an Acquity UPLC system (Waters Corp.). The autosampler and column were maintained at 4 and 60°C, respectively. Separation was achieved with an Acquity UPLC BEH C18 column (100 × 2.1 mm i.d., 1.7 μm; Waters Corp.). Mobile phase A consisted of water/1 mol/L ammonium acetate/5 mmol/L phosphoric acid/FA (990/10/1/1, v/v/v/v), and mobile phase B consisted of acetonitrile/IPA/1 mol/L ammonium acetate/FA (495/495/10/1, v/v/v/v). The flow rate for the separations was set at 0.4 mL/min. The gradient program for methods 1, 3, 4 and 5 was as follows: 0.00-8.00 min (increased from 35 to 80% B), 8.01-11.00 min (increased from 80 to 100% B) and 11.01-13.00 min (decreased from 100 to 35% B). The gradient program for method 2 was as follows: 0-10 min (increased from 25 to 80% B), 10.01-13.00 min (increased from 80 to 100% B) and 13.01-15.00 min (decreased from 100 to 25% B). A FCV-20AH2 system (Shimadzu, Kyoto, Japan) was used as a valve switch to allow for the introduction of the sample to the separation column (between 1-10 min for method 2, and 1-8 min for methods 1, 3, 4, and 5).

Mass spectrometry analyses of oxidized fatty acids were carried out on an API5000 Mass Spectrometer (AB SCIEX, Foster City, CA, USA) equipped with an ESI source. Nitrogen was used as the collision gas for the analysis of all of the metabolites. Oxidized fatty acids were detected in the negative ESI mode with the following source parameters: curtain gas, 15 psi; ion source gas 1, 50 psi; ion source gas 2, 60 psi; ion spray voltage, -4500 V; collision gas, 8 psi; temperature, 500°C; interface heater, on; entrance potential, -10 V; collision cell exit potential, -10 V; resolution Q1, unit; and resolution Q3, unit. The dwell time was set to 25 min for all of the molecules to obtain more than 10 data points per peak. The acquisition and processing of data were carried out with Analyst v1.4.2 (AB SCIEX). Selected reaction monitoring conditions for each molecule are summarized in Table 1.

Results and Discussion

LC-MS/MS has been used previously to allow for the analysis of fatty acids [10,14-17], and the methods described in these studies have been improved in our laboratory. However, despite our best efforts to improve these methods, there are still some issues in need of resolution. With this in mind, the principle aim of the current study was to further optimize the HPLC conditions and extraction method used for the analysis of fatty acids in biological samples.

Optimization of HPLC conditions

A number of issues can arise during the analysis of eicosanoids, including carryover and peak tailing with Cys-LTs and thromboxane B$_2$ (TXB$_2$) [18]. These problems can be caused by the adsorption of these molecules onto the metallic surfaces of the analytical system. Although phosphoric acid is generally unsuitable for LC-MS/MS analysis, it has been reported that the inclusion of a small amount of phosphoric acid (e.g., 5 μmol/L) can lead to significant improvements in the peak shapes of lipid mediators such as lysophosphatidic acid and lysophosphatidylserine by preventing the adsorption of these materials onto the metallic surfaces of the analytical system [30]. In this study, a small amount of phosphoric acid was included in the analysis of the oxidized fatty acids, and the subsequent impact on the analysis of Cys-LTs was evaluated in detail.

The extracted ion chromatograms of the Cys-LTs with and without phosphoric acid are shown in Figure 1. In the absence of phosphoric acid, the peak corresponding to leukotriene C$_4$ (LTC$_4$) was found to be very broad and the signal-to-noise ratio (S/N) was very low (Figure 1A). In the presence of phosphoric acid, there was a significant improvement in the USP tailing factor of LTC$_4$ from 12.5 to 1.25, and the S/N also increased from 4.4 to 1484.1 (Figure 1D). The S/N of leukotriene E$_4$ (LTE$_4$) also increased (374.2 to 878.6) following the addition of phosphoric acid (Figures 1C and 1F). Although the S/N of leukotriene D$_4$ (LTD$_4$) did not change following the addition of phosphoric acid, the peak intensity increased ($6.7e^4$ to $2.6e^5$) (Figures 1B and 1E). Similarly, the peak shape of TXB$_2$ improved following the addition of phosphoric acid (data not shown). Notably, the addition of phosphoric acid did not appear to have an adverse impact on the other molecules in the system. To the best of our knowledge, this work represents the first reported example of improvements in the USP tailing factors and S/N of Cys-LTs following the addition of phosphoric acid. With more than 10,000 measurements repeated under the same conditions to date, we have observed no problems with the sensitivity or the peak shapes of the oxidized fatty acids analyzed in this way.

Construction of mixed-mode SPE

RP conditions are generally used for lipid extraction and HPLC separation. However, in terms of analysis, it is generally accepted that the combination of different separation modes such as two-dimensional HPLC is more effective for the purification of biological samples than the combination of the same separation modes. Furthermore, the use of the same separation modes of lipid extraction and HPLC can lead to ion suppression effects resulting from the presence of specific types of lipid. Although anion-exchange SPE has been used in terms of the characteristic features of oxidized fatty acids, it is not suitable for the

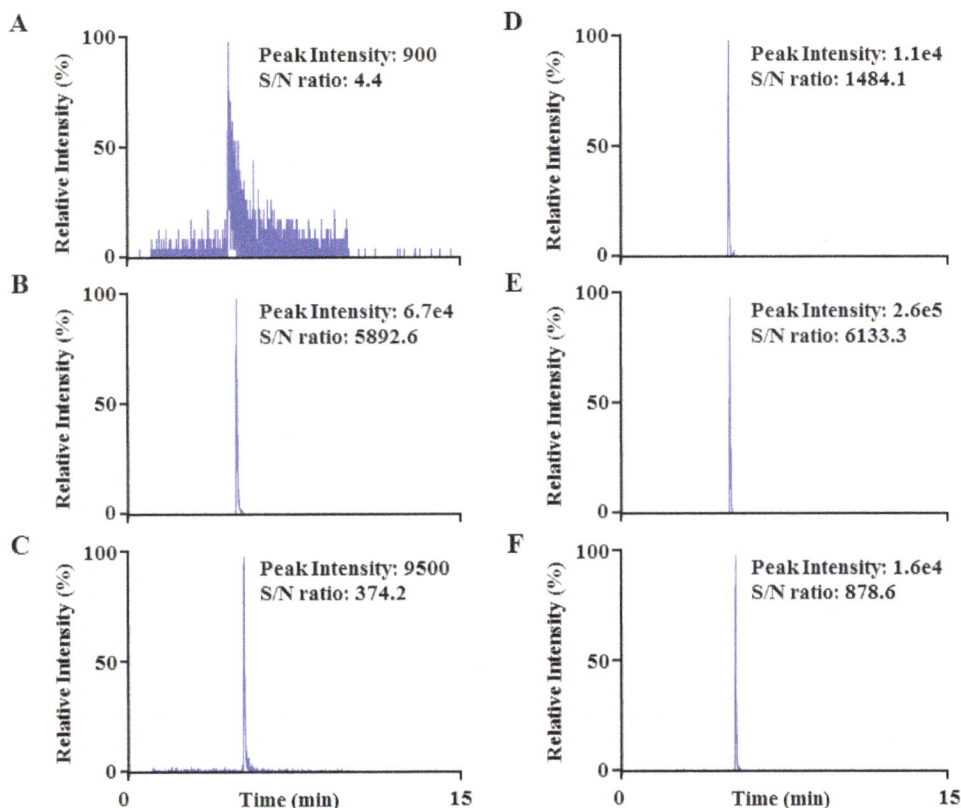

Figure 1: Extracted ion chromatograms of the leukotriene's (10 pg/μL). Leukotriene's were analyzed using measurement method 2 and the selected reaction monitoring transitions for leukotriene C$_4$, leukotriene D$_4$ and leukotriene E$_4$ were 625→272, 495→177 and 438→333, respectively. Extracted ion chromatograms of Leukotriene C$_4$, leukotriene D$_4$ and leukotriene E$_4$ without phosphoric acid (A, B and C, respectively) and with phosphoric acid (D, E and F, respectively) are shown.

extraction of polar lipids such as prostaglandins and thromboxanes [26]. Indeed, the results of our preliminary experiments using an anion-exchange SPE process revealed that the extraction efficiencies of prostaglandins and thromboxanes were <40 % (data not shown). To overcome these issues, we examined the lipid extraction conditions using mixed-mode SPE, which has been used in lipid biology to analyze fatty acid metabolites [27,28]. Although this approach is very attractive, its application has been limited because it can only be applied to a small number of lipids. Furthermore, the details of this method have never been fully clarified. Hence, the versatility of this approach is poor.

The method developed process used in the current study is shown in Figure 2. Using this method under neutral conditions, the dissociated oxidized fatty acids were adsorbed on to the column, which contained octadecyl and trimethylaminopropyl groups that were chemically bonded to monolithic silica. Furthermore, because a spin column was used in the current study, all of the handling procedures (e.g., sample loading, as well as the washing and elution of the oxidized fatty acids) were conducted by centrifugation. Using the spin column, the extraction time was 10 min per sample and the organic solvent requirements were <1 mL per sample.

To verify the utility of this method, the extraction efficiencies of oxidized fatty acids were tested using 62 lipids. The extraction efficiency of each lipid was evaluated in triplicate by LC-MS/MS analysis at concentrations of 3, 10 and 30 pg/μL (containing AA, EPA and DHA all at 30, 100 and 300 pg/μL, respectively) (Table 2). The results revealed that extraction efficiencies of ≥70 % were obtained for 61 of the lipids and Relative Standard Deviations (RSDs) of <15% were obtained for 60 lipids at the three different concentrations tested in the current study. Notably, the extraction efficiencies of prostaglandins and thromboxanes, which were insufficiently extracted by anion-exchange SPE, increased from <40% to >87% compared with anion-exchange SPE (No. 17-24, Table 2). These results therefore demonstrate that the combination of the RP and anion-exchange functions led to a significant improvement in the extraction efficiencies of polar lipids. Only the extraction efficiency of 5,6-epoxyeicosatrienoic acid (5,6-EET) was relatively low (<65%) (No. 11, Table 2). A detailed investigation of the extraction protocol resulted in an increased extraction efficiency

for 5,6-EET (≥85 %) following a change in the eluent from 5% NaCl/MeOH (1/9, v/v) to 2% formic acid in MeOH or 5% NaCl/MeOH/isopropanol (2/9/9, v/v/v) (data not shown). This result suggests that the low recovery of 5,6-EET was caused by the adsorption of this material onto the column, and that the extraction efficiencies for molecules with an otherwise poor recovery could be improved by changing the elution conditions. This newly developed extraction method therefore showed good extraction efficiencies for a range of different compounds.

Matrix effect and recovery from lung homogenates

The ME and RE values of this newly developed method were examined using mouse lung homogenates. The ME values for some of the lipids were found to be dependent on the weight of the lung homogenate in the sample. For deuterated LTC_4 (LTC_4-d5), an enhancement of >150% was observed in the ME value when the mass of lung homogenate was increased from 12.5 and 50 mg (No. 12, Table 3). Similar increases in the ME value of >150% were also observed for deuterated 15-hydroxyeicosatetraenoic acid (No. 2, Table 3) and deuterated prostaglandin E_2 (No. 6, Table 3) at 50 mg compared with 12.5 mg. Although enhancements of >130% were observed in the ME values of deuterated AA (AA-d8) (No. 1, Table 3) and deuterated DHA (DHA-d5) (No. 9, Table 3), these increases were found to be independent of the lung homogenate weight. Given that the IS peaks were not observed in samples without IS (data not shown), it was concluded that these results were not being influenced by the interfering peaks derived from the samples. RSD values of <15% were obtained for all lipids at each weight.

For the RE values, a weight-dependent reduction was observed for some of the lipids. However, the RE values were >70% for almost all of the lipids at each weight (Table 4). With the exception of AA-d8 (No. 1, Table 4) and DHA-d5 (No. 9, Table 4) at 50 mg, RSD values of <15% were obtained for all lipids at each weight. Thus, adequate RSD values for the ME and RE values were obtained for each weight using our newly developed method. Consideration of the ME and RE values for the lipids, as well as the concentration of oxidized fatty acids in lung tissue, meant that the weight of lung homogenate used in this study was set to ≤12.5 mg.

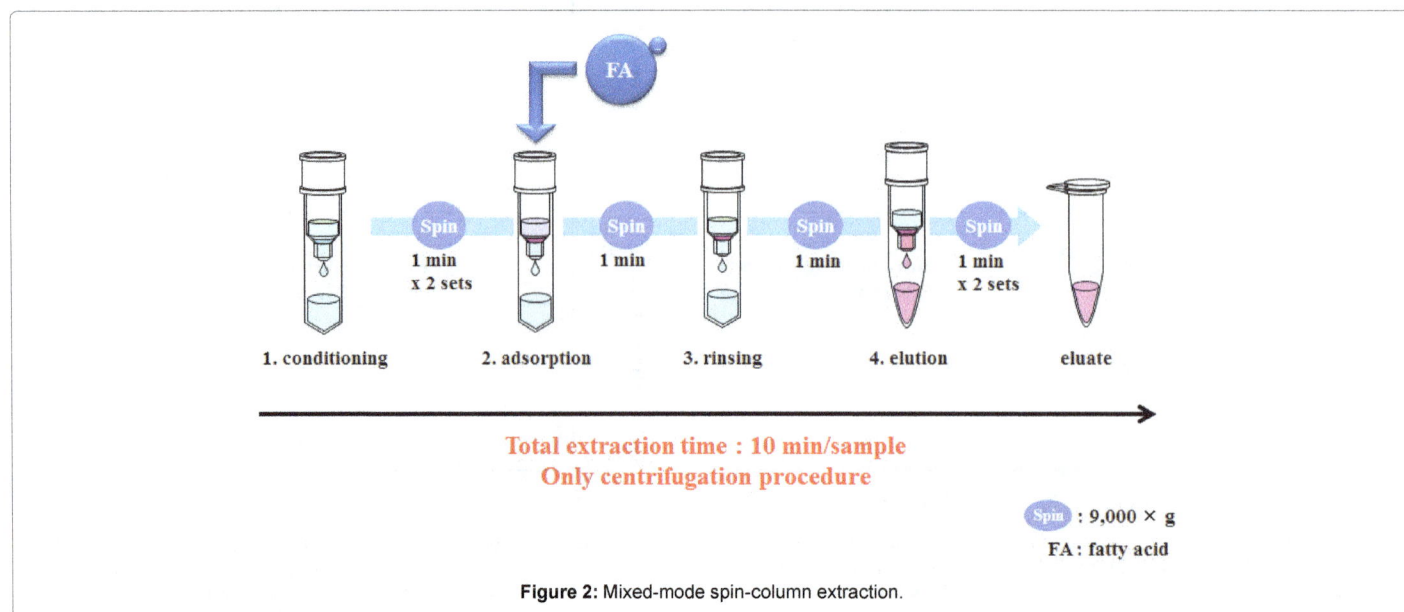

Total extraction time : 10 min/sample
Only centrifugation procedure

Spin : 9,000 × g
FA : fatty acid

Figure 2: Mixed-mode spin-column extraction.

No.	Lipid	Calibrated range (pg/µL)	r^{2a}	Extraction efficiency[b] (%)			RSD (%)		
				3 pg/µL	10 pg/µL	30 pg/µL	3 pg/µL	10 pg/µL	30 pg/µL
01	AA[c]	10-3000	0.99	74.99 ± 4.59	79.09 ± 5.26	101.28 ± 7.95	6.12	6.65	7.85
02	5-HETE	0.3-100	0.99	99.79 ± 4.37	84.99 ± 2.29	88.95 ± 0.37	4.37	2.69	0.41
03	8-HETE	0.3-100	0.99	97.68 ± 5.73	90.98 ± 2.14	87.76 ± 1.02	5.87	2.35	1.16
04	9-HETE	0.3-100	0.99	97.62 ± 1.20	89.21 ± 0.71	92.20 ± 1.49	1.23	0.80	1.61
05	11-HETE	0.3-100	0.99	93.89 ± 4.26	87.18 ± 2.57	89.61 ± 1.02	4.53	2.95	1.14
06	12-HETE	0.3-100	0.99	93.04 ± 5.76	91.42 ± 2.63	91.31 ± 1.85	6.19	2.88	2.02
07	15-HETE	0.3-100	0.99	96.14 ± 5.60	89.10 ± 2.25	94.49 ± 5.82	5.82	2.53	2.33
08	16-HETE	1-100	0.99	101.72 ± 10.66	95.80 ± 5.53	97.06 ± 2.65	10.48	5.77	2.73
09	17-HETE	0.3-100	0.99	92.70 ± 9.57	91.12 ± 0.89	93.30 ± 2.01	10.32	0.97	2.14
10	18-HETE	0.3-100	0.99	99.94 ± 11.37	88.27 ± 2.18	92.97 ± 0.80	11.38	2.47	0.86
11	5;6-EET	0.3-100	0.99	58.70 ± 6.20	61.05 ± 4.37	62.93 ± 1.82	10.55	7.16	2.89
12	8;9-EET	1-100	0.99	94.86 ± 12.89	78.48 ± 2.60	86.70 ± 0.44	13.59	3.32	0.50
13	11;12-EET	0.3-100	0.99	82.46 ± 4.01	82.70 ± 2.61	85.77 ± 4.35	4.87	3.16	5.07
14	14;15-EET	0.3-100	0.99	94.66 ± 7.23	79.29 ± 1.10	77.35 ± 2.50	7.65	1.39	3.23
15	LTB$_4$	0.3-100	0.99	95.68 ± 2.81	100.21 ± 3.18	92.38 ± 1.05	2.94	3.18	1.14
16	LXA$_4$	0.3-100	0.99	106.56 ± 13.70	108.96 ± 0.86	84.62 ± 3.78	12.86	0.78	4.46
17	PGD$_2$	0.3-100	0.99	95.84 ± 6.71	97.34 ± 1.39	96.25 ± 4.82	7.00	1.43	5.01
18	PGE$_2$	0.3-100	0.99	96.19 ± 1.36	96.08 ± 6.23	93.36 ± 4.45	1.41	6.49	4.77
19	PGF$_{2\alpha}$	0.3-100	0.99	97.92 ± 6.70	98.55 ± 3.83	94.91 ± 5.43	6.85	3.88	5.72
20	6-keto-PGF$_{1\alpha}$	0.1-100	0.99	96.19 ± 4.51	95.66 ± 2.65	93.88 ± 1.11	4.69	2.77	1.18
21	TXB$_2$	0.3-100	0.99	100.78 ± 3.06	99.53 ± 3.39	95.91 ± 3.47	3.04	3.40	3.61
22	dhk-PGD$_2$	0.1-100	0.99	101.95 ± 3.73	101.31 ± 3.66	101.56 ± 0.66	3.66	3.61	0.65
23	dhk-PGE$_2$	1-100	0.99	105.41 ± 3.44	105.05 ± 1.88	99.81 ± 2.04	3.26	1.79	2.04
24	15-deoxy-PGJ$_2$	0.3-100	0.99	89.09 ± 3.23	99.15 ± 0.55	87.57 ± 4.10	3.62	0.56	4.68
25	EPA[c]	10-3000	0.99	80.55 ± 5.08	89.71 ± 10.11	110.52 ± 4.73	6.31	11.27	4.28
26	5-HEPE	1-100	0.99	99.77 ± 4.30	96.79 ± 6.73	93.57 ± 6.81	4.31	6.96	7.28
27	8-HEPE	1-100	0.99	109.80 ± 8.16	94.39 ± 4.89	93.18 ± 4.69	7.43	5.18	5.03
28	9-HEPE	1-100	0.99	96.57 ± 7.64	90.80 ± 4.38	92.36 ± 5.07	7.91	4.82	5.49
29	11-HEPE	0.3-100	0.99	99.19 ± 2.24	95.90 ± 5.93	94.78 ± 5.47	2.26	6.19	5.77
30	12-HEPE	1-100	0.99	99.93 ± 5.92	101.10 ± 6.47	97.23 ± 6.39	5.92	6.40	6.57
31	15-HEPE	1-100	0.99	100.64 ± 1.42	104.35 ± 13.61	97.55 ± 8.24	1.41	13.05	8.45
32	18-HEPE	0.3-100	0.99	100.33 ± 7.60	98.18 ± 3.18	98.17 ± 6.97	7.57	3.24	7.10
33	8;9-EpETE	0.3-100	0.99	90.51 ± 3.27	88.52 ± 3.72	86.10 ± 6.95	3.61	4.20	8.07
34	11;12-EpETE	1-100	0.99	94.64 ± 5.26	89.85 ± 6.32	81.35 ± 3.46	5.56	7.04	4.25
35	14;15-EpETE	1-100	0.99	92.12 ± 3.25	75.80 ± 13.27	79.61 ± 4.90	3.52	17.50	6.16
36	17;18-EpETE	1-100	0.99	96.72 ± 2.51	88.80 ± 6.54	87.59 ± 6.28	2.60	7.37	7.17
37	DHA[c]	10-3000	0.99	73.27 ± 2.85	83.30 ± 2.52	97.87 ± 4.64	3.89	3.02	4.74
38	4-HDoHE	0.3-100	0.99	96.00 ± 7.38	106.04 ± 10.56	106.68 ± 9.02	7.68	9.96	8.53
39	7-HDoHE	0.3-100	0.99	96.99 ± 6.10	87.40 ± 5.89	90.07 ± 7.30	6.29	6.74	8.11
40	8-HDoHE	0.3-100	0.99	85.98 ± 4.61	80.24 ± 6.31	85.22 ± 5.93	5.36	7.86	6.96
41	10-HDoHE	0.3-100	0.99	94.78 ± 1.96	92.81 ± 10.99	93.70 ± 6.01	2.07	11.84	6.41
42	11-HDoHE	0.3-100	0.99	98.59 ± 2.39	94.42 ± 7.52	95.94 ± 7.36	2.42	7.97	7.67
43	13-HDoHE	1-100	0.99	105.97 ± 2.32	96.76 ± 9.98	98.40 ± 6.76	2.19	10.32	6.87
44	14-HDoHE	1-100	0.99	97.99 ± 1.00	97.90 ± 10.22	100.60 ± 8.64	1.02	10.44	8.58
45	16-HDoHE	0.3-100	0.99	99.74 ± 2.30	100.80 ± 9.51	101.12 ± 8.18	2.30	9.43	8.09
46	17-HDoHE	1-100	0.99	99.43 ± 9.24	98.27 ± 7.84	101.61 ± 7.63	9.30	7.98	7.51
47	20-HDoHE	0.3-100	0.99	93.27 ± 0.87	99.47 ± 9.45	95.47 ± 7.70	0.93	9.50	8.07
48	16;17-EpDPE	3-100	0.99	97.13 ± 7.26	72.99 ± 6.32	95.56 ± 6.32	7.18	9.95	6.61
49	19;20-EpDPE	1-100	0.99	91.30 ± 8.08	76.37 ± 10.89	87.17 ± 5.90	8.85	14.26	6.77
50	17R-RvD1	0.3-100	0.99	108.10 ± 11.19	93.99 ± 7.27	86.98 ± 12.42	10.35	7.73	14.28
51	PD1	0.3-100	0.99	101.06 ± 2.87	106.55 ± 7.83	106.91 ± 9.91	2.84	7.35	9.27
52	7S-MaR	1-100	0.99	98.00 ± 4.95	101.33 ± 7.19	97.33 ± 10.27	5.05	7.10	10.55
53	9-HODE	0.3-100	0.99	91.84 ± 10.72	93.05 ± 3.12	86.57 ± 5.72	11.67	3.35	6.61
54	13-HODE	0.3-100	0.99	91.96 ± 3.95	93.10 ± 6.12	87.70 ± 7.67	4.30	6.57	8.74
55	9;10-EpOME	0.3-100	0.99	95.63 ± 5.30	93.33 ± 8.68	90.31 ± 7.18	5.54	9.30	7.96
56	12;13-EpOME	0.3-100	0.99	96.08 ± 4.32	94.62 ± 6.00	89.94 ± 5.59	4.49	6.35	6.21
57	9-HOTrE	0.3-100	0.99	87.16 ± 10.54	90.92 ± 7.14	83.39 ± 6.82	12.09	7.86	8.18

58	13-HOTrE	1-100	0.99	83.75 ± 8.43	91.81 ± 8.49	87.59 ± 7.12	10.06	9.25	8.13
59	13-HOTrEr	0.3-100	0.99	88.87 ± 5.59	96.36 ± 9.75	85.78 ± 5.63	6.29	10.12	6.56
60	5-HETrE	0.3-100	0.99	84.24 ± 5.49	90.80 ± 9.07	83.95 ± 4.70	6.52	9.99	5.60
61	8-HETrE	0.3-100	0.99	89.06 ± 7.19	95.02 ± 7.43	86.08 ± 7.43	8.07	8.77	8.64
62	15-HETrE	0.3-100	0.99	83.62 ± 6.36	96.04 ± 14.52	86.37 ± 6.08	7.61	15.12	7.04

[a]Linearity
[b]Extraction efficiency is the average ± SD (*n*=3)
[c]Extraction efficiency is evaluated at 30; 100 and 300 pg/μL.

Table 2: Calibration curve data and extraction efficiencies of oxidized fatty acids .

No.	Lipid	Matrix Effect (%)				RSD (%)			
		6.25 mg	12.5 mg	25 mg	50 mg	6.25 mg	12.5 mg	25 mg	
01	AA-d8	139.55 ± 5.28	136.21 ± 12.79	140.95 ± 4.66	135.96 ± 3.35	3.79	9.39	3.31	2.47
02	15-HETE-d8	106.94 ± 3.19	116.70 ± 6.95	148.11 ± 3.33	171.48 ± 6.34	2.99	5.95	2.25	3.70
03	5;6-EET-d11	99.71 ± 0.95	102.09 ± 3.16	112.58 ± 4.03	104.30 ± 3.57	0.95	3.10	3.58	3.43
04	LXA$_4$-d5	94.62 ± 2.43	97.62 ± 2.78	101.51 ± 2.65	102.29 ± 0.98	2.57	2.85	2.61	0.95
05	PGD$_2$-d4	98.26 ± 1.49	105.23 ± 2.03	114.35 ± 1.14	133.47 ± 7.72	1.52	1.93	1.00	5.78
06	PGE$_2$-d4	100.52 ± 1.51	111.17 ± 1.95	122.88 ± 2.20	153.30 ± 7.95	1.50	1.76	1.79	5.18
07	TXB$_2$-d4	98.34 ± 1.04	99.96 ± 1.08	103.64 ± 1.28	111.64 ± 1.74	1.06	1.08	1.24	1.56
08	15-deoxy-PGJ$_2$-d4	96.76 ± 0.96	100.25 ± 1.29	103.80 ± 2.13	112.41 ± 0.29	0.99	1.29	2.05	0.26
09	DHA-d5	147.15 ± 1.98	145.54 ± 8.99	151.32 ± 4.33	152.55 ± 3.23	1.35	6.18	2.86	2.12
10	9-HODE-d4	103.91 ± 2.04	106.60 ± 3.35	116.88 ± 2.16	122.97 ± 3.74	1.97	3.15	1.85	3.04
11	14;15-DHET-d11	95.77 ± 0.29	100.89 ± 3.50	110.57 ± 2.00	129.04 ± 5.13	0.31	3.47	1.81	3.98
12	LTC$_4$-d5	100.33 ± 10.67	159.32 ± 19.74	179.57 ± 11.56	182.37 ± 22.49	10.63	12.39	6.43	12.33
13	LTD$_4$-d5	88.28 ± 4.12	94.93 ± 3.91	95.76 ± 3.30	84.90 ± 6.18	4.67	4.12	3.44	7.28

Table 3: Matrix effects for internal standards in mouse lung samples.

No.	Lipid	Recovery (%)				RSD (%)			
		6.25 mg	12.5 mg	25 mg	50 mg	6.25 mg	12.5 mg	25 mg	50 mg
01	AA-d8	106.47 ± 7.44	99.69 ± 9.21	83.03 ± 7.12	104.12 ± 22.09	6.99	9.23	8.58	21.2
02	15-HETE-d8	91.64 ± 2.08	89.00 ± 0.91	90.27 ± 7.12	102.92 ± 12.65	2.27	1.02	7.89	12.29
03	5;6-EET-d11	86.82 ± 3.95	72.07 ± 1.16	62.59 ± 1.80	56.29 ± 4.79	4.55	1.61	2.87	8.50
04	LXA$_4$-d5	92.68 ± 8.08	87.31 ± 1.27	90.48 ± 5.41	94.15 ± 3.44	8.72	1.46	5.98	3.66
05	PGD$_2$-d4	89.89 ± 2.60	78.36 ± 3.16	78.04 ± 5.15	67.87 ± 0.66	2.89	4.03	6.60	0.98
06	PGE$_2$-d4	94.91 ± 6.28	90.30 ± 1.35	90.53 ± 8.96	85.21 ± 3.36	6.61	1.49	9.89	3.95
07	TXB$_2$-d4	99.39 ± 2.45	92.85 ± 1.76	88.36 ± 8.03	82.84 ± 2.42	2.47	1.90	9.09	2.93
08	15-deoxy-PGJ$_2$-d4	63.74 ± 3.84	62.14 ± 0.50	69.41 ± 2.83	83.95 ± 4.56	6.03	0.81	4.08	5.43
09	DHA-d5	105.02 ± 11.45	101.57 ± 8.29	84.87 ± 7.65	102.92 ± 17.07	10.90	8.17	9.01	16.67
10	9-HODE-d4	95.48 ± 9.07	89.61 ± 3.06	88.92 ± 5.02	99.04 ± 7.19	9.50	3.42	5.64	7.26
11	14;15-DHET-d11	92.35 ± 3.73	73.72 ± 2.30	83.92 ± 4.49	93.42 ± 5.11	4.04	3.12	5.35	5.47
12	LTC$_4$-d5	100.74 ± 4.99	87.81 ± 6.59	93.22 ± 3.92	90.86 ± 2.36	4.96	7.50	4.21	2.60
13	LTD$_4$-d5	83.86 ± 4.20	72.80 ± 4.58	80.32 ± 0.72	83.32 ± 6.18	5.01	6.28	0.90	7.42

Table 4: Recoveries of internal standards in mouse lung samples.

Comparison of mixed-mode SPE with RP SPE using mouse lung homogenate samples

To evaluate our extraction method, the mixed-mode SPE system was compared with RP SPE system using mouse lung homogenate samples. In this study, MonoSpin™ C18 and Sep-Pak tC18 were used as RP SPE devices. Given that the MonoSpin™ C18 column consists of octadecyl groups that are chemically bound to a monolithic silica spin column, it is possible to compare the mixed-mode SPE spin column with the MonoSpin™ C18 column based entirely on the differences in their functional groups. In contrast, Sep-Pak tC18 cartridges are widely used in the field of lipid biology, where they have been reported to exhibit good performance characteristics for the analysis of a broad range of oxylipins in plasma [26]. With this in mind, we evaluated the possibility of using a mixed-mode SPE system by comparison with a Sep-Pak tC18 cartridge. The ratio of the process efficiency values between the mixed-mode SPE and RP SPE systems were calculated

for 12 of the ISs using mouse lung homogenate samples. With the exception of 5,6-EET-d11, the process efficiency ratio between the mixed-mode SPE and MonoSpin™ C18 systems was <100% in all cases (Figure 3). Notably, the process efficiency ratio was <60% for five of the ISs. These results therefore suggest that the mixed-mode SPE system was more efficient for the extraction of the lipids from the lung homogenate than the MonoSpin™ C18 spin column device. The mixed-mode SPE system was then compared with Sep-Pak tC18 cartridge. Interestingly, the process efficiency ratio between the mixed-mode SPE and the Sep-Pak tC18 was almost 100% for nearly all of the ISs (Figure 3), which suggested that the mixed-mode SPE system had the same degree of potential as the Sep-Pak tC18 system. In contrast, the process efficiency ratio of LTC$_4$-d5 was <5% for the MonoSpin™ C18 system and <25% for the Sep-Pak tC18 system. Since the process efficiency ratio of LTC$_4$-d5 was found to be <5% using the Oasis® HLB cartridge (60 mg/3 cc; Waters Corp.) (data not shown), our results suggested that the extraction efficiency of LTC$_4$ using the RP SPE system was generally

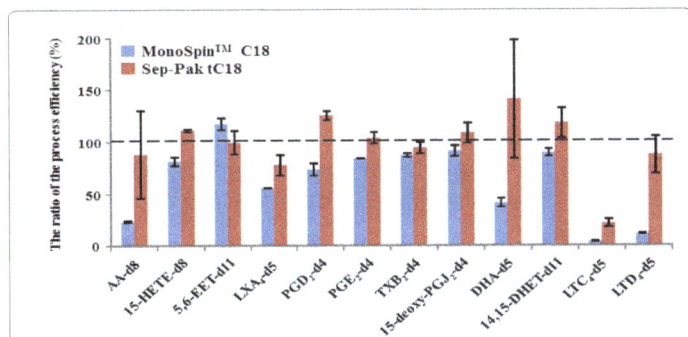

Figure 3: The ratio of process efficiency between mixed-mode SPE and RP SPE. Lipid extractions using the mixed-mode SPE, MonoSpin™ C18 and Sep-Pak tC18 systems were carried out in triplicate, and 12 ISs were analyzed by LC-MS/MS. The peak areas of the 12 ISs were calculated for each method and the ratio of the process efficiencies between the mixed-mode SPE and RP SPE system were calculated according to the "Comparison of mixed-mode SPE with RP SPE using mouse lung homogenate samples" described in the Materials and Methods section. The results for the MonoSpin™ C18 and Sep-Pak tC18 systems are indicated by blue and red bars, respectively.
HETE: Hydroxyeicosatetraenoic Acid; LXA_4: Lipoxin A_4; PG: Prostaglandin; DHET: Dihydroxyeicosatrienoic Acid

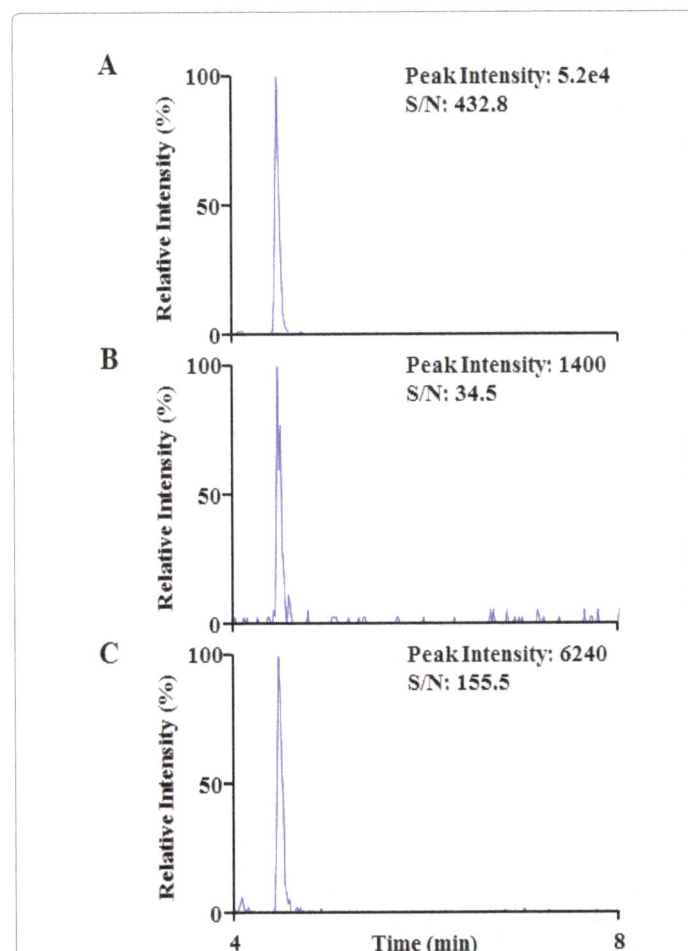

Figure 4: Comparison of leukotriene C_4 in mixed-mode SPE and RP SPE. Leukotriene C_4 in mice lung tissue samples was extracted using mixed-mode SPE, MonoSpin™ C18 and Sep-Pak tC18, and analyzed by LC-MS/MS. Extracted ion chromatograms of leukotriene C_4 in mixed-mode SPE (A), MonoSpin™ C18 (B) and Sep-Pak tC18 (C) are shown.

low. The extracted ion chromatograms of the endogenous LTC_4 extracted from the mouse lung homogenate samples using the different extraction methods are shown in Figure 4. Although the LTC_4 peak could be detected in full using the mixed-mode SPE and the S/N was 432.8, this peak was not sufficiently detected using the MonoSpin™ C18 system and the S/N was about 12-fold lower than the value achieved using the mixed-mode SPE system (432.8 to 34.5). Although there was a 5-fold increase in the S/N using the Sep-Pak tC18 system (34.5 to 155.5) compared with the MonoSpin™ C18 system, the S/N achieved using the Sep-Pak tC18 system was still three times lower than that of the mixed-mode SPE system (155.5 to 432.8). These results suggest that LTC_4 was not being sufficiently purified by the RP SPE system compared with mixed-mode SPE system. The poor performance of this system could be attributed in part to ion suppression from contaminants in the sample matrix. For conventional methods, simply eliminating contaminants that show similar retention behaviors to the compound of interest does not always allow for the resolution of this problem because the lipid extraction and HPLC separation processes are invariably carried out under the same RP conditions. Compared with conventional methods, potential contaminants are removed more effectively using our newly developed method, which involves the combination of mixed-mode and RP separation conditions for the lipid extraction and HPLC separation stages, respectively. Taken together, these results demonstrate that mixed-mode SPE is more effective for the lipid extraction of LTC_4 than RP SPE.

The correlation coefficients for the results measured between the mixed-mode SPE and RP SPE systems using mouse lung homogenate samples are shown in Figures 5 and 6. The proportion of variance was >0.99 for all of the lipids and >0.95 for all of the lipids present at concentrations of <500 pg/mg of tissue. For lipids present at concentrations <100 pg/mg of tissue, the proportion of variance was >0.86 (data not shown). These results therefore show that mixed-mode SPE performed much more effectively than RP SPE. Furthermore, more improvements were achieved for this method compared with the Sep-Pak tC18 system. Using a spin column, the extraction time was dramatically reduced from 1 h per 6 samples to 10 min per 6 samples, and the use of organic solvents was greatly reduced from >20 mL per sample to <1 mL per sample. In addition, the number of samples that could be simultaneously extracted was increased from 10 to 48 samples by changing the manifold to a centrifugal system. By eliminating time-consuming tasks such as those associated with the control of the manifold, the efficiency of the entire operation was markedly improved and the extraction time for 48 samples was less than 1 h. Therefore, this method is simpler, more readily available and more user-friendly than the conventional methods currently available for the analysis of lipids in biological samples.

Conclusions

A new approach has been developed for the analysis of oxidized fatty acids using LC–MS/MS combined with mixed-mode extraction with a spin column. The addition of a small amount of phosphoric acid to the mobile phase suppressed the adsorption of the Cys-LTs onto the metallic surfaces of the analytical system, which led to a dramatic improvement in the USP tailing factor and S/N of the Cys-LTs. This modified extraction method is suitable for various oxidized fatty acids including prostaglandins and thromboxanes. In addition, this method can be adapted to molecules that show poor recovery by changing the elution conditions. This new method was also compared with a conventional method using mouse lung homogenate samples, and the results showed that molecules exhibiting a poor recovery using conventional methods, such as LTC_4, were detected mush more

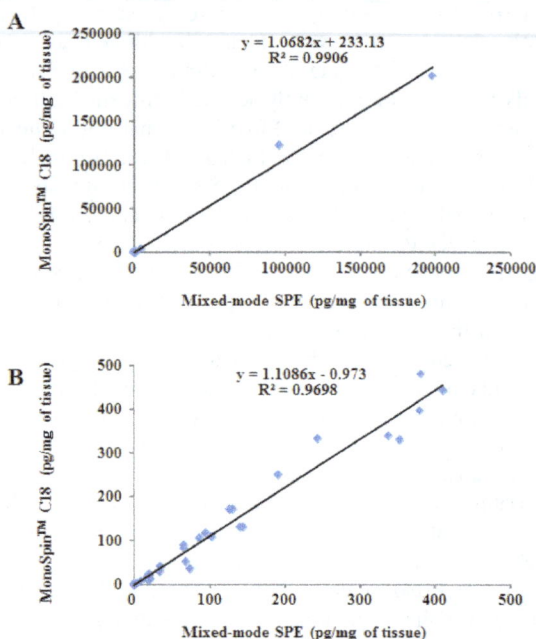

Figure 5: Scatter plot of the quantitative values of the mixed-mode SPE and MonoSpin™ C18 systems. Forty-five different metabolites derived from DHA and AA in mice lung tissue samples were analyzed using mixed-mode SPE and MonoSpin™ C18. The average values of the measurements (n=3) for each method were calculated for all lipids. All lipids (A) and lipid present at <500 pg/mg of tissue (B) are shown. R^2: proportion of variance.

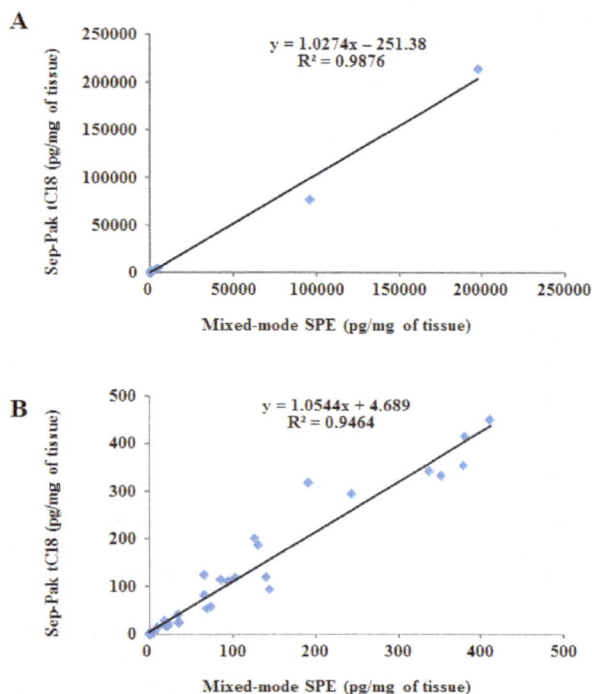

Figure 6: Scatter plot of the quantitative values of the mixed-mode SPE and Sep-Pak tC18 systems. Forty-five different metabolites derived from DHA and AA in mice lung tissue samples were analyzed using mixed-mode SPE and Sep-Pak tC18. The average values of the measurements (n=3) for each method were calculated for all lipids. All lipids (A) and lipid present at <500 pg/mg of tissue (B) are shown. R^2: proportion of variance.

effectively using our newly developed method because of its improved sample purification characteristics. We have already used this system in this context for lung, brain, kidney, plasma, and urine samples. Further work is needed to expand on the target lipid molecules, validate the analytical method and investigate the application of this method to a wide variety of tissue samples. We believe that this newly developed protocol is a useful tool for lipid biology and drug discovery research.

Acknowledgments

We thank Dr. Makoto Arita (RIKEN IMS) for guidance and valuable discussions, and Shigenori Ota (GL Sciences) for providing a schematic representation of the spin-column extraction.

References

1. Ruggieri S (1962) Separation of the methyl esters of fatty acids by thin layer chromatography. Nature 193: 1282-1283.

2. Michalec C, Sulc M, Mestan J (1962) Analysis of cholesteryl esters and triglycerides by thin-layer chromatography. Nature 193: 63-64.

3. Yamashita A, Watanabe Y, Hayaishi O (1983) Autoradiographic localization of a binding protein(s) specific for prostaglandin D2 in rat brain. Proc Natl Acad Sci U S A 80: 6114-6118.

4. Watanabe Y, Tokumoto H, Yamashita A, Narumiya S, Mizuno N, et al. (1985) Specific bindings of prostaglandin D2, E2 and F2 alpha in postmortem human brain. Brain Res 342: 110-116.

5. Miyazaki H, Ishibashi M, Yamashita K (1978) Use of a new silylating agent for separation of bile acids and cholesterol by selected ion monitoring with the computer controlled intensity matching technique. Biomed Mass Spectrom 5: 469-476.

6. Harada Y, Tanaka K, Uchida Y, Ueno A, Oh-Ishi S, et al. (1982) Changes in the levels of prostaglandins and thromboxane and their roles in the accumulation of exudate in rat carrageenin-induced pleurisy--a profile analysis using gas chromatography-mass spectrometry. Prostaglandins 23: 881-895.

7. Jörg A, Henderson WR, Murphy RC, Klebanoff SJ (1982) Leukotriene generation by eosinophils. J Exp Med 155: 390-402.

8. Arita M, Ohira T, Sun YP, Elangovan S, Chiang N, et al. (2007) Resolvin E1 selectively interacts with leukotriene B4 receptor BLT1 and ChemR23 to regulate inflammation. J Immunol 178: 3912-3917.

9. Xu ZZ, Zhang L, Liu T, Park JY, Berta T, et al. (2010) Resolvins RvE1 and RvD1 attenuate inflammatory pain via central and peripheral actions. Nat Med 16: 592-597.

10. Isobe Y, Arita M, Matsueda S, Iwamoto R, Fujihara T, et al. (2012) Identification and structure determination of novel anti-inflammatory mediator resolvin E3, 17,18-dihydroxyeicosapentaenoic acid. J Biol Chem 287: 10525-10534.

11. Spite M, Clària J2, Serhan CN3 (2014) Resolvins, specialized proresolving lipid mediators, and their potential roles in metabolic diseases. Cell Metab 19: 21-36.

12. Serhan CN (2014) Pro-resolving lipid mediators are leads for resolution physiology. Nature 510: 92-101.

13. Mirakaj V, Dalli J, Granja T, Rosenberger P, Serhan CN (2014) Vagus nerve controls resolution and pro-resolving mediators of inflammation. J Exp Med 211: 1037-1048.

14. Miyata J, Fukunaga K, Iwamoto R, Isobe Y, Niimi K, et al. (2013) Dysregulated synthesis of protectin D1 in eosinophils from patients with severe asthma. J Allergy Clin Immunol 131: 353-360.

15. Morita M, Kuba K, Ichikawa A, Nakayama M, Katahira J, et al. (2013) The lipid mediator protectin D1 inhibits influenza virus replication and improves severe influenza. Cell 153: 112-125.

16. Kubota T, Arita M, Isobe Y, Iwamoto R, Goto T, et al. (2014) Eicosapentaenoic acid is converted via ω-3 epoxygenation to the anti-inflammatory metabolite 12-hydroxy-17,18-epoxyeicosatetraenoic acid. FASEB J 28: 586-593.

17. Yokokura Y, Isobe Y, Matsueda S, Iwamoto R, Goto T, et al. (2014) Identification of 14,20-dihydroxy-docosahexaenoic acid as a novel anti-inflammatory metabolite. J Biochem 156: 315-321.

18. Kita Y, Takahashi T, Uozumi N, Shimizu T (2005) A multiplex quantitation method for eicosanoids and platelet-activating factor using column-switching reversed-phase liquid chromatography-tandem mass spectrometry. Anal Biochem 342: 134-143.

19. Kiss L1, Röder Y, Bier J, Weissmann N, Seeger W, et al. (2008) Direct eicosanoid profiling of the hypoxic lung by comprehensive analysis via capillary liquid chromatography with dual online photodiode-array and tandem mass-spectrometric detection. Anal Bioanal Chem 390: 697-714.

20. Dumlao DS, Buczynski MW, Norris PC, Harkewicz R, Dennis EA (2011) High-throughput lipidomic analysis of fatty acid derived eicosanoids and N-acylethanolamines. Biochim Biophys Acta 1811: 724-736.

21. Arita M (2012) Mediator lipidomics in acute inflammation and resolution. J Biochem 152: 313-319.

22. Rago B, Fu C (2013) Development of a high-throughput ultra performance liquid chromatography-mass spectrometry assay to profile 18 eicosanoids as exploratory biomarkers for atherosclerotic diseases. J Chromatogr B Analyt Technol Biomed Life Sci 936: 25-32.

23. Le Faouder P, Baillif V, Spreadbury I, Motta JP, Rousset P, et al. (2013) LC-MS/MS method for rapid and concomitant quantification of pro-inflammatory and pro-resolving polyunsaturated fatty acid metabolites. J Chromatogr B Analyt Technol Biomed Life Sci 932: 123-133.

24. Colas RA, Shinohara M, Dalli J, Chiang N, Serhan CN2 (2014) Identification and signature profiles for pro-resolving and inflammatory lipid mediators in human tissue. Am J Physiol Cell Physiol 307: C39-54.

25. Puppolo M, Varma D, Jansen SA (2014) A review of analytical methods for eicosanoids in brain tissue. J Chromatogr B Analyt Technol Biomed Life Sci 964: 50-64.

26. Ostermann AI, Willenberg I, Schebb NH (2015) Comparison of sample preparation methods for the quantitative analysis of eicosanoids and other oxylipins in plasma by means of LC-MS/MS. Anal Bioanal Chem 407: 1403-1414.

27. Abromeit H, Wu F, Scriba GK (2013) SPE of 5-lipoxygenase metabolites and the effect of head-column field-amplified sample stacking in MEKC. J Sep Sci 36: 3592-3598.

28. Lebold KM, Kirkwood JS, Taylor AW, Choi J, Barton CL, et al. (2014) Novel liquid chromatography-mass spectrometry method shows that vitamin E deficiency depletes arachidonic and docosahexaenoic acids in zebrafish (Danio rerio) embryos. Redox Biol 2: 105-113.

29. Matuszewski BK, Constanzer ML, Chavez-Eng CM (2003) Strategies for the assessment of matrix effect in quantitative bioanalytical methods based on HPLC-MS/MS. Anal Chem 75: 3019-3030.

30. Ogiso H, Suzuki T, Taguchi R (2008) Development of a reverse-phase liquid chromatography electrospray ionization mass spectrometry method for lipidomics, improving detection of phosphatidic acid and phosphatidylserine. Anal Biochem 375: 124-131.

34

Crystal Structure and Hirshfeld Surface Analysis of 1,2-Bis((2-(Bromomethyl)Phenyl)Thio)Ethane and Two Polymorphs of 1,2-Bis((2-((Pyridin-2-ylthio)Methyl)Phenyl)Thio)Ethane

Simplicio González-Montiel[1]*, Saray Baca-Téllez[1], Diego Martínez-Otero[2], Alejandro Álvarez-Hernández[1] and Julián Cruz-Borbolla[1]

[1]*Área Academic Chemistry, Chemical Research Center, University of the State of Hidalgo, 4.5 km. Road Pachuca-Tulancingo, City of Knowledge, CP 42184, Mineral de la Reforma, Hidalgo, Mexico*
[2]*Joint Research Centre in Sustainable Chemistry, UNAM UAEM, Toluca-Atlacomulco Highway 14.5 km., C.P. 50200, Toluca, State of Mexico, Mexico*

Abstract

1,2-Bis((2-(bromomethyl)phenyl)thio)ethane (**1**) and 1,2-bis((2-((pyridin-2-ylthio)methyl)phenyl)thio)ethane (**2**) were prepared and characterized by IR and NMR spectroscopy and single-crystal X-ray crystallography. X-ray diffraction studies shown that compound **1** crystallizes in a monoclinic space group $P2_1/n$ with crystal parameters a=8.3970(3) Å, b=12.4566(2) Å, c=8.9251(3) Å; β=117.911(3)°, V=824.96(5) Å^3 and z=2, and compound **2** exists in two monoclinic polymorphs (**2a** and **2b**). Polymorph **2a** crystals are in space group $P2_1$, with unit cell parameters a=5.3702(2) Å, b=14.4235(6) Å, c=15.4664(7) Å, β=119.97(9)°, V=1197.97(9) Å^3 and z=2, while polymorph **2b** crystals are in space group $P2_1/c$ with unit cell parameters a=7.8312(3) Å, b=9.6670(4) Å, c=16.2962(5) Å, β=121.219(3)°; V=1210.12(7) Å^3 and z=2. Variations in the crystal packing help to distinguish these two polymorphs via π-π and C–H•••π interactions. The 3D Hirshfeld surfaces and the associated 2D fingerprint plots have been performed to gain insight into the behavior of these interactions in compound **1** and polymorphs **2a** and **2b**.

Keywords: Polymorphs; 1,2-bis((2-((pyridin-2-ylthio)methyl)phenyl)thio)ethane; Crystal packing; Hirshfeld surface; π-π and C–H•••π interactions

Introduction

Polymorphism is the ability of a particular molecule to exist in more than one crystal structure and it has great importance in pharmacology, solid-state chemistry, and material science since different polymorphs may have different physicochemical properties such as thermal behavior, stability, solubility, melting point, and bioavailability, among others [1-6]. Variations in structural units in a crystal leading to polymorphs occur through different intermolecular interactions such as D–H•••A hydrogen bonding, π-π stacking, CH•••π, halogen•••π, halogen•••halogen and anion•••π. These crystal structure modification in each polymorph result in different thermodynamic stability, i.e., the free energy of the crystals, kinetics of nucleation and crystal growth promoted by crystallization conditions (solvent, temperature, concentration, cocrystallization, pressure, etc.) [7-15]. These molecular Hirshfeld surfaces, so named because they derive from Hirshfeld's stockholder partitioning, divide the crystal into regions where the electron distribution of a sum of spherical atoms for the molecule (the promolecule) dominates the corresponding sum over the crystal (the procrystal). Hirshfeld surface analysis has gained prominence as a powerful tool to explore and describe a wide variety of intermolecular interactions within a crystal, [16-18] and it is almost always related to its corresponding 2D fingerprint-plot. This plot provides a convenient way to quantify intermolecular interactions within crystal structures and helps to reveal important information both about close contacts and also about more distant interactions and areas where contacts are weak [19-25].

Herewith we report the synthesis and structural study of two polymorphs of 1,2-bis((2-((pyridin-2-ylthio)methyl)phenyl)thio) ethane (**2a** and **2b**) and 1,2-bis((2-(bromomethyl)phenyl)thio)ethane (**1**) (Scheme 1) as part of our studies concerning the construction of novel metallomacrocycles based on ligands that contain in their structure two 2-mercaptopyridyl groups [26]. The intermolecular interactions that exist in the crystal structure of the two polymorphs have also been investigated by Hirshfeld surface analysis [23-25].

Experiment

Materials and instrumentation

All reagents are commercially available and were used without further purification. Melting points were measured in a Mel-Temp II instrument and are not corrected. Elemental analyses of the compounds were determined on a Perkin–Elmer Series II CHNS/O Analyzer. IR spectra were recorded on the 4000-400 cm^{-1} range on a Perkin–Elmer 2000 FTIR spectrometer as KBr pellets. ^{1}H NMR and ^{13}C{^{1}H} NMR spectra were recorded on a Varian Inova 400 NMR spectrometer at 20°C in CDCl$_3$ solutions; ^{1}H, 399.78 MHz and residual protio-solvent signal were utilized as reference. Chemical shifts are quoted in the δ scale (downfield shifts positive) relative to tetramethylsilane (^{1}H). 1,2-bis-(2-hydroxymethylphenylthio)ethane (HOCH$_2$Ph-S-CH$_2$-CH$_2$-S-PhCH$_2$OH) was prepared according to a previously reported protocol [27].

Synthesis of 1,2-bis((2-(bromomethyl)phenyl)thio)ethane (1) and 1,2-bis((2-((2-pyridinylthio)methyl)phenyl)thio)ethane (2)

Preparation of 1,2-bis((2-(bromomethyl)phenyl)thio)ethane (**1**): 1,2-bis-(2-hydroxymethylphenylthio)ethane (8.15 mmol, 2.5 g)

*Corresponding author: Simplicio González-Montiel, Área Académica de Química, Centro de Investigaciones Químicas, Universidad Autónoma del Estado de Hidalgo, km. 4.5 Carretera Pachuca-Tulancingo, Ciudad del Conocimiento, C.P. 42184, Mineral de la Reforma, Hidalgo, México, E-mail: gmontiel@uaeh.edu.mx

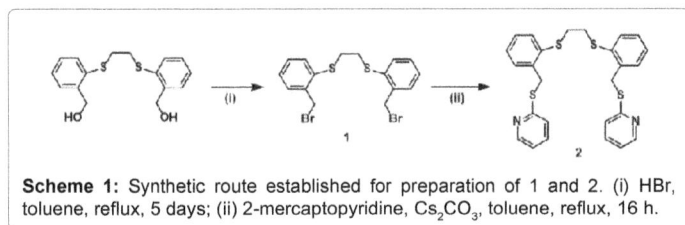

Scheme 1: Synthetic route established for preparation of 1 and 2. (i) HBr, toluene, reflux, 5 days; (ii) 2-mercaptopyridine, Cs_2CO_3, toluene, reflux, 16 h.

was dissolved in 50 mL of toluene and hydrobromic acid, 48% (25 mL) was added; the mixture was refluxed for 5 days. After cooling to room temperature, the two layers were separated and the organic layer was dried with $NaSO_4$, then filtered through a bed of Celite and the solvent was removed under reduced pressure to give a white solid. Yield: 3.15 g (90%). m.p.=111-115°C. *Anal.* Calc. for $C_{16}H_{16}Br_2S_2$: C, 44.46; H, 3.73. Found: C, 44.16; H, 3.83%. ^{1}H NMR (399.78 MHz, $CDCl_3$): δ=7.45-7.42 (2H, m, H1), 7.37–7.34 (2H, m, H4), 7.30-7.24 (4H, m, H2 and H3), 4.70 (4H, s, CH_2-Br), 3.18 (4H, s, S-CH_2) ppm. $^{13}C\{^1H\}$ NMR (100.53 MHz, $CDCl_3$): δ 138.68 (C6), 135.05 (C5), 131.26 (C4), 130.99 (C1), 129.43 (C2), 127.49 (C3), 33.81 (S-CH_2), 32.12 (CH_2-Br) ppm. IR (KBr): 3058, 2966, 2920, 2852, 1588, 1567, 1466, 1443, 1431, 1218, 1203, 1194, 1063, 1038, 818, 754, 727, 704, 672, 602, 567, 497, 445 cm^{-1}.

Preparation of 1,2-bis((2-((2-pyridinylthio)methyl)phenyl)thio) ethane (2): 1,2-bis((2-(bromomethyl)phenyl)thio)ethane (4.63 mmol, 2.0 g) and 2-mercaptopyridine (9.26 mmol, 1.03 g) were dissolved in 50 mL toluene, and then Cs_2CO_3 (4.63 mmol, 1.51 g) was added directly into the solution; the mixture was then refluxed for 16 h. After cooling, the resulting suspension was filtered through Celite and the solvent was removed under reduced pressure to give a white solid. Yield 2.10 g (92%). m.p.=82-87°C. *Anal.* Calc. for $C_{26}H_{24}N_2S_4$: C, 63.38; H, 4.91; Found: C, 62.98; H, 4.83%. ^{1}H NMR (399.78 MHz, $CDCl_3$): δ=8.46 (2H, ddd, ^3J=4.96 Hz, ^4J=1.84 Hz, ^5J=0.94 Hz, H12), 7.49 (2H, dd, ^3J=6.18 Hz, ^4J=2.96 Hz, H1), 7.45 (2H, ddd, $3J$=8.05 Hz, ^4J=7.41 Hz, ^5J=1.88 Hz, H10), 7.31 (2H, dd, ^3J=6.48 Hz, ^4J=2.66 Hz, H4), 7.16 (4H, m, H2, H3), 7.13 (2H, dd, ^3J=7.14 Hz, ^4J=0.98 Hz, H9), 6.98 (2H, ddd, ^3J=7.34 Hz, ^4J=4.95 Hz, ^5J=1.02 Hz, H11), 4.62 (4H, s, H7), 3.14 (2H, s, H8) ppm. 13C NMR ($CDCl3$): δ=158.80 (C8), 149.42 (C12), 139.01 (C5), 136.11 (C10), 134.79 (C6), 130.82 (C1), 130.61 (C4), 128.10 (C2), 127.07 (C3), 122.36 (C9), 119.69 (11), 33.99 (C13), 32.91 (C7) ppm. IR (KBr): 3056, 2992, 2924, 2847, 1578, 1556, 1453, 1414, 1281, 1242, 1147, 1122, 1062, 1043, 985, 956, 820, 757, 619, 580, 479 cm^{-1}.

X-ray diffraction

Suitable crystals of **1** were grown in toluene by slow evaporation while crystals of **2** were grown in dimethylsulfoxide and were separated by fractional crystallization; the polymorph **2a** crystallized first after 2 weeks stored at room temperature and then crystals of polymorph **2b** were obtained after 4 weeks at the same solution. All crystal structures were determined by X-ray analysis, and their crystallographic details and structure refinements are presented in Table 1. X-ray diffraction data were collected at room temperature on an Oxford Diffraction Gemini CCD diffractometer, using graphite-monochromated Cu Kα radiation (λ=1.54184 Å) for **1** and Mo Kα radiation (λ=0.71073 Å) for **2a** and **2b**. Data were processed using the Crysalis software package [28]. Using Olex2 [29], the structures were solved with the XS program [30] employing direct methods and refined with the XL refinement package using Least Squares minimization [30]. All non-hydrogen atoms were refined anisotropically and hydrogen atoms were located at calculated positions and refined using a riding model with isotropic

thermal parameters fixed at 1.2 times the Ueq value of the appropriate carrier atom (Table 1).

Computational details

The Hirshfeld surfaces and fingerprint plots were calculated using the Crystal Explorer (version 3.1) software [31].

Results and Discussion

Synthesis of compounds 1 and 2

Compound 1,2-bis((2-((2-pyridinylthio)methyl)phenyl)thio)etane (**2**) was prepared in good yield using an established route (Scheme 1). Thus, starting from 1,2-bis-(2-hydroxymethylphenylthio)etane [26] bromination using an excess of hydrobromic acid led to 1,2-bis((2-(bromomethyl)phenyl)thio)etane (**1**). 1,2-bis((2-((2-pyridinylthio) methyl)phenyl)thio)etane was obtained from the reaction between 1,2-bis((2-(bromomethyl)phenyl)thio)ethane and 2-mercaptopyridine in presence of Cs_2CO_3. Compounds **1** and **2** are soluble in chloroform, dichloromethane, toluene, benzene and acetonitrile, and are insoluble in methanol, ethanol, pentane and *n*-hexanes.

NMR spectroscopy

The ^{1}H NMR spectrum of compound **1** show three signals at high frequencies corresponding to *ortho*-substituted benzene rings, two single signals at low frequencies correspond to the bridge ethylene group and the bromomethylene group. The ^{1}H NMR spectrum of compound **2** shows eight signals at high frequencies due to two ABCD patterns corresponding to the *ortho*-substituted benzene rings and to the *ortho*-substituted pyridine rings; signals at low frequencies correspond to the methylene groups that link the phenyl and pyridine rings and to the ethylene brigde between both phenyl rings.

The ^{13}C NMR spectrum of compound **1** shows six signals at high frequencies that are attributed to *ortho*-substituted benzene rings and two single signals at low frequencies corresponding to the methylene group link bromide and the ethylene group that brigde both phenyl rings. The ^{13}C-NMR of compound **2** shows eleven signals at high frequencies which six are attributed to *ortho*-substituted benzene and pyridine rings, and two signals at low frequencies corresponding to methylene and ethylene groups. In solution, the two $-CH_2$-S-C_6H_4S-CH_2-Br and $-CH_2$-S-C_6H_4S-CH_2-C_5H_4N units are equivalent.

Molecular and crystal structures

Molecular structures: The molecular structure of **1**, **2a** and **2b** was confirmed by X-ray diffraction studies and their structure are depicted in Figures 1 and 2, and selected bond lengths, angles and torsion angles are given in Table 2.

In all compounds **1**, **2a** and **2b** the aromatic rings on the ethylene group adopt *anti*-conformation, i.e. the $-PhCH_2Br$ fragments in **1** and $-Ph$-CH_2-S-C_5H_5N fragments in **2a** and **2b** are positioned on opposite sides of the ethylene bridge with C1−C7---C7^i−C1^i torsion angles in **1** and **2b** equal to 180° and C1−C7---C8−C9 torsion angle in **2a** equal to 177.16°. In compounds **1** and **2b** the phenyl rings [Cg_A=C1/C2/C3/C4/C5/C6 and Cg_A=C1'/C2'/C3'/C4'/C5'/C6', symmetry code: (i) 2-x, 1-y, 1-z for **1** and -x, 1-y, -z for **2b**] are in an antiparallel manner in relation to each another with centroid-to-centroid distance Cg_A•••Cg_A. of 8.367 Å and 8.149 Å for **1** and **2b**, respectively. While in **2a** the phenyl rings (Cg_A=C1/C2/C3/C4/C5/C6 and Cg_B=C9/C10/C11/C12/C13/C14) are nearly coplanar with the dihedral angle between the two phenyl rings being 9.76° with centroid-to-centroid distance Cg_A•••Cg_B of 9.407

Compound	1	2a	2b
Empirical formula	$C_{16}H_{16}S_2Br_2$	$C_{26}H_{24}N_2S_4$	$C_{26}H_{24}N_2S_4$
Formula weight	432.23	492.71	492.71
Temperature [K]	301.3(8)	293(2)	293(2)
Crystal system	monoclinic	monoclinic	monoclinic
Space group	$P2_1/n$	$P2_1$	$P2_1/c$
a [Å]	8.3970(3)	5.3702(2)	7.8312(3)
b [Å]	12.4566(2)	14.4235(6)	9.6670(4)
c [Å]	8.9251(3)	15.4664(7)	16.2962(5)
α [°]	90	90	90
β [°]	117.911(3)	90.215(4)	101.219(3)
γ [°]	90	90	90
Volume [Å³]	824.96(5)	1197.97(9)	1210.12(7)
Z	2	2	2
ρ_{calcd} [mg/mm⁻³]	1.740	1.366	1.352
μ [mm⁻¹]	8.483	0.414	0.410
F(000)	428.0	516.0	516.0
Crystal size [mm³]	0.48 × 0.33 × 0.33	0.24 × 0.09 × 0.05	0.27 × 0.21 × 0.19
Radiation wavelength [Å]	CuKα (λ=1.54184)	MoKα (λ=0.71073)	MoKα (λ=0.71073)
2Θ range for data collection [°]	11.952 to 134.152	5.978 to 52.78	6.604 to 52.742
Abs. correction	Analytical	Analytical	Analytical
Index ranges	-10 ≤ h ≤ 9 -14 ≤ k ≤ 14 -10 ≤ l ≤ 10	-6 ≤ h ≤ 6 -18 ≤ k ≤ 18 -19 ≤ l ≤ 19	-9 ≤ h ≤ 9 -12 ≤ k ≤ 12 -20 ≤ l ≤ 20
Reflections collected	3971	10460	17858
Unique reflections, R_{int}	1374, 0.0384	4690, 0.0302	2467, 0.0261
Data/restraints/parameters	1374/0/92	4690/238/289	2467/0/145
Goodness-of-fit (GOF) on F²	1.068	1.015	1.048
R_1, wR_2 [I>=2σ (I)]	0.0416, 0.1108	0.0376, 0.0649	0.0327, 0.0773
R_1, wR_2 [all data]	0.0441, 0.1135	0.0554, 0.0707	0.0414, 0.0819
Largest diff. peak/hole [e·Å⁻³]	0.56/-0.60	0.22/-0.18	0.19/-0.25

Table 1: Details of crystal data and structure refinement parameters for 1 and 2a and 2b.

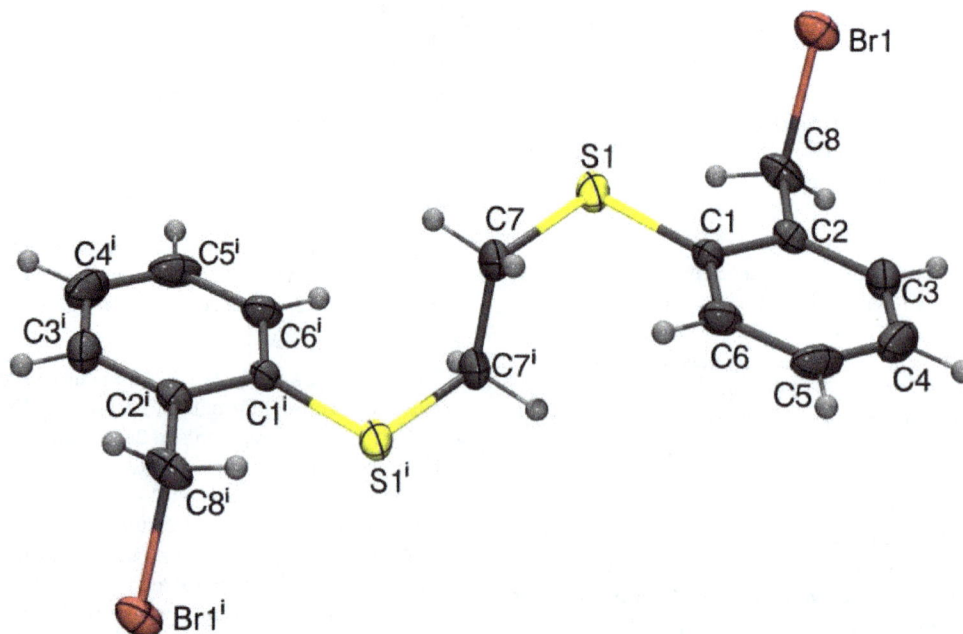

Figure 1: Molecular structure of compound 1, showing the atom labeling. Displacement ellipsoids are drawn at the 30% probability level. Symmetry code (i) 2–x, 1–y, 1–z.

Compound	1	2a	2b
Bond length			
S1-C1	1.772(3)	1.770(3)	1.7870(16)
S1-C7	1.806(4)	1.805(4)	1.8165(17)
C7-C7^i	1.512(8)		1.505(3)
C7-C8		1.510(4)	
C8-Br1	1.963(4)		
S2-C8		1.808(4)	1.8130(18)
S2-C9		1.768(3)	1.7704(17)
S3-C15		1.823(3)	
S3-C16		1.760(4)	
S4-C21		1.818(3)	
S4-C22		1.766(4)	
N1-C9			1.324(2
N1-C13			1.347(2)
N1-C16		1.318(5)	
N1-C20		1.345(6)	
N2-C22		1.321(4)	
N2-C26		1.344(5)	
Bond angle			
C1-S1-C7	104.92(19)	102.72(16)	101.32(7)
C2-C1-S1	117.6(3)	117.9(3)	120.30(12)
C6-C1-S1	123.3(3)	122.8(3)	119.74(13)
C7^i-C7-S1	113.8(4)		112.82(15)
C2-C8-Br1	110.0(3)		
C8-C7-S1		108.3(2)	
C9-S2-C8		103.14(16)	103.62(8)
C22-S4-C21		101.76(18)	
S2-C9-N1			120.96(13)
N1-C16-S3		120.6(3)	
N2-C22-S4		118.9(3)	
Torsion angle			
S1-C1-C2-C3	176.2(3)	176.6(3)	-176.02(12)
S1-C1-C2-C8	-2.7(5)		4.3(2)
C3-C2-C8-Br1	-86.6(4)		
C1-C2-C8-Br1	92.4(4)		
C3-C2-C8-S2			113.55(16)
C7-S1-C1-C2	164.3(3)	178.1(2)	
C1-S1-C7-C7^i	-71.7(5)		71.55(18)
C7-S1-C1-C6	-19.0(4)	-0.7(3)	62.25(15)
C13-N1-C9-S2			177.97(13)
S1-C1-C2-C15		5.9(4)	
S1-C1-C6-C5		177.4(3)	
S1-C7-C8-S2		-178.20(18)	
C1-S1-C7-C8		-175.9(2)	
S2-C9-C10-C21		-1.6(4)	
C8-S2-C9-C10		-171.9(2)	
C9-C10-C21-S4		-79.4(3)	
C15-S3-C16-N1		-1.5(4)	

Symmetry code: (i) 2–x, 1–y, 1–z for 1 and –x, 1–y, –z for 2b.

Table 2: Selected bond lengths (Å) and angles and torsion angles (°) for 1 and 2a and 2b.

Å. The fragments –CH$_2$Br in **1** and –CH$_2$-S-C$_5$H$_4$N in **2a** and **2b** are displayed in a *trans* arrangement (Figures 1 and 2).

In polymorph **2a** phenyl rings are nearly coplanar while in **2b** phenyl rings are positioned in an antiparallel manner and centroid-to-centroid distance increases when phenyl rings are nearly coplanar. In this context the polymorph **2a** presents a larger centroid-to-centroid distance compared to polymorph **2b** [$\Delta d_{(Cg\cdots Cg)}$=1.258 Å].

Crystal structures: The crystal lattice of **1** exhibits a supramolecular assembly that has a polymeric array via C–H•••π, π•••π interactions of the offset face to face stacking and Br•••π interactions. Thus, in this crystal lattice there is a C–H•••π interaction between one hydrogen atom of the methylene group bonded to bromine with a phenyl ring $d_{(H8a\cdots Cg)}$=2.947 Å and $\sphericalangle_{(C8-H8a\cdots Cg2)}$=111.78°; also a π•••π offset-stacked interaction is observed with $d_{(H6\cdots Cg)}$=3.465 Å, $d_{(Cg\cdots Cg)}$=4.655 Å and $\sphericalangle_{(Cg-H6\cdots Cg)}$=105.81°; finally Br•••π interactions are also observed with $d_{(Br1\cdots Cg)}$=3.465 Å, $d_{(Br1\cdots C6)}$=3.600 Å and $\sphericalangle_{(Br1-C6\cdots Cg)}$=92.96° (Figure 3).

Crystal analysis of polymorphs **2a** and **2b** suggests the presence of different π–π and C–H•••π interactions. The crystal lattice of polymorph **2a** exhibits the formation of a supramolecular, polymeric array via C–H•••π interactions between one hydrogen atom of the methylene group linking a phenyl ring with a pyridine ring and other interactions between one hydrogen atom of the methylene group that bridges two phenyl rings [C–H•••π interactions with $d_{(H7b\cdots Cg1)}$=2.708 Å and $\sphericalangle_{(C7-H7b\cdots Cg1)}$=142.92°; $d_{(H8a\cdots Cg2)}$=2.868 Å and $\sphericalangle_{(C8-H78a\cdots Cg2)}$=143.02°; $d_{(H15b\cdots Cg3)}$=3.116 Å and $\sphericalangle_{(C15-H15b\cdots Cg3)}$=145.68°; where C1/C2/C3/C4/C5/C6=Cg1, C9/C10/C11/C12/C13/C14=Cg2 and N1/C16/C17/C18/C19/C20=Cg3, respectively] (Figure 4). Polymorph **2b** presents a supramolecular, tetrameric assembly via π•••π offset-stacked and C–H•••π T-shaped interactions (π•••π offset-stacked interactions between two pyridinic rings with $d_{(H10\cdots Cg2)}$=3.434 Å, $d_{(Cg2\cdots Cg2)}$=4.446 Å and $d_{(interplanar)}$=3.380 Å. C–H•••π T-shaped interactions are due to contacts between one hydrogen atom of one phenyl ring and another phenyl ring with $d_{(H3\cdots Cg1)}$=3.302 Å, $d_{(Cg1\cdots Cg1)}$=5.397 Å and $\sphericalangle_{(C3-H3\cdots Cg)}$=148.16°, where C1/C2/C3/C4/C5/C6=Cg1 and N1/C9/C10/C11/C12/C13=Cg2, respectively) (Figure 5).

Hirshfeld surface analysis

The intermolecular interactions of crystal structures of 1, 2a and 2b were quantified using Hirshfeld surface analysis and fingerprint plots which are illustrated in Figure 6. Surfaces that have been mapped over a norm are shown. The relative contributions of different interactions of compounds 1, 2a and 2b are presented in Figure 7.

In general, the fingerprint plots of compounds **1**, **2a** and **2b** shown that dominant interactions are C•••H (19.2-31.8%) and H•••H (35.5-48.6%), and especially in compound **1** it is observed a Br•••H contribution (28.8%), other significant S•••H interactions have been observed varying from 10.8-17.5%. Hirshfeld surfaces of polymorphs **2a** and **2b** exhibit significant differences between intermolecular interactions. Thus, in **2b** C•••H contributions are 7.6% larger than those for **2a**; H•••H contributions in **2b** are smaller by 5.1% than in those in **2a**, S•••H contributions are quite similar in both polymorphs (17.0% in **2a** and 17.5% in **2b**), N•••H contributions are also similar (5.8% in **2a** and 4.0% in **2b**). Finally, polymorph **2b** does not present C•••C contributions while in **2a** this contribution is 3.3%. These differences in the contributions of all types of interactions found in the crystal structures of compounds **2a** and **2b** evidence that they are structurally different.

Conclusion

The crystal structure of the compound **1** and polymorphs **2a** and **2b** show different intermolecular interactions that lead to different types of supramolecular arrays. Quantification of the intermolecular interactions present in compounds **1**, **2a** and **2b** was realized by Hirshfeld surface analysis and 2D fingerprint plots. This analysis of intermolecular contacts present in the crystal packing of **2a** and **2b** leads to conclude these compounds are true polymorphs.

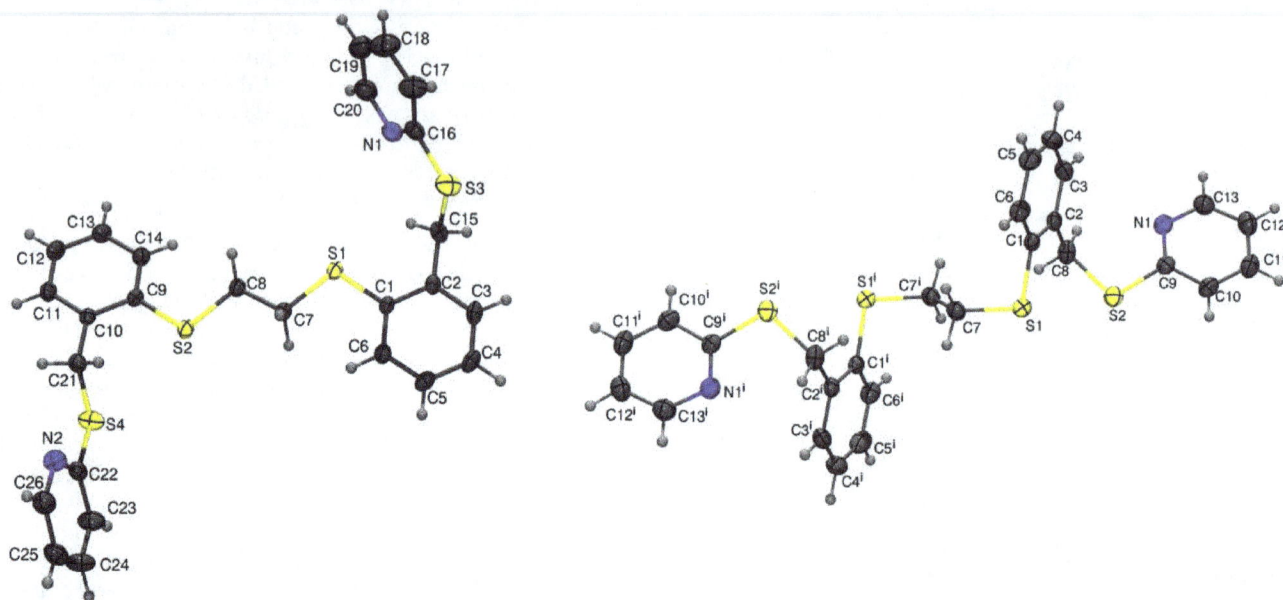

Figure 2: Molecular structure of the polymorphs 2a (left) and 2b (right), showing the atom labeling. Displacement ellipsoids are drawn at the 30% probability level. Symmetry code (i) −x, 1−y, −z.

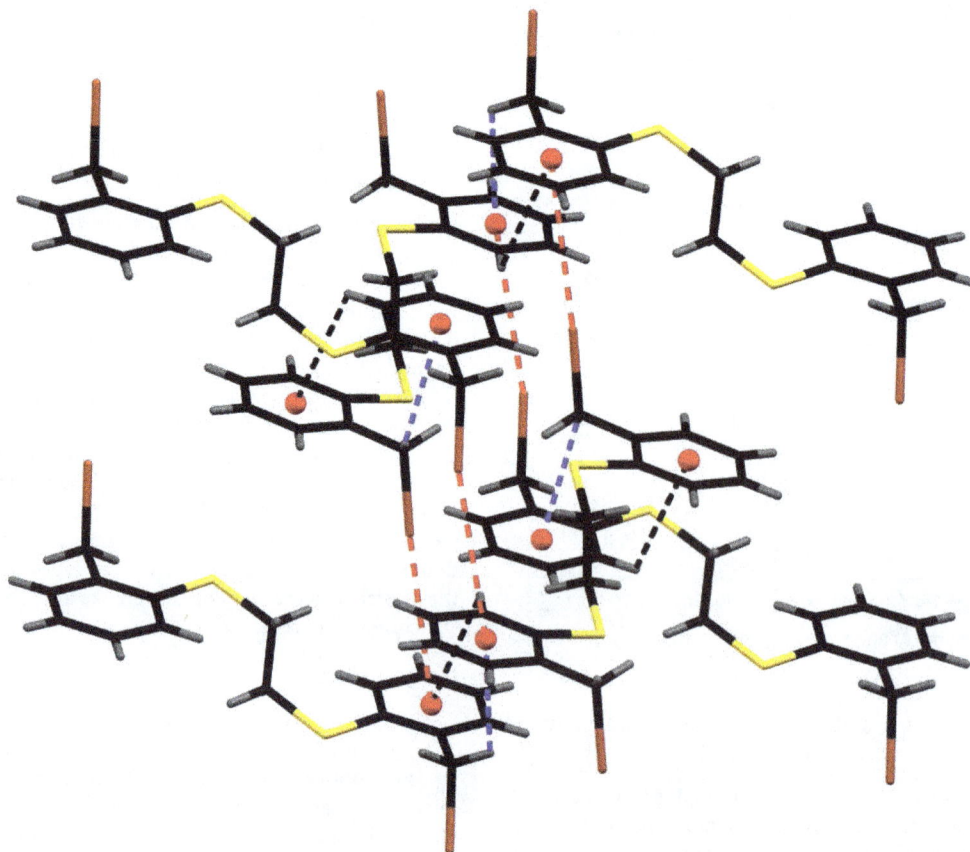

Figure 3: View of π•••π offset-stacked, C–H•••π and Hal•••π interactions present in the crystal packing of 1. (C: black; Br: brown; H: grey; S: yellow; π•••π offset-stacked, black; C–H••• π, blue; Hal•••π, red).

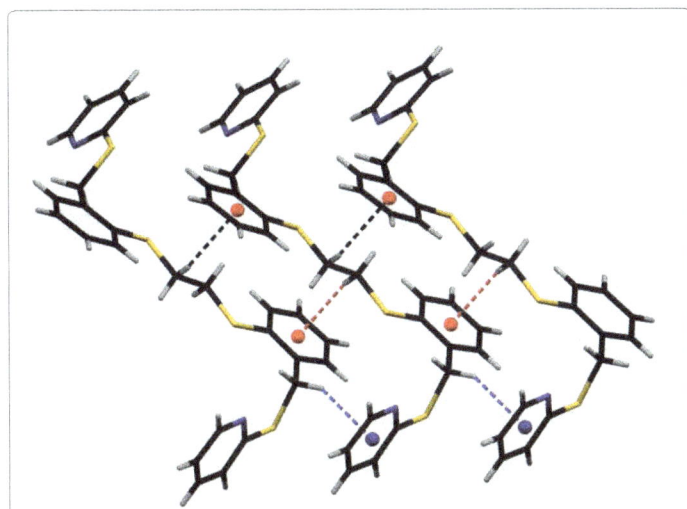

Figure 4: View of C–H•••π interactions present in the crystal packing of 2a. (C: black; N: blue; H: grey; S: yellow; C–H••• π, blue, red, black).

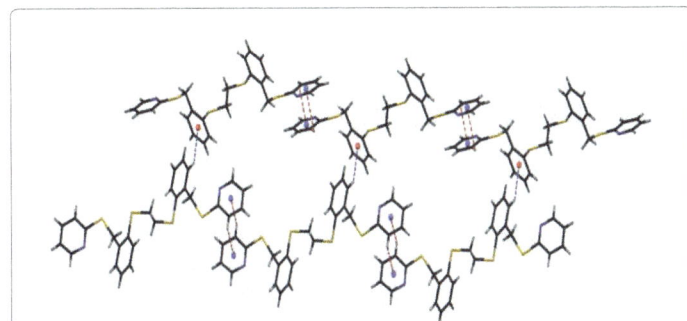

Figure 5: View of π•••π offset-stacked, C–H•••π interactions present in the crystal packing of 2b. (C: black; N: blue; H: grey; S: yellow; π•••π offset-stacked, red; C–H•••π, blue).

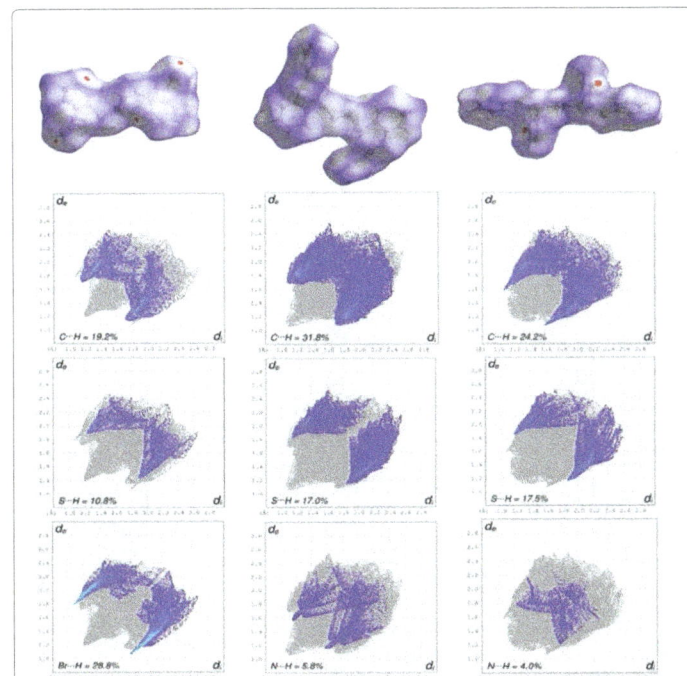

Figure 6: Hirshfeld surfaces and 2D fingerprint plots of compounds 1 (left), 2a (middle) and 2b (right) showing percentage contribution of contact molecules.

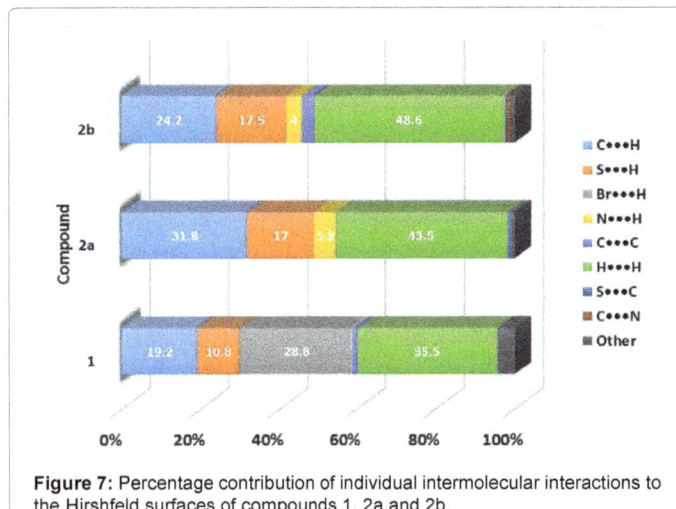

Figure 7: Percentage contribution of individual intermolecular interactions to the Hirshfeld surfaces of compounds 1, 2a and 2b.

Supplementary Material

Crystallographic data for the structural analyses have been deposited with the Cambridge Crystallographic Data Center, CCDC Nos. 1037743 for **1**, 1037744 for **2a**, and 1037745 for **2b**. The data can be obtained free of charge from The Director, CCDC, 12 Union Road, Cambridge CB21EZ, UK (Fax: +44-1223-336-033; E-mail: deposit@ccdc.cam.ac.uk or http://www.ccdc.cam.ac.uk).

Acknowledgements

S.G.M. thankfully acknowledges financial support from CONACyT "México" (Grant CB-2011-01-167873) and S.B.T. is grateful for a graduate fellowship from CONACyT (352507).

References

1. Threlfall TL (1995) Analysis of organic polymorphs. A review. Analyst 120: 2435-2460.

2. Berntein J, Henck JO (1998) Disappearing and reappearing polymorphs-an anathema to crystal engineering? Cryst Eng 1: 119-128.

3. Cruz-Cabeza AJ, Bernstein J (2014) Conformational polymorphism. Chem Rev 114: 2170-2191.

4. Rubcic M, Galic N, Halasz I, Jednacak T, Judaš N, et al. (2014) Multiple Solid Forms of 5-Bis(salicylidene)carbohydrazide: Polymorph-Modulated Thermal Reactivity. Cryst Growth Des 14: 2900-2912.

5. Kumar VS, Addlagatta A, Nangia A, Robinson WT, Broder CK, et al. (2002) 4,4-Diphenyl-2,5-cyclohexadienone: four polymorphs and nineteen crystallographically independent molecular conformations. Angew Chem Int Ed Engl 41: 3848-3851.

6. Yu V, Shavit E, Weissbuch I, Leiserowitz L, Lahav M (2005) Control of Crystal Polymorphism by Tuning the Structure of Auxiliary Molecules as Nucleation Inhibitors. The β-Polymorph of Glycine Grown in Aqueous Solutions. Cryst Growth Des 5: 2190-2196.

7. Roy S, Nangia A (2007) Kinetic and Thermodynamic Conformational Polymorphs of Bis(p-tolyl) Ketone p-Tosylhydrazone: The Curtin-Hammett Principle in Crystallization. Cryst Growth Des 7: 2047-2058.

8. Zhang W, Tang X, Ma H, Wen-Hua S, Janiak C (2008) {2-[1-(2,6-Diisopropylphenylimino)ethyl]pyridyl}palladium Dibromide Polymorphs Originating from Different Br•••π and C–H•••Br Contacts. Eur J Inorg Chem 2008: 2830-2836.

9. Ueda K, Oguni M, Asaji T (2014) Halogen Bond as Controlling the Crystal Structure of 4-Amino-3,5-Dihalogenobenzoic Acid and Its Effect on the Positional Ordering/Disordering of Acid Protons. Cryst Growth Des 14: 6189-6196.

10. Priimagi A, Cavallo G, Metrangolo P, Resnati G (2013) The halogen bond in

the design of functional supramolecular materials: recent advances. Acc Chem Res 46: 2686-2695.

11. Mukherjee A, Tothadi S, Desiraju GR (2014) Halogen bonds in crystal engineering: like hydrogen bonds yet different. Acc Chem Res 47: 2514-2524.

12. Jentzsch AV, Hennig A, Mareda J, Matile S (2013) Synthetic ion transporters that work with anion-π interactions, halogen bonds, and anion-macrodipole interactions. Acc Chem Res 46: 2791-2800.

13. Aitipamula S, Chow PS, Tan RBH (2014) Crystal Engineering of Tegafur Cocrystals: Structural Analysis and Physicochemical Properties. Cryst Growth Des 14: 6557-6569.

14. Landenberger KB, Matzger AJ (2012) Cocrystals of 1,3,5,7-Tetranitro-,3,5,7-tetrazacyclooctane (HMX). Cryst Growth Des 12: 3603-3609.

15. Sarcevia I, Orola L, Veidis MV, Podjava A, Belyahov S (2013) Crystal Structures and Density Functional Theory Calculations of o- and p-Nitroaniline Derivatives: Combined Effect of Hydrogen Bonding and Aromatic Interactions on Dimerization Energy. Cryst Growth Des 12: 3603-3609.

16. McKinnon JJ, Spackman MA, Mitchell AS (2004) Novel tools for visualizing and exploring intermolecular interactions in molecular crystals. Acta Crystallogr B 60: 627-668.

17. McKinnon JJ, Mitchell AS, Spackmann MA (1998) Hirshfeld Surfaces: A New Tool for Visualising and Exploring Molecular Crystals. Chem Eur J 4: 2136-2141.

18. Spackmann MA, McKinnon JJ (2002) Fingerprinting intermolecular interactions in molecular crystals. Cryst Eng Comm 4: 378-392.

19. Bakavoli M, Rahimizadeh M, Feizyzadeh B, Kaju AG, Takjoo R (2010) 3,6-Di(p-chlorophenyl)-2,7-dihydro-,4,5-thiadiazepine: Crystal Structure and Decoding Intermolecular Interactions with Hirshfeld Surface Analysis. J Chem Crystallogr 40: 746-752.

20. Seth SK, Saha I, Estarellas C, Frontera A, Kar T, et al. (2011) Supramolecular Self-Assembly of M-IDA Complexes Involving Lone-Pair···π Interactions: Crystal Structures, Hirshfeld Surface Analysis, and DFT Calculations [H_2IDA = iminodiacetic acid, M = Cu(II), Ni(II)]. Cryst Growth Des 11: 3259-3265.

21. Moggach SA, Allan DR, Parsons S, Sawyer L (2006) Effect of pressure on the crystal structure of alpha-glycylglycine to 4.7 GPa; application of Hirshfeld surfaces to analyse contacts on increasing pressure. Acta Crystallogr B 62: 310-320.

22. Tarahhomi A, Pourayoubi M, Golen JA, Zargaran P, Elahi B, et al. (2013) Hirshfeld surface analysis of new phosphoramidates. Acta Crystallogr B Struct Sci Cryst Eng Mater 69: 260-270.

23. Wong HL, Allan DR, Champness NR, McMaster J, Schröder M, et al. (2013) Bowing to the pressure of π···π interactions: bending of phenyl rings in a palladium(II) thioether crown complex. Angew Chem Int Ed Engl 52: 5093-5095.

24. Titi HM, Patra R, Goldberg I (2013) Exploring supramolecular self-assembly of tetraarylporphyrins by halogen bonding: crystal engineering with diversely functionalized six-coordinate tin(L)2-porphyrin tectons. Chemistry 19: 14941-14949.

25. Tawfiq KM, Miller GJ, Al-Jeboori MJ, Fennell PS, Coles SJ, et al. (2014) Comparison of the structural motifs and packing arrangements of six novel derivatives and one polymorph of 2-(1-phenyl-1H-1,2,3-triazol-4-yl)pyridine. Acta Crystallogr B Struct Sci Cryst Eng Mater 70: 379-389.

26. González-Montiel S, Baca-Téllez S, Martínez-Otero D (2015) Construction of 18-membered monometallic macrocycles by a trans-spanning ligand. Polyhedron 92: 22-29.

27. Taylor MK, Trotter KD, Reglinski J, Berlouis LEA, Kennedy AR, et al. (2008) Copper N_2S_2 Schiff base macrocycles: The effect of structure on redox potential. Inorg Chim Acta 361: 2851-2862.

28. Oxford Diffraction (2009) CRYSALIS software system, version 1.171.33.31 Oxford Diffraction Ltd., Abingdon, UK.

29. Dolomanov OV, Bourhis LJ, Gildea RJ, Howard JAK, Puschmann H (2009) OLEX2: a complete structure solution, refinement and analysis program. J Appl Cryst 42: 339-341.

30. Sheldrick GM (2008) A short history of SHELX. Acta Crystallogr A 64: 112-122.

31. Wolff SK, Greenwood DJ, McKinnon JJ, Jayatilaka D, Spackman MA (2007) Crystal Explorer 2.0. University of Western Australia, Perth.

Permissions

List of Contributors

Jason Olbrich and Joel Corbett
Poly-Med, Inc., Anderson, SC, USA

Cheryl E Green and Sylvia A Mitchell
The Biotechnology Centre, Faculty of Science and Technology, University of the West Indies, Mona Campus, Kingston 7, Jamaica, West Indies

Ahmed Khudhair Hassan
Environment and Water Research and Technology Directorate, Ministry of Science and Technology, Baghdad, Iraq

Najma Sultana
Research Institute of Pharmaceutical Sciences, Faculty of Pharmacy, University of Karachi, Karachi, Pakistan

Saeed Arayne M and Amir Haider
Department of Chemistry, University of Karachi, Karachi, Pakistan

Hina Shahnaz
Department of Environmental Science, Sind Madressatul Islam University, Karachi, Pakistan

Adrian D Allen
Department of Comprehensive Sciences, Howard University, Washington, DC, USA
NOAA Center for Atmospheric Sciences, Howard University, Washington, DC, USA

Folahan O Ayorinde
Department of Chemistry, Howard University, Washington, DC, USA

Broderick E Eribo
Department of Biology, Howard University, Washington, DC, USA

Carlos E Crespo-Hernández, Aaron Vogt R and Briana Sealey
Department of Chemistry and Center for Chemical Dynamics, Case Western Reserve University, 10900 Euclid Avenue, Cleveland, Ohio 44106, USA

Prasanna Datar
Sinhgad Institute of Pharmacy, Narhe, Pune, Maharashtra, India

Saeed M Arayne
Department of Chemistry, University of Karachi, Karachi, Pakistan

Najma Sultana
Research Institute of Pharmaceutical Sciences, Faculty of Pharmacy, University of Karachi, Karachi-75270, Pakistan

Hina Shehnaz
Department of Environmental Science, Sind Madressatul Islam University, Karachi, Pakistan

Amir Haider
Arysta Life Science Pakistan, Horizon Vista - 3rd Floor, Commercial 10, Block 4, Clifton, Karachi-75600, Pakistan

Wilma P Rezende, Leonardo L Borges, Danillo L Santos and José R Paula
Natural Products Research Laboratory, Federal University of Goiás, Brazil

Nilda M Alves
School of Health Sciences, Department of Pharmacy and Biochemistry, University of Rio Verde, Brazil

Blake Bonkowski, Jason Wieczorek, Mimansa Patel, Chelsea Craig, Alison Gravelin and Tracey Boncher
Department of Pharmaceutical Sciences, College of Pharmacy, Ferris State University, 220 Ferris Drive Big Rapids MI 49307, USA

Norshahidatul Akmar Mohd Shohaimi, Wan Azelee Wan Abu Bakar, Jafariah Jaafar and Nurasmat Mohd Shukri
Department of Chemistry, Universiti Teknologi Malaysia, Malaysia

Taha A and Ahmed HM
Faculty of Education, Ain Shams University, Roxy, Cairo, Egypt

Francisco Solá
NASA Glenn Research Center, Materials and Structures Division, Cleveland, OH 44135, USA

Urmi Roy, Izabela Sokolowska, Alisa G Woods and Costel C Darie
Biochemistry and Proteomics Group, Department of Chemistry and Biomolecular Science, Clarkson University, USA

Abdulhamid Alsaygh and Ibrahim Al-Najjar
Petrochemicals Research Institute, King Abdulaziz city for Science and Technology, P.O.Box 6086, Riyadh, 11442, Kingdom of Saudi Arabia

Jehan Al-Humaidi
Chemistry Department, College of Science, Princes Nora Bent Abdulrahman University, Riyadh, Saudi Arabia

Snezana Agatonovic-Kustrin and David W. Morton
School of Pharmacy and Applied Science, La Trobe University, Australia

Ahmad Pauzi Md. Yusof
Physiology Department, Medical School, Universiti Teknologi Mara, Selangor, Malaysia

Bablu Kumar, Gupta GP, Sudha Singh and Kulshrestha UC
School of Environmental Sciences, Jawaharlal Nehru University, New Delhi 110067, India

Lone FA
Centre for Climate Change, Mountain and Agriculture, SKUAST, Shalimar, Srinagar, 191123, Jammu & Kashmir, India

Rabya Aslam
Institute of Separation Science and Technology, Friedrich-Alexander-Universität Erlangen-Nürnberg, Germany
Institute of Chemical Engineering and Technology, University of the Punjab, Lahore, Pakistan

Karsten Muller
Institute of Separation Science and Technology, Friedrich-Alexander-Universität Erlangen-Nürnberg, Germany

Wanderley da Costa
Department of Metallurgical Engineering and School of Polytechnic, USP, Sao Paulo, Brazil

Edith MM Souza
Department of Metallurgical and Materials Engineering of the Polytechnic School, USP, São Paulo, Brazil

Leonardo GA Silva
Centro Technology Radiation, IPEN, São Paulo, Brazil

Helio Wiebeck
Departament of Metallurgical and Materials Engineering of the Polytechnic School, USP, São Paulo, Brazil

Mbugua JK, Mbui D and Kamau GN
University of Nairobi, Nairobi, Kenya

Sherrif SS, Madadi V, Mbugua JK and Kamau GN
Department of Chemistry, University of Nairobi, Nairobi, Kenya

Abdul Rauf Raza and Aeysha Sultan
Ibn e Sina Block, Department of Chemistry, University of Sargodha, Sargodha-40100, Pakistan

Nisar Ullah
Chemistry Department, King Fahd University of Petroleum and Minerals, Dhahran-31261, Saudi Arabia

Muhammad Ramzan Saeed Ashraf Janjua
Ibn e Sina Block, Department of Chemistry, University of Sargodha, Sargodha-40100, Pakistan

Chemistry Department, King Fahd University of Petroleum and Minerals, Dhahran-31261, Saudi Arabia

Khalid Mohammed Khan
HEJ Research Institute of Chemistry, International Centre for Chemical & Biological Sciences, University of Karachi, Karachi-75270, Pakistan

Bouhadiba K and Djennad MH
Laboratory of SEA2M, Department of Chemistry, Faculty of Sciences and Technology, University of Mostaganem, Mostaganem, Algeria

Hammadi K
Pedagogical Laboratory of Microbiology, Department of Biology, Faculty of Life and Natural Sciences, University of Mostaganem, Mostaganem, Algeria

Liu X and Yang G
Institute for Sport Research, Nanyang Technological University, Singapore

Lipik VT
School of Material Science and Engineering, Nanyang Technological University, Singapore

Lastovka AV, Fadeeva VP and Bazhenov MA
NN Vorozhtsov Novosibirsk, Institute of Organic Chemistry, Siberian Branch of Russian Academy of Sciences, Russia
Novosibirsk National Research State University, Russia

Tikhova VD
NN Vorozhtsov Novosibirsk, Institute of Organic Chemistry, Siberian Branch of Russian Academy of Sciences, Russia

Suresh DB, Jamatsing DR, Pravin SK and Ratnamala SB
School of Chemical Sciences, North Maharashtra University, Jalgaon, Maharashtra, India

Mohammad Shraitah and Malek MS Okdeh
Department of Chemistry, Techreen University, Lattakia, Syria

P Ponnambalam
Department of Chemistry, Bharathiar University, Coimbatore-641046, Tamilnadu, India

S Kumar
Department of Chemistry, Bharathiar University, Coimbatore-641046, Tamilnadu, India
Department of Chemistry, Thiruvalluvar College of Engineering and Technology, Tamilnadu, India

P Ramanathan
Department of Chemistry, Thanthai Hans Roever College (Autonomous), Perambalur, Tamilnadu, India

Yahya S, Beheshtiha SH, Majid M and Dehghani M
Department of Chemistry, School of Science, Alzahra University, Vanak, Tehran, Iran

Mohauman Mohammad Al-Rufaie
Department of Chemistry, College of Science, Kufa University, Iraq

Elsadig HK and Abdalfatah MB
Department of Pharmaceutical Chemistry, College of Pharmacy, Prince Sattam Bin Abdulaziz University, Riyadh, Al-kharj, Saudi Arabia

Amir Alhaj Sakur
Department of Analytical and Food Chemistry, Faculty of Pharmacy, University of Aleppo, Syria

Banana Al Fares
Department of Analytical and Food Chemistry, Faculty of Pharmacy, University of Aleppo, Syria
Department of Analytical Chemistry, Faculty of Science, Teshreen University, Lattakia, Syria

Malek Okdeh
Department of Analytical Chemistry, Faculty of Science, Teshreen University, Lattakia, Syria

Takao Sanaki, Takuji Fujihara, Takeshi Yoshioka, Kenichi Higashino, Toru Nakano and Yoshito Numata
Shionogi Innovation Center for Drug Discovery, Shionogi & Co., Ltd., Sapporo 001-0021, Japan

Ryo Iwamoto
Business-Academia-Collaborative Laboratory, Graduate School of Pharmaceutical Science, The University of Tokyo, Tokyo 113-0013, Japan

Simplicio González-Montiel, Saray Baca-Téllez, Alejandro Álvarez-Hernández and Julián Cruz Borbolla
Área Academic Chemistry, Chemical Research Center, University of the State of Hidalgo, 4.5 km. Road Pachuca-Tulancingo, City of Knowledge, CP 42184, Mineral de la Reforma, Hidalgo, Mexico

Diego Martínez-Otero
Joint Research Centre in Sustainable Chemistry, UNAM UAEM, Toluca-Atlacomulco Highway 14.5 km., C.P. 50200, Toluca, State of Mexico, Mexico

Index

www.ingramcontent.com/pod-product-compliance
Lightning Source LLC
Chambersburg PA
CBHW080647200326
41458CB00013B/4756